ENERGY IN NATURE AND SOCIETY

Also by Vaclav Smil

VACLAV SMIL

ENERGY IN NATURE AND SOCIETY

General Energetics of Complex Systems

The MIT Press
Cambridge, Massachusetts
London, England

For information about special quantity discounts, please email special_sales@mitpress.mit.edu

This book was set in Galliard, Bodoni, and Futura on 3B2 by Asco Typesetters, Hong Kong.
Printed on recycled paper and bound in the United States of America.

Library of Congress Cataloging-in-Publication Data

Smil, Vaclav.
Energy in nature and society : general energetics of complex systems / Vaclav Smil.
    p.   cm.
Includes bibliographical references and index.
ISBN 978-0-262-19565-2 (hardcover : alk. paper)—
978-0-262-69356-9 (pbk.: alk. paper)
1. Bioenergetics. 2. Energy budget (Geophysics)
I. Title.
QH510.S63   2008
577′.13—dc22   2007002801

10 9 8 7 6 5 4 3 2 1

CONTENTS

PREFACE

Energy is the only life and is
from the Body;
and Reason is the bound or outward
circumference of Energy.
Energy is Eternal Delight.
William Blake, *The Marriage of Heaven and Hell* (1793)

Energy's definition in Blake's poem comes from the Devil, as he corrects what he feels to be one of the principal errors of sacred codes, namely, "That Energy, call'd Evil, is alone from the Body; & that Reason, call'd Good, is alone from the Soul." This book is preoccupied with more pragmatic and less contentious propositions, but it too owes its existence to fascination with energy's elusive qualities.

I wrote this opening paragraph in 1987 as I was beginning to work on what was intended to be both a comprehensive and an analytically unified survey of energy. The first goal was met by offering a more sweeping treatment of energy sources, flows, conversions, uses, and their consequences in the Earth's biosphere and throughout the history of civilizations than was available at that time in any single volume. The second goal was pursued by using fundamental unifying metrics, most notably power density (W/m^2) and energy intensity (J/g) values. By far the most difficult task was to make sure that the forest of grand energy principles would not be obscured by the necessary focus on specific trees: inevitably, the book had to have thousands of numbers, and their torrent could be overwhelming.

General Energetics: Energy in the Biosphere and Civilization was published in 1991. Although it did not make waves compared to several of my "bestsellers," it found appreciative readers. My greatest satisfaction came from the judgment by Philip Morrison, for decades my model of clear and captivating scientific writing, who in *Scientific American* called the book "a work of tightly controlled audacity.... The pleasure and stimulation of the book come from its critical display of... unruly realities: its importance from the fact that serious argument on any of these matters must take into account this army of decisive magnitudes Professor Smil has marshaled."

For me *General Energetics* was an unprecedented project with multiple challenges and rewards. It took much longer to complete than any other book I have written, and the necessary research, endless calculations, and intermediate drafts turned up a great deal of fascinating information that had to be left out of that single volume. Consequently, *General Energetics* led directly to *Energy in World History* (1994) and a few years later to *Energies* (1998), and the work on several of its specific aspects also helped to inform parts of *Cycles of Life* (1997), *Feeding the World* (2000), *The Earth's Biosphere* (2002), and *Energy at the Crossroads* (2003).

As soon as the original book was published, I hoped that one day I could prepare a revised edition, but the continuing flood of new energy studies made it clear that a mere revision was not the way to go. My intent then became to retain the book's original divisions (combining chapters 5 and 6 was the only major change) but with deletions, rewrites, elaborations, expansions, introductions of new subjects, better illustrations, and more extensive references.

The fundamental goal has not changed: to produce a comprehensive, systematic, revealing (and hence interdisciplinary and quantitative) treatment of all major aspects of energy in nature and society. I believe the case for a book like this is even more compelling now than it was two decades ago. During the intervening years, energy has become even more prominent as the subject of scientific inquiry and public policy debates, and the consequences of its use raise even greater concern and anxiety, but approaches to its understanding and rational management continue to suffer from inadequate integration and insufficient understanding of complex wholes. My

objective thus remains the same as for *General Energetics*, so I end this preface by amending only slightly the closing paragraphs that I wrote in 1987.

I see a clear need for a book that not only embraces all essential energy sources, storages, flows, and conversions in a unified and systematic manner but that does so by combining an encyclopedic sweep and richness of detail with an evolutionary overview and analytical crispness, and that does not shy away from grand synthesizing generalizations or from acknowledging the inability to offer such statements. This is no modest goal, but I believe that Li Bo's ancient acute observation—"When the hunter sets traps only for rabbits, tigers and dragons are left uncaught"—justifies the attempt.

At the same time, this is also a very personal book. My fascination with energetics has been going on for more than 40 years, and new topics keep coming into closer focus. My interest in planetary energy flows and in bioenergetics started with university studies of climatology, geomorphology, zoology, and ecology, and these fields were soon augmented by work on energy economics, coal mining, and power plant engineering. Principal themes of my published energy research have been, chronologically, thermal electricity generation, acid deposition, coal mining, and internationalization of energy supply in the 1960s; atmospheric CO_2 modeling, evolution and forecasting of energy techniques, energy-economy correlations, and energy in China and other poor populous countries in the 1970s; biomass energies, grand biospheric cycles, energy analysis in agriculture, energy in agriculture and in human nutrition, energy in world history, and global dimensions of energetics in the 1980s and 1990s; and during the first years of a new millennium, energy's role in the creation and transformation of modern civilization.

A strong commonality tying these interests has been my fascination with unruly and fuzzy realities in preference to abstract models and dubious generalizations. Complexities and peculiarities of the real world and counterintuitive outcomes of many of its processes have seemed to me always more appealing than theoretical models. This preference requires a from-the-ground-up approach where gradual understanding of details and cumulative acquisition of the widest possible scope of information precedes any attempts at generalization. Inevitably, this book is marked by these interests and biases, but I have done my best to offer a balanced and comprehensive treatment free of excessive indulgence in favorite topics.

The aims are broad interdisciplinary coverage, richness of detail, clear analyses, syntheses rooted in the presented information, and commonsense generalizations. The means is a systematic and evolutionary account, but one whose boundaries and flavors are also clearly influenced by my scientific background, preferences, and lifelong fascinations and inclinations. The book takes just one of many possible routes to better understanding of energy in nature and in human societies, and its inherent sweep and complexity mean that it cannot be devoid of lapses and errors. But even if it were to fall far short of its ambitious mark, there is, as always, consolation in the wisdom of an ancient sage. Lao Zi, noting that it is void space that makes bowls and houses useful, wrote,

So advantage is had
From whatever is there;
But usefulness rises
From whatever is not.

ENERGY IN NATURE AND SOCIETY

1

THE UNIVERSAL LINK

Energetics, Energy and Power

*And the things best to know are first principles and causes.
For through them and from them all other things may
be known but not they through the things covered by
them.... But these things, the most universal, are perhaps
the most difficult things for men to grasp, for they are farthest removed from the senses.*
Aristotle (384–322 B.C.E.), *Metaphysics*

The Aristotelian preference for understanding what is
most universal finds the largest realm of inquiry in energetics, the study of transformations that have created
(and are incessantly changing) the inanimate universe
and that have sustained nearly four billion years of life's
evolution on the Earth. Nothing needs to be excluded
from this realm: any process can be analyzed in terms of
its underlying energy conversions; any object, as well as
any bit of information, can be valued for its energy content and for its potential contribution to future energy
transformations. On the Earth these conversions are

overwhelmingly transformations of solar radiation. The
other two prime energizers of the Earth, its internal heat
and its gravitational forces, are qualitatively indispensable
but quantitatively much less important.

Such disparate phenomena as multiarmed lightning,
long-stalked gladioli, crying babies, rotating steam
turbines, or elegantly designed books are ultimately
traceable to identical energetic origin in the Sun's thermonuclear reactions. A quest for the underlying energetic commonalities is only a part of the fascination of
studying energy conversions in nature and in society. A
no less exciting endeavor is to uncover, compare, and explain countless specific expressions of these commonalities that shaped evolution and history, and that create
the enormous heterogeneity of the world we live in. A
lifetime's fascination with this unified heterogeneity led
me to write this book. But before proceeding with systematic topical analyses it is essential to appreciate the
milestones in the study of energies and to understand

basic variables and general approaches used in these inquiries.

1.1 Evolution of Energetics: From Aristotle to Einstein

Energetics has no rivals in its interdisciplinary nature and reach, a reality reflecting not only the vastness of the potential subject matter but also the inquisitiveness and achievements of generations of scientists and engineers who contributed to filling in so many intellectual blanks. Studies of energy phenomena have always proceeded within many disciplines, sometimes in isolation, often in parallel, later with profitable transfers and fusing ideas, and since the mid-nineteenth century with increasingly frequent attempts at finding general patterns and offering grand syntheses from perspectives as diverse as physics and historical economy.

Many outstanding creators of classical and modern science were students of energetics. But this fundamental yet inherently diffuse field of inquiry—spanning not only physical, life, and social sciences but also many branches of engineering and management—has never acquired the much clearer identity of many reductionist disciplines, so their contributions have rarely been seen as constituent pieces of an impressive mosaic that has been assembled since the end of the seventeenth century. The basic course of this intellectual progress is chronicled in Mach (1896), Stallo (1900), Brody (1945), Cardwell (1971), Lindsay (1975), Martinez-Alier (1987), and C. Smith (1999). I offer just a brief account of advances in formulating the most fundamental explanations, including the laws of thermodynamics, the nature of metabolism, and the place of energy in economic advancement and civilization's survival.

The first isolated brushstrokes of this vast canvas were put down long before those decades of astonishing intellectual ferment at the close of eighteenth century. The word *energy* is, as so many of our abstract terms, a Greek compound. Aristotle left its first known written record in his *Metaphysics* (Düring 1966). He gave it a primarily kinetic meaning by joining εν (in) and έργον (*ergon*, work) to form ενέργεια (*energeia*): "The word 'actuality' (*energeia*) which we connect with 'complete reality' (*entelechia*), has, strictly speaking, been extended from movements to other things; for actuality in the strict sense is identified with movement." According to Aristotle, every object's existence is maintained by *energeia* that is related to the object's function.

But the word's etymology reveals as much as it hides. For Greeks the word and its cognates filled a much larger conceptual niche, qualitatively and figuratively, than does its modern scientific counterpart. In Aristotle's *Ethics*, *energeia* stands in opposition to mere disposition, *hexis*; in his *Rhetoric*, it carries the vigor of the style. The verb *energein* meant to be in action, implying ceaseless motion, work, production, change. The classical concept of *energeia* was a philosophical generalization, an intuition embracing the totality of transitory processes, the shift from the potential to the actual. The perception was clearly holistic and qualitative. These concepts remained unrefined for nearly two millennia. Roman civilization, Islam, dynastic China, and medieval Europe solved ingeniously many everyday energetic challenges, and their remarkably complex societies were energized not only by humans and animals but increasingly also by machines driven by water and wind (see chapter 7).

But there were few concurrent advances in systematic understanding of energy because even many founders of modern science had very faulty concepts of energy. To Galileo Galilei (1564–1642) heat was an illusion of the senses and the outcome of mental alchemies. Francis Bacon (1561–1626) thought heat could not generate motion, and vice versa. And the fundamental ideas of mass, momentum, and force used by Isaac Newton (1643–1727) had initially little relevance for men building better steam engines or pondering the energetic basis of living organisms. But these mental and manual experiments laid down the foundations of energetics as a systematic science. Scientific understanding was first sharpened on the basis of particularistic definitions and quantitative assessments. Only later came laws and theories as the burgeoning nineteenth-century science pursued its inquiries into fuels, engines, heat, motion, radiation, electricity, nutrition, metabolism, work, photosynthesis, and evolution.

The practical roots of this new knowledge are self-evident in contributions made by James Watt (fig. 1.1). His steam engine revolutionized industrial production, and his invention of a simple indicator, a miniature recording steam gauge, opened the way for detailed studies of engine cycles, which contributed immeasurably to the emergence of thermodynamics during the following century. In contrast, Sadi Carnot (1796–1832) pursued a purely abstract approach to the understanding of working engines, and this explains why his contributions were fully appreciated only decades later (Truesdell and Bharatha 1977). In order to explain production of kinetic energy from heat, Carnot (1824) set down the principles applicable to any imaginable heat engine, regardless of its working substance.

Carnot defined the maximum efficiency (e) of an ideal (reversible) heat engine as equal to 1 minus the quotient

1.1 James Watt (1736–1819).

of the machine's lowest (T_l) and highest (T_h) operating temperatures (expressed in degrees Kelvin):

$$e = 1 - \frac{T_l}{T_h}.$$

Consequently, a 100% efficient engine could be realized only if its cold end were at absolute zero, a practical impossibility. After Carnot's death, Emile Clapeyron (1799–1864) presented the ideal engine cycle in terms of Watt's indicator diagram, and this usage remains the standard in all thermodynamic studies (fig. 1.2). Antoine Lavoisier's (1743–1794) suggested equivalence between the heat output of animals and humans and their feed

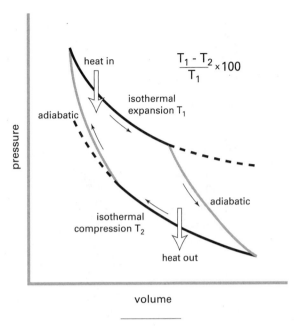

$$\frac{T_1 - T_2}{T_1} \times 100$$

heat in

adiabatic

isothermal
expansion T_1

adiabatic

isothermal
compression T_2

heat out

pressure

volume

1.2   Ideal Carnot cycle.

and food intake provided the foundation for studies of heterotrophic metabolism. On the other hand, Lavoisier's insistence on the caloric, an imponderable fluid carrier of heat, was an intellectual cul-de-sac (R. Fox 1971).

In contrast to Carnot's ideas, the work of Jean-Baptiste Dumas and Jean-Baptiste Boussingault (1842) was immediately influential, and their explanations of grand biospheric cycles underpinning the planetary bioenergetics were essentially correct. At the same time, Justus Liebig (1803–1873) was the first scientist to ascribe the generation of CO_2 and water to food oxidation (Liebig 1843), offering a fundamentally correct view of heterotrophic metabolism. He also introduced the powerful concept of limiting nutrients that is applicable

equally well to autotrophs and heterotrophs. Edward Smith (1857) collected about 1,200 samples of expired air and proved that respiration is a function of one variable, the ingested food, and one constant, a distinct individual metabolic rate. During the 1840s the advances in human energetics were closely associated with the genesis and progress of general energetic principles in the work of a German physician, Julius Robert Mayer (1814–1878).

Mayer's ideas had their origin in a voyage he made in 1840 as a ship's doctor. When bleeding his patients in Java (the ancient practice of bleeding as a cure for many ailments persisted well into the nineteenth century), he noticed that their venous blood was much brighter than the blood of patients in Germany (Caneva 1993). Mayer's correct explanation was that blood in the tropics does not have to be oxidized as much as in the temperate regions because less energy is needed for body metabolism in warm climates. But if less heat is lost in the tropics because of radiation, how about the heat lost as a result of physical work (that is, expenditure of mechanical energy), which obviously warms its surroundings regardless if done in Europe or tropical Asia? Unless we posit some mysterious origins for it, that heat must also come from the oxidation of blood, and hence heat and work must be equivalent and convertible at a fixed rate.

Mayer published the first quantitative estimate of the equivalence in 1842, and eventually he extended the idea of energy conservation to all natural phenomena, including electricity, light, and magnetism. He gave details of his calculation based on an experiment with gas flow between two insulated cylinders (Mayer 1842; 1851). His result was considerably below the real value, mainly because of incorrect numbers for the specific heat of air at constant pressure and constant volume that he

1.3   *Left,* James Prescott Joule (1818–1889); *right,* William
Thomson, Lord Kelvin (1824–1907).

took from published sources. This error had no effect on the validity of Mayer's law or on his conclusion that muscles are the equivalents of heat engines and that mammals convert food or feed energy to work (mechanical energy) with an efficiency of about 20%. The law of conservation of energy is now commonly known as the first law of thermodynamics. All of its many restatements, ranging from precise formulations (heat flowing into a system is equal to the increase of the system's internal energy and the work done by the system) to catchy summations (you can't get something for nothing) make it clear that energy can be neither created nor destroyed.

The correct value for the equivalence of heat and mechanical energy was found, contemporaneously with Mayer's work but entirely independently, by the English physicist James Prescott Joule (fig. 1.3) after he conducted a large number of carefully executed churning experiments. Joule used very sensitive thermometers to measure the temperature of water that was being churned by an assembly of revolving vanes driven by descending weights; this arrangement made it possible to measure fairly accurately the mechanical energy invested in the churning process. Joule reported his first results to the British Association meeting at Oxford in 1847, and the full paper finally appeared three years later (Joule 1850). Joule's later revision set the equivalence with an error of less than 1%. The third independent formulation of the law was the work of a German physiologist, Hermann von Helmholtz (1821–1894), who, like

Mayer, wrote about forces rather than energies (Helmholtz 1847).

Soon afterwards William Thomson, Lord Kelvin (fig. 1.3), identified the Sun as the principal source of kinetic energies available to humans and wrote about a universal tendency in nature toward the dissipation of mechanical energy (Thomson 1853). In 1850 a German theoretical physicist, Rudolf Julius Emanuel Clausius (1822–1888), published his first (and still most famous) paper on the mechanical theory of heat, in which he proved that the maximum work obtainable from a Carnot cycle depends solely on the temperatures of the heat reservoirs, not on the nature of the working substance, and that there can never be a positive heat flow from a colder to a hotter body (Clausius 1850).

Clausius continued to refine this fundamental idea. In 1865 he coined the term *entropy*—from the Greek τροπή (transformation)—to measure the degree of disorder in a closed system, and formulated the second law of thermodynamics: the energy content of the universe is fixed, but its distribution is uneven, and thus its conversions seek uniform distribution, and the entropy of the universe tends to maximum (Clausius 1867). A system changes its entropy (ΔS) in proportion to the amount of heat (Q) applied at temperature (T, in absolute terms), and ΔS is positive when a system's entropy increases:

$$\Delta S = \frac{Q}{T}.$$

In practical terms this means that in any closed system (without an external supply of energy) the availability of useful energy can only decline. A lump of coal is a high-quality, highly ordered (low-entropy) form of energy; its combustion will produce heat, a dispersed, low-quality, disordered (high-entropy) form of energy, and the sequence is irreversible: diffused heat (and emitted combustion gases) can never be reconstituted as a lump of coal. Heat thus occupies a unique position in the hierarchy of energies. All other energies can be completely converted to it, but the conversion of heat into other forms of energy can never be complete because only a portion of the initial input ends up as kinetic energy or as electricity. The second law of thermodynamics, the universal tendency toward disorder, can be seen as the scientific counterpart of the futility cries in *Ecclesiastes*. Only at absolute zero ($-273°C$), in the absence of any movement, is the entropy nil. This is the third law of thermodynamics, initially formulated in 1906 as Walther Nernst's (1864–1941) heat theorem.

The second law is perhaps the grandest of all cosmic generalizations, yet one of which most nonscientists remain unaware, as Charles Percy Snow (1905–1980) eloquently pointed out in his Rede lecture on the two cultures (Snow 1959). Soon after the precise formulation of the thermodynamic laws, Josiah Willard Gibbs (1839–1903), a U.S. physicist, applied their concepts to chemistry and introduced the important notion of free energy, G (Gibbs 1906). Change of this energy (ΔG)—basically the maximum amount of work that can be derived from a reaction proceeding at constant temperature and pressure—is determined by subtracting the product of temperature and entropy change from the change of enthalpy (H, the heat content of a system) energy that enters a chemical reaction:

$$\Delta G = \Delta H - T\Delta S.$$

Formation of compounds from elements requires inputs of G: at standard conditions (298 K, 101.3 kPa)

1.4 *Left to right*: Wilhelm Ostwald (1853–1932), Nobel prize in chemistry, 1909; Frederick Soddy (1877–1956), Nobel prize in chemistry, 1921; Albert Einstein (1879–1955), Nobel prize in physics, 1921. All photographs © The Nobel Foundation.

it takes about 235 kJ/mol to form H_2O and about 911 kJ/mol to synthesize $C_6H_{12}O_6$. In a spontaneous process ΔG is negative. For the two critical reactions sustaining life—hydrolysis of ATP to ADP and inorganic phosphate, and oxidation of glucose—ΔG is, respectively, about −31 kJ/mol and −2870 kJ/mol. The second law exercised a powerful influence on scientists thinking about the energetic foundations of modern civilization. Edward Sacher (1834–1903), a little-known Austrian science teacher, viewed economies as systems for winning the greatest possible amount of energy from nature, and he tried to correlate stages of cultural progress with per capita availability of fuels (Sacher 1881).

He ascribed 12.5 GJ/a to foragers, 25 GJ/a to nomadic societies, roughly 60 GJ/a to agriculturalists, and close to 85 GJ/a to contemporary Central Europe.

In 1892, Wilhelm Ostwald, a leading German chemist and Nobel laureate (fig. 1.4), began his manifesto on the *Fundamentals of General Energetics* (1892) by stressing energy's unique position:

The concepts that find application in all branches of science involving measurement are space, time, and energy. The significance of the first two has been accepted without question since the time of Kant. That energy deserves a place beside them follows from the fact that because of the laws of its

THE UNIVERSAL LINK

transformation and its quantitative conservation it makes possible a measurable relation between all domains of natural phenomena. Its exclusive right to rank along space and time is founded on the fact that, besides energy, no other general concept finds application in all domains of science. Whereas we look upon time as unconditionally flowing and space as unconditionally at rest, we find energy appearing in both states. In the last analysis everything that happens is nothing but changes in energy. (Translated in Lindsay 1976, 339)

The contrast between the rising demand for coal and the inexorable thermodynamic losses during coal's conversion led first to greatly exaggerated fears about the fuel's exhaustion (Jevons 1865) and eventually to reasoned arguments in favor of energy conservation. Ostwald's energetic imperative—Waste no energy but value it—is relevant as humankind makes the inevitable transition to a permanent economy based exclusively on solar radiation. Another Nobel laureate, Frederick Soddy (fig. 1.4), was the first scientist to make the analogy between utilizing natural energy flows and relying on fossil fuels: "The one is like spending the interest on a legacy, and the other is like spending the legacy itself." He also anticipated "a period of reflection in which awkward interviews between civilization and its banker are in prospect" (Soddy 1912, 139). Concerns about the necessity to start living again on the interest were rekindled with recent forecasts of the imminent peaking of global crude oil extraction (C. J. Campbell 1997; Deffeyes 2001).

After World War I, Soddy looked closer at the relation between energy, evolution, economics, and human prospects, and he generalized the links by observing, "From the energetic standpoint progress may be regarded as a successive mastery and control over sources of energy ever nearer the original source" (1926, 48). The twentieth century brought a fundamental extension of the first law of thermodynamics with Albert Einstein's (fig. 1.4) follow-up of his famous relativity paper (Einstein 1905). Soon after its publication Einstein, writing to a friend, noted, "The principle of relativity in conjunction with Maxwell's fundamental equations requires that the mass of a body is a direct measure of its energy content—that light transfers mass" (cited in A. I. Miller 1981, 353). During the next two years Einstein formalized this "amusing and attractive thought" in a series of papers firmly establishing the equivalence of mass and energy.

In the last of these papers he described a system behaving like a material point with mass,

$$M = \mu + \frac{E_0}{c^2},$$

and noted that this "result is of extraordinary theoretical importance because a physical system's inertial mass and energy content appear to be the same thing. An inertial mass μ is equivalent with an energy content μc^2" (Einstein 1907, 464). The world's most famous equation requires any mass releasing energy to diminish, a fact of no practical importance in chemical reactions. For example, the combustion of 1 kg of hard coal (requiring 3 kg of O_2 and releasing ~30 MJ of energy) will diminish the mass of the two reactants by about 10^{-10}, a reduction too small to measure. In contrast, in nuclear reactions the reduction is obvious. Einstein, aware of this difference, noted that a practical demonstration of the law was difficult but that for radioactive decay the quantity of free energy is enormous. Fissioning 1 kg of U^{235} will release about 8.2 TJ of energy—about 270,000 times more than the same mass of coal—and diminish the uranium mass by 1 g, or 0.1%.

The last decades of the nineteenth century and the first half of the twentieth century also brought great gains in bioenergetics. Numerous calorimetric experiments performed by Max von Pettenkofer, Carl von Voit, Graham Lusk, and Wilbur Atwater determined fairly accurate energy balances of living organisms. Consequently, Max Rubner (1902) was able to offer an excellent systematic account of human metabolism that included all the essentials of modern understanding, including determination of energy values of various foodstuffs; the distinction between food intake and energy consumption; the realization that carbohydrates, proteins, and lipids can all be converted to work and heat; correlations between environmental conditions, individual circumstances, and metabolic rates, and between basal metabolism and body surface area; and appreciation of the dynamic effect of food digestion.

Max Kleiber (1893–1976) uncovered the fundamental allometric relation between the basal metabolic rate and the body mass of heterotrophs (Kleiber 1932), a beginning of fascinating studies of energetic scaling (see chapters 3–5). And Samuel Brody (1890–1956) published his monumental synthesis of bioenergetics analyzing and summarizing a century of scientific progress (Brody 1945). An important theoretical advance in bioenergetics came with Alfred Lotka's (1880–1949) formulation of a law of maximum energy. For biota it is not the highest conversion efficiency but the greatest flux of useful energy, the maximum power output, which is most important for growth, reproduction, and maintenance, and

species radiation. Consequently, living organisms and ecosystems do not convert energy with the highest supportable efficiencies but rather at rates optimized for the maximum power output (Lotka 1925). Odum and Pinkerton (1955) demonstrated that the efficiencies are always less than the possible maxima: they never surpass 50% of the ideal rate.

During the 1930s came rapid advances in understanding nuclear energy that included not only the epochal discovery of the neutron (Chadwick 1932) and the first laboratory demonstration of fission (Hahn and Strassman 1939) but also the first correct explanation of energy processes in stars by Hans Bethe (1939) (fig. 1.5). At the same time, an inconspicuous but fundamental wave of change began to affect the scientific method. After centuries of progressive reductionism and compartmentalization came a gradual formulation of general system approaches based on recognition of underlying commonalities, multidimensional complexities, nonlinear feedbacks, and probabilistic outcomes. Vladimir Ivanovich Vernadsky (fig. 1.5), with his powerful idea of the biosphere (1926), and Ludwig von Bertalanffy (1901–1972), with his systematic theoretical look at biology (1932–1942), were the early pioneers of the approach in life sciences. And Arthur Tansley's (1935) introduction of the concept of the ecosystem enriched bioscience with one of its most important cognitive tools.

Just before the end of World War II, Edwin Schrödinger (1887–1961) addressed the thermodynamic oddity of living systems that create and maintain exquisite order by using disordered elements dominated by carbon (1944). He explained this apparent violation of the second law by introducing the idea of "nonequilibrium thermodynamics" of open systems (the nonequilibrium

1.5 *Left*, Hans Bethe (1906–2005), Nobel prize in physics, 1967; *right*, Vladimir I. Vernadsky (1863–1945). Bethe's photograph © The Nobel Foundation; Vernadsky's photograph courtesy of Eric Galimov, V. I. Vernadsky Institute of Chemistry and Analytical Chemistry, Russian Academy of Sciences, Moscow.

conditions apply to material flows as well). Living systems, be they cells or civilizations, maintain themselves in highly improbable states of order and organization, temporarily defying the unidirectional entropic imperative by importing large amounts of free energy (and constituent nutrients and structural elements) from outside and processing it to generate a lower entropy state within, that is, by thriving on negentropy. These ideas, often embedded in the newly flourishing general system studies, opened up the debate about the behavior of

open systems ranging from cells to civilizations, and about the thermodynamic fundamentals of bioenergetics (Prigogine 1947; Bertalanffy 1968; Morowitz 1968; Brooks and Wiley 1986; Weber, Depew, and Smith 1988; Haynie 2001; Urry 2004).

But outside of bioenergetics there was much less interest in general energy studies. Perhaps the best explanation of this neglect lies in the increasing abundance and decreasing real cost of principal commercial energies between 1945 and 1973; in those circumstances there was little need to heed Ostwald's energetic imperative or to do complex energy studies. As a result, synthesizing approaches to energy were rare during the first two post–WW II decades. Works by Ubbelohde (1954), Cottrell (1955), and Thirring (1958) were the exceptions. M. King Hubbert's (1903–1989) innovative analysis of the cycle of mineral production was limited to forecasting the course of nonrenewable resource recovery, but his model (Hubbert 1962) became especially influential when it correctly predicted the peak of crude oil production in the contiguous United States. And a new generalizing path was opened up by Howard T. Odum (1924–2002) with his ecoenergetic approach to environment, power, and society (Odum 1971). He subsequently refined this approach and extended it to include the concept of emergy (embodied energy) expressed in units of one type of energy, often in terms of solar emjoules (Odum 1996; Odum and Odum 2000).

The first oil price "crisis" (1973–1974) gave rise to an unprecedented interest in energy affairs and to a flood of new publications. Even leaving aside naive or poorly researched publications by instant experts, most of this attention still remained very particularistic (dominated by traditional sectoral studies of coal, hydrocarbons, and electricity), too beholden to the rapidly changing perceptions of the day (hence the widespread anticipation of continuously rising prices), preoccupied with immediate and improperly interpreted problems, and as result highly error-prone in its ignorance of broader settings and implications. Virtually all forecasts produced at that time became obsolete almost instantly (Smil 2003). At the same time, our understanding benefited from syntheses that ranged from delineating the history of prime movers (Needham 1954–1986; L. White 1978) to pioneering accounts of primary biospheric productivity (Lieth and Whittaker 1975) and the intricacies of human nutritional requirements (FAO 1973).

The second "energy crisis" (1979–1981), triggered by the fall of Iran's Pahlavi dynasty, finally led to a number of syntheses that appraised realistically both the prevailing global energy situation and the technical opportunities, and economic and environmental implications, of possible solutions (Gibbons and Chandler 1981; Rose 1986; Smil 1987). The 1980s also brought many interdisciplinary surveys and syntheses of geoenergetics (Verhoogen 1980), atmospheric energetics (Kessler 1985), systems ecology (Odum 1973), animal scaling (Schmidt-Nielsen 1984), and human nutrition (FAO 1985). This multifaceted interest has continued with contributions ranging from analyses of energy efficiency (Schipper and Meyers 1992; Sorrell 2004) to surveys of principal resources and conversion techniques (Odell 1999; Sørensen 2004; Zahedi 2003; Miller 2005; da Rosa 2005) and explorations of their environmental and social consequences (ExternE 2001; Goldemberg 2000).

This enormous heterogeneity of energy studies—be it in disciplinary coverage, preferred concepts, or practical preoccupations—is fascinating but makes synthesis difficult. There is a need for defining and imposing a manageable set of variables that would apply across the vast field of general energetics. Such a quantification would support meaningful attempts at grand generalizations, which must almost always be modified by important qualitative caveats. An understanding of basic SI units (see appendix) is indispensable before embarking on systematic inquiries in general energetics. The remainder of this chapter outlines all the common measures I use in this book in order to reveal the grand processes and patterns of general energetics.

THE UNIVERSAL LINK

1.2 Approaches to Understanding: Concepts, Variables, Units

Energy and power are the key variables of energetics, the first measured in joules J, the other in watts W. The nouns *energy* and *power* are often used to express attributes that have little to do with their scientifically defined meanings. As a result, it is not surprising that incorrect understanding and improper use of these key terms are so widespread. This is not a new frustration: the term *energy* became practically indistinguishable from *power* or *force* centuries ago. In 1748, David Hume complained in *An Enquiry Concerning Human Understanding* (sec. VII, 49), "There are no ideas, which occur in metaphysics, more obscure and uncertain, than those of power, force, energy or necessary connexion, of which it is every moment necessary for us to treat in all our disquisitions."

And in 1842 the seventh edition of *Encyclopaedia Britannica* offered only a brief entry, which described energy as "the power, virtue, or efficacy of a thing. It is also used figuratively, to denote emphasis in speech." Potential energy represents yet another semantic problem: *potential* commonly signifies a possibility or latency, but in a strict physical sense, potential energy represents a *change* in the spatial setting of a mass or its configuration. Gravitational potential energy (the product of the elevated mass, its mean height above ground, and the gravitational constant) is the consequence of a changed position in the Earth's gravitational field. Springs that have been tensioned by winding exemplify the practical use of elastic potential energy: it is stored through the springs' deformation and released as useful work when the coil unwinds.

At the outset of the twenty-first century the scientifically specific terms *energy* and *power* continue to be used imprecisely even by scientists and engineers (for instance, in such ingrained phrases as "total consumption of energy," although energy cannot be consumed, only transformed).

There is no difficulty defining power simply as the rate of flow of energy:

$$W = J/s.$$

One can provide an equally elegant derivative for energy as the integral of power

$$E = \int P \, dt,$$

but this formulation does not make for an intuitive understanding of the phenomenon.

Richard Feynman (1918–1988) summed up the problem in his famous *Lectures on Physics*: "It is important to realize that in physics today, we have no knowledge of what energy is. We do not have a picture that energy comes in little blobs of a definite amount. It is not that way" (1963, 4-2). I prefer Rose's (1986, 5) solution, which acknowledges the difficulty by being appropriately evasive: Energy "is an abstract concept invented by physical scientists in the nineteenth century to describe quantitatively a wide variety of natural phenomena." The leading choice of science textbooks and encyclopedias is "the capacity for doing work," a definition that is too reductionist in its inevitable mechanical connotation and that fills the intended space only when work is not understood in the everyday sense of invested labor but as the generalized physical act that produces a change of configuration in a system in opposition to any force that resists such a change (Maxwell 1872). Defining energy as the ability to transform a system, a process that can involve any kind of energy (fig. 1.6), is thus much more helpful.

The standard physical derivation of the basic energy unit via Newton's second law of motion—1 joule is the force of 1 newton (mass of 1 kg accelerated by 1 m/s^2) acting over a distance of 1 m—certainly provides an impeccable definition, but it pertains only to kinetic energy and, simple as it is, it hardly fosters an intuitive understanding of the elusive entity. The classical derivation of calorie, a common non-SI unit of thermal energy, as the amount of heat needed to raise the temperature of 1 g of water from 14.5°C to 15.5°C, describes a process that is easy to imagine, but it is not easily related to other energies. As already explained, Mayer's and Joule's experiments made it possible to express heat in dynamic units: 1 cal = 4.1855 J.

Heat content (enthalpy, H) of a thermal energy source equals the sum of the internal energy (E, the measure of the molecular activity in the absence of motion, external action, and elastic tension) of a system plus the product of the pressure-volume work done on the system:

$$H = E + pV.$$

Heat of combustion (or specific energy, J/kg) is the difference between the bonds in initial reactants and the bonds in a newly formed compound. Heat can be transferred in three distinct ways: by conduction (direct molecular contact, most commonly in solids), by convection (moving liquids or gases), and by radiation (emission of electromagnetic waves).

Latent heat is the amount of energy needed to effect a physical change with no temperature change: changing water to steam (latent heat of vaporization) at 100°C requires exactly 6.75 times more energy than does the changing of ice into water at 0°C. Heating of water also accounts for most of the difference between gross (or higher) heat of combustion and net (lower) heating value. The first rate is the total amount of energy released by a unit of fuel during combustion with all water condensed to liquid (and hence the heat of vaporization is recovered); the second rate subtracts energy required to evaporate the water formed during combustion. The difference between the two rates is negligible for charcoal or anthracite (both being virtually pure carbon) but huge for fresh (green) wood.

Equivalence of energies makes it theoretically very easy to add up disparate flows, and various common denominators have been used to offer aggregate supply or utilization accounts in terms of SI units (J, Wh) or other measures commonly used in energy studies, such as hard coal or crude oil equivalents or barrels of oil. But these quantitatively impeccable reductions to a common denominator mislead in several ways. On the most general level, the common denominators ignore the fundamental distinction between low-entropy stores or flows and high-entropy states. Yet the practical difference between equivalent totals of available energy (in fuels, steam, or electricity) and dissipated energy (in heat) is obvious. The concept of exergy (the maximum work possible in an ideal process) has been introduced to quantify this loss of quality: unlike energy, exergy is not conserved but is destroyed in every real process because of the increase of entropy (Ahern 1980; Sciubba 2004; Sciubba and Ulgiati 2005; Hermann 2006).

Energy units also do not capture qualitative differences that determine actual modes of use. Three kilograms of freshly cut wood are not qualitatively equivalent to 1 kg of bituminous coal, or to a day's intake of food of three adults, or to solar radiation absorbed in 1 s by 1 ha of

from / to	electro-magnetic	chemical	thermal	kinetic	electrical	nuclear	gravitational
electro-magnetic		chemi-luminescence	thermal radiation	accelerating charge phosphor	electro-magnetic radiation electro-luminescence	gamma reactions nuclear bombs	
chemical	photo-synthesis photo-chemistry	chemical processing	boiling dissociation	dissociation by radiolysis	electrolysis	radiation catalysis ionization	
thermal	solar absorption	combustion	heat exchange	friction	resistance heating	fission fusion	
kinetic	radiometers	metabolism muscles	thermal expansion internal combustion	gears	electric motors electro-strictions	radioactivity nuclear bombs	falling objects
electrical	solar cells photo-electricity	fuel cells batteries	thermo-electricity thermionics	conventional generators		nuclear batteries	
nuclear	gama-neutron reactions						
gravitational				rising objects			

1.6 Matrix of energy conversions.

CHAPTER 1

Iowa cornfield on an overcast July day, although each of these energies amounts to 30 MJ. A common denominator does not differentiate between renewable and fossil fuel energies, yet this distinction has many fundamental implications for an energy system. Where animate labor remains an important source of mechanical energy, there is no easy way to compare such energy inputs with those contributed by fuels and electricity. Should the animate energy inputs be counted in terms of all the feed consumed by draft animals (including those kept just for reproduction) or just in terms of the work they actually accomplished? Total food intake should not be counted as an energy input for working humans (they spend most of it on basal metabolism and a part of it on nonlabor tasks), but multiple assumptions are then needed to quantify the labor-related energy input.

And how to account for mental work, the intellectual input whose direct metabolic costs are negligible but whose development (years of education) is clearly very energy-intensive? This question becomes especially vexing in a modern high-energy society, where human labor is overwhelmingly invested in nonphysical tasks. Even when the comparisons are limited to fossil fuels, there is a considerable loss of information in conversion to a common denominator. Some qualitative differences are important even before these fuels are burned: they determine the ease, costs, and impacts of extraction, transportation, and distribution. And during combustion, maximum flame temperatures obtainable from different fuels set the limits on the availability of useful energy and the flexibility of final conversions. Moreover, equal amounts of available energy may be accompanied by vastly different environmental effects. Odum (1996) tried to take these important qualitative differences into account by calculating specific energy transformities based

on sequential upgrading of solar radiation, but these largely arbitrary and often questionable constructs do little to resolve the irresolvable (see section 12.1).

Differences of energy form, availability, density, extractability, transportability, ease of conversion, pollution potential and convenience, and safety of use cannot be subsumed by a single denominator. Equally intractable are comparisons of different electricity generation modes. In the case of fossil-fueled generation, the energy equivalent of electricity is clearly the heat content of the consumed fuel. But what is the primary energy content of nuclear electricity when the fission reactors use only a small portion of uranium's thermal potential? And should hydroelectricity be converted by using its thermal equivalent (1 kWh = 3.6 MJ) or the prevailing rate of thermal generation whereby 1 kWh = 9–15 MJ? A country that generates only hydroelectricity will appear to have a relatively small primary energy use if the first conversion is used, and a highly inflated one if the second conversion is applied, because it is most unlikely that a nation would produce as much electricity if all of it were to be generated by burning fossil fuels. Both the United Nations and BP employ a hybrid solution, using the thermal equivalent for hydroelectricity and the prevailing efficiency rate (about 33%) of fossil-fueled generation for nuclear electricity.

Energy efficiency is captured by yet another set of measures hiding considerable complexity (Lovins 2004). There is no single or best yardstick to assess the performance of energy transformations; the most commonly used ratio is not necessarily the most revealing one; the quest for the highest rate is not always the most desirable goal; and inevitable preconversion energy losses may be far greater than any conceivable conversion improvements. The ratio of energy output or transfer of the desired kind achieved by a converter to the initial energy input to the device, organism, or system is by far the most commonly used value. This first-law efficiency (e_1) has at least three important drawbacks: its maximum values may be less than, equal to, or greater than 1 (in the last case, it is usually called a coefficient of performance); it does not capture the efficiency limitations imposed by the second law; and it is not readily applicable to systems whose desired output is a combination of work and heat.

The second-law (or exergy) efficiency (e_2) is the ratio of the least available work that could have performed the task to the available work actually used in performing the job with a given device or system (Ford et al. 1975). Clearly, this efficiency cannot surpass unity. It offers direct insight into the quality of performance relative to the ideal (minimization of inputs as the goal of energy management), and it focuses attention on the desired task, not on a device or a system currently used for that purpose. For example, overall energy and exergy efficiencies will both be about 40% for the best coal-fired electricity-generating plants, but while steam generation has very high e_1 (with some 95% of the input energy transferred to water), it is only about half as efficient in exergy terms, e_2 (M. A. Rosen 2004). Similarly, the e_1 of standard household natural gas furnaces is about 75%, but the e_2 will be only about 10% because the equivalent heating could have been done much more efficiently by a heat pump. All conversions where high-temperature combustion is used to provide low-temperature heat will be exposed as similarly wasteful using e_2.

Calculations of e_2, always more difficult than appraisals of e_1, often run into social and behavioral concerns. For example, the e_2 of cars will not be calculated with respect to the broadly defined task of moving people speedily and comfortably around but strictly with respect to the narrowly defined task of supplying kinetic energy at the drive wheels. How then to factor in the unoccupied seats, or what is indeed the true meaning of e_2 in the case of completely frivolous cruising? And using an equivalent to e_2 in comparing the efficiencies of photosynthetic or heterotrophic conversions is even more questionable. Because of different photosynthetic pathways, wheat is a less efficient converter than sorghum, but sorghum does not produce bread-making flour. And conversion of grasses to beef shows inferior e_2 compared to bleeding the animals, but are Americans willing to behave like Maasai?

As for the quest for maximum efficiency, practical limits were already recognized in Carnot's writings, and the subsequent history of engineering design of both prime movers and converters has been always marked by compromises taking into account bearable costs, acceptable reliability, and desired power outputs. Infinitesimally slow conversions produce maximum possible efficiencies, but transformations done at socially and economically rewarding rates inevitably carry high heat waste penalties. Similarly, Lotka's (1925) law of maximum energy recognized that optimal efficiencies are considerably lower than the maxima that would limit the power output necessary for growth and maintenance of biota. Moreover, in many instances lower efficiencies are preferable to practices whose energy conversion efficiencies are higher but whose utility, manageability, and ensuing productivity or comfort are greatly inferior. And what is the efficiency of hydroelectric generation? Is the initial input the potential energy of the precipitation in the entire watershed, of the water behind the dam, of the storage above the penstock intakes (including all evaporation losses), or only of the water actually used in generation?

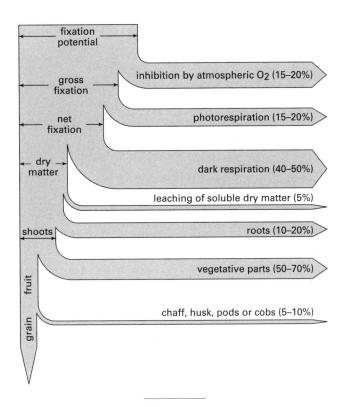

1.7 Energy losses along the photosynthetic, respiratory, and harvesting pathways in a temperate grain field.

Similarly, very different efficiencies result when tracing the fate of solar radiation incident in a temperate grain field: no more than 0.1% of solar irradiance may be converted into chemical energy in grain, but the photosynthesis is actually energized with only about 10% of all incident radiation owing to a sequence of large prefixation losses (fig. 1.7): redefinition of inputs boosts the efficiency by 1 OM. And why should all residual phytomass (roots, shoots) be excluded from efficiency calculations when it provides excellent livestock feeds as well as irreplaceable ecosystem services after its return to the soil?

Photosynthetic efficiency can be at least doubled by counting this phytomass. In any case, higher photosynthetic efficiency may actually be counterproductive if the grain is fed to cattle whose inefficient metabolism produces fatty meat, the consumption of which promotes obesity and cardiovascular mortality. Clearly, worshipping higher energy efficiency as an abstraction is a dubious faith.

The primary analytical tools used in this book (in addition to the measures already noted) make matters even more complicated because they have no universally accepted, binding definitions. Perhaps the best example is the treatment of power density in *Encyclopedia of Energy* (2004). In this comprehensive, authoritative treatment of energy matters, four authors define this rate in four different ways: as "power per unit of volume" (W/m^3) in the first volume (Thackeray 2004); as "power per unit of weight" (W/kg) in the third volume (German 2004); as "power per unit of land area" (W/m^2) also in the third volume (Smil 2004b); and as "energy harnessed, transformed or used per unit area" (J/m^2) in the sixth volume (Grübler 2004).

All of these rates are used in practice by different disciplines. Nuclear engineers use volume power density in order to express the energy release in reactor cores: it ranges, expressed usually in kW/dm^3, from just 1–5 for MAGNOX and advanced gas reactors to 70–110 for the most widespread pressurized water reactors (Zebroski and Levenson 1976). Both volume (W/m^3) and weight (W/kg) power density are commonly used to quantify the performance of batteries. Radio engineers and builders of wind turbines calculate power density from both isotropic and directional antenna with area in the denominator; for an isotropic antenna the density in W/m^2 is simply a quotient of the transmitted power and the surface area of a sphere at a given distance, $P_t/4\pi r^2$. I should also add that the official list of SI-derived units with special names calls W/m^2 "heat flux density" or "irradiance," has no special name reserved for W/m^3, and calls J/m^3 "energy density" (BIPM 2006).

Consequently, readers should be forewarned that the definitions of key rates used in this book cannot coincide with every definition found elsewhere, but (in the absence of universal agreement) all measures are unambiguously defined and used in a logical and consistent manner. My leading choices for universal measures are applicable to all segments of energetics. No matter if the processes are animate or inanimate, natural or anthropogenic, they can be profitably studied in terms of energy and power densities and intensities, the rates that relate energy and power to space and mass (or volume), making it possible to express key physical realities in a few simple yet revealing measures.

Energy density (J/m^3) and *specific energy* (J/kg) convey critical information about fuels and foodstuffs, telling us how concentrated sources of energy they are. This attribute is decisive for all portable applications or where space and weight are at premium: airplanes cannot cross oceans powered by natural gas, a fuel whose volume density is only 1/1,000 that of aviation kerosene, and Himalayan climbers do not subsist on carrots, whose specific energy is 1.7 MJ/kg, one-tenth that of power bars. Specific energy is also used to sum up the amount of fuels and electricity that have been invested in the extraction of minerals, the production of food and manufactures, the distribution of goods, and the provision of services. This quantity, also called *energy cost* or *energy intensity* (MJ/kg or GJ/t), reveals relative energy needs (aluminum needs 1 OM more energy than steel) and informs about low-energy-intensity substitutions.

Energy intensity of energy (J/J), expressed simply as a fraction of gross energy content, appraises the net energy gain of commonly used fuels and electricity. It is very high for natural gas from giant Persian Gulf fields but actually negative for early versions of corn-based ethanol fermentation. *Energy intensity of an economy* (J/unit of currency, J/$ for international comparisons, and J/$ in constant monies for long-term series) is a valuable indicator (albeit one that needs careful interpretation taking into account national peculiarities) of both the stage of economic development and the overall efficiency of economic performance. Secular and global comparisons of energy intensities of national economies lead to many interesting questions about the magnitudes and causes of indicated disparities.

Energy concentration (J/m^2) reveals the spatial density of resources (be they fossil fuels or trees), a critical determinant of extraction or harnessing methods and costs and of associated infrastructure needs. When applied to natural ecosystems the measure is a fine surrogate for specific richness and diversity. In heavily managed ecosystems, such as crop fields or fast-growing tree plantations, it informs about harvest possibilities. In urban and industrial areas it expresses the levels of intensification either in terms of habitation densities or as energy incorporated in structures and infrastructure of these areas. And in all instances it reminds us of inexorable limits to future growth.

Power density (W/m^2) is perhaps the most revealing variable in energetics, and in this book it is used in two different ways: either with the generated power prorated over the area of an energy converter (perpendicular to energy flow, and hence either vertical or horizontal), or (much more often) with harnessed, generated, or used power over the land or sea surface (prorated only horizontally).

In the first instance, the measure is of fundamental physical importance in determining the maximum performance of individual energy converters in all cases where the transformation of energy can be envisaged as proceeding within a certain volume with one form of energy supplied into this volume across its surface and the transformed energy leaving that surface. The flux of energy through the working surface of such a converter is then the product of the velocity of propagation of a distur-

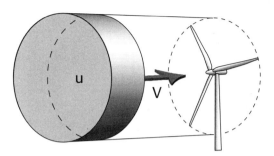

Umov-Poynting vector = uV

1.8   Derivation of the Umov-Poynting vector for a wind turbine. Based on Kapitsa 1976.

bance (v, be it a kinetic or a heat flow) and its energy density (U, measured in J/m^3), which is limited by the physical properties of the working medium. Figure 1.8 illustrates the concept for wind turbines. The vector $\mathbf{v}U$ can be used to assess the limits of any energy converter.

The vector was originally suggested in 1874 by a Russian physicist, Nikolai A. Umov (1846–1915), and a decade later an English physicist, John Henry Poynting (1852–1914), described the propagation of electromagnetic waves as a product of electric and magnetic fields expressed in W/m^2 (Poynting 1885). Application of the Umov-Poynting vector was favored by Pyotr L. Kapitsa (fig. 1.9), a student of Rutherford and a Nobel laureate in physics (1978), for being a fundamental and convenient characteristic of energy converters. Kapitsa (1976, 10) argued that ignoring the restrictions of various energy fluxes revealed by the vector "results in wasting money on projects that can promise nothing" in terms of supplying high-power energy production needed by modern societies.

Kapitsa recalled how the use of this rather simple measure refuted the Russian Academician A. F. Ioffe's idea of

1.9 Pyotr Kapitsa (1894–1984), Nobel prize in physics, 1978. Photograph © The Nobel Foundation.

of temperature, and a constant dependent on the molecular composition of the gas. Its use also illustrates the limits of high conversion efficiencies. Transformation of chemical energy into electricity in fuel cells is more than 70% efficient, but the low diffusion rates in electrolytes limit the power density to some 200 W/m^2 of the electrode, too low for a centralized base load supply in a high-energy society (Brandon and Thompsett 2005).

The second and broader meaning of power density—the rate of energy flow per unit of surface area (rather than per unit of the working surface of a converter)—is perhaps the most critical parameter determining both the structure and the operating modes of energy conversion systems. I use it for systematic comparisons of all important inanimate fluxes (from solar constant to geothermal energy), for autotrophic and heterotrophic conversions (from photosynthetic performance to predator feeding), and for all principal modes of past, present, and contemplated anthropogenic energy conversions (from biomass to electricity generation and including waste heat rates).

Specific power (W/g) is used to characterize intensities of autotrophic and heterotrophic metabolism (adult basal metabolic rate is ~1 mW/g) as well as those of inanimate transformations (car engines develop ~1 W/g). When looking at the performance of prime movers, from steam engines to rockets, it is better to use the reverse (g/W) in order to emphasize the importance of lightweight conversion devices in transportation. For their first flight the Wright brothers built a four-stroke engine (with aluminum body and steel crankshaft) that weighed 8 g/W; the most advanced turbofan engines that power widebodied jets weigh less than 0.1 g/W. Finally, *metabolic intensities* (g/J), whether of recyclable nitrogen in domestic animals or SO$_2$ emissions by different combustion

replacing electromagnetic generators by the electrostatic ones for large-scale electricity production. Two clear advantages would be simpler construction and the possibility of direct feeding of high voltage to transmission lines. But to avoid sparking, the electrostatic field is restricted by the dielectric strength of the air, and to produce 100 MW the electrostatic rotor would have to cover about 400,000 m^2, a clear impossibility compared to about 10 m^2 for an identically rated electromagnetic generator with a power density vector of about 1 kW/cm^2.

Where combustion is involved, the Umov-Poynting vector is the product of gas pressure, the square root processes, call attention to the material requirements of energy fluxes and their environmental impacts.

Using these general measures, and augmenting them by more specific quantifications, this book proceeds in an evolutionary sequence. Chapter 2 examines terrestrial energy fluxes, starting with solar radiation and concluding with the energetics of geomorphic processes. This sets the stage for reviewing both the bioenergetic fundamentals and the specific and ecosystemic peculiarities of the plant and animal kingdoms. Chapter 3 on primary productivity and chapter 4 on heterotrophic conversions are followed by a closer look at human energetics in chapter 5. From that point on, the book concentrates on the energetic exploits and limit of our species. Brief notes on the energetics of foraging societies are followed by more extensive analyses of traditional (solar) farming and preindustrial complexification based largely on animate power and biomass fuels (chapters 6 and 7).

The energetics of modern civilization is discussed, focusing first on its resource foundation and key conversion techniques, that is, by appraising stores and combustion of fossil fuels, generation of electricity, and development and ratings of inanimate prime movers (chapter 8). Then the perspective widens to embrace general patterns and trends of energy use in modern societies (chapters 9 and 10) and to survey environmental implications and socio-economic complexities that accompany this high-energy way of life (chapters 11 and 12). The book closes with a juxtaposition: chapter 13 contrasts grand energetic generalizations concerning planetary flows, the biosphere, and the advancement of civilization with thoughts on the inadequacies inherent in approaches that, though fundamental and universal, are still insufficient to be the measure of all things.

THE UNIVERSAL LINK

2

PLANETARY ENERGETICS

Atmosphere, Hydrosphere, Lithosphere

... consider first of all
Seas, lands, and sky; those threefold essences,
Those bodies three, those threefold forms diverse,
Those triple textures vast ...
Lucretius (c. 99–c. 55 B.C.E.), *De Rerum Natura*

Even when compared to its closest planetary neighbors
the Earth is unique. Its improbable atmosphere is dominated by molecular nitrogen and oxygen and contains
many trace gases (CH_4, N_2O, NH_3) whose concentrations violate the rules of equilibrium chemistry to an
infinite extent (Lovelock 1979). This atmosphere is a
creation of life. The atmosphere of a lifeless Earth would
resemble those of Venus and Mars: 98% CO_2, less than
2% N_2, and a mere trace of O_2; the surface temperature
would be about 563 K (290°C), above the melting point
of tin (232°C); and the surface pressure would be a few
MPa rather than just 101.325 kPa (1 atmosphere). Only
a small share of solar radiation is needed to power the

planet's atmospheric circulation and its massive water
cycle, the two processes that redistribute heat and moisture and that are responsible for atmospheric and hydrospheric fluxes ranging from ephemeral (lightning) to
continuous (rivers), from highly localized (tornadoes)
to very extensive (monsoon). And only a tiny part of the
received radiation is converted by photosynthesis into
phytomass, whose production sustains an enormous variety of heterotrophs.

But photosynthesis needs much more energy indirectly. Without surface temperatures high enough to
keep water liquid (at least seasonally) and to allow for
rapid rates of biochemical reactions, there would be no
excitation of chlorophyll, no absorption of nutrients
through roots, no decomposition of dead organic matter. And solar energy transformed into winds and
flowing water has indispensable indirect roles in photosynthesis, ranging from pollination of tens of thousands
of gymnosperm and angiosperm species to denudation

of continents and mobilization of needed plant nutrients.

The Earth's uniqueness goes deep below its surface. The planet's lithosphere is constantly recycled into the underlying mantle as the incessant motions of giant tectonic plates create new ocean floor, form mountain belts, and generate earthquakes, volcanic eruptions, and tsunamis.

The uniqueness of these tectonic arrangements results in a giant material cycle: an internal heat engine in the mantle extrudes magma to form laterally rigid plates that are pushed away from mid-ocean ridges, and the floor is eventually subducted back into the mantle in order to be reprocessed into new magma (Eiler 2003). Without these tectonic processes there would be no re-creation of continents and oceans, and the Earth's surface would be flattened by erosion. Such an undifferentiated world would not have been conducive to evolutionary diversification of life. Accounts of planetary energy balances and of thermal and kinetic energy flows in the atmosphere, hydrosphere, and lithosphere could thus be seen as setting the grand stage for all subsequent inquiries into the energetics of life and civilization.

2.1 Sun: The Star and Its Radiation

Not everything on the Earth is energized by solar radiation: it does not keep the planet on its orbital path, it does not drive the plate tectonics, the process that constantly reshapes oceans and continents (see section 2.5), and it does not power the metabolism of chemoautotrophic bacteria, most notably those that live in total darkness at the bottom of deep ocean near hot vents, where they oxidize H_2S and support many larger organisms, including white clams, crabs, and giant tube worms. But chemoautotrophs aside, the whole pyramid of life

rests on photosynthetic conversion of solar radiation to phytomass while precipitation, ice, and wind—all just thermal and kinetic transformations of solar radiation that was absorbed by the atmosphere, land, and waters—keep reshaping the planetary surfaces and are the key determinants of the food-producing potential of every civilization.

The source of this radiation, the Sun, is one of roughly 100 billion radiant bodies in our galaxy, a star located in one of the spiral arms of the Milky Way galaxy, about 33,000 light-years (ly) from its center and near the galactic plane (Phillips 1992). The Sun's closest neighbors are the binary Alpha Centauri (4.3 ly distant) and Epsilon Eridani (10.8 ly distant), and it is visible with naked eye from a distance of 20 parsecs, that is, about 620 Pm, or roughly 4.1 million astronomical units (AU, the mean distance between the Earth and the Sun). Its size (class V) makes it a dwarf star (like Sirius or Vega); its spectrum is G2, yellow stars with characteristic lines of ionized calcium and metals. Its mass is 1.991×10^{33} g (5 OM larger than that of Earth); the radius of its visible disk spans 696.97 Mm (compared to Earth's $\sim$6.5 Mm); its average density is 1.41 g/cm^3; and its steady radiation corresponds to a surface temperature of nearly 5800 K.

The Sun is almost perfectly in the middle of the main sequence of the Hertzsprung-Russell diagram, which plots the distribution of absolute visual magnitude against the spectral class for stars of known distance (fig. 2.1). Hydrogen makes up about 91% of its huge mass, helium all but about 0.1% of the rest; C, O, and N are the most abundant elements among the minor constituents. All of the Sun's energy is produced within the innermost quarter of its radius, which encloses a mere 1.6% of its volume, and most of it (nearly 80%) comes from the proton-proton (p-p) cycle, the net result of

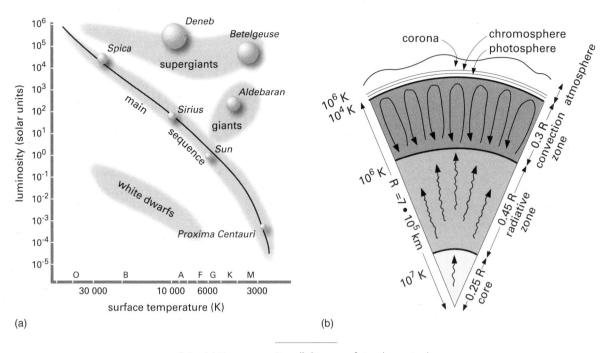

2.1 (a) Hertzsprung-Russell diagram of visual magnitude and spectral classes of stars. (b) A section through the Sun's interior. Based on Kaler (1992).

which is the fusion of four protons to form a helium nucleus (Phillips 1992):

$$^1H + {}^1H \rightarrow {}^2D + \text{positron} + \text{neutrino}$$

$$^1H + {}^2D \rightarrow {}^3He + \gamma \text{ rays}$$

$$^3He + {}^3He \rightarrow {}^4He + {}^1H + {}^1H$$

The probability of the first step is extremely low, and random collisions at a speed sufficient to overcome the electrical barrier take place only because of the enormous quantity of protons in the Sun's core. The second step is accomplished almost instantaneously; the fusion of two ^3He atoms returns two free protons available for another p-p reaction. This cycle provides the bulk of thermonuclear energy in all stars with core temperatures of up to about 17 MK; above that level the much less important carbon-nitrogen cycle becomes dominant (Cox, Livingston, and Matthews 1991; Kaler 1992; Phillips 1992). Approximately 9.2×10^{37} thermonuclear reactions take place every second (every one liberating 4.2×10^{-27} J), and they consume 4.4 t of the Sun's core every second. According to the Stefan-Boltzman law, the radiant flux (F, the total energy radiated per unit of area) is proportional to the fourth power of temperature,

$F = \sigma T^4$,

with σ (Stefan-Boltzman constant) equal to 5.67×10^{-8} $W/m^2/K^4$. The Sun's isotropic radiation thus produces an average flux of 63.2 MW/m^2 of the photosphere's surface, corresponding to the effective temperature of 5778 K. Total luminosity is the product of the star's surface and total flux:

$$F = 4\pi r^2 \sigma T^4.$$

The Sun's immense energy flux of 3.85×10^{26} W is transported outward through the radiation zone ($\sim$45% of the star's radius) and then through the outermost convection zone to the photosphere. Temperature declines from 10^7 K in the core to 10^6 K in the radiative zone and 10^4 K in the outer layer of the convection zone. The thin ($\sim$200 km), finely granulated, and opaque photosphere radiates the bulk of both visible and infrared wavelengths (fig. 2.1). The photosphere has numerous sunspots, regions of reduced temperature that persist for hours to weeks and whose frequency is subject to an 11-year cycle during which their locations converge toward the solar equator (Radick 1991). The chromosphere separates the photosphere from an extremely hot corona ($\sim$2 MK, and $\leq$29 MK in flares), which emits X-rays (1–10 nm) and extreme (10–100 nm) ultraviolet (UV) radiation. Coronal flares, marked by terrestrial magnetic disturbances and high-latitude auroral displays, follow the 11-year cycle.

Because the conversion of H to He increases Sun's density, and hence the star's core temperature and its rate of thermonuclear reactions, luminosity of the young Sun was about 30% lower than the present rate. This faint young Sun would not have allowed liquid water for the first 2 or perhaps even 3 Ga of the Earth's evolution, yet the oldest sedimentary rocks can be reliably dated to as far back as 3.8 Ga before the present time. Enhanced greenhouse gas effect is the best solution of this paradox, and Sagan and Mullen (1972) proposed that this was achieved by about 10 ppm of atmospheric ammonia. This assumption was questioned owing to the short lifetime of NH_3, caused by photochemical decomposition, and Sagan and Chyba (1997) suggested that UV absorption by high-altitude organic solids produced from CH_4 photolysis may have shielded ammonia from such decomposition. But higher atmospheric CO_2 levels (equilibrated by feedbacks within the carbonate-silicate cycle and later amplified by biota) provide a much more likely explanation (Kasting and Grinspoon 1991).

After some 5 Ga of radiation, hydrogen will have been depleted from 75% to 35% of the Sun's core mass. As the core continues to contract, the star's radiation will increase, and its diameter will expand. Eventually the heat will evaporate the oceans as the lifeless Earth continues orbiting the brightening Sun. Then, after some 5 Ga, the yellow star will leave the main sequence and be transformed into a red giant, whose diameter will eventually be 100 times larger than it is now (the star's photosphere will reach beyond the orbit of Mars) and whose luminosity will be about 1,000 times greater than today's. Finally, after it has swallowed the four inner planets, the Sun will become a brilliant, Earth-size, white dwarf. But it is the Sun's remarkably stable past, due in large part to the star's fortuitous but fortunate galactic location, that has made life possible.

The Sun's orbit around the galactic center is less elliptical than those of similar nearby stars, and this prevents the solar system's passage through the inner galaxy with its many destructive supernovas. A small inclination of the

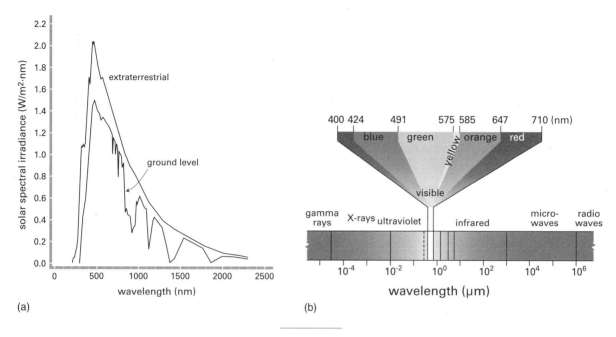

2.2 (a) Extraterrestrial and ground-level irradiance.
(b) Electromagnetic spectrum. From Smil (2002).

star's orbit prevents the Sun from an abrupt crossing of the galactic plane that would stir up the Oort cometary cloud. Moreover, the Sun, unlike most of the stars, is solitary and hence much more likely to generate suitable planetary orbits (Donnison and Mikulskis 1992). And unlike the more massive stars with their very high rates of thermonuclear reactions, the Sun has been sufficiently long-lived to allow for billions of years of life's evolution. The energy output of much smaller and long-lasting stars could be too low, or it could be almost totally in infrared (IR) and too variable.

In contrast, the solar spectrum ranges from wavelengths less than 0.1 nm (γ rays) to more than 1 m (fig. 2.2), resembling radiation from a blackbody (an object

with a continuous radiation spectrum in all wavelengths that absorbs all radiant energy reaching it) at about 5800 K. According to the Wien's displacement law,

$$\lambda_{max} = \frac{0.002898}{T\ (K)},$$

the maximum wavelength corresponding to that photospheric temperature is almost exactly 500 nm, near the lower limit of green light (491–575 nm). According to Planck's law, which expresses the distribution of energy emitted from a black body as a function of its wavelength and temperature, the energy flux density emitted at this maximum wavelength is about 2 W/m$^2 \cdot$ nm.

PLANETARY ENERGETICS

As a result, nearly 40% of all solar energy is carried by visible wavelengths, only about 8% by UV (<400 nm), and 53% by IR frequencies (>700 nm). Longer UV (UVA, 320–400 nm) is not very harmful to organisms, but shorter wavelengths (UVB, 290–320 nm) burn skin, and the highest frequencies are lethal to most organisms, particularly marine phytoplankton (de Mora, Demers, and Vernet 2000). Visible wavelengths, from 400 nm (the deepest violet) to 700 nm (the darkest red), energize photosynthesis and are sensed by heterotrophs (fig. 2.2). The peak sensitivity of human eyes is in green (491–575 nm) and yellow (576–585 nm) light, and maximum visibility is at 556 nm (3.58×10^{-19} J/photon). Most of the heat supplied to the biosphere comes from wavelengths shorter than 2 μm (less than 1×10^{-20} J/photon).

There is hardly any attenuation of the Sun's radiation as it travels through interplanetary space, and the irradiance at the top of the Earth's atmosphere is simply the quotient of the Sun's total energy flux (3.89×10^{26} W) and the area of the sphere with the radius equal to the Earth's mean orbital distance from the star (149.6 Gm), or about 1370 W/m². This rate is known as the solar constant, and atmospheric interference makes it impossible to detect its subtle changes by ground-based monitoring. During the 1960s and the early 1970s, NASA used a solar constant of 1353 W/m² as the design value for its space vehicles, although unsystematic pre-1980 measurements from high-altitude balloons, planes, rockets, and spacecraft indicated an increase of 0.029%/a after 1967 (Fröhlich 1987).

Systematic extraterrestrial studies of solar irradiance began with the Earth Radiation Budget (ERB) radiometer on Nimbus 7 in 1978, and they were followed by the Active Cavity Radiometer Irradiance Monitor (ACRIM I) in 1980, the Earth Radiation Budget Satellite (ERBS) in 1984, ACRIM II in 1991, Variability of Solar Irradiance and Gravity Oscillations (VIRGO) monitoring on the Solar and Heliospheric Observatory (SOHO) in 1996, and ACRIM 3 in 2000 (Foukal 1990; de Toma et al. 2004). A composite record of 24 years of solar irradiance reveals the average of 1366 W/m². Short-term dips of up to 0.2–0.3 W/m² correspond to the passage of large sunspots across the Sun, and long-term undulations are due to the 11-year solar cycle with peaks about 1.3 W/m² above the minimum of 1365.6 W/m² (de Toma et al. 2004). Spectral irradiance during the 11-year cycle varies at least 100% for wavelengths shorter than 100 nm, but the total declines only by about 0.1% between the cycle's peak and trough (NRC 1994). Variations in solar luminosity have not had a significant influence on global warming since the seventeenth century, but additional climate forcing due to enhanced solar UV output cannot be ruled out (Foukal et al. 2006).

A reconstruction of the sunspot frequency during the past 11,400 years (based on dendrochronologically dated radiocarbon) shows that the level of solar activity has been exceptionally high since about 1940 but that this episode (with more than 70 sunspots a year compared to an average of about 30 during the past millennium) is unlikely to continue for more than two or three decades (Solanki et al. 2004). Satellite observations show that the Sun's shape and temperature vary with latitude in a complex way, and indicate that variations of radius and luminosity do not originate in the star's inner depths but rather in its outer layers (Emilio et al. 2000). None of this changes the mean value of about 1366 W/m² for the solar constant, the base for tracing the conversions of solar radiation within the planet's interacting spheres.

CHAPTER 2

2.2 Energy Balance of the Earth: Radiation Fluxes

Although the disk with the Earth's diameter (12.74 Mm) catches only a tiny fraction of total solar output (4.5×10^{-10}), the rate of this intercept, 174.26 PW, and its annual aggregate, 5.495 YJ, are enormous: at the beginning of the twenty-first century the global consumption of all fossil fuels was just above 10 TW, or about 0.006% of the solar irradiance. Total resources of fossil fuels are perhaps as large as 200 ZJ, but even this generous estimate would be no more energy than in the solar radiation intercept by the Earth in only about 13 days. Average insolation (radiation received per m^2 of the planet's surface) is considerably lower: even without any atmospheric interference it would be only a quarter of the solar constant, or about 342 W/m^2 (a sphere's area is four times larger than that of a circle of the same radius). And because the Earth's clouds and surfaces reflect about 30% of the incoming shortwave (SW) radiation, the total actually absorbed by the atmosphere and by the ground is 240 W/m^2 (122 PW, 3.85 YJ/a) and the mean flux reaching the surface is about 173 W/m^2.

Albedo, α, the share of SW radiation that is reflected and scattered back to space without any change of wavelength, can be commonly as high as 0.8–0.9 for thick cumulonimbus clouds and only 0.02–0.03 for wispy cirrus. Fresh snow is as efficient a reflector as thick clouds, and hence the snow-covered parts of the Northern Hemisphere greatly affect the planetary energy balance. Satellite observations indicate average winter albedos of 0.6 in Eurasia and 0.56 in North America, with a Northern Hemisphere mean of 0.59 (Robinson and Kukla 1984). Albedo of older snow falls below 0.50, light sandy deserts rate 0.3–0.4, green meadows 0.1–0.2, and coniferous forests just 0.05–0.15. Albedos of water

surfaces are just 0.04 for the 90° sun angle but 0.6 for the 3° angle (and the Moon averages just 0.07, Mars 0.16).

The Earth's albedo is thus variable on a seasonal basis, and it changes faster on regional and local scales with cloud cover, aerosols in the atmosphere, and the extent of snow and ice. Its presatellite estimates ranged between 0.28–0.42. Remote sensing by satellites constrained the range, but because the planetary albedo cannot be measured directly, theoretical models are needed to derive it from the monitored parameters. Six standard models used in 2005 had annual averages between 0.29–0.31, most had the maximum interseasonal amplitude of less than 0.02, and all of them shared the expected minimum global albedo in September, when the Northern Hemisphere's radiation-absorbing vegetation cover is at its peak (Charlson, Valero, and Seinfeld 2005). The global albedo of 0.3 is largely due to cloud reflection (returning 20% of the incoming SW), with back scattering and surface reflection sending outward, respectively 6% and 4% of all SW input.

Deforestation has been the single largest anthropogenic cause of about 0.01 albedo increase during the historic era. With average irradiance at 342 W/m^2, albedo change of just 0.01 means a global energy balance shift of 3.4 W/m^2, the rate similar to that caused by the doubling of current atmospheric CO_2 level. Unfortunately, the evidence regarding the recent changes in Earth's albedo is inconsistent. Pallé et al. (2004) used earthshine (the sunlight reflected from the Earth's bright side to the Moon and then back to an observer on the Earth's dark side) to conclude that the albedo decreased steadily from 1984 to 2000 (by an equivalent of ~10 W/m^2) and then rose by 1.7% (~6 W/m^2) between 2000 and 2004. Satellite observations indicate a small decrease of 0.6%

($\sim$2 W/m^2) during the same period (Wielicki et al. 2005). Whatever the actual direction and the rate of change, changing albedo affects average global temperatures as well as global ocean heat storage.

All SW radiation that is not reflected gets absorbed and transformed into longwave (LW) flux. The process starts in the stratosphere and continues throughout the troposphere, but most of it takes place on the Earth's surface. Stratospheric O_3 contributes by far the largest share to the drastic reduction of UV waves (<300 nm). Visible light passes through the atmosphere largely unaffected, but water and CO_2 and, to a lesser extent, CH_4, N_2O, and O_3 absorb much of the infrared flux. The total absorption profile shows two peaks in the near infrared and three extended blocks, 1.5–2, 2.5–3.5, and 4–8 μm (fig. 2.3). Changing concentrations of absorptive gases alter both the quantity and the quality of surface insolation.

The fate of stratospheric O_3 is of the greatest concern: its destruction by such anthropogenic pollutants as chlorofluorocarbons and N_2O would have progressively injurious effects on biota, including higher frequencies of erythemal damage, skin cancers, and genetic malformations (see section 11.4). Sky color is due to molecular scattering that diffuses the incoming radiation with efficiency inversely proportional to the fourth power of the wavelength: the blue light at 400 nm is scattered more than nine times as much as the red one at 700 nm. Natural and anthropogenic aerosols (above all, black carbon and sulfates) are a variable but often important cause of SW absorption in the troposphere.

The sources of radiation extinction (O_3-dominated absorption, Rayleigh scattering, and turbidity) can be combined in a single attenuation coefficient (extinction optical thickness) and inserted, together with the depth of the air mass, as the negative exponent in the Beer-Bouguer-Lambert formula for calculating the fraction of the solar constant reaching the surface. Values for these extinction factors are available for individual wavelengths for the standard atmosphere (Thekaekara 1977). Combinations of different air masses (solar zenith angles) and extinction optical thicknesses result in very different surface irradiances. For one air mass and very clear air, the rate is nearly 960 W/m^2 or 70% of the solar constant, whereas with seven air masses and heavy pollution, the rate is merely 133 W/m^2, less than 10% of the solar constant. Fractions of the total energy carried by UV, visible, and IR waves also differ: 5%, 47%, and 48% in the first instance, 0.1%, 30%, and 70% in the other.

The UV disparity is by far the greatest, the reason for the skin-burning effect of noontime summer radiation in high mountains and a key factor in the prevalence of rachitis in heavily polluted nineteenth-century northern cities under low winter sun. Typical attenuation ranges can be also combined to give the daily maxima of clear sky sunlight penetration. Absorption can eliminate 11%–23% of incoming radiation; scattering can return 1.1%–11% and send 5%–15% of the beam downward as diffuse sunlight, leaving 56%–83% in the direct beam. Total mid-latitude peak insolation would then be 970–1203 W/m^2. Typical shares of radiation received under different cloud types are (all for solar altitude of 65°) 85% for cirri, 52% for altocumuli, 35% for stratocumuli, and 15% for nimbostrati. The percentage of possible sunshine ranges seasonally from less than 10% in the world's cloudiest regions (North Pacific and Atlantic) to around 90% in subtropical deserts (peaks in Chile's Atacama, southern Egypt, and northern Sudan).

Because of the selective absorption by atmospheric gases the SW radiation that reaches the Earth's surface has a notably different wavelength profile than does the

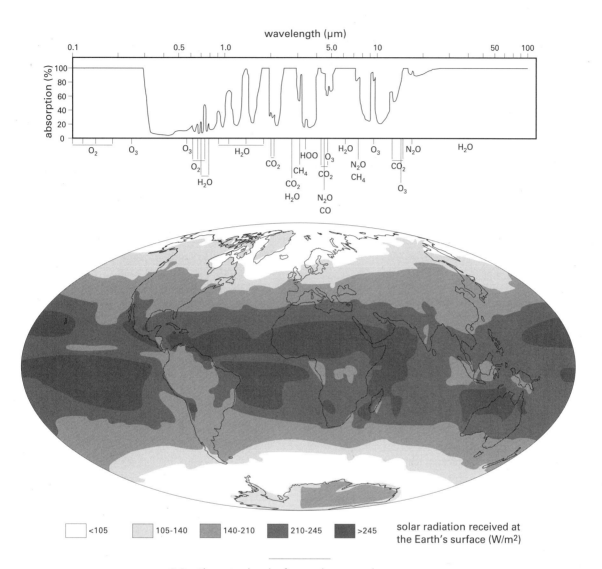

2.3 Absorption bands of atmospheric greenhouse gases;
bottom, annual means of solar radiation received at the Earth's
surface based on 96 months of satellite measurements between
1983 and 1991. From NASA (2005).

PLANETARY ENERGETICS

irradiance at the top of the atmosphere. Total insolation is primarily the function of cloudiness: the highest values, in excess of 250 W/m^2, are in the cloudless high-pressure belts of subtropical deserts (fig. 2.3). A fairly regular poleward decline is unmistakable, as are the relatively low values in the tropics: insolation in the U.S. Corn Belt, the Great Plains, or the Southeast is higher than throughout most of southern Nigeria and the Amazon Basin. But tropics have little seasonal or daily variation. Manaus on the Amazon (3° S) has a daily average of 195 W/m^2, with the highest values in August just 16% above, and the lowest ones in April 13% below, the mean. Ames, Iowa (42° N) averages 167 W/m^2, just 15% less than Manaus, but its December flux is 41% below, and its July insolation 55% above, the mean.

Implications for agriculture are obvious: plants maturing in two to six months will, when provided with adequate moisture and nutrients, perform no worse, and those requiring long days will do much better, in temperate regions. Radiative impoverishment of the tropics is seen even in the peak midday fluxes. Jakarta's 550–580 W/m^2 is no different from the summer fluxes in Edmonton or Yakutsk. Local cloudiness can also create large differences within short distances. Oahu's Koolau Range has an annual mean of 150 W/m^2, and the Pearl Harbor, just 15 km downwind, averages 250 W/m^2. Insolation averages just short of 170 W/m^2 for the oceans and about 180 W/m^2 on land, a flow totaling about 88 PW. Measurements of surface solar radiation found that this flux was decreasing at 0.23%–0.32%/a between 1958 and 1992, and cloud changes, anthropogenic aerosols, and volcanic eruptions were suggested as the major cause of this global dimming, which amounted to 6–9 W/m^2 over land (Pinker, Zhang, and Dutton 2005). This trend did not persist; brightening has been observed since the late 1980s. Wild et al. (2005) put it at 0.68 W/m^2 per year for the clear-sky flux, and Pinker, Zhang, and Dutton (2005) used satellite records to compute an overall annual rise of 01.6 W/m^2 (0.1%) between 1983 and 2001.

All radiation that is absorbed by the Earth's diverse surfaces must be returned to space in order to maintain planetary thermal equilibrium. The most notable characteristic of this outgoing longwave (LW) radiation is its shifted wavelength. Whereas the Sun radiates as a blackbody of about 5800 K with most of the flux between 0.1 μm and 4 μm and the peak at 500 nm, the terrestrial radiation ranges between 4 μm and 100 μm with the peak flux at 10 μm, far into the IR zone. If the Earth were a perfect blackbody radiator, its effective temperature (T_E) would be simply a function of its albedo (α) and its orbital distance (a expressed in AU)

$$T_E = 278 \frac{(1 - \alpha)^{0.25}}{a^{0.5}},$$

and it would radiate at 255 K, the temperature that would leave all of its water frozen. This is not the case, because several atmospheric gases selectively absorb some of the outgoing radiation and reradiate it both downward and upward.

This absorption of IR radiation is commonly known as the greenhouse effect, and Svante Arrhenius (1896) published its first detailed elucidation. In the absence of water vapor on the Earth's two neighbors, it is the presence of CO_2 that generates a greenhouse gas (GHG) effect—a very strong one on Venus, resulting in average surface temperature of 750 K (477°C), and a very weak one on Mars (surface at 220 K). Thanks to GHG absorption the Earth's actual average surface temperature of

288 K (15°C) is 33 K higher than its blackbody temperature. Water vapor is by far the most effective GHG, accounting for almost two-thirds of the 33 K difference; it has five major absorption bands between 0.8 μm and 10 μm, the broadest ones centered at 5–8 μm and beyond 19 μm. Other major GHGs are CO_2, accounting for nearly one-quarter of the forcing (absorption peaks at ∼2.6 μm and ∼4.5 μm, and between 12 μm and 18 μm); O_3 (4.7 μm and 9.6 μm); N_2O (4.5, 7.8, and 8.6 μm); and CH_4 absorbing at about 3.5 μm and 7.6 μm (Kondratyev 1988; Ramanathan 1998). Minor natural contributions come mostly from NH_3, NO_2, HNO_3, and SO_2.

Radiation balance of the Earth at the top of the atmosphere (Q_{ET}) requires that the outgoing LW radiation (Q_{LWT}, counted as negative) be equal to the incoming SW stream (Q_{SWT}) corrected for the planetary albedo (α):

$$Q_{ET} = Q_{SWT}(1 - \alpha) + Q_{LWT} = 0.$$

Global annual value for the net radiation at the Earth's surface (Q_{ES}) must also be equal to 0 because the incoming direct (Q_{DSW}) and scattered (diffused, Q_{SSW}) SW radiation, diminished by the albedo, must be balanced by the LW radiation streaming upward (Q_{LWU}) and downward (LW counterradiation, Q_{LWD}):

$$Q_{ES} = (Q_{DSW} + Q_{SSW})(1 - \alpha) - Q_{LWU} + Q_{LWD}.$$

As with so many scientific concepts, the ideas of planetary energy budgets were turned into systematic and impressively complete accounts during the last decades of the nineteenth century (Voeikov 1884; Homen 1897; D. H. Miller 1969). Advances in our understanding of the Earth's energy budgets are reviewed by Budyko

(1963), Kessler (1985), Kiehl and Trenberth (1997), and R. D. Rosen (1999). Results of global radiation balance observations from satellites are assessed by J. T. Houghton (1984); Ramanathan, Barkstrom, and Harrison (1989); Hatzianastassiou and Vardavas (1999); and Wielicki, Wong, Allan et al. (2002), and solar energy flows at the surface of the Earth have been evaluated by D. H. Miller (1981); Rosenberg, Verma, and Blad (1983); Dabberdt et al. (1993); and NASA (2005).

Individual components of the Earth's mean annual radiation balance have been quantified by dozens of studies during the twentieth century with the extreme differences for some shares being on the order of 50%. Global monitoring from satellites constrained some of these uncertainties: NASA's Earth Radiation Budget Experiment (ERBE) began in 1984; the first component of Clouds and the Earth's Radiant Energy System (CERES) became operational in 1997; and the satellite carrying the first sensors of the European Geostationary Earth Radiation Budget Experiment was launched in 2002 (ERBE 2005; CERES 2002; GERB 2005). But significant differences remain even when comparing only the budgets that have been published since 1990.

For some models the differences between SW and LW fluxes at the top of the atmosphere give radiative surpluses greater than 5 W/m^2; others have small deficits. At the Earth's surface, modeled annual global averages for net SW and LW fluxes range, respectively, from 142–172 W/m^2 to 40–68 W/m^2, and the net all-wave (AW) radiation ranges between 99 W/m^2 and 128 W/m^2 (Hatzianastassiou and Vardavas 1999). For comparison, global July and January means of SW and LW surface fluxes derived from NASA's Surface Radiation Budget Project, conducted between 1983 and 1991, are, respectively, 158 W/m^2 and 162 W/m^2, and 47 W/m^2

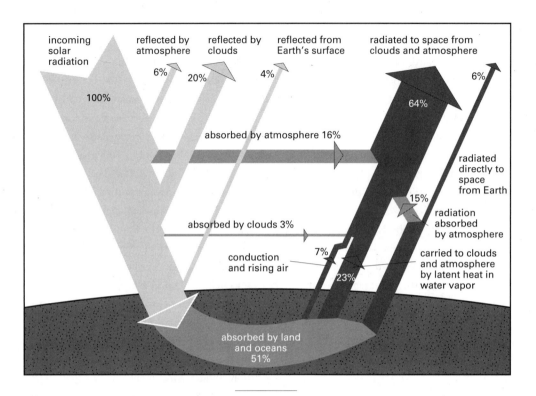

incoming solar radiation

reflected by atmosphere

reflected by clouds

reflected from Earth's surface

radiated to space from clouds and atmosphere

6%

20%

4%

100%

64%

6%

absorbed by atmosphere 16%

radiated directly to space from Earth

15%

radiation absorbed by atmosphere

absorbed by clouds 3%

conduction and rising air

7%

23%

carried to clouds and atmosphere by latent heat in water vapor

absorbed by land and oceans 51%

2.4   Earth radiation balance. From ERBE (2005).

and 48 W/m², and the values of net AW radiation, 111 W/m² and 114 W/m² (NASA 2005).

Moreover, two decades of accurate data from satellite observations have shown that the decadal variability of both SW and LW fluxes at the top of the atmosphere is much larger than previously thought. Since the early 1990s tropical anomalies of emitted LW flux as large as 5–8 W/m² have been associated with a decrease in tropical cloudiness (due to El Niño events), and SW-reflected flux showed a large increase (nearly 10 W/m²) between 1991 and 1993 (due to aerosols from the Mount Pina-

tubo eruption) and decreased subsequently (Wielicki, Wong, Allan et al. 2002). Chen, Carlson, and Del Genio (2002) ascribed these changes primarily to decreased convection, cloudiness, and upward motion above Indonesia, whereas at other longitudes along the intertropical convergence zone these processes actually increased.

Graphic representation of the global radiation balance (fig. 2.4) charts the multiplicity of flows, with the mean values given as percentages of the mean incident solar radiation at the top of the atmosphere. Latent heat flux is determined by the net radiation at the surface and over the land areas by the availability of water to be evaporated. Sensible heat flux is relatively unimportant as long as there is sufficient water for evaporation. Peaks of its annual means, around 80 W/m², coincide with the areas of high aridity, while most of the oceans have yearly means below 10 W/m². About 60% of the global mean of 17 W/m² derives from the land, and some 45% of radiation absorbed by the continents goes into this turbulent exchange with the atmosphere. Daytime means in dry inland locations can be up to 200 W/m², and the maxima can reach up to 400–500 W/m². Relatively humid mid-latitudes have peaks around 150 W/m² and averages of 50–60 W/m².

Only the central areas of large cities go much above these levels, especially where the absorbed radiation and the latent heat in moist air inside high-rise buildings are removed by air conditioning into the surrounding space. There the sensible heat fluxes may rival those above the deserts, reaching over 300 W/m² in the early afternoon. Sensible heat flux is usually of minor importance in all well-watered ecosystems, but it is a critical defense against overheating during dry spells when plants translocate heat from the soil to leaves and from exposed leaves to the shaded ones to maintain tissue temperature within optimum range. The Bowen ratio relates sensible and latent heat fluxes: the quotient is mostly about 1 for the continents and about 0.1 for the oceans, resulting in a global mean of about 0.3.

As for the incoming (downward) LW radiation emitted between 4 μm and 50 μm by the triatomic atmospheric molecules, stratospheric O_3 contributes 15–20 W/m², CO_2 adds 70–75 W/m², and water vapor emits 150–300 W/m². This flux is just an internal subcycle, a temporary delay in the outward flow of reradiated heat,

but it is a critical determinant of tropospheric temperatures and the largest supplier of energy to nearly every ecosystem (Miller 1981). The annual global mean of the flux is around 320 W/m², with mid-latitude continents receiving around 300 W/m² and cloudy equatorial regions up to 400 W/m². Diurnal ranges are mostly between 20 W/m² and 50 W/m², and the flow weakens little (25%–30%) even in winter. Momentary variations can be fairly large: a passing cloud can boost the flux by 25 W/m², and changing levels of greenhouse gases are the major source of its intensification.

Finally, on the second most important extraterrestrial influence on the Earth, its encounters with extraterrestrial bodies. Objects up to 220 m in diameter produce only air blast (Bland and Artemieva 2003), but there are about 1,200 near-Earth asteroids with diameters of 1 km or more, and collision with such objects would have global consequences (Stuart 2001). Fortunately, by the beginning of 2007 the international telescopic Spaceguard Survey had already identified about 700 of these bodies, and it found that none of them is on a trajectory that would lead to collision with the Earth during the twenty-first century (NEOP 2007). Impact energy of a 1-km body would be equivalent to the release of close to 100 Gt TNT (1 t TNT = 4.184 GJ), nearly 1 OM more of energy than would have been expended by an all-out thermonuclear war in 1980 (Sakharov 1983). If such an object were to enter the ocean, the impact would generate a large tsunami, and the principal global effect of a continental impact would be due to an immense mass of shattered material lifted high into the atmosphere, resulting in a drastic drop of temperature, extensive deposits of dusts, and long-term reduction of plant productivity.

2.3 Hydrosphere and Atmosphere: Thermal and Mass Fluxes

Oceans cover just over 70% of the Earth (to an average depth of 3.8 km), store 96.5% of the Earth's water, are the source of about 86% of all evaporation, and receive 78% of all precipitation. But water's importance for the planet's energy budget rests more on its extraordinary properties than on its area and mass. A mere 0.0009% of the Earth's water is in the atmosphere (where it accounts for just 0.3% of mass and 0.5% of volume), but this relatively tiny amount suffices to cover always about 60% of the planet by clouds, and it is the decisive absorber and radiator of incoming and outgoing radiation. In soils (which contain 0.001% of the Earth's water) and plants (0.0001% of the Earth's water), it is the carrier of latent heat; in the ocean it is the planet's largest reservoir of warmth; and everywhere it is an unsurpassable thermal "flywheel," an ideal absorber and a gradual emitter of large quanta of energy. All of this is possible only because of water's peculiar properties (M. W. Denny 1993).

Because of its intermolecular hydrogen bonds, water has an unusually high boiling point (100°C, whereas all other similar H_2X compounds boil at less than 0°C). Water's specific heat (4.18 J/g · °C) is 2.5–3.3 times that of common land surfaces (soils, rocks). Water's heat of fusion, 334 J/g, is larger than among similarly structured compounds. Its anomalously high heat of vaporization (2.45 kJ/g) makes it an ideal transporter of energy as latent heat, and it helps to retain plant and soil moisture in hot environments. Finally, its relatively low viscosity makes it an excellent medium for swimming and an outstanding carrier of heat in countless eddies and currents.

Oceans dominate the planetary energy balance. About 80% of all radiation intercepted by the Earth (total of 173.5 PW) enters the atmosphere above oceans. With 50% reaching the surface and with average oceanic albedo of 0.06, oceans thus receive about 65 PW, nearly twice as much energy as is absorbed by the whole atmosphere and four times as much as the continents. Oceans also absorb about two-thirds of the downward LW radiation (~110 PW), so their global annual heating rate is about 175 PW. Because of oceans' great average depth, the air-sea interactions cannot directly affect the entire water column. Water is densest at about 4°C, and hence the deep ocean, with temperature stable near that point, is isolated from the atmosphere by a relatively thin mixed layer that is agitated by winds and that experiences both daily and seasonal temperature fluctuations. Solar energy absorption takes place in the top 100 m, and water's high specific heat restricts the range of temperature amplitudes. Equatorial waters and the southern seas between 30°S–50°S are the main zones of heat storage.

Principal vertical transport to the surface is upwelling: molecular conduction, convection, and turbulence are less important. Meridional overturning circulation (MOC)—a downward flux followed by lateral flow—is driven by a severe heat loss to the atmosphere, water cooling, and ice formation in high latitudes (Wunsch 2002). Colder (higher-density) water sinks in high-latitude ventilation areas, and Macdonald and Wunsch (1996) concluded that there are two nearly independent overturning cells, one connecting the Atlantic Ocean to other basins through the Southern Ocean, and the other connecting the Indian and Pacific basins through the Indonesian archipelago. Complex pathways of the Pacific water flowing into the Indian Ocean in the Indonesian

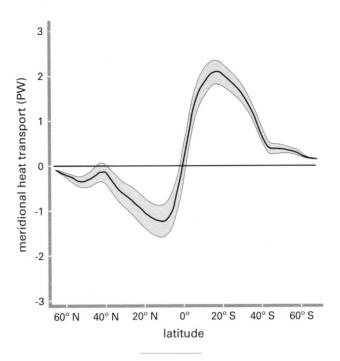

2.5 Global ocean heat transport. Meridional rates based on
Ganachaud and Wunsch (2003).

seas were traced only recently: low-salinity water moves through the Makassar and Lombok Straits, and more salty water goes through the Banda Sea (Gordon and Fine 1996; Gordon, Susanto, and Vranes 2003).

MOC requires about 2 TW in order to turbulently warm the abyssal waters and warm them enough so they can upwell (Alford 2003). Hydrographic records indicate that the ventilation of the deep North Atlantic has steadily changed over the past four decades and that the subpolar seas have become fresher and warmer, suggesting a slackening of the MOC (Dickson et al. 2002). Weakening of this circulation could have undesirable consequences for major ocean currents, marine ecosystems

(because of reduced upwelling of nutrient-rich water), and regional climate (Clark et al. 2002; Schmittner 2005). Measurements taken during the World Ocean Circulation Experiment have improved our understanding of large-scale ocean heat transport. While considerable uncertainties remain (particularly in the Southern Ocean with its paucity of data), grand patterns are clear (fig. 2.5) (Ganachaud and Wunsch 2003).

The Atlantic Ocean shows positive northward meridional heat transport across all of the studied hydrographic sections between 45° S and 47° N, with maxima of 1.26 and 1.27 PW at 7.5° N and 24° N. In contrast, the Indian and Pacific oceans show pronounced southward heat flows throughout the Southern Hemisphere (as much as −1.6 PW at 18° S) and appreciable northward positive transport at about 20° N. Globally these patterns result in a considerable asymmetry across the equator: the vigorous northward transport in the South Atlantic nearly cancels the net southward flow in the Indian and Pacific oceans. Estimated heating totals 2.3 PW in the tropics, and 70% of the corresponding cooling (−1.7 PW) takes place north of 24° N.

Warm seawater can be employed to generate electricity by using thermal gradients greater than 20°C between the surface and deep cold layers (<1 km). The warm water can be used either to evaporate a fluid with a low boiling point, or the seawater can be evaporated in a vacuum to generate low-pressure steam. Actual efficiencies of these conversions would not most likely surpass 2%. The theoretical limit of ocean thermal energy conversions is given by cooling the surface waters because a change greater than 10°C would have widespread climatic consequences. The limit of about 10 TW would prorate to 200 mW/m² over the world's warmest tropical seas. Only the sites with the highest thermal gradients very close to the shore can be seriously considered for OTEC (ocean thermal energy conversion), but the commercial viability of these conversions is yet to be demonstrated.

Energy release from the ocean to the atmosphere is dominated by thermal radiation (~80%) and by evaporation as latent heat (~16%), and its global pattern shows the highest fluxes generated by strong northward ocean currents. With water's extraordinarily high heat of vaporization it takes 28–29 W/m² in order to evaporate 1 mm/day. This much energy is available as the annual mean even at 70° N in the Atlantic and up to about 60° S in the Antarctic waters. The highest latent heat flux, close to 200 W/m², is associated with the warm waters of the Kuroshio (off Japan's Pacific coast) and Gulf streams (along the U.S. seaboard). The global mean of the flux is close to 90 W/m², which means that the Earth's water cycle is driven at a rate of about 45 PW, corresponding to daily evaporation of about 3 mm, or an annual total of 1.1 m.

Evaporation exceeds precipitation in the Atlantic and Indian oceans; the reverse is true in the Arctic Ocean; and over the Pacific the two processes are nearly balanced. Irregular patterns of oceanic evaporation (maxima up to 3 m/a) and precipitation (maxima up to 5 m/a) require substantial compensating flows from the regions with excess rainfall in order to maintain sea level (Schmitt 1999). The North Pacific, particularly its eastern tropics, is the largest surplus region (and hence its water is less salty), whereas evaporation dominates the Atlantic waters. Cyclones, low-pressure systems ranging from thunderstorms to monsoons, are characterized by strong, often destructive winds, but they carry much more energy in latent heat.

Thunderstorms that leave behind just 1 cm of rain in 20–40 min over 100–200 Mm² release 2.5–5 PJ (1–4 TW) of heat, 10–100 times the kinetic energy total. Most of this energy goes into heating the atmosphere. An average year sees about 80 tropical cyclones that develop by heat transfer from ocean surfaces warmer than 26°C (Emanuel 2003). When maximum winds reach 33 m/s, cyclones are classed as hurricanes (in the western North Atlantic) or typhoons (in the eastern North Pacific). Mature tropical cyclones have axisymmetric flows whose energy cycle (with nearly isothermal expansion and compression) resembles that of an ideal Carnot engine (Bister and Emanuel 1998). Large cyclones discharge less than 0.5% of the planetary latent heat flux, with rates for individual events mostly between 10 EJ and 50 EJ (120–580 TW); the kinetic energy dissipated by an average Atlantic hurricane is about 3 TW and by a Pacific supertyphoon ten times as much (Emanuel 1998).

The Earth's largest seasonal landward water-borne transfer of latent heat is the Asian monsoon that originates in the heating of the Pacific Ocean waters (Gadgil 2003). Its course affects the lives of about 2 billion people in an area totaling nearly 10 Gm². Continental precipitation averages about 10 Gm³ during six months, a latent heat transfer amounting to almost 1.5 YJ at a rate of 1.5 PW and land power density of some 150 W/m². All but about 10% of water evaporated from the ocean is precipitated back onto sea surfaces. In the long run oceanic evaporation is also influenced by the mean sea level: during the last glacial maximum, 18,000 years ago, that level was 85–130 m lower than it is now (CLIMAP 1976).

Direct estimates from tide gauges indicate a global sea level rise of 1.5–2 mm/a during the twentieth century, although indirect estimates based on changes in mass and volume suggest much lower (<0.5 mm) annual rates. Miller and Douglas (2004) found that gauge measurements are not biased by any above-average local warning. For comparison, measurements by satellite altimeters for the period 1993–2003 indicate a very similar annual rate of about 2.5 mm. Shoreline retreat is the most obvious consequence of sea level rise, with coastal plains typically experiencing losses of 30–100 cm/a (Pilkey and Cooper 2004). Longer-term effects, related to more substantial global warming–driven sea level rise, would be much more worrisome.

There is no way to ascertain directly the share of planetary heating that powers the atmospheric motion in order to offset energy dissipated in turbulence (aloft) and friction (at the surface). The best estimate derived from the parameters of general circulation theory puts the share at about 2% (Lorenz 1976), or roughly 7 W/m² and 3.5 PW for the whole planet, but Peixoto and Oort (1992) estimated that energy transferred to wind and dissipated as friction is no more than 870 TW. With the average density of 1.2 kg/m³ and mean speed of about 10 m/s, the kinetic energy of 1 m³ of moving air equals about 60 J. For the whole atmosphere (5.1 Em³) it adds up to roughly 300 EJ (9.5 TW), a total smaller than annual global commercial energy use in 2000. But this figure is not an equivalent of a resource base to be used for estimates of eventual extraction. Rather the entire solar recharge (at least 870 TW) is the absolute theoretical limit on any utilization of wind energy. More practically, the downward flux of kinetic energy, averaging about 1.5 W/m² over the land surface (~220 TW in total), puts the limit on extractable power (Best 1979).

2.4 Water and Air in Motion: Kinetic Fluxes

Kinetic energies of ocean water include the currents that redistribute the absorbed heat, wind-generated waves, seismic waves (including tsunamis and seiches), and tides. Aggregate energies of these flows are large, but average power densities are relatively low (although those of catastrophic waves are obviously enormous), and any commercial conversions will have to focus on areas of exceptional fluxes encountered in strong currents, in stormy seas with large waves, and in unusually high tides. The total power of ocean currents has been estimated at 100 GW (Isaacs and Schmidt 1980), that is, merely 0.3 mW/m² of ice-free ocean surface. Major currents are obviously much more powerful: for example, the Florida Current between Bimini and Miami rates about 20 GW, or 1.6–2.2 kW/m².

Wind-generated waves abound on the ocean. Their total power was put at as much as 90 TW and the renewal rate at no less than 1 TW (Isaacs and Schmidt 1980). Assuming an average height of 1.5 m and a period of 6 s yields about 40 TW for the whole ice-free ocean; this would be about 0.1 W/m^2 of the ocean surface or 1/1000 of the global average of the net radiation flux. Linear power density is much more relevant for potential energy utilization than estimates of area totals. As the power goes up with the square of the wave height (Salter 1974), a 4-m-high wave with a 6-s period will have nearly twice the power of a 3-m-high wave (45.9 kW/m vs. 25.8 kW/m). The world's highest reported waves, encountered in the Pacific by the USS *Ramapo* on February 7, 1933, had a height of at least 33 m and a period of 14.8 s (Bascom 1959), containing enormous power, 7.7 MW/m.

In contrast, wavelets in largely placid inland seas may carry no more than 1 W/m. Measurements of linear power densities in such stormy waters as the North Atlantic show rates between 25 kW/m in the Irish Sea and 91 kW/m west of Scotland (Ross 1979). When the waves break on the shore, their kinetic energies are rapidly dissipated. The kinetic energy of a 2-m wave is about 5 kJ/m^2 of beach, or about 1 kW/m^2, a tremendous amount of work when considering the incessant nature of the pounding. Plunging waves trapping and compressing air against cliffs can generate pressures greater than 600 kPa for 0.01 s and do a great deal of rock destruction. Tsunamis (erroneously called tidal waves) are generated by underwater earthquakes and volcanic eruptions, much less frequently by massive rockslides, and rarely by impacts of extraterrestrial bodies. The speed of seismic tsunamis (in m/s) equals $\sqrt{gh}$ (g = 9.8 m/s^2, ocean depth h, in m).

This means they can travel across deep ocean for thousands of kilometers at speeds between 500 km/h and 800 km/h with little loss of power and with a hardly discernible (<1 m) amplitude (NOAA 2005). As they slow down in shallow coastal waters, the waves rise up to tens of meters; the recorded modern high was 38.2 m at Shirahama on Honshū in 1896. The energy of the tsunami generated by the Sumatra-Andaman earthquake of December 26, 2005, was estimated at 4.2 PJ, or less than 0.5% of the strain energy released by the faulting (Lay et al. 2005), but the waves—up to 24 m high in Banda Aceh and higher than 10 m even on the southern coast of Sri Lanka, about 1500 km from its epicenter (Liu et al. 2005)—were responsible for more than 200,000 casualties. This toll greatly surpassed the more than 36,000 people who died in the tsunami caused by the Krakatau eruption in 1883.

Large tsunamis impact vertical surfaces (cross-sectional power densities equal to $0.5\rho v^3$) with power of 1–2 MW/m^2, or twice that of tornadoes classed in violent category and 2 OM higher than strong thunderstorm winds or river floods. Tsunamis generated by giant landslides and asteroids can be much more powerful. Landslides that recur roughly once every 100,000 years and create waves in excess of 100 m high were first documented in the early 1960s as massive hummocks of debris on the seafloor surrounding Hawaii (Moore, Normark, and Holcomb 1994). An eruption of the Cumbre Vieja volcano at La Palma (Canary Islands) could cause a catastrophic failure of the mountain's western flank (Ward and Day 2001). The resulting landslide (up to 500 km^3) would generate a mega-tsunami that would cross the Atlantic at up to 350 km/h and hit the eastern coast of North America with repeated walls of water up to 25 m high.

CHAPTER 2

If an asteroid with the diameter of 400 m were to hit the ocean at 20 km/s, it would lift the surface by about 2.1 km, and the maximum amplitude of the tsunami generated by this impact would be about 50 m at the distance of 100 km, and nearly 250 m only 20 km away. A near-shore impact off California or eastern Honshū would thus instantly devastate a core region of one of the world's two leading economies, and unlike a tsunami generated by a distant earthquake, it would not give sufficient time for mass evacuation of affected regions (Smil 2005b). Like tsunami, seiches are also generated by seismic motions (as well as by other mechanisms), but they develop primarily in closed or semiclosed bodies of water, where their rhythmic sloshing (before the undisturbed level is regained) can exceed 3 m (Korgen 1995).

Waves generated by the gravitational power of the Moon and the Sun have the great advantage of accurate predictability because there are two tidal cycles roughly every 24 hours. There is also considerable regularity of tidal ranges, with variations ranging from differences between the two successive cycles to monthly and annual fluctuations. Average day lengthening of 1.5 ms/century, caused by almost linear decrease of the Earth's rotation since the Paleozoic era, represents an input of about 3 TW into the tidal friction. Long-lasting debate about the principal sites of this energy dissipation—in oceanic tides or in imperfectly elastic mantle—has been resolved overwhelmingly (∼80% of the total flux) in favor of the ocean. Satellite altimeter studies show that 25%–30% of the total dissipation of about 2.5 TW occurs in the deep ocean; the rest is due to bottom friction in shallow seas (Egbert and Ray 2000).

The maximum energy extractable in a tidal cycle goes up with the square of the tidal range and linearly with the impoundment's area. A 10-m tide filling a basin of 10 km² would deliver about 227 MW, or nearly 23 W/m². A quick formula to approximate this density is to multiply the square of the tidal range by 0.2. Only combinations of high tides (at least 5 m) and coastal features allowing for suitable impoundments can provide sites for commercial conversions of tidal energy (WEC 2001). The Earth's highest tidal ranges are in Nova Scotia's Bay of Fundy (6.47–11.71 m mean, 7.50–13.30 m spring tides), in Alaska's Cook Inlet (average 7.5 m), in southern Argentina (5.9 m), along the coast of Normandy (5.0–8.4 m), and in the White Sea bays (up to 11.4 m). The theoretical power of the 28 best potential sites is 360 GW (Merriam 1978).

The global water cycle is driven by annual evaporation of some 430 Gt of water (fig. 2.6), but compared to its latent heat, the kinetic energy of precipitation is quite small. Even a fairly heavy rainfall of 2 cm/h, with raindrops falling with terminal velocity of 6 m/s, will have impact energy of just 360 J/m², that is, a power density of only 100 mW/m², but its effect may be large as it breaks up topsoil unprotected by vegetation. Although the resulting runoff may look far more damaging than does the impact of raindrops, their kinetic energy and hence their erosive power is far stronger. Terminal raindrop velocity is primarily a function of drop diameter (2 m/s for diameters of 0.5 mm, about 9 m/s for the largest sizes about 5 mm), although strong driving winds, boosting the drop velocity by the reciprocal of the cosine of the rain's angle of inclination off the vertical, can make a substantial difference (Wischmeier and Smith 1978). Experimental measurements show that between 5% and 22% of a drop's impact energy is spent on cratering and that the relation between crater volume and kinetic

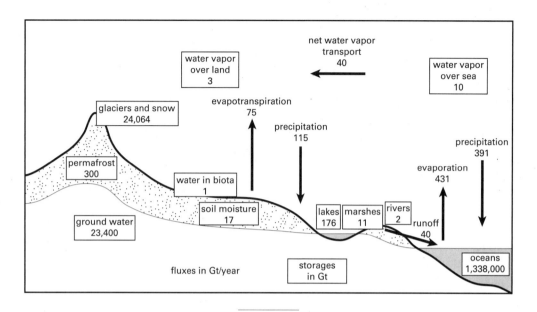

2.6   Global water cycle with annual flows. According to Oki
(1999).

energy of a raindrop is curvilinear; the efficiency of impact energy in removing target material declines as the energy increases (Ghadiri 2004).

Even ordinary velocities of 6 m/s convey kinetic energy of 180 J/m^2 for every centimeter of rain; should the runoff equal one-third of all precipitation, its kinetic energy, with an average speed of 1 m/s, would be 1.65 J/m^2, a rate 2 OM smaller than the rain's kinetic energy density. Not surprisingly, field studies found the strongest correlation between soil erosion and the kinetic energy of the rain and its intensity. The importance of plant cover in controlling erosion is obvious: relative soil losses are at least 100 times higher in fields planted to row crops than in forests. With extensive row cropping (corn, soybeans), these losses are a worrisome threat to long-term crop productivity in many regions (Larson, Pierce, and Dowdy 1983; Smil 2000b).

Hailstones are much more damaging. Even the largest raindrops (diameter 5 mm), hitting with a velocity of about 9 m/s, have just 2.6 mJ of kinetic energy. In contrast, hailstones with diameters of 2 cm, not uncommon in heavy storms, will have kinetic energy about 75 times higher (200 mJ). Northern Colorado, the area of heaviest damage in the United States, averages about 2,200 hailstones/m^2 with mean diameter of about 1.25 cm (Frazier 1979). The kinetic energy of such an event adds up to about 500 J/m^2, and the resulting power densities of 10–20 W/m^2 are 100–200 times those of a heavy rainfall. At the point of impact, kinetic power releases are easily equivalent to 10^3–10^4 W/m^2, enough for heavy damage. These rates are surpassed by the power of avalanches. A slab of dry snow measuring 10 m × 10 m × 20 m, weighing about 500 t and hanging 500 m above the lower slopes of a mountain valley contains potential energy equal to nearly 2.5 GJ. Kinetic energy of such a falling mass, assuming average speed of 30 m/s, would be 225 MJ, resulting in vertical power densities around 110 kW/m^2, a rate directly comparable to the vertical wind density of tornadoes. Not surprisingly, the effects of these two sudden energy releases are devastatingly similar.

Once the precipitation reaches elevated ground, the gradual release of its potential energy is usually a subdued affair. For example, meltwater from 1 m of snow accumulated during winter in a high mountain valley, when dropping on the average 1 km to reach the main valley channel within one month, would reduce its potential energy by just 380 mW/m^2. Total potential energy of 2 m of precipitation in maritime mountains (3 km above sea level) would be just short of 60 MJ/m^2 a year (1.9 W/m^2). The importance of this potential energy is much greater than the power densities can indicate. This quiet and unrelenting flux imparts a common direction to many ecosystem functions (D. H. Miller 1981). Moreover, since it wears away rocks and moves the products of weathering, falling water is a key agent of geomorphic processes.

Surface runoff returns the precipitated water rather rapidly: average residence times of fresh water range from just two weeks in river channels and weeks to months in soil to years in lakes and swamps. Annual river runoff ranges typically between 200 mm/a and 300 mm/a, but it is as high as 800 mm/a for the Amazonian South America and as low as 25 mm/a for Australia. Not surprisingly, the Amazon alone carries about 16% of the planet's river water, and the world's five most voluminous streams (Amazon, Ganges-Brahmaputra, Congo, Orinoco, and Yangtze) carry 27% of all river runoff (Shiklomanov 2003). Estimates of global surface runoff range from 33,500 km^3 to 47,000 km^3 a year. The most likely average, excluding the Antarctic ice flow but including the runoff to continental interiors, is 44,500 km^3 (Shiklomanov 2003). Assuming the mean continental elevation of 840 m (Ridley 1979), the potential energy of this flow is about 367 EJ. Waterfalls are the sites of the highest concentrated releases of potential energy of flowing water (Czaya 1981). Maxima are 16.25 GW for the Inga Falls on the Congo; 5.2 GW for the Sete Quedas on the Paraná; and 4.9 GW for the spectacular 72-m-tall Iguassú; Niagara's potential is 3.4 GW.

Stream flows are indispensable for transporting huge loads of sediments, accreting nutrient-rich alluvial plains that became the core areas of all high cultures, and also providing humankind's first practical inanimate source of mechanical energy (see section 8.2) that has been exploited since the 1880s for generation of inexpensive electricity (see section 9.2). Stream velocities are unevenly distributed throughout a channel, but the flows are typically around 0.5 m/s in wide lowland rivers and between 2 m/s and 3 m/s in floods; none exceed 9 m/s. Cross-sectional kinetic power density of a flooding stream can be up to about 13 kW/m^2, less than in many thunderstorms and 1 OM below tornado or avalanche impacts, but considerably longer durations of typical floods and their ability to weaken the foundation soils have much greater impacts than the power density alone would indicate (threshold for structural damage is ~18 kW/m^2).

The biggest known catastrophic flood was not caused by a stream but by the discharge of glacial Lake Missoula that created spectacular erosional and depositional land-scape in the loess and basalt of the Columbia Plateau of eastern Washington commonly known as the channeled scabland about 15,000 years ago (Bretz 1923; V. R. Baker 1981). Maximum discharge was up to 21 Mm^3/s (for comparison, global discharge of all rivers is 1.2 Mm^3/s) with speeds averaging 20 m/s. Total kinetic energy of this unparalleled breach was about 400 PJ, its power most likely between 1 TW and 2.5 TW, and its vertical kinetic power density was up to 13.5 MW/m^2, ten times the impact of the fastest tornado gusts. These enormous flows reshaped about 8000 km^2, moving boulders up to 10 m in diameter and creating such impressive features as Columbia's Grand Coulee in a matter of hours or days. Another enormous flood took place some 8,450 years ago when the glacial Lake Agassiz (containing some 163,000 km^3 and extending from Manitoba to Quebec) emptied within less than a year into the Hudson Bay with the maximum discharge rates of 5–10 Mm^3/s (Clarke et al. 2003).

Sediments carried by rivers include dissolved compounds, a fine load of suspended clay, and silt particles; and coarse bedload (gravel, stones, boulders) that requires high velocities for downstream transport (Goudie 1984). Stream competence, the maximum movable weight of individual bedload pieces, varies with the sixth power of water velocity (a flow of 4 m/s can carry stones 64 times more massive than one of 2 m/s), and the stream capacity (ability to move total bedload) goes up with the cube of the velocity (doubled speed moves an eight times larger load). Average shares of the three components are difficult to estimate, but a ratio of 4:5:1 may be a good global approximation. The best available reconstruction of the global prehuman sediment discharge adds up to about 14 Gt/a; humans have increased this

flux through accelerated erosion by about 2.3 Gt/a but at the same time cut the mass of the sediment reaching the ocean by 1.4 Gt/a because of retention in reservoirs (Syvitski et al. 2005). Sediment flux of 16.3 Gt/a would be equal to roughly 130 PJ of lost potential energy, or less than 0.1% of the total loss in global water runoff.

Uneven heating of the atmosphere creates distinct pressure belts (equatorial low, subtropical highs, mid-latitude lows) that give rise to predictable steady or semi-permanent prevailing winds (trade winds blowing toward the equator, prevailing westerlies in mid-latitudes) on a planetary scale as well as to smaller rapidly moving atmospheric systems (thunderstorms, tornadoes, and hurricanes). These cyclonic flows are irregular, short-lived (mostly between tens of seconds and a few hours), and often very intense (wind speed commonly in excess of 20 m/s in heavy thunderstorms, up to 130 m/s in tornadoes). Kinetic energy of common thunderstorms sweeping typically 50–150 Mm^2 with winds of 15–25 m/s during 8–15 min will be between 30 TJ and 300 TJ, and their total power will range from 75 GW to 600 GW.

Simply prorated, their power density would be 1.5–4 kW/m^2, but because most thunderstorms release their vast kinetic energies aloft, the impact in the lowermost 100 m above the ground is equal to just between 30 W/m^2 and 100 W/m^2, and the vertical surfaces have to withstand briefly power densities up to 20 kW/m^2, enough to do scattered damage. The United States has about 100,000 such events a year, only 3,000 of them severe (Allaby 2004). Kinetic energy of hurricanes, whose 30–50 m/s winds (maxima up to 90 m/s) can affect up to 1 Tm^2 for many hours, is small when prorated over the entire impact area (mostly just 200–500 W/m^2), but the eye of the cyclone produces brief energy releases as intense as 0.5–1 MW/m² of vertical surface: few structures can remain undamaged under such onslaught.

The impact of tornadoes is frequently even worse. Analyses of 10,826 U.S. tornadoes indicate the mean width of their path to be 126.7 m, average length 9.6 km, speed 60 m/s, and duration 160 s (Schaeffer, Kelly, and Abbey 1980). Total energy in the average 100-m-tall tornado funnel is about 275 GJ, its power is 1.7 GW, and its power density is about 1.4 kW/m². Wind cross-sectional densities average about 135 kW/m², and maximum wind gusts release up to 1.35 MW/m² against the structures. The largest tornadoes (classified as "incredible") will not surpass this highest localized impact, but their total energies in the lowest 100 m may be up to 150 TJ, with power up to about 100 GW, about 60 times more powerful than the average event, and their destruction path can be as wide as 1.5–2 km and extend over more than 150 km.

Kinetic energies of water and wind are the most widespread agents of geomorphic change. Some of them can produce major changes in a matter of hours (flooding streams, catastrophic rains), but most of the erosional and sedimentation processes proceed at very slow rates and are revealed on timescales of 10^3–10^6 years; new methods of measuring cosmogenic nuclides (^{10}Be) in exposed rocks (Bierman and Nichols 2004) can be used to quantify this gradual denudation. And in some instances erosion rates are, counterintuitively, unrelated to precipitation. Burbank et al. (2003) found no measurable difference in erosion rates across the Greater Himalayas despite a fivefold range of precipitation (tectonic uplift is the key factor affecting erosion). Denudation rates are often expressed in Bubnoff units, with 1 B = 1 mm/1000 years or 1 m/Ma, an equivalent of 1 μm/a

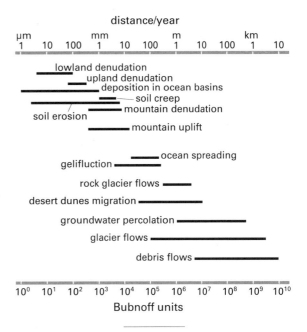

2.7 Typical ranges of annual denudation rates. Plotted from data in Saunders and Young (1983), Selby (1985), Clayton (1997), and W. W. Hay (1998).

or 1 m³/km² · a. Figure 2.7 summarizes the rates of all important slow geomorphic processes.

Considerable ranges for most of them are not surprising: erosion rates are very difficult to quantify, and differences in relief, effective precipitation (the net after subtracting evapotranspiration), and ground cover combine to give rates ranging over 3 OM. Perhaps the highest recorded rate, 19 kB, was from a volcano in New Guinea (Ollier and Brown 1971); erosion in the Himalayas can be sustained at nearly 10 kB; and the Indus river incises through the bedrock at extremely high rates of 2–12 mm/year, or 2–12 kB (Burbank et al. 1996). The British Isles are being denuded at rates ranging mostly between 20 B and 100 B. In many lowlands the rates are no more than 10 B, in closed canopy forests they may be merely 4–5 B, but in badly damaged croplands they may reach nearly 2 kB. Actual field measurements of various landscape denudation rates have been infrequent (Rapp 1960; Caine 1976; Clayton 1997; Bierman and Nichols 2004). Their results indicate that most geomorphic processes proceed at power densities of just 10^{-7} to 10^{-6} W/m^2. Maximum reported rates per square meter are 200 nW for surface wash, 500 nW for snow avalanche debris and rockfalls, 2 µW for soil creep and solifluction, 20 µW for solute transport, and 50 µW for earth slides and mudflows.

Given this variability, it is not surprising that the total detrital sediment flux from continents to oceans remains uncertain (W. W. Hay 1998). A widely quoted rate of 20 Gt/a (Milliman and Syvitski 1992) would prorate (with average density of 2.5 g/cm^3) to about 65 B for ice-free land (glacial denudation rates are extremely difficult to estimate). This rate implies erosion of some 160 t/$km^2 \cdot$ a and (assuming, once again, the mean continental elevation of 840 m) is equivalent to a change of roughly 165 PJ of potential energy, or about 5.2 GW, a mere 35 $\mu W/m^2$ of continental surfaces. (If recent erosion rates were roughly equal to the increase in potential energy of continents, then the sum of 5.2 GW also represents the power going into the uplift of terrestrial formations.) This is a minuscule fraction of potential energy loss in runoff or geothermal flux, and an even smaller fraction of solar heat absorbed by soils. Geomorphic power densities are tiny, but their durations are immense. The Romans knew it well: *gutta cavat lapidem, non vi, sed saepe cadendo* (the drop hollows out the stone by frequent dropping, not by force).

Finally, a few facts about lightning, perhaps the most spectacular atmospheric energy conversion. Cloud con-vection and a proper mix of water and ice particles determine the electrification of clouds (Black and Hallett 1998). High current strokes (10–100 kA) originate mostly in cumulonimbus clouds, their power going up with the fifth power of the cloud size (doubling the cloud increases the total power 30-fold). More than half of all discharges are within these clouds. Cloud-to-ground strikes are a major cause of forest fires, and electricity outages and return strokes are their main visible event. The average cumulonimbus contains about 4 GJ of energy that can be dissipated in lightning. Cloud-to-ground discharges dissipate 10^9–10^{10} J, and estimates of input energy in return strokes range from less than 5 kJ/m to more than 100 kJ/m (Rakov and Uman 2003). The entire flash lasts usually no more than 200 ms, but the bulk of energy is dissipated by a return stroke within less than 100 µs, resulting in enormous power of 10^9–10^{12} W.

Globally lightning dissipates merely about 4×10^{-7} of the atmosphere's convective energy, and spectral measurements of visible stroke light indicate typical releases of 200–2000 J/m, or less than 5% of all flash energy and (with a 5-km stroke) total optical power of up to 1 TW. Most of the discharge goes into heating the atmosphere (peak temperatures are ~30,000 K, more than five times the Sun's photosphere) and into producing the acoustic energy of thunder, which is typically three times greater for ground flashes than for in-cloud discharges. Data from the Optical Transient Detector (MicroLab-1 Satellite) indicate 1.4 billion flashes worldwide per year (44/s), with the highest frequency in the Congo Basin, ten times more activity over land than over sea, and nearly 80% of all discharges taking place between 30°S and 30°N (Christian et al. 2003).

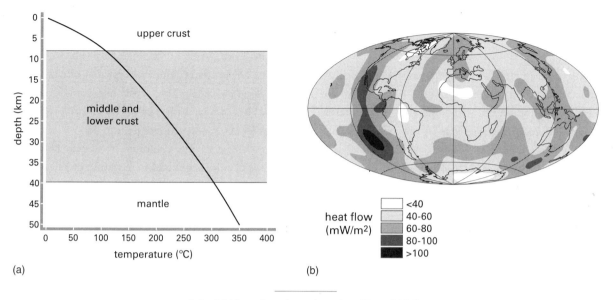

2.8 (*a*) Normal geothermal gradient (Sleep 2005).
(*b*) Earth's heat flow map (based on Pollack and
Chapman 1977).

2.5 Geoenergetics: Heat, Plate Tectonics, Volcanoes, Earthquakes

A celestial body without any internal heat has its surface shaped only by external forces; the Earth's grand features are sculpted primarily by a heat flux whose power density is 3 OM lower than the mean insolation but whose incessant action is refashioning the ocean floor, breaking up and reassembling continents, creating ocean ridges and mountain ranges, and generating earthquakes, volcanic eruptions, and tsunamis. The planet's early thermal history remains speculative, and there are also major uncertainties regarding the relative contributions of basal cooling of the Earth's core and of heat-producing isotopes of ^{235}U, ^{238}U, and ^{232}Th. The role of ^{40}K is particularly intriguing: its decay generates a mere 3.6 pW/g

(97 nW/g for ^{235}U), but experimental evidence suggests that it could be a substantial radioactive heat source not only in the crust but also in the core (Murthy, van Westrenen, and Fei 2003).

The geothermal gradient in the Earth's upper crust is typically 25°C–30°C/km; it decreases with depth through the middle and lower crust, and it is difficult to extrapolate into the mantle (fig. 2.8) (Sleep 2005). Its variability, as demonstrated by hot springs and volcanic regions, is large. Measurements of heat flow conducted through the rocky crust (gradients are established by taking temperatures at intervals in bore holes on land and by using hollow probes in the soft sediments at the ocean bottom) began only in 1939 on land and in 1952 in the ocean. The total number of measurements rose from 47 in 1954 to nearly 25,000 by the year 2000, and there is now a readily accessible global heat flow database (IHFL 2005).

The measurement sites have been concentrated in Europe, the United States, and Japan, and are relatively rare in southern oceans. Early results led to a mistaken conclusion about the near-equality of continental and oceanic values (Pollack and Chapman 1977). Subsequent revisions raised the oceanic contributions, and reviews by Davies (1980); Sclater, Jaupart, and Galson (1980); and Pollack, Hurter, and Johnson (1993) make it also possible to offer some detailed subdivisions by region and crustal age. Terrestrial heat flow generally declines with crustal age. Where thicker sediments prevent major losses owing to hydrothermal circulation, the heat flow decays uniformly from rates in excess of 250 mW/m^2 for ocean floors younger than 4 Ma to 46 mW/m^2 for those older than 120 Ma, reaching an equilibrium value of 38 mW/m^2 after about 200 Ma.

On the continents the time of the last orogenic event, distribution of heat-producing elements, and the speed of erosion are the key determinants of heat flow rates. The youngest regions average 77 mW/m^2, the oldest shields (>800 Ma) just 44 mW/m^2, with the non-radiogenic share declining to a constant rate of 21–25 mW/m^2 after 200–400 Ma. Sclater, Jaupart, and Galson (1980) calculated the total heat loss through the oceans at 30.4 TW, through the continental shelves at 2.8 TW, and through the continents at 8.8 TW, for a grand total of 42 TW. With 24–38 TW, radioactivity alone could supply at least 55% and up to 90% of this flux. Pollack, Hurter, and Johnson (1993) assigned averages of 65 mW/m^2 to the continents (including marine continental shelves) and 101 mW/m^2 to the ocean, for the planetary means of 87 mW/m^2 and a grand total of about 44 TW,

of which some 70% is lost through the oceans and 30% through the continents. Heat flow minima (<50 mW/m^2) are characteristic of old, thick Precambrian shields; maxima are associated with the formation of new ocean floor along the oceanic ridges, particularly in the western Pacific (Nazca ridge with >240 mW/m^2) and southern Indian Ocean (>180 mW/m^2).

Aggregate heat release through the ocean floor and ridges adds up to about 60% of the global heat loss, with nearly half of the total taking place in the Pacific (fig. 2.8). About 9 TW, one-fifth of the global flow, are transported by hydrothermal circulation in the oceanic crust, with 2–4 TW coming from axial flows (Elderfield and Schultz 1996). Exploratory dives found that relatively small openings of hydrothermal vents associated with ocean ridges eject water with temperatures up to 350°C at rates between 25 MW and 330 MW (10^6–10^7 W/m^2). In nature these levels of heat generation are equaled or surpassed only by ephemeral volcanic explosions. Hydrothermal megaplumes are produced on much larger scales (diameters >10 km, rising up to 1 km above the sea floor) by the cooling of hot (1200°C) pillow basalts (Palmer and Ernst 1998). Sixty megaplumes would produce heat loss of 230 GW, equal to less than 10% of the total hydrothermal heat loss from young (<1 Ma) oceanic crust.

Global mean flux of less than 90 mW/m^2 is minuscule in comparison with the mean insolation of about 170 W/m^2, but when acting over huge areas and across long time spans it provides much more power than is needed to prime the processes of global geotectonics whose understanding was revolutionized by the formulation and maturation of the plate tectonic theory that unified the earlier ideas of continental drift (Wegener 1924), sea-floor spreading (Hess 1962), and slab subduction

(Coats 1962). The genesis of this new science is detailed in Le Grand (1988), Oreskes (1999 and 2001), and D. M. Lawrence (2002), and its advances are reviewed in Bebout et al. (1996), Richards, Gordon, and Hilst (2000), Stein and Freymueller (2002), and Eiler (2003).

The Earth's lithosphere consists of irregularly shaped, semirigid oceanic and continental plates that are separated by divergent (constructive), convergent (destructive), and transform boundaries, which meet at triple junctions. The seven largest plates cover 94% of the Earth's surface, but there are twenty plates in total, an equal number of broad zones of deformation, and areas with accreted terranes (crustal blocks whose composition is different from their surroundings). The African plate may actually consist of two blocks, the Nubian and the Somalian, which are splitting along the East African Rift (Djibouti to Mozambique). Continental plates are thick (typically 10 km and up to 200 km) and nearly stationary, with parts composed of rocks more than 3 Ga old. Relatively thin (~10–15 km) oceanic plates are created at the ocean ridges as basaltic magma rises from the mantle and as plate divergence (spreading) produces new ocean floor.

Fast-spreading ridges (8–18 cm/a) have low axial highs, slow-spreading ridges (<5.5 cm/a) have deep rift valleys, and some 20,000 km of the roughly 55,000 km global ridge system belong to an ultraslow-spreading class with annual divergence of less than 2 cm/a (Dick, Lian, and Schouten 2003). Annual extrusion of oceanic basalts amounts to nearly 20 km^3 (~56 Gt), and because their latent heat of fusion is almost 400 kJ/kg, their production requires only about 700 GW of heat. An average spreading rate of 3 cm/a creates nearly 2 km^2 of new ocean floor, sufficient to form the current floor of about 310 million km^2 in only about 175 Ma. Indeed, there is no sea floor older than about 200 Ma because the ocean crust is repeatedly recycled by subduction into the mantle.

Plate collisions result in massive deformation as well as in subduction (fig. 2.9). Major mountain ranges (Himalayas, Andes, Tianshan) are the most spectacular results of continental deformation. The Himalayas, the highest chain, began to form about 55 Ma ago when India collided with Asia, and this process also gave rise, in a stepwise manner, to the high Tibetan plateau (Tapponnier et al. 2001). Deep ocean trenches (Mariana, Kuril-Japan, Tonga, Sunda, Alaska) mark the most active subduction zones where cooler and denser slabs penetrate deep into mantle, and there are also zones of diffuse oceanic transformation (mainly in the Indian Ocean) and relatively small areas of ridge-transform systems, above all, along the mid-Atlantic Ridge and the Pacific-Antarctic Rise. According to Kreemer, Holt, and Haines (2002), 63% of the minimum tectonic moment rate within all plate boundary zones (totaling 7×10^{21} Nm/a) are associated with subduction, 14% with continental transformation, 19% are expected for ridge-transform systems, and just 4% for diffuse ocean zones.

Compression keeps the plates together, and once the stress changes, plates may get reconfigured (D. L. Anderson 2002). Every new grand tectonic cycle begins with the breakup of a supercontinent (as the mantle heat trapped beneath it eventually rises to the surface in the form of massive injections of basaltic magma that produce juvenile crust) and ends with renewed continental amalgamation (Murphy and Nance 2004). During the last billion years the Earth has experienced the formation and breakup of three supercontinents: Rodinia (assembled ~1.1 Ga ago, broke up ~760 Ma ago), Pannotia (formed ~650 Ma ago, broke up ~540 Ma ago),

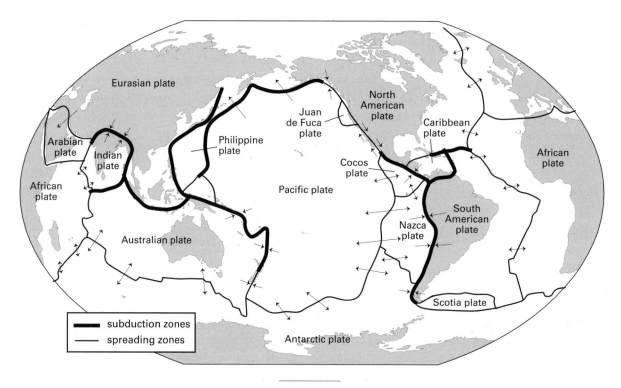

2.9 Earth's tectonic plates, the directions and approximate rates of their annual motion, and their boundaries. Based on USGS (2003) and Kreemer, Holt, and Haines (2002).

and Pangea (formed 300–250 Ma ago, began to break up about 200 Ma ago). Pangea spanned the planet latitudinally, from the high Arctic to Antarctica. The latest continental rifting is now underway in the Red Sea and along the Great Rift Valley of East Africa.

The early model of a relatively few rigid plates separated by narrow boundaries has been replaced by a more nuanced understanding (Richards, Gordon, and Hilst 2000). Large parts of continent-bearing plates have diffuse deformation zones, and at least 15% of the Earth's surface does not behave in expected plate tectonic fash-

ion, as rigid plates with localized boundaries; the African plate has not moved during the last 30 Ma. At the same time, the dynamics and energetics of plate tectonic processes remain poorly understood: plate tectonics and mantle convection are obviously two aspects of the same coupled system, but there is no consensus on the thermal convection patterns within the mantle, on the coupling between the two processes, and on the forces that drive plate motions and create stresses within plates. The mantle, the layer of silicates and oxides that underlie the lithosphere, extends to a depth of about 2900 km and accounts for about 70% of the planet's mass (Jackson 1998; Helffrich and Wood 2001). The transfer of heat through the mantle takes place mostly by convection, heat-induced motion of solid yet (on geological time scales) flowing rock.

But the dynamic link between the convecting mantle and the overlying lithosphere remains conjectural. Plate tectonics have commonly been seen as nothing but the surface manifestation of mantle convection, but a more accurate view may be to see the convection patterns as the result of plate tectonics (D. L. Anderson 2002). Oceanic plates have negative buoyancy because they are cooler and hence heavier that the underlying mantle and descend into it more rapidly than the heat conduction warming the slab (Peacock 2003). As a result, the subducting lithosphere remains cooler and heavier than its surroundings even after more than 100 Ma. The gravitational pull of subducted slabs is seen as the prime mover of plate tectonics, but the coupling between slabs and plates is unclear, and the motion is best explained by invoking both slab pull and slab suction forces (Conrad and Lithgow-Bertelloni 2002). We are also not sure about the way the mantle convection takes place. Is it a whole-mantle process or a layered convection separated into two giant cells by a discontinuity at a depth of 660 km (O'nions, Hamilton, and Evensen 1980; M. W. Kellogg Co. 1998; Hofmann 2003)?

Geochemists, basing their arguments on compositional differences within the mantle, have favored the layered model; seismologists, pointing to cold subducting slabs that penetrate deep below the 600-km boundary, have argued against any layered mixing. A new hypothesis has tried to bridge this gap by proposing a slow-rising deep mantle and a fast-rising upper mantle, which are separated by a thin discontinuity at 410 km, where the material undergoes dehydration-induced partial melting that explains the difference in composition (Bercovici and Karato 2003). And, going even deeper, we are uncertain about the degree to which the lower mantle interacts with the Earth's core (C. A. Lee 2004) and now believe that the core-mantle boundary may be structurally as complex as the Earth's crust (Garnero 2000).

Another controversial topic has been the genesis, persistence, and stability of hotspots. J. T. Wilson (1963) chose the term for those volcanic phenomena (ridges or subduction zones) that are not associated with plate boundaries and hence appear to represent a convection mode independent of plate tectonics. By far the best known example of a massive hotspot is the one that created the Hawaii islands and the chain of seamounts in the North Pacific. Other well-known island hotspots include Iceland, Azores, Galapagos, Samoa, and Tahiti; the best examples of continental hotspots are the Yellowstone and Africa's Afar and Hoggar. Morgan (1971) attributed hotspots to fixed, deeply anchored (near the mantle-core boundary), and fairly narrow (50–100 km across) plumes of hot rock whose relatively high speed (up to 50 cm/a) and buoyancy carry them through the upper mantle and the overlying (and much slower moving) plates. The worldwide total of putative hotspots eventually grew to more than 5,000, but only about 50 volcanic locales deserve to be in this category (fig. 2.10) (Courtillot et al. 2003).

The very existence of plumes feeding volcanic hotspots is disputed. Nataf (2000) argued that direct evidence for actual plumes is weak, and any plumes in the lower mantle would be difficult to detect by seismic means. Heat flow measurements do not show highly elevated temperatures for eruptive plumes, and the plumes are not fixed relative to one another: relative motion between the

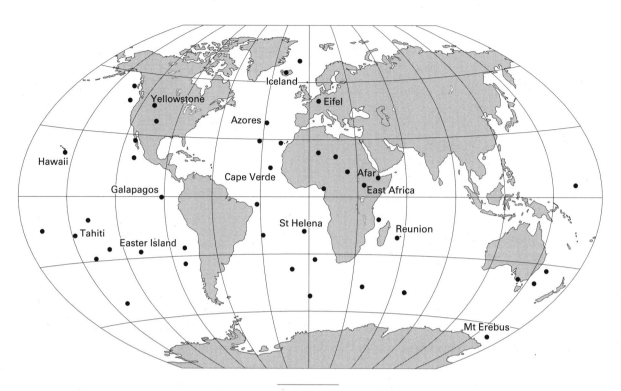

2.10   Earth's principal hotspots. Plotted from coordinates in
Courtillot et al. (2003).

Indo-Atlantic (including Iceland) and Pacific hotspot families is up to 20 mm/a (Norton 2000). Compositional arguments are not strong either: Humayun, Qin, and Norman (2004) used higher Fe/Mn ratios of Hawaiian lavas (compared to those from mid-ocean ridges) as proof that massive plumes are rising all the way up from the lower mantle or core-mantle boundary, but the difference (only 15%) is well within the scatter for the upper-mantle ratios (C. A. Lee 2004).

Consequently, Foulger and Natland (2003) concluded that the plume hypothesis has survived largely as a belief and that hotspots are by-products of plate tectonics that require a source of melt in the shallow mantle and stretching and cracking of the crust. Only a single mode of convection is required and plate tectonics becomes an all-encompassing explanation of the Earth's volcanism. In contrast, Montelli et al. (2004) presented seismic tomographic evidence for the existence of at least six well-resolved plumes (including the Azores, Canaries, Samoa, and Tahiti) that reach into the lowermost mantle. But interpretation of seismic data remains in dispute, and more time is needed before the existence of plumes is unequivocally established (Kerr 2006). Hirano et al. (2006) provided evidence for an alternative hypothesis: hotspot volcanoes can be fed by a crack caused by elastic bending of a plate.

One conclusion is clear: these grand geotectonic processes had an enormous impact on the evolution of life. For the past 3 Ga they have made it possible to keep a significant share of the Earth's surface well above sea level, allowing for the evolution of complex terrestrial forms of life. Diffusion and diversification of land biota have been influenced by changing locations and sizes of the continents. These changes have also created different patterns of global oceanic and atmospheric circulation, the two key determinants of climate. Two massive land features that affect the climate for nearly half of humanity (the Himalayas and the Tibetan Plateau) are direct consequences of the continuing breakup of Pangea (An et al. 2001). So is the warm northward flow of the Gulf Stream, which moderates the climate in Western Europe.

And plate tectonics generates volcanic eruptions and earthquakes. The overwhelming concentration of both active volcanoes and the areas of the most frequent and most powerful earthquakes along the subduction edges of tectonic plates leaves no doubt about their genesis. Volcanic eruptions have been the most important natural source of CO_2 and hence a key variable in the long-term balance of biospheric carbon. Intermittently they have been also by far the largest source of aerosols that can be injected all the way to the stratosphere and whose high atmospheric concentrations can produce a hemispheric or global cooling detectable for months or years after an eruption (Lamb 1970; Briffa et al. 1998; Soden et al. 2002; Robock and Oppenheimer 2003). During the twentieth century no other natural disaster has been responsible for more deaths than earthquakes, whose devastating effects recur in regions that are now inhabited by roughly one half of humanity.

As spectacular and devastating as they often are, volcanic eruptions account for a surprisingly low share of the total geothermal release. A very liberal assumption of 30 Pm^3 of lavas ejected since the Cambrian period (ended 505 Ma ago) would imply the total cooling and crystalization loss (averaging 1.7 MJ/kg) of some 170 YJ (about 11 GW), or a mere 0.025% of the total planetary heat loss during the same period (at least 7×10^{29} J even when assuming the current rate of 44 TW). Obvious measures of the intensity of volcanic eruptions are the volume of ejecta and the height of the ash column. Newhall and Self (1982) modified Tsuya's 1955 scale using both these criteria to construct a volcanic explosivity index (VEI). Fedotov's more quantitative scale relates eruptions to the rate of ejection (kg/s), the logarithm of thermal power output, and the height of the eruption column.

Indices less than 4 include eruptions that take place somewhere on the Earth daily or weekly and that produce less than 1 km^3 of tephra (airborne fragments ranging from fairly large blocks to very fine dust) with maximum plume heights below 25 km. Mount St. Helens (1980) had VEI 5 (paroxysmal eruption, the same magnitude as Vesuvius in 79 C.E.), producing just 1 km^3 of ejecta (Lipman and Mullineaux 1981). Krakatau's (1883) VEI was 6 (colossal eruption with 18 km^3), and Tambora's (1815) supercolossal eruption that ejected 150 km^3 of solids was the largest historic event. On Fedotov's scale, eruptions with intensity XI (equivalent to VEI 6) have columns reaching 28–47 km and releasing thermal energy at the rate of at least 100 TW.

Analysis of a fairly complete set of VEI for continental eruptions since 1500 shows that the mean number of events in each category increases in the same proportion as the energy released by each eruption of that magnitude decreases. This means that over long periods of time randomly occurring eruptions have a nearly constant release of thermal and kinetic volcanic energies for each energy magnitude category; De la Cruz-Reyna (1991) calculated it to be equal to about 36 PJ a year. Invariably, heat is the dominant energy form, at least 10 and up to 1,000 times greater than the other releases, and hence a careful estimate of heat gives a good idea of the proper order of magnitude of an eruption's total energy.

Hedervari (1963) expressed the thermal energy (E_{TH}) as

$$E_{TH} = Vd(CT + B),$$

where V is the rate of extrusion of rocks, d their mean density, C the specific heat of lava (1.25 $J/g \cdot C$), T its temperature above the ambient level, and B its heat of fusion (207.8 J/g). Other energies include the change in the height of the magmatic column (potential energy), kinetic and thermal energy of lavas and pyroclastic flows (mixtures of hot rocks and gases whose high temperature, up to 700°C, and high speed, more than 80 km/h, are highly destructive), and three more forms of kinetic energy releases: seismic waves of the accompanying earthquakes, power of the associated tsunamis or air shock waves, and energy fractioning or deforming the surrounding crust.

The world's best-monitored volcanic eruption, that of Mount St. Helens, Washington, on May 18, 1980, provided an excellent illustration of the dominance of thermal energies (Decker and Decker 1981): heat amounted to 96% (1.73 EJ) of the total energy release (fig. 2.11). Ash cloud, accounting for nearly half of all thermal energy, had enough buoyancy to rise to more than 20 km.

In contrast, a virtually identical amount of energy was released by the 1950 Mauna Loa eruption (1.4 PJ), but there was no high ash column, just massive flows of hot lavas characteristic of the Hawaiian type of eruptions. The Bronze Age Minoan eruption in the Aegean Sea, about 3,650 years ago, was the largest release of volcanic energy during the historic period. Its 100 EJ created the great Santorini caldera (surrounded by the islands Thera, Therasia, and Aspronisi) by ejecting about 70 km^3 (Friedrich 2000). Iceland's Laki released 86 EJ in 1783, Tambora 84 EJ in 1815, Nicaragua's Coseguina 48 EJ in 1835, Krakatau 30 EJ, and Alaska's Katmai 20 EJ in 1912.

Historic eruptions are dwarfed by VEI 8 (megacolossal) events, most recently the creation of the giant Toba caldera (an oval roughly 30 km by 100 km filled by a lake) in northern Sumatra about 75,000 years ago, which produced about 2800 km^3 of ejecta (Rose and Chesner 1990). This catastrophe is perhaps the best explanation for the late Pleistocene population bottleneck, when small and scattered groups of humans were reduced to fewer than 10,000 individuals and when our species came very close to ending its evolution (Rampino and Self 1992; Ambrose 1998). And in terms of total lava flow, Toba was a small event in contrast with the formation of the massive Indian and Siberian basalts. Deccan Traps (formed 65–60 Ma ago) contain more than 500,000 km^3 of basalt, Siberian Traps (formed ~250 Ma ago) about 1.6 million km^3 (Renne and Basu 1991). These massive effusions were not formed by volcanic eruption but by prolonged floods of basaltic lavas accumulating in layers.

Assumptions must be made in order to derive the annual rate of energy released by volcanic eruptions. Elder (1976) estimated the total annual rate of volcanic heat

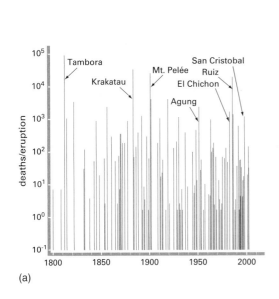

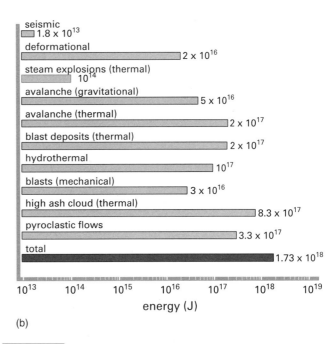

2.11 (a) Volcanic eruptions and fatalities, 1800–2000 (plotted from data at Volcano Live). (b) Energies released by Mount St. Helens eruption on May 18, 1980 (plotted from Decker and Decker 1981).

loss at 200 GW, less than 0.5% of the global geothermal flux. In contrast, Verhoogen (1980) put the maximum volume of continental lava outpouring at less than 1 km³/a and estimated the highly uncertain rate of oceanic eruptions at close to 4 km³/a. A global total of less than 5 km³/a corresponds to heat loss of about 800 GW, still no more than 2% of planetary heat flow. But power ratings of individual explosions are huge: those of Tambora and Pinatubo were put at about 400 TW, and that of Vesuvius in 79 C.E. at 200 TW (Ritchie and Gates 2001). The main eruption of Mount St. Helens (1.9 EJ) on May 18, 1980, lasted 9 h, prorating to 52 TW. An eruption releasing 80 EJ in 10 h rates over 2 PW, roughly equal to a magnitude 8 earthquake lasting 30 s.

Modern volcanologists have repeatedly used power comparisons with nuclear bombs, whose energy is expressed in equivalents of TNT. Mount St. Helens's 1.9 EJ equals 450 Mt TNT, or 27,000 Hiroshima bombs. Such comparisons ignore different ways and durations of energy release. Nuclear explosions produce blinding light within 10^{-3} s and emit deadly IR radiation 0.2–3 s after the eruption, and their blast wave, containing about 50% of all latent energy, travels at initial velocities of 10^2 m/s (CCMD 1981). In contrast, the median duration of volcanic eruptions is about 7 weeks; 10% last no longer than 1 day, most less than 100 days; a few go on for more than a decade (Simkin 1993). Moreover, as is the case in Hawaii, eruptions may consist of slow upwellings of lavas without any explosions and with energy releases dispersed along fracture lines or over molten lava pools.

About half a billion people live within a 100-km radius of a volcano that has been active during the historical era, but the number of fatalities and the extent of material damage caused by volcanic eruptions is highly variable (fig. 2.11). Hot lava usually spreads only over several km^2, ballistic projectiles fall on an area of up to 10 km^2, substantial tephra deposits affect areas of 10^2–10^6 km^2, tsunamis generated by large eruptions can cross oceans, and volcanic dust is transported worldwide. Consequently, it is impossible to calculate any typical impact power densities. Ejection power densities can be enormous: with tephra clouds rising up to 80 km and with initial ejection velocities up to 500 m/s (Kittleman 1979), they can be on the order of 10^8–10^9 W/m^2, unequaled by any other natural terrestrial fluxes.

There is no clear correlation between spectacular ash plumes and the total energy released by a volcanic eruption, but there is a well-established link between the mass of dust ejected all the way into the lower stratosphere and hemispheric or even global temperature declines during the following months or years (Angell and Korshover 1985; Robock and Oppenhemier 2003). Lamb (1970) formulated the dust veil index (DVI) in order to classify these climatic impacts: among the nineteenth- and twentieth-century eruptions Nicaraguan Coseguina (1835) had DVI 4000, Tambora (1815) 3000, Krakatau (1883) 1000, El Chichon (1982) 800, and Mount Pinatubo (1991) about 2400. As for the frequency of erup-

tions, it rose from fewer than 20/a before 1800 to more than 60/a by the late twentieth century, largely because of improved reporting. Ammann and Naveau (2003) analyzed sulfate spikes in polar ice and discovered a strong 76-year cycle of tropical explosive volcanism during the last six centuries.

Loss of life and property depends on the prevailing form of energy release. Kilauea eruptions, with their slow-flowing, glowing lavas, give plenty of time to evacuate houses. Pyroclastic flows that swept down Vesuvius in 79 C.E. and buried Pompei and Herculaneum, and those from Mount Pelée in 1902, which killed all but two of 28,000 people in St. Pierre on Martinique, are the two most famous examples of instant mass burials (Sigurdsson et al. 1985; Heilprin 1903). Because of larger populations the frequency of eruptions that proved fatal rose from fewer than 40 per century before 1700 to more than 200 during the twentieth century (Simkin 1993; Simkin, Siebert, and Blong 2001). Nearly 30% of the roughly 275,000 fatalities between 1500 and 2000 were due to pyroclastic flows, 20% to tsunamis. The four greatest disasters (fatalities in parentheses) were Tambora (92,000), Krakatau (36,000), Mt. Pelée (28,000), and Colombian Nevado de la Ruiz in 1985 (23,000).

The creation, collision, and subduction of plates are accompanied by frequent earthquakes: about 95% of all tremors occur along the plate boundaries, 90% of all destructive seismic energy is released in subduction zones in the Circum-Pacific Belt, and all the great earthquakes recorded since 1900 have been caused by underthrust subduction in South American, Alaska-Aleutian, and Kamchatka-Kuriles-Japan zones, whereas the numerous subduction zones in the Southwestern Pacific have experienced no large interplate earthquakes (Kanamori and Boschi 1983; Ruff 1996). On December 26, 2004, the world's second strongest recorded earthquake had its epicenter just west of Aceh in northern Sumatra, where the Indian Ocean plate subducts the Asian continental plate (Lay et al. 2005).

Stick-slip frictional instability, long recognized as the prime mover of earthquakes, explains not only seismogenesis but also pre- and post-seismic phenomena (Scholz 1998). During an earthquake a rupture usually propagates along the fault plane in just a few tens of seconds. During the Sumatra-Andaman earthquake the rupture extended (at ~2.8 km/s) more than 1300 km along the Andaman trough for about 8 min, the longest on record (Ammon et al. 2005; Ishii et al. 2005). But there are also very slow-moving ruptures, which cause silent earthquakes without any sudden, strong tremors or waves on seismometers (Hirn and Laigle 2004). As for the temporal variation, the twentieth-century record shows a clear cluster of events during the 1950s and 1960s, but a detailed analysis of wait times for all of the 40 great interplate events of the twentieth century shows them to conform to a model of random occurrence, with average wait time of 2.3 years between the successive events.

Energies dissipated during an earthquake include heat produced by friction and kinetic energy of new cracks in the crust and seismic waves radiated through the Earth. Only the waves, felt as the earthquake tremors and recorded by seismographs, can be measured. The earliest earthquake intensity scales (De Rossi in 1874, Sekiya in 1885, Murashi in 1887) were highly subjective. The first fundamental approach to strength-based classification of tremors was taken by Richter's (1935) magnitude scale defined by decadic logarithms of the largest trace amplitude (in μm) recorded with a standard Wood-Anderson torsion seismograph at a distance of 100 km from the epicenter. This measure eventually became known as local magnitude (M_L). Other measures, including surface wave magnitude (M_S), followed during the next 50 years.

Earthquake magnitude was first correlated with the release of seismic energy (E_S) by Gutenberg and Richter (1942). The conversion (E_S originally in ergs) had the form

$$\log_{10} E_S = A + BM_L.$$

A and B, two empirical coefficients, were initially 11.3 and 1.8, and later various authors used values between 5.8 and 14.2 for A and between 1.21 and 4 for B (Howell 1990). This conversion is a gross oversimplification that yields only very approximate equivalents. The same is true for the conversion of M_S into seismic energy because M_S is calculated from a bandwidth that is too narrow to capture all the radiated frequencies. Inherent inaccuracies aside, relative differences remain unchanged: a magnitude increase of 0.2 doubles the seismic energy release, and a unit increase multiplies it 32 times.

The M_L and M_S scales were eventually replaced by a more consistent scale, not based on an instrumental recording but on the measurement of the ruptured area and of the average slip across the affected fault:

$$M_0 = \mu AD,$$

where M_0 is moment, μ rigidity of the material around the rupture zone, A the area displaced, and D the average displacement across the fault. Moment magnitude (M_W), introduced by Hanks and Kanamori (1979), is defined (in SI units) as

$$M_W = \tfrac{2}{3}(\log_{10} M_0 - 9.1),$$

and E_S is proportional to the seismic moment:

$$E_S = 5 \times 10^{-5} M_0.$$

Modern seismographs record digitally across a broad bandwidth (0.01–5 H), and this makes it possible to calculate radiated seismic energy by direct integration of records (Boatwright and Choy 1986). Energy magnitude, M_e, in J, can be calculated (Choy and Boatwright 1995) as

$$M_e = \tfrac{2}{3} \log_{10} E_S - 2.9,$$

with E_S in Nm.

Seismic energies derived from the recordings of broadband waves and those calculated from M_W (or M_S) will not be identical: the measures capture two different quantities, shaking from higher-velocity spectra vs. low-frequency displacement spectra due to the area of rupture and the slip across the fault (USGS 2005). Magnitude 5 corresponds to E_S of 710 GJ from broadband measurements and to 2 TJ when derived from M_W (or M_S). But all estimates of E_S become difficult for great earthquakes with long duration because the established empirical conversions do not work in the case of the strongest events. The Sumatra-Andaman earthquake had M_0 of 4.0×10^{22} Nm (maximum slip of about 20 m) or M_W of 9.0 for the main shock (empirically derived E_S is as high as 2 EJ). This is precisely the value used by the USGS (2005), but Lay et al. (2005) assigned M_W of 9.2 and offered 1.1 EJ as the best estimate of seismic energy. Banerjee, Pollitz, and Bürgmann (2005) concluded that M_W did not exceed 9.2.

These discrepancies, amounting to a twofold difference in overall energy discharge, illustrate the difficulties of quantifying very large tremors. In addition, there is no universal ratio between the seismic wave energy and the total energy released by an earthquake. Total energy release by the Sumatra-Andaman earthquake was put by Bilham (2005) at 4.3 EJ, roughly four times the seismic energy flux. An illustrative global aggregate derived by using average annual frequency of earthquakes in the three highest categories (much more common, smaller earthquakes account for less than 1% of all seismic energy)—1 of M_W 8.0 or higher, 17 between 7 and 7.9, 134 between 6 and 6.9—and applying conservative E_S averages adds up to releases of about 800 PJ (USGS 2005). Quintupling this to account for strain energy accumulated in irreversible deformations and for friction-generated heat along the faults would yield roughly 125 GW, less than 0.3% of the Earth's conductive heat flow.

Verhoogen (1980) estimated that the average seismic power may be about 300 GW and that the addition of accumulated strain energy and of friction-generated heat may raise the total to about 1 TW. The Sumatra-Andaman earthquake was equivalent to a 1-Gt bomb, or as many as 68,000 Hiroshima bombs. Earthquakes of M_W 8–9 last usually 30–90 s (and have a serious impact over areas with diameters of 80–160 km), and those with magnitudes 7–7.9 last only 2–50 s with strong ground shaking within radii of 50–120 km. A 30-s M_W 9 earthquake rates nearly 67 PW, and its power density (prorating over the area with radius of 80 km subjected to strong ground shaking) could be as high as 3.3 MW/m². A much more common M_W 6 tremor of the same duration with shocks felt over 100 km² would have a surface power density of about 21 kW/m². In contrast, the Sumatra-Andaman earthquake lasted about 500 s, and the strongest recorded earthquake in Chile in 1960

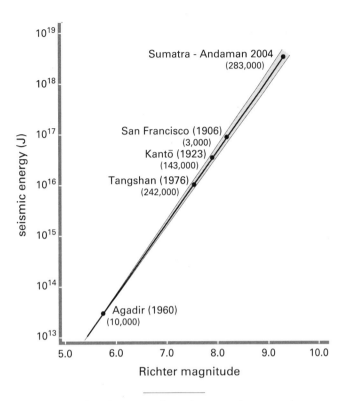

2.12 Earthquake magnitudes and energy releases. Fatalities in parentheses. Plotted from Ruff (1996).

(M_W 9.5) was felt for 340 s (Ni, Kanamori, and Helmberger 2005).

Magnitudes and approximate seismic energies of major earthquakes that struck between 1900 and 2005 do not correlate with fatalities (fig. 2.12): the population densities of the stricken areas and the quality of construction are far more important. The September 1, 1923, Kantō earthquake that set Tokyo on fire claimed more than 40 times as many lives as did the April 18, 1906, San Francisco earthquake, which released roughly four times as much energy. The highest death toll caused by an earth-

quake was in Tangshan on July 28, 1976: the Maoist regime reported officially 242,219 fatalities in that northern Chinese coal-mining city, but estimates of the actual death toll are as high as 655,000 (Huixian et al. 2002).

Even with low fatalities, urban densities guarantee costly material damage and massive homelessness: the January 17, 1995, Kobe (Hyōgoken Nanbu) quake killed 5,470, displaced 310,000 people, and cost at least $110 billion in damage (Chung et al. 1996). During the twentieth century, earthquakes caused more deaths than volcanic eruptions, floods, and cyclones combined. Because even a small rupture can potentially extend its length, and there are 10^5 times more M_W 2 events than M_W 7 tremors, predictions of which small events become major tremors are impossible. Such chaotic events can be predicted only in a statistical sense (Kanamori, Hauksson, and Heaton 1997). But in California, progress has been made with seismic hazard models that forecast probabilities of strong shaking within the next 24 h (Gerstenberger et al. 2005).

3

PHOTOSYNTHESIS

Bioenergetics of Primary Production

The Sun appears to be poured down, and in all directions indeed it is diffused, yet it is not effused. For this diffusion is extension.

Marcus Aurelius (121–180), *Meditations*

And what an extension! All complex life on this planet, all its incredible diversity, all our hopes and worries, are but transmutations of the Sun's light, and photosynthesis is the agent of this miracle. Absorption of sunlight and the subsequent sequence of photochemical and thermochemical reactions in the chloroplasts of photosynthesizing bacteria and green plants are the most important energy conversions on the Earth. Plants provide (directly or after being eaten by animals) all our foods; their immediate harvests (as wood and crop residues) or the extraction of their fossilized remains (as coals and hydrocarbons) supply all our fuels. All the richness of heterotrophic life and all the intricacies of human civilizations are thus energized by photosynthesis (with help from primary electricity).

The origins of photosynthesis remain speculative. If the early Archean (4–3.5 Ga bp) atmosphere had no oxygen, then the first phototrophic prokaryotes had to assimilate carbon in ways akin to those of the still extant anoxygenic bacteria. These were eventually marginalized by high levels of atmospheric O_2 produced by water-splitting photosynthesis (Peschek 1999). The sequencing of genes involved in photosynthesis and phylogenetic analyses indicates that green nonsulfur bacteria (heliobacteria) are the last common ancestors of all photosynthetic lineages (Xiong et al. 2000). But if the Archean atmosphere already had more than a trace of oxygen, then the earliest phototrophs might have been more like today's cyanobacteria. Many strains of these prokaryotes can perform anoxygenic photosynthesis in hypoxic or anoxic sulphide-rich environments (hot springs, ocean sediments) and then shift to O_2-releasing photosynthesis in aerobic niches (Whitton and Potts 2000).

The shift to water-cleaving photosynthesis freed the bacteria from their dependence on limited amounts of reduced S, Fe^{2+}, Mn^{2+}, and H_2 and CH_4. Des Marais (2000) estimated that microbes dependent on hydrothermal energy could sustain annual fixation of less than 25 Mt C, whereas oxygenic photosynthesizers, able to tap a virtually unlimited supply of hydrogen from water, could eventually fix more than 100 Gt C/a. Descendants of the earliest cyanobacteria continue to fill almost every aquatic and terrestrial niche. Unicellular *Prochlorococcus*, the smallest and the most abundant photosynthesizer in the ocean, discovered only in 1988 (Chisholm et al. 1988), contributes 30%–80% of total primary production in oligotrophic waters.

Grypania, the first fossil alga, dates to 2.1 Ga bp, which means that eukaryotic phototrophs had to evolve at some time during the preceding 500 Ma. Their phototrophic ability was not an original evolutionary achievement but an import (Nitschke et al. 1998): algal chloroplasts were derived directly from cyanobacteria through a primary endosymbiosis. The fossil record shows microbial mats declining, and green and red algae increasing in abundance only about 1–0.9 Ga bp. Both bryophytes (nonvascular liverworts, hornworts, and mosses without distinctive water-conducting tissues) and tracheophytes (vascular lycopods, horsetails, ferns, and seed plants) evolved from charophytes, freshwater green algae whose fossils are documented from more than 600 Ma bp. Liverworts, the earliest land plants, may have appeared during the mid-Ordovician period, after 450 Ma bp, and fungus-plant symbioses were essential in colonizing nutrient-poor and desiccation-prone environments (Blackwell 2000).

Following rapid diversification and diffusion, land plants became the dominant photosynthesizers by the end of the Devonian period, 360 Ma bp (Bateman et al. 1998). They acquired stems with complex fluid transport, structural tissues that enabled them to reach unprecedented heights, roots and leaves with stomata for respiratory exchange of gases, and specialized sexual and spore-bearing organs and seeds; cellulose, a microfibrillar polysaccharide, emerged as the dominant structural compound that now accounts for about half of all phytomass. Angiosperms (flowering plants) have been dominant since the mid-Cretaceous period, about 90 Ma bp. Photosynthesis of complex organic compounds is energized by photons absorbed by a small group of plant pigments. The synthetic sequence always shares the core of an intricate multistep reductive pentose phosphate (RPP) cycle; its rates are determined by the same limiting factors in all terrestrial environments; and the bulk of the newly formed phytomass has a highly uniform energy density. After describing the fundamentals of the photosynthetic process I appraise the productivities of major biomes and the magnitudes of standing phytomass in principal ecosystems.

3.1 Photosynthetic Pathways

The standard scientific description of photosynthesis is as a process of CO_2 fixation and O_2 evolution. As Tolbert (1997) noted, it is not easy to overcome this dogma, which goes back to the mid-nineteenth century. Photosynthesis is actually a complex process of O_2 and CO_2 exchange that is energized by the absorption of specific wavelengths of solar radiation (D. O. Hall et al. 1993; Raghavendra 1998; Hall and Rao 1999; Lawlor 2001; Kê 2001). Every photosynthetic sequence starts with light absorption by pigments in disk-like thylakoid membranes inside chloroplasts in specialized leaf cells (in some species also stem cells). Excitation of pigments (chlorophylls *a* and *b*, bacteriochlorophyll, and carotenoids) takes place in two different reaction centers: photosystem

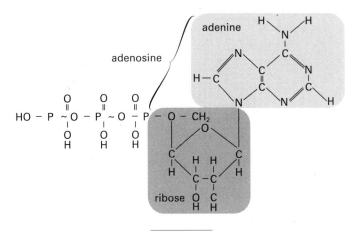

3.1   ATP molecule is composed of adenine, ribose, and a
triphosphate tail.

I is excited by wavelengths of 700 nm, and photosystem II works best at 680 nm. Photosynthesis is thus energized overwhelmingly by red light. In total, the photosynthetically active radiation (PAR), the wavelengths between 400–700 nm, is less than half (43%) of total insolation.

Light-driven enzymatic transfer of electrons from water produces nicotinamide adenine dinucleotide phosphate (NADPH) and converts adenosine diphosphate (ADP) and inorganic phosphate to ATP. The electron transfer splits water in a single reaction step within a $MnCa_4$ cluster and release of O_2 (Yachandra et al. 1993). ATP and NADPH are then used to energize the reduction (fixation) of CO_2 to carbohydrate, which can proceed without any light (dark reaction). ATP, the principal energy carrier in the biosphere, powers a vast variety of enzymatic reactions via the hydrolysis of its phosphate bond (fig. 3.1). Evolution selected this phosphate because its enzymatic hydrolysis (loss of the endmost phosphate group) releases a large amount of energy, 31 kJ/mol (Westheimer 1987). A reverse reaction, catalyzed by ATP synthase, takes place when energy is not immediately needed. It reforms the compound by adding a third phosphate molecule to ADP. ATP can thus be used again and again.

The sequence of this multistep reductive carbon cycle was first elucidated by Melvin Calvin and Andrew Benson during the early 1950s (Bassham and Calvin 1957; Calvin 1989). They introduced ^{14}C-labeled CO_2 (for periods ranging from fractions of a second to minutes) to photosynthesizing *Chlorella*, and after preselected periods they stopped the enzymatic reactions by alcohol and extracted materials for analysis. In the early 1950s, once they began using newly invented paper chromatography, they could clearly demonstrate that the first stable compound produced by photosynthesis was phosphoglyceric acid (PGA), a compound containing three carbons (3-phosphoglycerate is a synonym). The RPP

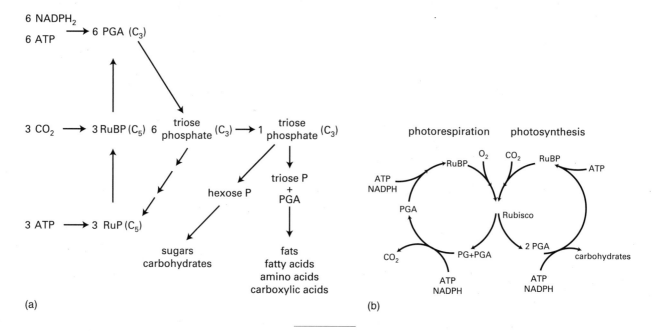

3.2 (*a*) Basic reactions of the photosynthetic carbon reduction (Calvin) cycle (Smil 2002). (*b*) Dual role of Rubisco, as a carboxylase in photosynthesis and as an oxygenase in photorespiration (based on Tolbert 1997).

pathway is made up of 13 enzyme-catalyzed reactions; its three key sequences are carboxylation, reduction, and regeneration.

During carboxylation CO_2 (1-C compound) reacts with the 5-C sugar ribulose-1-5-bisphosphate (RuBP) to produce two molecules of 3-C PGA. The enzyme that catalyzes this transformation is Rubisco, ribulose-1-5-biphosphate carboxylase, a large water-soluble molecule that makes up 16% of all protein in chloroplasts and hence is the most abundant protein in the biosphere. The resulting PGA is first phosphorylated with ATP and then reduced, in a reaction using NADPH, to 3-phosphoglyceraldehyde (PGAL, $C_3H_7O_6P$), a 3-C (tri-

ose) sugar phosphate, the end product of the RPP cycle (fig. 3.2). PGAL can be immediately metabolized in the chloroplast or dimerized to produce stable 6-C sugars (hexoses). Two of these sugars, fructose and glucose, form a 12-C disaccharide sucrose. This is by far the most important transport metabolite, the compound that plants use to distribute most of their fixed carbon from the leaves to be used by subsequent organic syntheses in roots, stems, or flowers. Five out of every six molecules of PGAL are converted, first to pentose phosphate, then to ribulose-5-phosphate (R5P), whose phosphorylation with ATP regenerates three molecules of RuBP, closing and sustaining the cycle.

Rubisco's abundance is not only a matter of its ubiquity but also of the slowness of its reactions (catalyzing just 3 molecules/s compared to 1000 molecules/s for a typical enzymatic process) and its reversibility. The enzyme acts as a carboxylase (as just described) and, once the CO_2 concentrations inside the leaf drop to about 50 ppm, as an oxygenase. In the latter role Rubisco is the catalyst for the binding of O_2 to RuBP to produce not only PGA but also 2-phosphoglycolate, a 2-C compound, during the C_2 oxidative photosynthetic cycle that releases CO_2 (this photorespiration is an entirely different process from night-time mitochondrial respiration in leaves). Additional energy is needed to remove 2-phosphoglycolate by converting it first to glycolic acid and then to glycine and serine. This oxygenation appears to be of no benefit for the plant, yet photosynthesis and photorespiration are inextricably linked: there is just one CO_2 and one O_2 pool, and the C_3 and C_2 cycles create a necessary balance for net exchange of the gas (fig. 3.2).

Because of the relatively low CO_2 and high O_2 levels in today's atmosphere, some species use about half of photosynthetic energy in the C_2 cycle (Tolbert 1997; Hall and Rao 1999). Only a drastic reduction of atmospheric O_2 (to about 2%) or greatly elevated ambient CO_2 levels would eliminate C_2-cycle losses in C_3 plants. But some plants can avoid the photorespiration losses. Kortschak tried to replicate the Calvin-Benson sequence with sugarcane, but his first product was not PGA but rather the 4-C acids malate and aspartate, which he assumed to have a common precursor in oxaloacetate, another 4-C acid. Hatch and Slack then unraveled the details of this distinct process. Instead of reducing CO_2 with Rubisco, they used phosphoenol pyruvate (PEP) carboxylase in the mesophyll cells to form oxaloacetate (fig. 3.3) (Hatch 1992).

This acid is reduced to malate, transported into chloroplasts of the bundle sheath cells where CO_2 is regenerated, and only then used in the Calvin-Benson cycle (Sage and Monson 1999). These C_4 plants also differ structurally from C_3 species. The latter have no significant differentiation in mesophyll and bundle sheath, and the vascular conducting tissue of C_4 species is surrounded by a bundle sheath of large thick-walled cells containing chloroplasts (fig. 3.3). PEP carboxylase has a greater affinity for CO_2 than Rubisco; moreover, O_2 levels in the bundle sheath are low, and CO_2 concentrations are near what is required to saturate Rubisco, whose oxygenating action (causing photorespiration) is practically eliminated. But structural differences (Kranz anatomy) are not obligatory: Voznesenskaya et al. (2001) discovered that a species of Chenopodiaceae (*Borszczowia aralocaspica*) uses C_4 pathway through spatial compartmentation of enzymes and separation of two types of chloroplasts.

A C_4 pathway needs more energy than the Calvin cycle alone. Additional ATP is required to energize the regeneration of pyruvate to PEP, but in the absence of photorespiration its overall net conversion efficiencies are considerably higher. Maximum daily growth rates of C_3 and C_4 species are, respectively, 34–39 g/m^2 and 50–54 g/m^2, a 40% difference that is especially significant in food production (Monteith 1978; Edwards and Walker 1983). Daily maxima averaged over the whole growing season show still greater difference: with 22 g/m^2, C_4 plants are about 70% ahead of C_3 species that fix 13 g/m^2. C_4 species also do not have any light saturation (C_3 plants saturate at irradiances around 300 W/m^2), and their optimum photosynthetic temperature is 30°C–45°C (15°C–25°C in C_3 plants). The C_4 pathway thus appears to be an adaptation to hot climates and aridity,

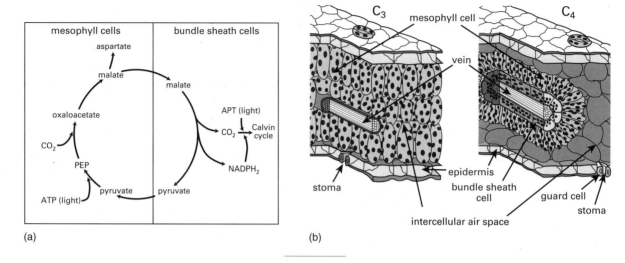

3.3 (a) Carbon fixation in C$_4$ plants. (b) Sections through
leaves of C$_3$ and C$_4$ species showing their different structures.
From Smil (2002).

and the distribution of C$_4$ species is associated with relatively high temperatures and low moisture availability.

Corn, sugarcane, and sorghum are the most important C$_4$ crops; some of the worst weeds, including crab grass (*Digitaria sanguinalis*), also follow this path. A global comparison of annual phytomass accumulation shows C$_4$ plants taking 11 out of the 12 top places (Hatch 1992). Maximum daily growth rates reported for field crops (all in g/m^2) range from just over 50 for corn and sorghum to around 40 for sugar cane, above 35 for rice and potatoes, 20–25 for most legumes, and just short of 20 for wheat. Atmospheric CO$_2$ decline during the past 7 Ma favored the C$_4$ species, whereas rising levels of CO$_2$ are tilting the balance once again in favor of C$_3$ plants. And the line between C$_3$ and C$_4$ species is not that definite: tobacco, a typical C$_3$ plant, has features of C$_4$ photosynthesis in cells of stems and petioles that are supplied with carbon from the vascular system and from stomata (Hibberd and Quick 2002).

Crassulacean acid metabolism (CAM) is the other important modification evolved to minimize H$_2$O losses. Succulents use it to absorb large volumes of CO$_2$ during night and, as in C$_4$ species, to convert it initially into C-4 acids. During the day, with stomata closed, sunlight energizes decarboxylation of these acids and carbon refixation via the RPP cycle. These processes, unlike in C$_4$ species, are not spatially separated; they take place at different times in the same cells. Many CAM plants can totally suspend any gas exchange for weeks, even months. CAM metabolism has worldwide distribution, with Crassulaceae being the most important group in northern hemisphere and Cactaceae in the Americas. Throughout the tropics there are numerous epiphytic Orchidaceae and Bromeliaceae, some of them able to shift between C_3 and CAM in response to environmental stresses. Pineapple, aloe, and opuntia are the only notable CAM crops.

In energy terms the simplest equation describes photosynthesis as an endothermic reaction that requires 2.8 MJ of radiant energy to synthesize one molecule of glucose from six molecules of CO_2 and H_2O. A more realistic black box description is as follows: $106CO_2 + 90H_2O + 16NO_3 + PO_4 +$ mineral nutrients $+ 5.4$ MJ of radiant energy $= 3258$ g of new protoplasm (106C, 180H, 46O, 16N, 1P, 815 g of mineral ash) $+ 154O_2 + 5.35$ MJ of dispersed heat. Where light is a limiting factor of photosynthesis, individual plants follow two basic strategies: shade-intolerant species redirect their development into internodal extension or stem growth at the expense of leaf development; shade-tolerant species cope by increasing photosynthetic efficiency and cutting down respiration losses (H. Smith 1982). Solar tracking by plants is also widespread, either to boost the photosynthetic rates (with leaves perpendicular to the sun's rays) or, with parallel leaves, to reduce leaf temperature and transpiration water losses (Ehleringer and Forseth 1980). And although plants reject near IR wavelengths and thus avoid overheating, the heat absorbed by plants in the far IR is essential for the initiation and progression of thermochemical reactions of the RPP cycle.

The energy efficiency of actual carbon assimilation is very high. To reduce one molecule of CO_2 requires three molecules of ATP and two molecules of NADPH in the RPP pathway. Free energies of the two compounds are, respectively, about -29 and -216 kJ/mol. The reacting compounds contribute 519 kJ, and the difference between the broken (in H_2O and CO_2) and newly formed (in sugars) bonds during carbohydrate formation is about 465 kJ/mol. Theoretical efficiency of the process is almost exactly 90%. Actually measured performance is 80%–85%, a great contrast with the much lower efficiency of the whole photosynthetic sequence. The minimum quantum requirement for the synthesis of the three ATP molecules needed for the reduction of one molecule of CO_2 depends on the H^+/ATP ratio. Theoretically, there should be an efflux of $2H^+$/ATP; the actual ratio is up to 4. With $2H^+$/ATP the minimum quantum requirements would be 6, but synthesis of two molecules of NADPH would raise this to 8 quanta for each molecule of CO_2 assimilated, and 10 may be a more realistic total.

Energy content of a quantum depends on the light frequency, varying inversely with the wavelength. Assuming the mean PAR wavelength at 550 nm, the energy content of an average-sized quantum is 3.61×10^{-19} J (Planck's constant, 6.62×10^{-34} J, multiplied by the light frequency, a quotient of the light speed and the mean wavelength). One einstein (1 mol, or Avogadro's number, 6×10^{23}) of green photons would have energy of 217 kJ and 8 einsteins would supply 1.736 MJ of radiant energy. The overall maximum theoretical efficiency of photosynthesis would be almost 27% (465 kJ/1.736 MJ). Alternatively, using 680 nm (the peak for chlorophyll), 1 einstein of red photons carries 176 kJ, and with the more realistic requirement of 10 quanta per fixed CO_2 molecule, the total radiant energy input would be 1.756 MJ, virtually identical with the result of the first calculation. A sequence of adjustments brings this theoretical maximum to realistic levels.

Adjustment for PAR (43% of total insolation) reduces the maximum theoretical efficiency to about 12%. Some of the incident light is reflected by plants, and some of it is transmitted through canopy leaves. These losses of green light, as well as minor forms of inactive absorption of blue and red, total at least 10%, reducing the photosynthetic efficiency maximum to around 11%. An ideal leaf exposed at a 90° angle to direct sunlight would then reduce about 250 mg $CO_2/dm^2 \cdot h$ and synthesize 170 mg/dm^2 of new carbohydrates, converting the radiant energy to chemical bonds with a power density of 8 mW/m^2. With an average leaf weight of 2.85 g/dm^2, the metabolic intensity would be about 280 mW/g. During a sunny day this leaf, absorbing the total of 250 kJ/dm^2, would produce about 1.7 g of new photosynthate. A field of such ideal leaves would fix daily 1.7 t/ha of phytomass, and where growth could continue for the whole year, 620 t/ha would be added.

Actual short-term increments of new phytomass are at best 50%, more likely just 33% of these rates. The top seasonal or annual additions are between 20% and 25% of the ideal rates, and long-term, large-scale averages are merely 10% and all the way down to just 2% of the best hypothetical performance. The two main reasons for these disparities are the respiration costs and the inevitable losses that go with rapid rates of photosynthetic reactions. In order to conserve as much light as possible during the limited hours of intensive insolation, the rates must be quite fast, but this rapidity results in two kinds of considerable inefficiencies. Unless the plant's enzymes can keep up with the radiation flux coming into the excited pigments, the absorbed energy will be reradiated as heat. Utilization must be immediate because the chlorophyll molecules cannot store sunlight. Only at very low light intensities, when radiation would be the only factor limiting the rate of the terrestrial photosynthesis, is there such a perfect match.

Rapid photosynthesis maximizes growth rates and hence improves the chances of early survival and competitive maturation, but it is paid for by large irreversible losses that lower the photosynthetic efficiency to roughly 8%–9%. Fixed carbon dissipated in metabolic processes and in the maintenance of the photosynthetic system and its supporting structures is highly variable, ranging from less than 20% in high-yielding crops to virtually 100% in old-growth forests. With respiratory losses at 40%–50%, the peak plant growth efficiencies would be around 5%. Only at this point do theoretical calculations and actual performance meet; the highest recorded values of net photosynthesis (for highly productive plants under optimum conditions during short periods of time) are between 4% and 5% (see fig. 1.7). For most plants, even the 4% efficiency is impossible because their potential performance is limited by environmental factors (Nemani et al. 2003).

The two most widespread limiting factors in carbon fixation are the availability of water, dominant on some 40% of land, and temperature, the strongest productivity-limiting factor on a third of land (fig. 3.4). Photosynthesis is impossible without an extremely lopsided trade-off between CO_2 and H_2O. The difference between internal (inside the leaves) and external water vapor is 2 OM higher than the difference between external and internal CO_2 levels. As a result, C_3 plants need 900–1200 mol, and some up to 4000 mol of H_2O, to fix 1 mol of CO_2, whereas C_4 plants can manage with 400–500 mol H_2O/mol of CO_2 fixed, and the rates in CAM species are as low 50, with typical values between 70–150 mol H_2O/mol CO_2, losses of 1 OM lower than in C_3 plants. As the kinetic energies of enzyme molecules and substrates increase with rising temperatures, thermochemical reactions proceed faster, and conversion efficiencies eventually reach maximum rates and then decline as higher temperatures denature the enzymes.

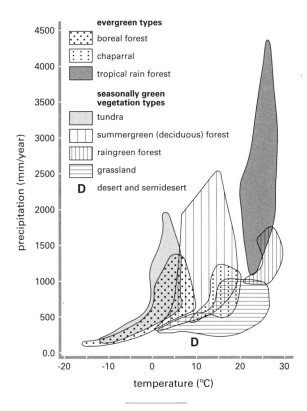

precipitation (mm/year)

evergreen types

⬚ boreal forest

⬚ chaparral

⬛ tropical rain forest

**seasonally green
vegetation types**

⬚ tundra

⬚ summergreen (deciduous) forest

⬚ raingreen forest

⬚ grassland

D desert and semidesert

temperature (°C)

3.4 Average annual temperature, precipitation, and major biomes. Based on Lieth and Whittaker (1975).

Temperate deciduous ecosystems need at least 0°C and photosynthesize best at about 25°C; in the tropics both the minima and maxima are about 10°C higher.

Insolation is the primarily limiting factor of photosynthesis over 27% of the world's vegetated area, particularly in cloudy tropical lowlands (see section 2.2). Nutrient shortages limit the productivity of grasslands and woody phytomass in many arid regions as well as forest growth in such well-watered and thermally ideal environments as the Amazon basin, where 90% of soils are deficient

in N and P, nearly 80% in K, and about 60% in Ca and S (Nicholaides et al. 1985). In many ecosystems two or more of these factors operate concurrently or sequentially: low temperatures limit the productivity of boreal forests in winter, high temperatures, in summer; many tropical rain forests are co-limited by heavy cloud cover and poor soils.

3.2 Global Primary Productivity

Primary productivity, the rate of synthesis of new phytomass during a specified period of time (usually a year), is expressed in three different ways: as assimilated carbon, as dry matter (DM), or as stored energy. Because of a large range of moisture contents in fresh (wet) biomass (from >98% in phytoplankton to <5% in mature seeds) it is necessary to use DM values for all comparisons. I use the carbon metric, and the following conversions should be used for dry matter and energy: 1 t C $\approx$ 2 t of dry woody phytomass and 2.2 t of dry herbaceous phytomass; 1 t of dry phytomass (average of woody and herbaceous species) $\approx$ 17.5 GJ. The rate is quantified on three levels of diminishing content.

The gross primary productivity (GPP) is the amount of phytomass (carbon, energy) fixed (after photorespiration losses) in a year by all photoautotrophs (chemotrophic prokaryotes that do not energize their biosynthesis by sunlight are a negligible addition). GPP is reduced by autotrophic respiration (R_A) to yield the net primary productivity (NPP): NPP = GPP – R_A. NPP is the most frequently used rate in modern bioenergetic studies concerned with the photosynthetic performance of ecosystems. There are two interpretations of NPP: one posits that the rate must be $\geq$0; the other allows NPP to be negative during the periods when R_A > GPP (Roxburgh et al. 2005). I use the first convention.

Plant respiration is responsible for a major loss of fixed carbon, but (unlike wasteful photorespiration) this loss is a function of fundamental gains: the released CO_2 is an inevitable by-product of biochemical reactions that support plant growth (including the transport of photosynthates and biosynthesis of complex compounds) and maintenance. Respiration is thus a metabolic bridge from photosynthesis to plant structure and function (Amthor and Baldocchi 2001). Respiration rates (usually expressed as the quotient R_A/GPP) are generally lowest in crops (often as low as 0.3; agriculture could be defined as a quest for maximized GPP and minimized respiration). They range between 0.3 and 0.65 in grasslands and between 0.45 and 0.70 in young forest, and can even be over 0.9 in mature forests. The rate of 0.5 is a good first-order approximation.

NPP is a fundamental concept, but its value cannot be measured directly. A variety of methods (including recurrent harvesting with corrections for litter fall and heterotrophic losses, gas-exchange techniques, and mathematical treatments relating plant growth to various environmental variables) have been used to determine NPP. The task is obviously most difficult with forests. Until recently highly accurate gas-exchange studies were practical only on a very small scale (10^0–10^1 m^2), and reliable complete destructive harvesting is limited by logistics and cost to 10^3 m^2. Good approximations of above-ground NPP can be based on short-term measurements of fixation rates. Typical daily means sustainable for several weeks of the most rapid growth are about 20 g/m^2 for C_4 plants and 5–15 mg (average 13) for C_3 species. All the latest NPP estimates rely on increasingly complex models that take into account many relevant environmental variables (Cramer et al. 1999).

Phytomass is subject to heterotrophic consumption that ranges from rapid microbial decomposition to seasonal grazing by large ungulates. Subtracting heterotrophic respiration (R_H) by microbes, invertebrates, and vertebrates from the NPP of a particular ecosystem yields the net ecosystem production (NEP, the annual rate of storage), whose ultimate aggregate is the net primary biospheric production. NEP is the phytomass that is potentially available for human harvests, but because of difficulties in quantifying R_H, the NPP has become the preferred measure of primary production. It has received a great deal of research attention, particularly with respect to the anthropogenic releases of CO_2 affecting the carbon cycle.

Estimates of the Earth's primary productivity have a long history. In 1862 one value was based on the seemingly unrealistic assumption of all land entirely covered with a green meadow and yielding annually 5 t/ha (Liebig 1862). Surprisingly, the result, about 63 Gt C, falls within the range of estimates published since the 1970s: the Earth's mean primary productivity does indeed resemble that of temperate grassland (fig. 3.5). Most of the values offered during the intervening generations were either serious under- or overestimates of the most likely total. A clearer consensus emerged only during the 1970s, when appraisals using empirically derived relations between climatic variables and NPP yielded totals between 40 Gt C and 60 Gt C (Smil 2002).

The first estimate of global phytoplankton production—25.4 Gt C/a for the open ocean and 3.2 Gt C/a for coastal seas—was based on just a few values for relatively unproductive areas (Noddack 1937). Riley (1944) used productivity values for seven western Atlantic sites to produce a range of 44–208 (mean 126) Gt C.

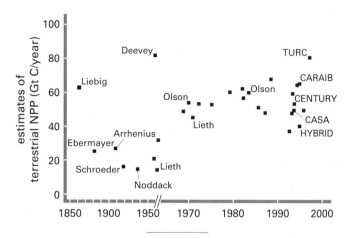

3.5   Nearly 150 years of estimates of global NPP. From Smil (2002).

Koblents-Mishke and colleagues (1968) based their aggregate on data from 7,000 sites around the world, but their total of 23 Gt C/a excluded all benthic production. Whittaker and Likens (1975) estimated 55 Gt C, and De Vooys (1979) put the total marine NPP at 46 Gt C/a. The best estimates of global productivity are now derived from satellite and surface data in combination with models of basic ecological processes. They are made possible by remote monitoring on a planetary scale and by a better understanding of the fundamental relations determining GPP, NPP, and R_A.

Chlorophyll reflects less than 20% of the longest wavelengths of visible light, but about 60% of near IR. These differences are easily monitored from space; LANDSAT and SPOT offered relatively high resolution, but the high cost of their images led to a reliance on the Advanced Very High Resolution Radiometer (AVHRR), installed on polar-orbiting satellites. Its maximum resolution of only 1–4 km is sufficient to appraise large-scale patterns of vegetation coverage and to detect seasonal and statistically significant year-to-year variability of global plant conditions (Gutman and Ignatov 1995). The normalized difference vegetation index (NDVI) compensated for changing illumination conditions, surface slope, and viewing aspect.

The first NDVI calculations used reflectances in visible band (0.58–0.68 μm) and in the near IR (0.73–1.1 μm); since April 1985 they have also included brightness temperatures in the thermal IR (11 and 12 μm) and the associated observation-illumination geometry (Gutman et al. 1995). Sequential monthly averages of NDVI showed dramatically the seasonal ebb and flow of the Earth's photosynthesis, and the images provide an excellent tool for monitoring changes of land cover as well as a highly reliable input to global models of photosynthetic productivity. Similarly, satellite data from the Coastal Zone Color Scanner have revealed global patterns of surface chlorophyll (Antoine, André, and Morel 1996;

Falkowski, Barber, and Smetacek 1998; Behrenfeld et al. 2005).

The results of 16 models of global NPP were compared using standardized input variables (Cramer et al. 1999). Most of these models simulate carbon fluxes using a prescribed vegetation structure; some use data from the AVHRR sensor as their major input. After excluding the two extreme values, the results of 16 models that were reviewed in the end-of-the-century systematic comparison ranged from 44.3 Gt C to 66.3 Gt C, with a mean of 54.1 Gt C (Cramer et al. 1999). Assuming an average of 45% C, this translates to roughly 98–147 Gt (mean of 120 Gt) of dry phytomass, or assuming 15 GJ/t, to 1.8 ZJ. Two satellite-based models of marine NPP came up with very similar results: 37–46 Gt C/a (Antoine, André, and Morel 1996), and 47.5 Gt C/a (Behrenfeld and Falkowski 1997). The Pacific Ocean accounts for slightly more than 40% of the total, and the marine NPP of the two hemispheres is roughly equal because of much higher output (about 60%) per unit area in northern oceans.

The most likely range of global NPP at the end of the twentieth century was thus 100–110 Gt C, that is, 220–245 Gt of dry phytomass, or 3.3–3.6 ZJ, and annual flux of 105–115 TW. Two conceptually similar models of terrestrial and marine NPP, both with an emphasis on integrating large-scale satellite observations, combine to yield a global total of 104.9 Gt C, with 56.4 Gt C on land and 48.5 Gt C in the ocean (Field et al. 1998; Geider et al. 2001). Extrapolation of these totals results in power densities of about 450 mW/m^2 of ice-free land and 130 mW/m^2 of ocean. The NPP total of 105 Gt C also means that the global GPP during the late 1990s (using the standard assumption NPP = R_A) was on the order of 210 Gt C, that slightly more than a one-quarter of the

atmospheric CO_2 (at 370 ppm in 2000 = 787 Gt C) was drawn into photosynthesis, and that about 13% of it was annually incorporated into new phytomass. The carbon flux in NPP was thus roughly 16 times as large at the beginning of the twenty-first century as the annual emissions of carbon from the combustion of fossil fuels (6.67 Gt C in 2000).

The Global Primary Production Data Initiative, launched in 1994, made NPP measurements readily available in a standardized format. By the year 2005 the NPP database at the Oak Ridge National Laboratory contained data for 65 intensively studied sites (mainly grasslands and tropical and boreal forests) with geo-referenced climate and site characteristics data (ORNL 2006). But the greatest advance in productivity studies was the introduction of continuous satellite-derived measures of terrestrial GPP (issued weekly). This was made possible by combining the measurements of canopy reflectance by the Moderate Resolution Imaging Spectroradiometer (MODIS) on the Terra satellite, launched in 1999, with the information on biome type, fraction of PAR absorbed by vegetation (changing with growth and senescence), and daily surface climate conditions (Running et al. 2004). GPP and NPP totals were, respectively, 108.42 Gt C and 56.06 Gt C in the year 2000, and 107.5 Gt C and 54.8 Gt C in 2003 (NTSG 2006). Global NPP based on MODIS data shows the expected maxima (>1 kg C/m$^2 \cdot$ a) in the Congo Basin, parts of the Amazon, the highlands of Central America, and the wettest and warmest parts of monsoonal Southeast Asia (fig. 3.6).

But even these advances do not allow us to determine global NPP with a high degree of accuracy. Measurements of large-scale CO_2 flux between vegetation and the atmosphere indicate (after being corrected for R_H

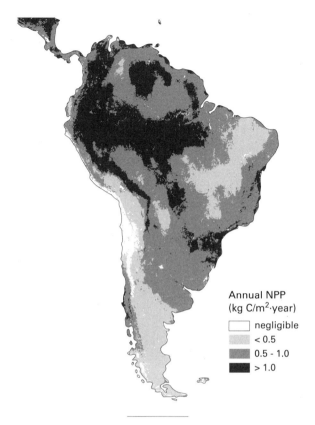

3.6 South America's NPP (based on MODIS data). As expected, the largest area of the highest productivities is in the Amazon basin (Running et al. 2004).

in plant litter. Meentemeyer, Box, and Thompson (1982) estimated the worldwide annual leaf fall at 35.1 Gt and the total litter production at 54.8 Gt, or about half of the global NPP. Studies in various ecosystems showed ranges of 5–15 t/ha (~9–26 MJ/m^2, or 0.3–0.9 W/m^2) in tropical forests (11 t/ha may be a good mean) and mostly 4–8 t/ha (8–16 MJ/m^2) in temperate and boreal biomes, where 4.5–5 t/ha may be a typical loss.

Two comparisons illustrate the degree to which the fossil-fueled civilization depends on accumulated and transformed products of photosynthesis and some recent globally significant changes of the process. Photoautotrophy is ancient: organic carbon in Archean sediments puts the first prokaryotic photosynthesizers 3.8 Ga ago (Schidlowski 1991), and massive deposits of fossil fuels attest to high intensities of NPP during the Phanerozoic eon. High carbon preservation rates mean that only 5–10 g C in paleophytomass were needed to produce 1 g C in anthracites and bituminous coals, whereas the typical rates were in excess of 10,000 for crude oil and natural gas (Dukes 2003). The weighted mean of about 7,000 implies that the combustion of about 7.5 Gt C in fossil fuels in 2005 required initially no less than 50 Tt C in ancient phytomass and was equivalent to about 500 years of the current NPP (105 Gt C); or, assuming terrestrial phytomass stores of 500–600 Gt C, we have been consuming every year fossilized phytomass equivalent to 80–100 times today's entire planetary stocks of phytomass carbon.

Advances in satellite monitoring and in the modeling of NPP have uncovered many fascinating changes of this key biospheric indicator, both on the global scale over relatively short periods of time and on regional or continental scales over extended time spans. During the first four years of the program, the MODIS-based continuous

flux) that the NPP of some ecosystems can be not just 20% but even 50% higher than indicated by the best current models. For example, Grace (2001) found that the NPP of a Brazilian rain forest near Manaus was as high as 15.6 t C/ha · a (nearly 3.5 kg/m^2), whereas the total that neglected fine root turnover was nearly 40% lower. Such adjustments could make the global NPP markedly higher than indicated by the recent modeling consensus. In any case, a large part of NPP is continuously discarded computations of global NPP showed annual fluctuations up to 3.8% (NTSG 2006). A longer comparison, using satellite-based models of global NPP, indicated that climate change weakened several constraints on plant productivity, resulting in a 6% (3.4 Gt C) rise of NPP between 1982 and 1999 (Nemani et al. 2003). Enhanced photosynthesis was also found as a result of lingering volcanic aerosols from the 1991 Mount Pinatubo eruption (because plant canopies use diffuse radiation more efficiently than the direct beam) and stronger monsoon winds over the western Arabian Sea. They increase nutrient upwelling and boost average summertime phytoplankton mass by more than 350% (Gu et al. 2003; Goes et al. 2005).

On the other hand, the European heat wave in 2003 reduced the continent's GPP by about 30% and was responsible for a strong (500 Mt C/a) anomalous net source of CO_2 (Ciais et al. 2005). Long-term averages of NPP and standing phytomass are bound to change with progressively more pronounced global warming, but the extent and pace of these changes remain highly uncertain, often even as to the direction of the sign (will boreal or tropical forests be a long-term sink or source of carbon?). A model by Cao and Woodward (1998) indicated much enhanced global NPP (by 25%) and substantially higher phytomass stocks (up by 20%) with doubled CO_2, but other studies predict no gains or minimal gains in some key biomes. Similarly, it is unclear to what extent the continuing deforestation of sub-Saharan Africa, parts of Latin America, and Asia will be balanced by reforestation and enhanced NPP in other regions.

3.3 Productivities of Ecosystems and Plants

Division of the global NPP shows the expected domination of forests. A great deal of research has been devoted to the productivity and structure of the tree-dominated biomes (Reichle 1981; Landsberg 1986; Perry 1994; Waring and Running 1998; Barnes et al. 1998; Roy, Saugier and Mooney 2001), and it has produced extensive databases that reveal universal patterns as well as numerous peculiarities. Tropical rain forests produce most of the biosphere's new phytomass. Field et al. (1998) ascribed about 48% of terrestrial NPP to forests and 32% to tropical rain forests alone. Reliable moisture, steady high temperatures, and very high leaf area index (LAI, the upper area of foliage per unit area of ground) explain the unmatched rates. While the LAI of grasslands and cereal crops is often no higher than 2–3, tropical rain forests have an LAI of 4–7.5 (Myneni, Nemani, and Running 1997; Scurlock, Asner, and Gower 2001).

Available data show NPP averaging around 2.5 kg/m^2 (about 1.3 W/m^2), almost equally divided between above- and below-ground phytomass, and the reported maxima reach about 1.7 W/m^2. As already noted, photosynthesis in the wet tropics is limited by insolation and nutrient-poor soils. Unlike the temperate rain forests (rooted in relatively nutrient-rich soils covered with thick litter mats), most tropical trees are shallowly rooted in highly weathered and leached nutrient-poor soils overlaid with rather thin litter layers. Nutrients necessary for photosynthesis reside in the phytomass itself, and the fertility of the forest depends on their constant rapid recycling. Nutrient-conserving adaptations include rapid direct absorption of scarce nutrients, extensive mycorrhizal symbioses, absence of denitrifying bacteria, leaves scavenging nutrients but resistant to rainfall leaching and heterotrophic attack, and quick regrowth in clearings (Jordan and Herrera 1981; Primack and Corlett 2005).

The efficient use of nutrients, moderately high productivity, high efficiency of stemwood production (as a share of total insolation of 0.42–1.81 compared to 0.20–0.50 for tropical forests), and the high proportion of phytomass in the stem make the temperate forests, especially their coniferous stands, the most suitable ecosystems for management. Another important advantage for management of these forests is their low nutrient-cycling intensity, resulting from relatively large litter falls and slow decomposition rates. Coniferous forests have commonly accumulated litter at rates of 30–50 t/ha (1 OM higher than the tropical stands), and a single application of fertilizers may be effective for 10–25 years. Their NPP averages around 1.5 kg/m^2 (about 40% below ground), and their LAI can be as high as 7. Environmental limits (insolation, temperature, nutrients) keep the average global NPP of boreal forests below 0.4 kg/m^2, but there are major differences between Europe (mean rate ~0.9 kg/m^2) and Siberia (average <0.25 kg/m^2), where after logging the forest remains a net source or just a very weak sink of carbon (Schulze et al. 1999).

But the knowledge of NPP (or even NEP) is of little interest to those foresters who manage temperate and boreal forests for timber. The concern is with the yields of harvestable timber, and since the late nineteenth century foresters have acquired a wealth of detail on wood growth in pure and mixed stands (Assmann 1970; Davis et al. 2001). Spruces have about 55% of their aboveground phytomass in steam and bark, 24% in branches, and 11% in needles, and their stumps account for about 20% of the whole-tree mass. In contrast, pines have 67% of phytomass in steam and bark, and broadleaved trees 78% (Lehtonen et al. 2004). Merchantable bole may thus amount to no more than half of the above-ground phytomass (fig. 3.7).

The disparities between concepts are easily illustrated. The annual NPP of a mixed temperate forest is most

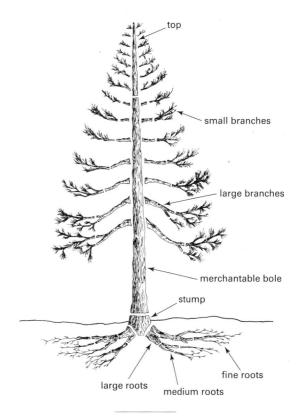

3.7   Division of phytomass in a coniferous tree. From Young (1979).

often about 1 kg/m^2, its NEP is about 0.5 kg/m^2, and a forester's net annual increment (NAI) may be no more than 150–200 g/m^2 (the latter rate is most commonly expressed volumetrically, as 2–2.5 m^3/ha). The NPP is an academic concept, and the NEP is still a substantial overestimate of a potential harvest unless every stump and leaf were removed. On the other hand, the last value is unduly conservative for assessing more complete wood harvests: whole tree utilization (including stump removal) is now a technical possibility, but it is the standard in commercial evaluations of forest productivity.

The global average of forest growing stock is about 110 m^3/ha. European rates are close to this figure, Latin America's rate tops 150 m^3/ha, and the highest rates in tropical rain forests are about 200 m^3/ha (FAO 2005). The latest survey of U.S. forest resources puts the nation-wide average at just over 113 m^3/ha (Smith et al. 2004). Reliable European data show nationwide NAI rates ranging from less than 0.9 m^3/ha in Greece to nearly 6 m^3/ha in France (Lehtonin et al. 2004). But these rates are determined by the density and composition of forest stands, and hence a better indicator of economically useful productivity is the ratio of NAI/1000 m^3 of growing stock. Finland, with its exemplarily managed boreal forests, has a very high NAI/1000 m^3 of growing stock, about 39 m^3; the rates are about 32 m^3 for France and Germany and just over 20 m^3 for Italy and Greece.

In energy terms, typical sustainable worldwide wood harvests thus range from a mere 4 GJ/ha in many tropical stands to over 20 GJ/ha in North American or Scandinavian coniferous forests (dry weight equivalents are ~525 kg/m^3 for hardwoods and 440 kg/m^3 for softwoods). The productivity of temperate and boreal forests shows a general decline with the stand age. The ratio of bole to leaf production, a good indicator of wood production efficiency and stand vigor, shows an especially striking decline in boreal coniferous forests, where cold climate shortens the period during which the trees can maintain positive phytomass and energy balance. Age-related decline is also conspicuous in even-aged stands of temperate conifers: a 70-year-old stand of Douglas fir produces annually the same mass of needles as its 20-year-old counterpart, but its stem storage will be only 40% of the earlier rate.

Savannas and other tropical and subtropical grasslands are, in aggregate, the biosphere's second most productive biome, accounting for nearly 30% of the global NPP. Highly rain-dependent NPP can be quite variable, averaging about 1 kg/m^2 with extremes ranging from less than 500 g/m^2 in semiarid locations to more than 4.5 kg/m^2 in subhumid tropics, and with above-ground production accounting for more than below-ground NPP in a minority of studied cases (Coupland 1979). Similarly, the NPP of temperate grasslands, averaging mostly 1–1.5 kg/m^2 (but surpassing 2 kg/m^2 in the richest meadows), comes mostly from often dense and deep root mats. Lawns account for a large part of temperate urban areas; their NPP is mostly 1–1.8 kg/m^2, comparable to the fixation of natural grasslands (Falk 1980).

At the beginning of the twenty-first century, cultivated land—according to FAO (2006) about 1.54 Tm^2, or about 12% of ice-free land, under annual and permanent crops—produced about 14% of the global NPP. The specific rates (given the great variety of crops, growing conditions, and agronomic practices) differ substantially, and they also vary with the availability of water and the intensity of fertilization. Conversion of readily available yields to NPP must take into account the changes of harvest index, the ratio of grain to total above-ground production (Donald and Hamblin 1976; R. K. M. Hay 1995). In 1900 harvest indices of cereals were 0.25–0.35 because the bulk of their DM phytomass was in long stalks and numerous leaves. Modern high-yielding cultivars have short stems with fewer and narrower leaves, and harvest indices 0.40–0.42 for wheat, 0.47–0.50 for corn, and about 0.5 for rice. With optimum water supply, fertilization, and effective weed and pest control, traditional varieties would produce no less phytomass than the modern cultivars, but the partitioning of their photosynthates is economically much less favorable.

In the year 2000 the mean global NPP for all crops (calculated by enlarging the total DM harvest of ∼7.5 Gt by 15% in order to account for root productivity and by another 15% to factor in the preharvest losses to heterotrophs) was about 10 Gt or about 7 t DM/ha (0.7 kg/m^2, or 12 MJ/m^2, or <0.4 W/m^2). Two studies of the NPP of U.S. agriculture (a satellite observation-based model and an estimate based on harvest data) found the annual range of 0.54–0.62 Gt C (Lobell et al. 2002). These totals prorate to 9.5–11 t DM/ha, 35–55% above the global mean but still no better than the NPP of a good lawn. This is not surprising because lawns have high LAI and longer growing periods (6–9 months even in winter climates, compared to 90–150 days for cereals).

Extreme crop NPP ranges from low-yielding cereals and legumes in arid regions (<2 t DM/ha) to tropical sugarcane (>50 t DM/ha). Among the leading commercial crops, Iowa grain corn and Dutch wheat (8–9 t grain/ha), would have whole-plant NPP of 19–22 t/ha, (assuming that pesticide applications reduce R_H to just 5% of NPP), identical to that of a dense lawn. Those C$_4$ crops that fix CO$_2$ year-round do as well as the best natural grasses, and when irrigated and fertilized, better than any other plants. Even the worldwide average of sugar cane NPP (with 15% of NPP from roots and R_H equal to at least 5% of NPP) is about 30 t/ha; the best national average (about 150 t/ha of fresh cane in Peru) translates into productivities of about 80 t/ha; and the highest recorded fixation in Java amounted to 94 dry t/ha. These record productivities (8–9.4 kg, or 135–160 MJ/m^2) put sugarcane well ahead of other crops, even after adjustments for the length of growing period (365 days for cane, 90–150 days for temperate crops).

Blue light penetrates farthest in the open ocean, and even in clean waters little red light will be available for pigments below 10 m. Macronutrient concentrations nearly always show a sharp decline near the surface and a remarkable stability below 1000 m, a result of dominant stratification of ocean waters. Even in relatively nutrient-rich coastal water, photic zone has much lower N and P concentrations. Areas of the highest NPP coincide with the zones of nutrient enrichment (continental runoff in near-shore waters or coastal upwelling). The northern Atlantic, the Pacific shelves of Asia, and the upwelling zones off Africa, the Americas, and in the northern Indian Ocean have the highest annual production (fig. 3.8). And because of a strong negative abundance-mass scaling, energy used by all phytoplankton cells in a given class size (regardless of species) equals that of all the cells in other size classes (Li 2002).

The most efficient ecosystems are wetlands, which benefit from a high influx of waterborne nutrients. Both tropical and temperate marshes convert commonly 1.5% of insolation into new phytomass, and the best sites do at least twice as well. Typical grassland conversions are much lower. US/IBP Grassland Biome sites show a range from 0.13% for a desert grassland to 1.2% for a mountain formation, with a mean of 0.46% for ungrazed sites and 0.57% for grazed sites (Coupland 1979). Alpine grasslands have efficiencies of 0.05%–0.13%, Arctic grasses below 0.09%. The best conversions for temperate forests are around 1.5%, for tropical rain forests about 1%, for rich, mature coniferous and deciduous forests 0.4%–0.9%, and in stressed locations 0.3%.

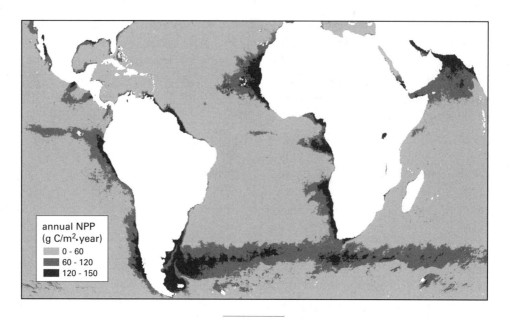

3.8   Annual oceanic NPP in parts of the Atlantic, Indian, and
Pacific oceans. Simplified from maps in IMCS 2000.

The global average for all terrestrial NPP is held low because of the huge extent of water- and cold-stressed ecosystems. With annual global NPP averaging only 15 MJ/m^2 and mean radiation reaching the Earth's surface at 5 GJ/m^2, the planetary efficiency of photosynthesis would be just 0.3%. Similarly, with oceanic NPP averaging 180 g, or roughly 3 MJ/m^2, the global mean of phytoplanktonic fixation efficiency would be a mere 0.06%. Detailed long-term studies measured efficiencies from less than 0.092% in the oligotrophic Sargasso Sea to 0.9% in Nova Scotia's coastal waters and in excess of 5% in Eniwetok Atoll (Parsons 1984). Thus both the terrestrial and marine efficiencies of NPP differ by 2 OM, and these differences are reflected, but far from perfectly mir-

rored, by the richness of standing phytomass as well as by the diversity of heterotrophic life.

3.4   Phytomass Stores

Fresh phytomass has a highly variable content of water, ranging from less than 5% for mature seeds to as high as 95% for young shoots (and values for phytoplankton cells are even higher). The only way to ensure the uniformity of comparisons is to express all masses in absolutely dry terms (after desiccation of samples at 104°C–105°C to constant weight). Their densities are then 0.2–0.8 g/ cm^3, with organic matter accounting for about 95% of most species but with inorganic substances making up as much as 50%, even 70% of the dry mass in many aquatic

3.9 A segment of cellulose, the biosphere's most abundant macromolecule, composed of about 3,000 units of glucose.

plants. Monosaccharides (40% C, 15.5 kJ/g) and disaccharides (42% C, 16.5 kJ/g) are present in plants in relatively limited quantities, and the bulk of the world's phytomass is stored as cellulose, a polysaccharide (44% C, 17.5 kJ/g) that makes up 66%–72% of softwoods and 74%–80% of hardwoods. Long-chained cellulose, consisting of about 3,000 units of glucose, is the most abundant biopolymer in the biosphere (fig. 3.9). Lignin (63% C, 26.4 kJ/g), the biosphere's most abundant aromatic polymer (composed of varying ratios of cross-linked alcohols), makes up about 20% of hardwoods and 30% of softwoods. Larger shares of energy-dense lipids (39 kJ/g) are usually present only in seeds.

The dominance of cellulose and lignin is a structural necessity because these polymers make it possible to build stiff yet flexible vertical (trunks, stems) and horizontal (cantilevered branches, boughs) supports that can expose myriads of perennial or seasonal photosynthesizing surfaces (leaves, needles) to the Sun. Standard conversion of dry phytomass to carbon is 0.45, corresponding to about 17.5 kJ/g, but 0.48–0.50 is a more appropriate rate for woody phytomass in temperate forests. However, Savidge's (2001) analyses of 40 North American tree species show that interspecific differences matter for accurate conversions. C in hardwoods ranged from 46.2% to 49.97%, in softwoods (conifers) from 47.21% to 55.2%. Lieth and Whittaker (1975) recommended the following conversions for the world's major biomes (in kJ/g): tropical rain forest 17.1, temperate mixed forest 19.7, boreal forest 20.1, grasslands 16.7, and cultivated land 17.1. Detailed chemical composition analyses and energy values are available for many hundreds of food and feed crops (Watt and Merrill 1963; NRC 1971), and forestry literature contains similar data on many tree and shrub species (Tillman 1978; NAS 1980; Davis et al. 2001).

Two major uncertainties complicate any large-scale estimates of phytomass (in small areas its above-ground components can be readily and accurately determined by destructive sampling): unreliable data on the extent of biomes and ecosystems, and the questionable validity of extrapolating detailed local assessments to larger areas. Errors in converting volume to mass and energy equivalents are another problem. Conversion of green volume of wood to dry matter phytomass ranges between 0.4 and 0.8, with typical rates between 0.55 and 0.60. As a result, there is less consensus on global phytomass than on global NPP. Vernadsky's (1926) estimate of the Earth's green matter in the first edition of his pioneering book on the biosphere, corresponding to 10^{13}–10^{14} t C, proved to be an enormous exaggeration. Whittaker and Likens (1975) estimated continental phytomass in 1950 at 1.837 Tt, or 827 Gt C.

Olson, Watts, and Allison (1983) improved the accuracy by subdividing continents into $0.5° \times 0.5°$ cells and by collecting the best available data on climatic factors, vegetated areas, and phytomass ranges on that scale. Their range was 460–660 (mean of 560) Gt C. Models of the global carbon cycle published during the 1990s contained total continental phytomass values as low as 486 Gt C (Amthor et al. 1998) and as high as 780 Gt C (Post, King, and Wullschleger 1997). Pilot Analysis of Global Ecosystems opted for a surprisingly broad range of 268–901 Gt C (Matthews et al. 2000), whereas a comprehensive review of global NPP offered 652 Gt C as its best estimate (Roy, Saugier, and Mooney 2001). Disparities in the categorization of land cover and differences in assumed phytomass densities explain these differences. Land cover assessments published since 1980 have used values as low as about 25 Gm^2 and as high as 75 Gm^2 for the total area of the Earth's forests (Emanuel, Shugart, and Stevenson 1985; Solomon et al. 1993; Cramer et al. 1999; FAO 2005a). In contrast, oceanic phytomass adds up to only 1–3 Gt C.

With 500–800 Gt C, the biosphere's phytomass binds an equivalent of 62%–99% of the element's current atmospheric content (809 Gt C in 2005). Uniform distribution of dry terrestrial phytomass over ice-free land would produce a layer about 1 cm thick; the same process in the ocean would add a mere 0.03 mm of phytoplankton (in both cases, I assume an average biomass density of 1 g/cm^3). I know of no better examples to illustrate the evanescent quality of life, but a different definition of phytomass would produce an even smaller terrestrial total. Most of the structural polymers that play essential supportive, protective, and conductive roles are not alive, and this reality makes it possible to argue for both drastically reducing and greatly expanding the definition of the Earth's phytomass.

The first course would be to restrict the phytomass definition to the living protoplasm but, as a closer look at tree phytomass illustrates, it is difficult to offer reliable large-scale corrective multipliers. The radial extent of the cambial zone, the generator of tree growth, is difficult to define because of the gradual transition to differentiating xylem and phloem. Most of the conducting tissue in trees is sapwood, with typically only 5%–8% of living cells in conifers and 10%–20% of living cells in hardwoods. Conversion of sapwood into nonconducting heartwood involves death of the cytoplasm of all living cells in softwoods, but in hardwoods some cells in axial and radial parenchyma remain alive for years and decades. And there are substantial specific differences in the shares of total phytomass made up of fresh leaves, buds, young branches, and rapidly growing fine roots (Gartner 1995; Waring and Running 1998). In addition, trees and shrubs have dead branches and dead roots. If, in a strict sense, no more than 15% of all forest phytomass were alive, then the terrestrial phytomass would be less than 80 Gt C, and the real total might be less than 50 Gt C. On the other hand, it can be argued that the definition of phytomass could be extended to include not only all the standing dead structural matter but also plant litter and the stores of organic soil carbon (the latter inclusion brings the problem of separating autotrophic and heterotrophic contributions). Such inclusions would multiply the aggregate storage. For example, the FAO's latest forest assessment put the average phytomass at 71.5 t/ha in living trees, 9.7 t/ha in dead trees, 6.3 t/ha in litter, and 73.5 t/ha in soil up to 30 cm deep (FAO 2005a).

Forests store the bulk of all phytomass (at least 80%–85%), with tropical rain forests dominant. Tropical rain forests can be monospecific, or dominated by a single family, such as Dipterocarpaceae in Southeast Asia and in the northeastern basin of the Congo River (Connell and Lowman 1989). But high biodiversity is the hallmark of this biome, with many large tree families entirely or largely restricted to the biome. Destructive sampling of an Amazonian site near Manaus ended up with nearly 95,000 plants/ha belonging to more than 600 species stratified in six distinct layers (Klinge et al. 1975). Nine-tenths of all plants were in the light-starved layers, and hence their phytomass contributed less than 1% of the total. In contrast, 50 emergents and more than 300 canopy trees, altogether just some 50 species dominated by Leguminosae and Euphorbiaceae, contained 85% of the standing phytomass.

Almost two-thirds of above-ground phytomass was in the stemwood of dicot species (palms contribute merely a fraction of 1%), just over one-quarter in branches and twigs, a mere 2.5% in leaves, and a small remainder in lianas, epiphytes, and parasitic plants. Total dry phytomass can be as high as 450–500 t/ha, but the global mean is just short of 400 t/ha, with about 22% of all phytomass in roots (Roy, Saugier, and Mooney 2001). The forest's enormous diversity means that a single tree species will store no more than 5% of all phytomass, most commonly just 2%–3%. High diversity increases the spacing among the adult trees of the same species, and this lowers the high mortality of juveniles caused by concentrated heterotrophic attack (Janzen 1970). Clark and Clark (1984) confirmed this hypothesis, finding the seedling survival positively correlated with distance to adult trees and negatively with local conspecific seedling density. Tropical trees also have such protective mechanisms as smooth barks and coated leaves, chemical defenses, and symbiotic myrmecophily (ants guarding the huge energy stores in stems).

In contrast, in boreal ecosystems a single species may store the bulk of the site's phytomass. This is most pronounced in the case of the world's highest accumulation of phytomass, the old-growth coniferous forests of western North America (Edmonds 1982). These ecosystems shelter the oldest living plants (bristlecone pines, *Pinus longaeva*, some 4,600 years old) as well as the biosphere's most massive living creatures (although most of their phytomass is dead wood): giant sequoias (*Sequoidendron giganteum*) growing over 100 m tall and able to live for over 3,200 years. Phytomass accumulations in these forests are enormous. A century-old forest dominated by Sitka spruce (*Picea sitchensis*) or noble fir (*Abies procera*) can store about 900 t/ha, older stands of Douglas fir (*Pseudotsuga menziesii*) and noble fir may have up to 1700 t/ha, and the maximum rates for coastal redwoods are at 3500 t/ha (~68 TJ/ha), with all of these values excluding roots. Even the richest tropical rain forests will store less than one-quarter of this mass.

PHOTOSYNTHESIS

The most diverse temperate deciduous forests have five distinct layers, ranging from a ground layer of mosses and lichens through a layer of herbs, shrubs, and small trees to the surmounting tree stratum, whose canopies are mostly 15–30 m above the ground. Some of them are also great accumulators of phytomass in various species of oak, beech, maple, chestnut, and, on sandy soils, pine. The richest cove forests of the Great Smoky Mountains store up to 600 t/ha, but their typical phytomass is around 250 t/ha, with 20% below ground. Energy limits on tree productivity are evident in the life cycle of foliage. In stressed ecosystems the investment in annual renewal of leaves would far surpass the costs of their prolonged retention. Life spans of expensive leaves are maximized in order to amortize the high cost of their formation; that is why all subxeric and boreal habitats are dominated by evergreens (Chabot and Hicks 1982). Their leaves have high specific weight (6.3–15 mg/cm^2, compared to the deciduous range of 2.9–7.8 mg/cm^2) because their photosynthetic tissues are diluted with structural supports and protective coatings (to reduce herbivory, leaching, and desiccation) to ensure their longevity.

With progressing deforestation the tropical grasslands have become the world's most extensive terrestrial biome, covering perhaps as much as 2.5 Gha, and the second largest reservoir of phytomass after the tropical rain forests. Their best-known regional formations are East African savannas, Brazilian *cerrado*, and Venezuelan *llanos*, and they exist either as pure perennial grass formations or as open woodlands with varying densities of appropriately adapted (drought-, fire- and browse-resistant) shrubs and trees. The presence of the woody phytomass indicates that tropical grasslands are not climax ecosystems. Their typical standing phytomass is 50–60 t/ha,

nearly one-third of it underground. Temperate grasslands (now largely converted to fields) were dominated by species of Compositae and Leguminosae. Short-grass formations consist of species growing no taller than 50 cm, and tall-grass prairies can exceed the height of 2 m. Temperate grasses, such as spear grass and tussocky iron grass, are admirably adapted to seasonal fire and drought, with most of their phytomass below ground.

Root to shoot ratios in temperate grasslands can be as high has 13:1, and bulky root mats retain water and bind soils so tightly that they virtually eliminate water and wind erosion. A representative mean for phytomass stocks is 7.5 t/ha, two-thirds of it in roots (Roy, Saugier, and Mooney 2001). Agricultural crops are the only vegetation category whose phytomass is known with a fairly high degree of certainty (Smil 1999a; Wirsenius 2000). During the late 1990s the aggregate annual phytomass of field crops was 3.5 Gt C (35% in harvested parts, 48% in crop residues, and the rest in roots and unharvested phytomass), or just over 5 t DM/ha. Because of the more frequent multicropping (the global mean near 1.5 harvests/ha) and staggered harvesting, the peak crop phytomass is no more than two-thirds of the annual aggregate, less than 1% of all terrestrial biomass.

The smallest oceanic monocellular autotrophs—ultrananoplankton of bacteria and blue-green algae—have diameters less than 2 μm. Nanoplankton, 2–20 μm, includes diatoms, coccolithophores, and silicoflagellates, and diatoms and dinoflagellates (20–200 μm in diameter) are the most common kinds of microplankton (Falkowski and Raven 1997). Phytoplankton densities peak near the coast and decline oceanward; the density difference between the continental shelf and the open ocean is up to 3 OM. Eutrophic waters, enriched by nutrient upwelling or flux from rivers, average 10^4–10^5 cells/L, and their standing phytomass is 100–1500 mg/m^3. In contrast, the oligotrophic regions of the central Pacific or Atlantic have as few as 100 cells/L. Because of short cell life the annual phytomass turnover rates are 300–400 in nutrient-rich waters and 40–50 even in the nutrient-poor open ocean.

Benthic autotrophs are mixtures of algae and a limited variety (only about 60 species) of vascular plants. Rocky coastal zones often support dense stands of macroalgae, above all, kelp and fucoid rockweeds. Reef-building corals live symbiotically with photosynthesizing dinoflagellates. Standing stock of intertidal phytomass is lowest in the tropics, highest on the boreal shores: measurements in the Sea of Japan show up to 6.5–7 kg/m^2 in grassy beds (Menzies, George, and Rowe 1973). The average density of algal beds and reefs is most likely no greater than 2 kg/m^2. Estuarine waters may have the same order of phytomass densities, whereas the mean for upwelling regions is 2 OM smaller, and for open ocean 3 OM smaller.

While we cannot reconstruct global phytomass totals that were characteristic of past geological eras, we have some revealing numbers regarding the historic trend, and increasingly accurate indications of recent changes. Bazilevich, Rodin, and Rozov (1971) reconstructed the world's potential vegetation cover (the extent of natural biomes of the preagricultural era) and came up with a continental phytomass of 2.4 Tt, or almost 1.1 Tt C. The terrestrial phytomass total of 500–600 Gt C in the year 2000 would be roughly half of the preagricultural storage, and about half of this loss took place during the last three centuries. The Earth's standing phytomass has declined by about 25% since 1700, and during the twentieth century the net loss of global plant mass amounted to about 15% of the 1900 total, with deforestation being the main reason.

In addition to deforestation there has been major phytomass loss in temperate grasslands. Deep and fertile soils, like chernozems, that gave rise to this extensive biome have also been the main cause of its enormous retreat because of the conversion to cropland of most of the U.S. Great Plains, Canadian Prairies, Ukrainian, Russian, and Kazakh steppes, Argentinian pampas, and South Africa's *veld*. Crops replacing the grasses may have similar NPP, but their phytomass is generally much lower and of short duration. The third major category of phytomass loss has been the destruction and drainage of wetlands. Reconstructions of global land use changes (Ramankutty and Foley 1999; Goldewijk 2001) offer the best available estimates of the historic progression of these processes, and they indicate cropland expansion from about 265 Mha in 1700 to nearly 1.5 Gha by the year 2000.

DeFries et al. (1999), Houghton and Hackler (2002), and R. A. Houghton (2005) expressed these losses in terms of declining phytomass: a cumulative phytomass decline of about 50 Gt C before 1850 and, depending on land use data and phytomass density used, losses of 125–200 Gt C by the year 2000, with more than 85% from deforestation and the rest from the conversion of temperate grasslands (fig. 3.10). This massive phytomass loss has been partially counterbalanced by post–WW II expansion of forests in Europe and the United States, massive post-1980 afforestation in China, and higher NPP due to inadvertent atmospheric deposition of nitrogen and better management of forests. Most notably, the standing phytomass of European forests has increased by more than 40% since 1950, and these forests have become a substantial (140 Mt C/a) carbon sink (Nabuurs et al. 2003).

PHOTOSYNTHESIS

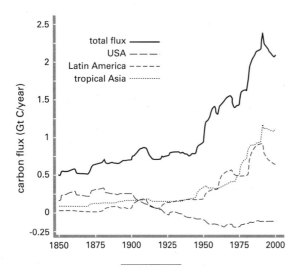

3.10 Carbon losses from land use changes, 1850–2000. From Houghton and Hackler (2002).

3.5 Autotrophic Scaling

Changes in size have both structural and functional consequences for otherwise very similar organisms. If there is no change in similarity between the studied variables across a wide range of values, then they scale isometrically and the plot of this relationship will be a straight line. If the similarity is not maintained, the scaling is allometric and the plotted lines are curved. Both autotrophs or heterotrophs rarely maintain geometric similarity relative to the mass of the whole body (M, the most commonly considered independent variable) or to the mass or size of its parts. Their allometric scaling is expressed by the general equation

$$y = ax^b,$$

where y is the variable of interest, a is a constant multiplier needed to express the result in particular units, x is the variable of size (most often M), and b is an exponent (isometry is maintained when $b = 1$). Restated logarithmically,

$$\log y = \log a + b \log x,$$

the exponent determines the slope of the straight line on a log-log graph; a is then the intercept of y (it represents the value of y when $x = 1$).

Allometric scaling has been a common tool in investigating functional (particularly metabolic) and structural regularities in heterotrophs (see chapter 4), but a series of scaling studies published since the late 1990s claimed that autotrophs share the same basic scaling exponent as far as the intensity of their metabolism (NPP) is concerned. Enquist et al. (1999) examined life history variation of 45 species of tropical trees that attain similar canopy sizes despite substantial differences in their rates of growth and ages of maturity, and found that their metabolism scales as $M^{3/4}$, much like the metabolism of many heterotrophs (see section 4.1). Niklas and Enquist (2001) extended this study of interspecific phytomass production rates and body size to photosynthesizers, including unicellular and multicellular autotrophs representing three algal phyla, aquatic ferns, aquatic and terrestrial herbaceous plants, and trees (monocots, dicots, and conifers). The examined set spanned over 20 OM of body size and over 22 OM of body length (cell diameter or plant height).

Annualized growth rates (in kg DM/plant) scaled as $M^{3/4}$, and plant body length scaled as $M^{1/4}$ (fig. 3.11). Light-harvesting capacity (pigment content of an algal cell or foliage phytomass of higher plants) also scaled as $M^{3/4}$, as did foliage photosynthesizing phytomass (foliage) in relation to nonphotosynthesizing plant mass.

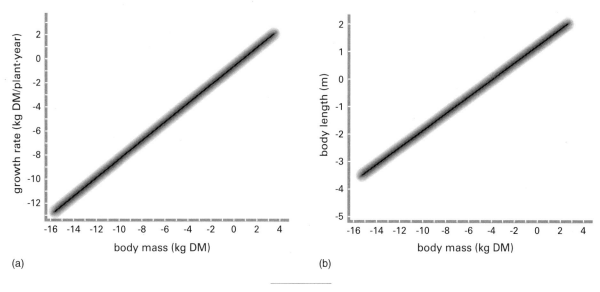

3.11 (a) Plant growth rates scale as $M^{3/4}$, and (b) plant body lengths scale as $M^{1/4}$, for species whose mass spans 20 OM. Based on Niklas and Enquist (2001).

And while heterotrophs have similar allometric exponents but a different normalization constant (different intercepts when graphed), photoautotrophs across the entire range of extant body masses conform to a single allometric pattern. This means that the relative growth rate decreases with increasing plant size as $M^{-1/4}$. This uniformity despite differences in phylogenetic affiliation and habitat is explained by shared hydrodynamics and biomechanics that govern the resource distribution through hierarchical branching networks of plants (West, Brown, and Enquist 1999).

Theoretical explanation of this functional unity rested on indispensable fractal-like distribution networks (in plants they have to transport photosynthate from leaves and transpire water through roots, stems, and leaves) that evolved to maximize metabolic capacity and effi-

ciency by maximizing exchange surfaces and throughputs while minimizing transport distances and transfer rates (West, Brown, and Enquist 2000). This uniformity means that NPP is largely insensitive to species composition: identical density of similarly massive plants fixes the same amount of carbon. Because the abundance of terrestrial plants per unit area scales as $M^{-3/4}$, and their individual annualized production rates scale as $M^{3/4}$, the rate of total community production (the product of the two variables) scales as M^0, and hence the phytomass production (and the overall claim on resources) should be invariant as to plant size or species composition.

Total energy flux through a plant community is thus not dictated by individual body size but limited by the resource supply. Belgrano et al. (2002) extended this relation also to marine phytoplankton, confirming that maximum rates of both terrestrial and marine primary productivity are subject to very similar body mass–related energetic limits. Allometric regularities were also found in the partitioning of phytomass in seed plants and the density of autotrophs. Standing leaf mass (M_L) scales as the 3/4 power of stem mass (M_S) and root mass (M_R), whereas M_S and M_R scale isometrically with respect to each other (Enquist and Niklas 2002). This means that the above-ground phytomass ($M_A = M_L + M_S$) will scale nearly isometrically with respect to roots and that the ratio of M_A/M_R (often called the shoot/rot ratio) declines rapidly for small plants and becomes nearly asymptotic for plants with $M > 10$ kg. This size-specific prediction makes it easier to estimate the below-ground phytomass from more common assessments of above-ground phytomass (Zens and Webb 2002).

At the same time, I must note that a high degree of conformity across many orders of magnitude hides substantial differences at many levels. For example, for a given M_S, there is a 2 OM difference in M_L among different species. And while the scaling exponent is the same for angiosperms and gymnosperms, their allometric constants (intercepts on the y axis) are quite different (respectively, 0.12 and 0.24) because conifers average 2.6 times more foliage (needles) than leafy trees of the same stem size. But these realities do not fundamentally challenge the remarkable invariance of scaling exponents across such autotrophic diversity and in such a range of habitats. A fundamental challenge of the universal validity of 3/4 allometric scaling in plants came only with direct measurements of plant respiration (Reich et al. 2006).

Some 500 observations of 43 perennial species—both field- and laboratory-grown, ranging in age from 1 month to 25 years, and spanning 5 of the roughly 12 OM of size in vascular plants—lent no support to 3/4 power scaling of plant night-time respiration (and hence its overall metabolism) and instead supported very strongly isometric scaling (exponent ~1). Moreover, the study found no single universal relation between R_A and M but uncovered such a link between R_A and total plant nitrogen content (reflecting the fundamental role of the nutrient in plant biochemistry), again with a scaling exponent of ~1 (fig. 3.11). This demonstration of near-isometric scaling of plant respiration eliminates the need for complex fractal explanations of 3/4 power scaling and makes it unlikely that there is a single size-dependent law of metabolism for plants and animals.

The last remarkable mass-dependent regularity that should not be omitted in this brief review of autotrophic scaling is the self-thinning rule, which describes plant mortality owing to competition in crowded even-aged stands of terrestrial plants. Total phytomass of a stand (M_{tot}) can be increasing independently of the stem density (ρ, numbers/m^2) until an intensive competition sets in, lowers the density, and limits the average mass per plant, in g) in a highly predictable way. Studies found that in the allometric relation

$$M = k \cdot \rho^{-a}$$

the exponent lies between -1.3 and -1.8 and has an ideal value of -1.5, and k varies between 3.5 and 5.0 (J. White 1985). Since $M = M_{tot}/\rho$, an alternative formula proposed by Westoby (1984) is $M_{tot} = k \cdot \rho^{-0.5}$ (fig. 3.12). The -1.5 rule has three fascinating features: time can be ignored because mortality depends only on phytomass accumulation; this makes the thinning rate slower

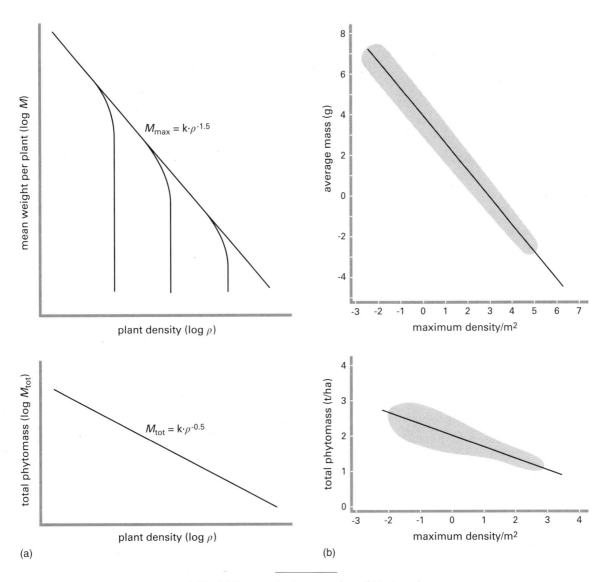

3.12   (a) Two graphs illustrating the self-thinning rule
(Westoby 1984). (b) Two graphs showing a strong correlation
between maximum plant density and average plant mass for
autotrophs whose mass ranges over 11 OM (Enquist, Brown,
and West 1998).

when conditions for growth are worse; and the location of the thinning line varies remarkably little for different species and growing conditions.

There is only about a sixfold range of intercepts, a minimum variation considering the enormous span of mean plant weights. The rule fits not only trees but also shrubs, herbs, ferns, and mosses, whose average masses span more than 10 OM. This means that a monospecific even-aged stand with 100,000 stems/ha has trees averaging no more than about 0.2–0.4 kg and adding up to a maximum of a bit over 40 t, whereas 100 trees/ha average 6 t/tree with a mass of more than 600 t/ha (fig. 3.12). Economies of scale mean that every tenfold concentration of photosynthesis brings roughly a 32-fold increase in the average plant size as more phytomass is accumulated in stands of fewer but larger plants, resulting in large, long-lasting structures that require durable construction (dense wood) and effective protection (barks or defensive chemicals) against heterotrophs.

The self-thinning rule has been recurrently questioned, and Lonsdale (1990) even concluded that there is no evidence to support it: the slope is much more variable than previously claimed, and the straight lines demarcating the M_{max} are the exception rather than the rule. In contrast, Enquist, Brown, and West (1998) found a strong correlation between maximum plant density and average plant mass for autotrophs whose mass ranged over 11 OM, from *Sequoia* to *Lemma* (duckweed). Their slope for the allometric relation between M_{tot} and ρ was -0.325 rather than -0.5. (fig. 3.12), but their analysis strongly confirms the existence of the self-thinning rule as plants grow and proliferate until checked by the availability of energy and nutrients.

4

HETEROTROPHIC CONVERSIONS

Consumer Bioenergetics

For the main factor in the nature of an animal is much more the final cause than the necessary material.... If any person thinks the examination ... of the animal kingdom an unworthy task, he must hold in like scorn the study of man.

Aristotle (384–322 B.C.E.), *Parts of Animals*

Heterotrophic life, precisely because of its total dependence on photosynthesis, has evolved countless adaptations to cope with environmental challenges, diffused into nearly every conceivable niche, and eventually resulted in the emergence of global civilization. Autotrophs take care of themselves by reducing atmospheric CO_2 and producing a vast variety of complex organic compounds. Heterotrophs, incapable of *ab initio* synthesis of complex molecules, use those compounds (directly as herbivores and detritivores, indirectly as carnivores) to energize their growth and activity. Heterotrophic cells usually absorb carbon in relatively simple molecules, and

this means that enzymes must first destroy the structure of biopolymers.

This is done by severing glycoside bonds of complex sugars (to produce the constituent monosaccharides, with glucose dominant), and amide bonds of proteins (amino acids are then used for protein synthesis), and by hydrolyzing triglyceride bonds in lipids (into glycerol and fatty acids). There are three basic metabolic strategies: aerobic glycolysis, anaerobic fermentation, and dissimilatory anaerobic oxidation. Oxygen is the most common electron acceptor. Most heterotrophs use the highly exergonic (-2870 kJ/mol) oxidation of biomass, which produces water and CO_2. Some anaerobic organisms convert carbon compounds to lactate or ethanol. These fermenters include both bacteria (e.g., yogurt-making *Lactobacillus*) and fungi (e.g., *Saccharomyces*, the yeast responsible for alcoholic fermentation and leavened bread), and they gain just 197 kJ/mol by those oxidations.

Other anaerobes use nitrates, nitrites, sulfates, and ferric iron as electron acceptors. *Pseudomonas* and *Clostridium* are common bacteria that reduce nitrate to nitrite. Thiopneute bacteria, including red *Desulfovibrio* common in muds and in estuarine brines, reduce sulfates to H_2S or S while incompletely oxidizing lactate and acetate. These dissimilatory reductions (less exergonic than anaerobic glycolysis, more exergonic than fermentations) are critical for the functioning of global biospheric cycles. Methanogens perform the final task of biomass degradation in those environments where oxygen, nitrate, sulfate, and ferric iron have been depleted. They use CO_2 as the final electron acceptor to produce CH_4 (Ferry 1993). Anaerobic fermentation proceeds naturally in marine and freshwater sediments, marshes, bogs, flooded soils, gastrointestinal tracts, and geothermal habitats.

Aerobic respiration is a way of life for 8 of 16 phyla of *Monera* (bacteria), ranging from myxobacteria to nitrogen fixers, for nearly all fungi, and for the whole kingdom of *Animalia*. Energy costs of this prevalent mode of metabolism are examined in this chapter. Evolutionary steps toward the origin of aerobic respiration are not difficult to postulate, but details and timings are elusive. The preplanetary matter was clearly anoxic, and the Earth's secondary atmosphere had only a limited amount of the gas formed by photolysis of water vapor by UV radiation. Eventually photosynthesis (see chapter 3) increased the partial pressure of oxygen to the point where some prokaryotes could use aerobic respiration to generate energy in the form of ATP more efficiently than by fermentation.

Finding a generally accepted division between plants and animals has been difficult because both phycologists and protozoologists claim taxonomic dominion over euglenids and trichomonads. Many eukaryotic protoctists can be classified in both groups. Many organisms have found it advantageous to evolve toward mobile protozoan existence, first by osmotrophy (absorbing dissolved nutrients through cell surfaces), then by phagotrophy (as active consumers, even predators). Evolution of heterotrophs remains to be clarified for the late Proterozoic eon, but it is fairly clear after 530 Ma bp. Metazoa evolved between 1100 and 600 Ma ago, and the fossil record from about 530 Ma ago documents a spectacular emergence of diversified skeletonized fauna in a span of just a few million years (Erwin, Valentine, and Jablonski 1997). Sponges are the most primitive surviving animal phylum; other simple phyla include Ctenophora (comb jellies), Cnidaria (jellyfish and sea anemones), and Plathelminthes (flat worms).

More complex animals are classified by the different fate of the initial opening of the primitive digestive tract in an embryo: Arthropoda, Annelida (earthworms), and Mollusca (snails, clams, squids) belong to protostomes; Echinodermata (star fish, sea urchins) and Vertebrata (fish to mammals) to deuterostomes. Fish-like animals from the early Cambrian (Shu et al. 1999) put vertebrates among the organisms of the Cambrian eruption of new life forms. Transition between fish and amphibians (tetrapods) took place during the Late Devonian, about 365 Ma bp (Janvier 1996). Some 310 Ma bp mammal-like reptiles split from bird-like reptiles, and molecular clocks confirm that modern orders of mammals go back to the Cretaceous period, more than 100 Ma bp, and that they diversified before the extinction of dinosaurs (Kumar and Hedges 1998).

4.1 Metabolic Capabilities

Lives of Metazoa require highly differentiated mouth and digestive organs, well developed circulation, efficient modes of locomotion, and complex nervous systems to control these processes. For them the energetic advantages of oxidation over anaerobic fermentation are clear. Lactic acid fermentation liberates 195 kJ for each molecule of glucose, alcoholic fermentation yields 232 kJ, but a complete oxidation of that sugar releases 2.8 MJ, a 12–14-fold gain. Three kinds of nutrients can be metabolized to yield energy: carbohydrates, lipids, and proteins (but proteins are used only if the other nutrients are in short supply). Energy released by their oxidation is partially conserved in ATP. As in photosynthesis, ATP is the principal energy carrier, the key link between cellular catabolism (degradation of nutrient substrates) and anabolism (biosynthesis of complex substances), locomotion (muscle contraction) and active transport of metabolites against the concentration gradient.

The biochemistry of these sequential, enzymatically catalyzed reactions is well understood (de Duve 1984; Nelson and Cox 2000). Glycolysis of glucose or glycogen takes place in the cytoplasm of all heterotrophs, and it produces, following the Embden-Meyerhof-Parnas pathway, pyruvic acid (fig. 4.1). Nicotinamide adenine dinucleotide (NAD) is the electron carrier (NADH), and pyruvic acid is the precursor compound for anaerobic respiration (the pathway that ends in lactic acid), alcohol fermentation (producing ethanol and CO_2), and aerobic respiration, the tricarboxylic acid (citric acid, Krebs) cycle (fig. 4.1). This cycle, taking place inside the mitochondria, converts a variety of organics (fatty acids and amino acids) to CO_2 and transfers the released electrons down the electron transport chain, producing large amounts of ATP and reducing oxygen to water. The maximum energy gain is 38 mol of ATP for each mole of glucose broken down in prokaryotic cells, an overall free energy change of about −2.8 MJ. With −31 kJ/mol available

from each ATP transformation to ADP, the overall efficiency of the whole sequence would be about 42%.

In eukaryotic cells the net ATP gain is a bit smaller—two moles are needed to move NADH from the cytoplasm—but because the free energy of the compound may be up to −50 kJ/mol in mammalian cells, the overall efficiency may be over 60% (a value of −31 kJ/mol is valid only for unimolar concentrations, neutral pH, and 25 C). Respiration of fatty acids yields a maximum of 44 ATP/mol, but since oxidized compounds have higher energy contents than glucose (around 3.4 MJ/mol), the peak efficiencies are about 60%. Only two molecules of ATP are gained during the breakdown of glucose to lactic acid (the overall free energy is −197 kJ), and the process has efficiency of about 30%; in vertebrates it can be sustained only briefly. Only invertebrates living in oxygen-poor environments evolved longer-lasting low-efficiency anaerobic pathways leading to alanine, succinate, and propionate.

The intensity of ATP generation is stunning (Broda 1975). A 60-kg man consuming daily about 12 MJ (~700 g) of food in carbohydrates would make and use no less than 70 kg of ATP (assuming production of 36 molecules of ATP for every digested hexose molecule), more than his total weight. This rate, roughly 3 g ATP for each gram of dry body mass, is minuscule compared to intensities achieved by respiring bacteria. *Azotobacter*, breaking down carbohydrates while fixing large amounts of N_2, produces 7000 g of ATP for each gram of its dry mass. While solar luminosity is immense (390 YW), so is the star's mass (1.99×10^{33} g). Consequently, the Sun's power intensity averages about 200 nW/g, but the daily metabolism of schoolchildren proceeds at a rate of 3 mW/g of body weight, 15,000 times the power intensity of the Sun, and respiring *Azotobacter* reaches up to 100

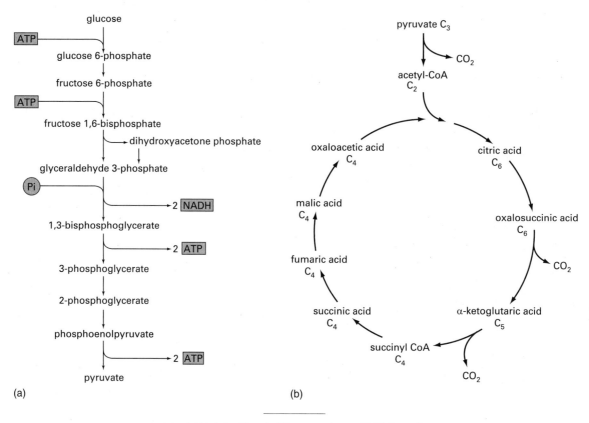

4.1 (a) Embden-Meyerhof-Parnas pathway. (b) Krebs cycle.

W/g, 500 million times the Sun's rate. ATP-driven energy conversions in heterotrophs are no less awesome in their intensity than the Sun's performance is in its overall magnitude.

Basal metabolic rate (BMR) informs about the minimum energy cost of steady-state existence, but it is not easily determined because it must be measured at rest, in a postabsorptive state, and in a thermoneutral environment. The most common approach to studying it across the enormous variety of heterotrophic organisms has

been allometric scaling (Pedley 1977; McMahon and Bonner 1983; Schmidt-Nielsen 1984; Brown and West 2000). The earliest scaling, based on just seven entries for dogs (Rubner 1883), confirmed logical expectations based on relations between BMR and body surface. A body's mass and area are proportional, respectively, to the cube and the square of a linear dimension: $M \propto L^3$ and $A \propto L^2$; by rearrangement, $A \propto M^{2/3}$. This means that doubling a body's length increases its surface fourfold and its volume eightfold. As the metabolic heat is lost through the body surface, it is logical to expect that an organism will adjust its BMR accordingly and that the rate will be proportional to $M^{2/3}$. Rubner's surface law of metabolism remained unchallenged for nearly half a century.

Then Kleiber (1932), working with a set of 13 data points (including two steers, a cow, and a sheep), showed that BMR goes up as $M^{0.74}$. Kleiber (1961) eventually recommended a rounded expression of $70M^{0.75}$ (in kcal/day) or $3.4M^{0.75}$ (in W). When plotted on double-log axes, this exponential relation became one of the most important generalizations in bioenergetics, the straight mouse-to-elephant line of the 3/4 law (fig. 4.2). Unlike the 2/3 law, the 3/4 law presented a challenge of causal interpretation. Maynard Smith (1978) explained the exponent as a compromise between the surface-related BMR (0.67) and the mass-related inputs needed to overcome the gravitation (exponent 1.0). McMahon (1973) based his explanation on the elastic criteria of limbs. The weight of these loaded members is a fraction of M, so their diameter will be proportional to $M^{3/8}$. The power output of muscles depends only on their cross-sectional area (proportional to d^2), and thus the maximum power output is related to $(M^{3/8})^2$, or $M^{0.75}$. If applicable to any particular muscle, the scaling should rule the total organism, and BMR should be a function of $M^{0.75}$.

West, Brown, and Enquist (1997) offered an explanation based on the geometry and physics of a network of tubes needed to distribute resources and remove wastes in organisms. They argued that the rates and times of life processes are ultimately limited by the rates at which energy and material flows are distributed between the surfaces where they are exchanged and the tissues where they are used or produced. This means that distribution networks must be able to deliver these flows to every part of an organism; that their terminal branches must have identical size because they have to reach individual cells; and that the delivery process must be optimized in order to minimize the total resistance and hence the overall energy needed for the distribution. A complex mathematical derivation shows that these properties require the metabolism of entire organisms to scale with the 3/4 power of their mass.

The first conclusion demands that many structures and functions of the delivery system (be they hearts or heart-beats) must be scaled according to the size of organisms. Because the obviously tightly interdependent components cannot be optimized separately, they must be balanced by an overall design that calls for a single underlying scaling. West, Brown, and Enquist (1997) showed that the networks providing these fundamental services have a fractal architecture that requires many structural and functional attributes to scale as quarter powers of body mass, and that this requirements applies equally well to heterotrophs and plants. The second conclusion regarding the invariant components of the delivery system seems to be convincingly demonstrated by the identical radius of capillaries in mammals, whose sizes span 8 OM (from shrews to whales), or by the identical number of heartbeats per lifetime, that is, by the identical total of energy needed to support a unit mass of any organism over its lifetime (Marquet et al. 2005). The third conclusion rests on the economy of evolutionary design: organisms develop structures and functions to meet but not exceed maximal demand.

But Makarieva, Gorshkov, and Li (2005) concluded that this model is logically inconsistent because its assumptions of the size-invariance of the rate of energy supply in terminal units contradicts its central prediction—namely that the mass-specific metabolic rate

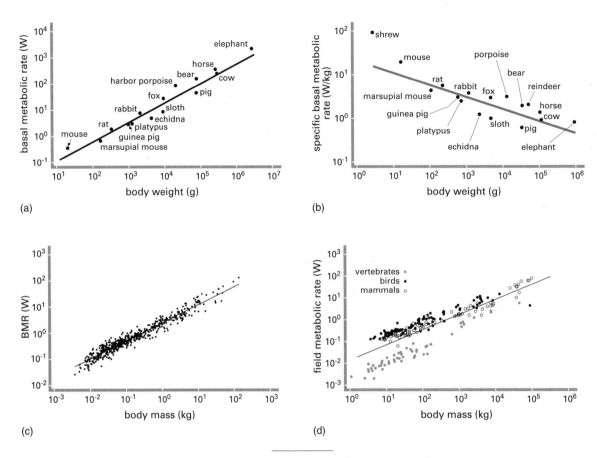

4.2 Basal metabolic rate scaling. (a) Kleiber's (1961) 0.75 slope, and (b) declining specific BMR for mammals with body masses spanning almost 6 OM (Smil 2002). (c) 0.67 slope (White and Seymour 2005), and (d) field metabolic rates (Nagy 2005).

should decline with body mass with a scaling exponent $-1/4$. According to them the best available evidence is consistent with a size-independent mean value of mass-specific metabolic rate, but one not equivalent to strict uniformity, with the range confined between 1 to 10 W/kg for most unicells, insects, and mammals.

Banavar, Maritan, and Rinaldo (1999) argued for a simpler explanation of a universal scaling expenent, believing that the $1/4$ power law is an attribute of every optimally efficient network used to distribute nutrients. Hence there is no need to invoke fractality, merely to consider the need to feed roughly L^3 sites in a network distributing nutrients over the circulation length L. Other explanations of the $3/4$ power law have been offered, but all of them may be misplaced because several careful reexaminations of BMR scaling (for mammals, birds, and fish) found exponents that differ significantly from 0.75. White and Seymour's (2003) analyses of the allometry of mammalian BMR (spanning 5 OM of M and including 619 species of 19 orders) found no support for the $3/4$ exponent (fig. 4.2). The main reason for this is that previous data sets were contaminated with non-BMR values: the inclusion of ruminants (overrepresented in the earliest sets) is particularly problematic because their enteric fermentation delays or entirely precludes postabsorptive measurements of their BMRs.

After eliminating all large herbivores from their data set, White and Seymour (2005) concluded that the exponent for true BMR is 0.686 and for the temperature-normalized standard metabolic rate, 0.675, essentially Rubner's exponent. Analyses for individual mammalian orders showed that exponents were not significantly different from 0.75 for carnivores (0.784) and primates (0.772) but were as low as 0.457 for insectivores and 0.629 for lagomorphs (Kozłowski and Konarzewski

2005). A careful selection of reliable BMRs for birds confirmed Rubner's exponent of 0.669 (McKechnie and Wolf 2004). Bokma (2004) argued that it is more informative to investigate intraspecific allometry, and his most powerful sample of sea trout individuals (with body sizes ranging from 0.1 g to 600 g) yielded the exponent 0.86. Based on his large sample of 113 species of fish, he concluded that there is no universal scaling exponent for BMR.

Perhaps most important, BMR is only one of many heterotrophic states of living (and a rather artificial one), and hence it may be much more useful to study the scaling of resting metabolic rate (RMR) or field metabolic rate (FMR), the longer-term average energy expenditure under normal living conditions, or the maximal rate that may have been a more pertinent goal of evolutionary selection (Hoppeler and Weibel 2005). This is evident when one considers the metabolic rates of common activities: walking and running mammals metabolize at rates that are 4–20 times their BMR, and the multiples are 15–100-fold for flying insects (R. F. Chapman 1998). How would a fractal-like network that creates a rate-limiting supply of oxygen (and hence is seen to impose the $3/4$ scaling on BMR) cope with these challenges?

Before the introduction of the doubly labeled water (DLW) technique—developed in the mid-1950s (Lifson, Gordon, and McClintock 1955) and used first to measure energy expenditure in small animals—there was no practical way to measure FMR. The DLW method uses water marked with stable heavy isotopes of hydrogen and oxygen ($^2H_2$ and ^{18}O). When the isotopes are washed out of the body and replaced with dominant $^1H_2$ and ^{16}O, the loss of deuterium measures water flux, and the elimination of ^{18}O traces not only the water loss but also the flux of CO_2 in the expired air (DeLany 1997). The technique's relative simplicity, high accuracy, and advances in its use (including implantable data loggers and externally mounted transmitters) have made it the method of choice for reliable FMR determinations (Butler et al. 2004).

Nagy's (2005) scaling of FMR was based on a set of 229 free-living mammals, birds, and reptiles whose metabolism was measured by the DLW method. Daily energy expenditures ranged nearly 6 OM, from 0.23 kJ for a gecko to 52.5 MJ for a seal, and the exponent for the entire set was 0.808, with extremes ranging from 0.59 for marsupials to 0.92 for lizards. Nagy's (2005) conclusions were that allometric slopes for FMR are not well represented by the 3/4 power law, and FMR slopes are not identical to BMR slopes (fig. 4.2). The analysis also confirmed the expected importance of thermoregulation: more than 70% of the variation was due to variation in body mass, but most of the rest was explained by differences in specific metabolic rates (W/g), with FMRs of mammals and birds being, respectively, 12 and 20 times those of equally massive reptiles.

FMRs of wild terrestrial vertebrates also reveal interesting differences in the specific metabolism below the class level (Nagy, Girard, and Brown 1999; Nagy 2005). FMRs of marsupials are about 30% lower than those of the eutherian mammals, and the rate for the primitive monotreme echidna is only about 20%–30% of an equally massive hare. In contrast, FMRs of Procellariiformes (storm-petrels, albatrosses, and shearwaters) are nearly 1.8 times as high as those of desert birds and about 4.3 times higher than those of desert mammals. The low FMRs of desert vertebrates reflect their adaptation to periodic food shortages and to recurrent or chronic scarcity of water. Some notable outliers, such as sloths, can be explained by environmental adaptations, but there are also many unexplained specific departures from the norm.

There may be no convincing evidence for any single universal exponent relating metabolic rates to body mass, but there is no doubt about the allometric nature of the relation, and its obvious corollary (whatever the actual scaling exponent may be) is the exponential decline of specific BMRs (total BMR divided by body mass). With BMR scaling as $M^{0.75}$, the decline scales as $M^{-0.25}$; with BMR scaling as $M^{0.66}$, the decline scales as $M^{-0.33}$. For the latter case this would mean that a 10-g kangaroo mouse metabolizes 16 mW/g, a 10-kg coyote uses just 1.6 mW/g, and a 1,000-fold jump in body mass reduces the specific metabolism by 90% (fig. 4.2). This scaling has another important implication: its exponents have a very similar range but the opposite sign when compared with the relation between lifespan and M (exponents 0.15–0.30). Consequently, the product of these exponents, the mass-specific expenditure of energy per lifespan, with mean values of about -0.07, is essentially independent of M.

This means that during its lifetime 1 g of animal tissue would process the same amount of energy regardless of whether it is in a shrew's or an elephant's tail. Speakman (2005) argues that this long-standing interpretation is incorrect because BMR or RMR are poor measures of total energy metabolism. His comparison, using daily energy expenditure (DEE), showed that lifetime energy expenditure in mammals is not independent of M and that smaller animals process more energy per unit mass than larger creatures. Bird data showed only an insignificantly negative trend, but independent of M, over its lifetime 1 g of bird tissue expends roughly 3.5 times more energy than 1 g of mammalian tissue; moreover, there is enormous variation of expenditures within each animal class. Finally, it is not surprising that an attempt to express the effect of temperature on metabolic rate by a single universal equation has not met with uncritical acceptance.

Gillooly et al. (2001) used a century-old Arrhenius equation relating temperature, reaction rate, and the equilibrium constant to formulate their expression for the variation of metabolic rate (B) of all organisms:

$$B \approx M^{3/4} e^{-E/kT},$$

where E is the activation energy, k is Boltzmann's constant, and T temperature in K. But while temperature governs metabolism because of its effect on the rate of biochemical reactions, such a simple mechanistic explanation ignores the fact that heterotrophic metabolism is influenced by a large number of interacting physiological processes (Clarke 2004). As a result, evolutionary adaptations have resulted in reaction rates that are relatively independent of the temperature at which an organism habitually metabolizes, and responses to temperature cannot be reliably predicted from first principles (Hochachka and Somero 2002; Clarke and Fraser 2004).

4.2 Ectotherms and Endotherms

The most obvious consequence of high specific BMRs in tiny organisms is the limit on the size of the smallest warm-blooded animals. Creatures lighter than shrews and hummingbirds would have to feed incessantly to compensate for rapid heat losses. In contrast, low specific BMRs make it considerably easier for larger creatures to cope with prolonged environmental stresses because they can draw on accumulated fat reserves for relatively long periods of time. Metabolic imperatives thus dictate two very different grand strategies used by heterotophs (with

some interesting transitory adaptations) to cope with nonoptimal temperatures: ectothermy and endothermy (or poikilothermy and homeothermy).

All higher ectotherms—arthropods, fish, amphibians, and reptiles—have very low specific BMRs and poor body insulation. Their thermoregulation is behavioral; their goal is to raise body temperatures into the preferred range (~10°C for salamanders, 35°C–40°C for heliothermic lizards) by seeking optimal microenvironments. In terrestrial ectotherms, body size is a critical determinant of sustainable body temperatures and hence of their behavior and niches (Stevenson 1985). Large ectotherms with high heat capacity warm up slowly, but their maximum body temperature ranges are narrow (3.5°C–5.5°C in giant tortoises, and 2°C would be expected for an ectothermic 3 t dinosaur). They are able to maintain a larger gradient between body and ambient temperature, and their considerable thermal inertia allows for longer periods of activity.

In contrast, small insects unable to raise their temperature above the ambient level control it by moving around; they can heat up quickly or retreat rapidly to protected microenvironments. Basking is very important and uses both direct absorption and reflected radiation. Butterflies, the most attractive basking ectotherms, either use spread wings to capture the radiation and transfer it to the body or hold their wings in a wide variety of angles above the body (Kingsolver 1985). Thin wings are poor conductors, so species living at higher altitudes have darker wing bases to maximize absorption. Baskers with very narrow angles can convey radiation to the body from a much larger area than more open baskers.

Remarkably, many flying insects can be endothermic during the short periods preceding takeoff, when they warm up the flight muscles by shivering. Winter moths, which at less than 200 mg are more than 1 OM smaller than hummingbirds, are the most accomplished practitioners of this art (Heinrich 1993). They can reach the thoracic temperature of 30°C required for flying even when the surrounding temperature is near 0°C by starting to shiver at temperatures as low as −2°C and persisting often for more than half an hour. They repeat the cycle after short flights that only accelerate their heat loss. Such energy-intensive endothermy is ephemeral, and the moths spend at least 99% of the winter inactive. In contrast, hummingbirds drop their temperature during cold nights to as little as half the waking level, a quasi-ectothermic torpor that reduces otherwise excessive heat loss. But when shivering themselves back to life, they require nearly as much energy as for hovering. Marine ectotherms regulate temperature by selecting waters that will support optimum growth rates and swimming speeds (∼30°C for young carp, ∼10°C for trout).

The portable environments of endotherms are highly uniform (36°C–40°C for most mammals, 38°C–42°C for birds). Endotherms survive in environments that remain for most of the year well below the freezing point and that can plunge repeatedly to below −40°C in absolute terms and to chill-factor equivalents of more than −60°C. Musk oxen and polar bears prove that both herbivores and carnivores can thrive in extremely cold environments, but their thermoregulatory achievement pales in comparison with that of small birds, which maintain a temperature gradient of about 80°C across less than 5 cm between the outside air and the core of their tiny bodies, warmed to just above 40°C. Heat stored in an animal should be equal to the sum of inputs generated by internal metabolism, radiation, conduction, convection, and evaporation. This balancing does not have to be instantaneous, but no endotherms can tolerate large and prolonged excursions of their core body temperatures.

In contrast, microbes and some invertebrates have the option of surviving extreme temperatures by becoming almost completely dehydrated and entering a death-like state of cryptobiosis. Cryptobiotic forms that can survive the greatest temperature extremes belong to the phylum Tardigrada (fig. 4.3). These tiny water bears (50 μm to 1.2 mm), related to arthropods and nematodes, and common in thin water films and on mosses, lichens, and algae, survive exposures to as much as 151°C and as little as −270°C, very close to absolute zero (Greven 1980). The highest temperatures compatible with cellular metabolism are clearly much lower, but studies of extremophilic organisms have raised them to levels that were thought previously impossible. Before 1960 the record belonged to *Bacillus stearothermophilus*, growing at 37°C–65°C; then it passed to hyperthermophilic strains of *Bacillus* and *Sulfolobus* that survive at 85°C (Herbert and Sharp 1992). During the 1980s came the discoveries of marine hydrothermal thermophiles, growing at 95°C–105°C, and *Pyrolobus fumarii*, an archaeon that grows in the walls of deep-sea vent chimneys and tolerates 113°C (Stetter 1998).

Some hyperthermophilic enzymes are effective up to 140°C–150°C, and autotrophic synthesis of all protein-forming amino acids is favored in 100°C submarine hydrothermal solutions compared to syntheses in 18°C warm seawater. We still do not understand why the hyperthermophilic enzymes have optimal catalytic activity above 100°C because they contain the same 20 amino acids as enzymes of other organisms, and there are no gross structural differences between them and the compounds in mesophilic prokaryotes (Zierenberg, Adams, and Arp 2000). Higher organisms are much less

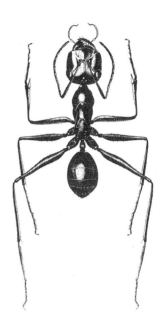

4.3   Three remarkable extremophiles. *Left to right: Echiniscus arctomys* (a tardigrade), *Alvinella pompejana* (a polychaete worm), and a Saharan *Cataglyphis* ant. From Smil (2002).

heat-tolerant. One marine eukaryotic organism comes close to the extreme heat tolerances of microbes. The Pompeii worm (*Alvinella pompejana*) is a 6-cm-long, shaggy creature that colonizes the sides of active deep-sea hydrothermal vents; temperatures within its tubes average 69°C, and spikes exceed 81°C (fig. 4.3) (Cary, Shank, and Stein 1998).

Even the most heat-tolerant Saharan desert ant of genus *Cataglyphis* (fig. 4.3), able to forage during the midday heat, has critical thermal maxima at 53°C–55°C (Gehring and Wehner 1995). Optimal body temperatures required for feeding, reproduction, and growth of terrestrial animals are restricted mostly to 36°C–42°C.

Mammalian enzymes pass their optima as body temperatures approach 50°C, lipids change, and cell membranes become increasingly permeable (Spotila and Gates 1975). That is why endotherms in hot environments have evolved remarkable adaptations. Sweating and panting are the best active responses; dealing with heat is closely related to the management of water balance (Schmidt-Nielsen 1972). Small desert rodents as well as camels reduce water loss by using countercurrent heat exchange in respiratory passages to cool the exhaled air to temperatures much lower than their body cores (Walsberg 2000).

Some desert rodents and birds resort to estivation (torpidation), a lethargic state similar to hibernation.

Reduced metabolism conserves water as well as scarce food. The ability to live on air-dried food without any access to water, the production of concentrated excreta, hiding in burrows, venturing out mostly at night, and tolerating salt water are other common thermoregulative adaptations among desert mammals. Camels have particularly effective adaptations (Gauthier-Pilters and Dagg 1981). They are able to forage on dry, thorny plants, do not drink during cooler months even when water is offered, tolerate long spells without water (up to 10–15 days at 30°C–35°C), can lose as much as 40% of their body weight, and have an almost instantaneous rehydration ability as they drink an equivalent of more than 30% of body weight within 10–20 min.

Cold environments are much more extensive on the Earth than extremely hot niches. Besides vast polar deserts there is also the ocean: except for the topmost 1200 m, all of it remains constantly below 5°C, and the underlying sediments are permanently at about 3°C. Microorganisms that prefer temperatures lower than 15°C are considered psychrophilic, and many of them live at close to 0°C (Russell and Hamamoto 1998). Living microbes are found in supercooled cloud droplets. Psenner and Sattler (1998) reported their occurrence at altitudes above 3000 m in the Alps at around −5°C. They are also found in brine solutions, which remain liquid at well below the freezing point. The most extreme psychroecosystem was discovered in the perennial Antarctic lake ice in the McMurdo Dry Valleys (Priscu et al. 1998). Solar heating of the sediment layer containing sand and microbes blown from the surrounding cold deserts and embedded in the 3–6-m-thick ice produces miniature pockets of water that contain bacteria and cyanobacteria.

In order to survive in the planet's ice-laden polar waters (−1.4°C to −2.15°C), Notothenioids (teleost fish) produce antifreeze proteins (glycopeptides), whose adsorption to minute ice crystals stops their growth and lowers the freezing point (Fletcher, Hew, and Davies 2001). Other marine organisms protecting their cells by secreting ice nucleators into the extracellular fluids are molluscs, both bivalves and gastropods, living in intertidal zones that are temporarily exposed to low temperatures (Loomis 1995). *Littorina littorea*, a snail from the northeastern United States, may survive with more than 70% of its total water content frozen when it is exposed to temperatures as low as −30°C; barnacles survive even with 80% of ice in their extracellular tissues. Glycine-rich antifreeze proteins are also found among many terrestrial arthropods (beetles, mites, spiders), but these organisms also survive by insulating their bodies by ice formed across their cuticles, and by extending the supercooling capacities of water, from which they remove ice nucleators (Duman 2001). Small volumes of pure water can be supercooled by as much as 40°C (the point of spontaneous nucleation), and antifreeze proteins can lower this temperature further. Alaskan willow cone gall fly larvae supercool to −56°C. Among metazoans, turtles and amphibians also use antifreeze proteins together with ice-nucleating proteins to initiate the formation of tiny extracellular ice crystals as they freeze solid while preserving their vital intracellular structures (Storey and Storey 1988).

Endotherms cope with cold climate by highly effective insulation. Arctic wolves and caribous have skin temperatures comparable to those for a well-clothed human even at −32°C. Experiments with shorn, shaven, or hairless animals demonstrate increased heat losses by radiation and the necessity of higher metabolic outputs, whereas raising the furry species in temperatures colder than their normal habitat results in the growth of thicker insulation. The plumage of birds is about one-third more insulative than the pelage of similarly sized mammals. Light-colored species reflect over 50% of visible wavelengths, dark-colored ones may reject just 15%, but actual effects on radiation balance are not that simple. White coats scatter incoming radiation both away from the animal and toward its skin. Individual hairs are actually colorless, appearing white only as their central core scatters incoming radiation; the hair shaft may also conduct scattered radiation to the skin. The white pelts of Arctic mammals absorb UV radiation, a desirable property in cold environments. In contrast, black covers absorb incoming radiation, and the amount of solar energy reaching the skin may be higher in a lighter-colored animal.

This apparent paradox is most pronounced in birds in windy environments, and it explains why dark colors of such desert species as corvids or vultures and light colors of arctic ptarmigans are not maladaptations: with high winds heat load on erected black plumage is below that on white feathers (Wolf and Walsberg 2000). Insulation quality (measured as conductance) of furs is a linear function of their thickness. Naturally, density of hair also matters, as does the degradation of the insulative layer by sweat or rain. At normal temperatures water conducts 24 times more than the air, and the fur of a polar bear swimming in ice water will lose virtually all of its outstanding insulation capacity. Consequently, all entirely marine mammals will have an especially difficult task of thermoregulating in cold waters. The smallest cetaceans (harbor and porpoise dolphins) are at the greatest disadvantage as they try to maintain temperature differences of nearly 40°C across their blubber; their BMRs are two to three times those of similarly sized terrestrial mammals (Kanwisher and Ridgway 1983), a high price for their ability to move and to feed in frigid waters.

Small endotherms are especially at risk during long winters. Because the capacity to store fat is directly proportional to body mass, large mammals can maintain constant temperatures for months without eating. Consequently, no hibernating animal heavier than about 5 kg needs to reduce its winter body temperature by more than a few degrees, but in small mammals it drops sometimes by more than 30°C until spring, when the animal's brown fat tissue is activated to rewarm it and energize it for mating (Lyman 1982; French 1988). And some animals can become repeatedly hypothermic in order sustain amazing physical feats. Antarctic king penguins dive so deeply (>500 m) and for so long (maxima >15 min) because they reduce their abdominal temperature to as low as 11°C during such deep, prolonged diving (Handrich et al. 1997). Similarly, Arctic mammals habitually resort to deep hypothermy in their feet.

Given these challenges, it is logical to ask why the homeotherms are thermoregulating at such high levels (approaching the level of protein decay), why not at lower levels, perhaps just around 20°C? Regulation at such levels would call for lower metabolic rates but for high rates of evaporative cooling; with surface body temperatures below the ambient level in warm climates and during summers there would be no conduction or convection heat loss. These high evaporative heat losses would pose excessive risks of desiccation and restrict the radiation of homeotherms in arid climates. Maximization of evaporative cooling would also require sparse insulation, which would restrict diffusion in cold environments. And lower temperatures would reduce the efficiency of metabolism, making such a low-level homeothermic effort hardly more useful than heterothermy.

An evolutionary compromise is thus in evidence as advantages of relatively high and stable body temperature (near the biochemical optimum) are set against the dangers of heat death, water loss, and the costs of metabolism and insulation (Spotila and Gates 1975). The link between endothermy and the evolutionary success of birds and mammals is obvious. Possession of a constant portable microenvironment required countless feeding and behavioral adaptations and hence very high metabolic rates, but it conferred competitive advantages in benign environments and opened up even the most inhospitable parts of the biosphere for colonization. Our species could not have succeeded without it; ectothermic sapience is unthinkable with life as we know it.

4.3 Locomotion

Much of the human fascination with animals arises from our admiration of flying geese, hovering hummingbirds, jumping salmons, swimming whales, or running horses. Many feats of animals in motion (from squirting scallops to gliding tree snakes) are extraordinary, but none more so than the annual long-distance migrations (Alerstam, Hedenström, and Åkesson 2003). Among the famous accomplishments are the annual 3600-km flight of Monarch butterflies from Canada to Mexico, the 11,500-km swims of loggerhead turtles from California to Japan, the up-to-1000-km countercurrent advances of Pacific salmon in Canadian rivers, and the unrivaled 19,000-km/a peregrination of the Arctic tern between Greenland and Antarctica. The modalities of long-distance animal navigation have yet to be satisfactorily explained (Alerstam 2006).

The energetics of animal locomotion has been surveyed by Pedley (1977), Schmidt-Nielsen (1984), Videler (1993), Alexander (1999a; 2003), and Biewener (2003). Here I present only the essentials, first by introducing cross-modal generalizations and then by focusing on some of the most remarkable performances in running, jumping, flying, and swimming. Comparisons of animal locomotion are done best in terms of the cost of transport (COT), the quotient of metabolic energy used, and the product of body mass and distance traveled (J/kg · m). This treatment eliminates speed as a variable and makes it possible to generalize energy requirements of unrelated species moving at speeds differing by more than 1 order of magnitude (fig. 4.4).

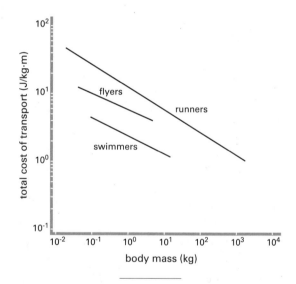

4.4 Energy cost of transport. Running is always the most, and swimming the least, energy-intensive mode of locomotion. Flying falls in between, but only a few flyers are heavier than 10 kg. Based on Schmidt-Nielsen (1972), Tucker (1975), and T. M. Williams (1999).

Comprehensive comparisons (Schmidt-Nielsen 1972; Tucker 1975) showed that swimming is the least energy-demanding form of animal locomotion; flying comes next, and running (by ectotherms or endotherms) is the most expensive. T. M. Williams (1999) offered a more nuanced appraisal, confirming that submerged fish swimming is by far the most energy-efficient way of animal locomotion, followed by bird and bat flight. But the submerged swimming of marine mammals has an energy cost comparable to, or even higher than, mammalian running; phylogenetic history, rather than mode of locomotion, determines the cost of transport in mammals. The most expensive mode of transport is the surface swimming of vertebrates ranging from ducks to humans (fig. 4.5).

All transport modes are limited by metabolic scopes, the ratios between peak metabolism and RMR or BMR. For running mammals the scopes are typically about 10 times RMR, for horses 20, and for coyotes, wolves, and dogs 31–32. For birds the scopes do not appear to go beyond 20 (but they start from higher specific BMRs), and 15 is a short-term scope in some fish. Reptilian and amphibian scopes are just 5–10, posing a severe limitation on the activity of these heterotrophs (Huey, Pianka, and Schoener 1983). Whereas the top aerobic power in small birds and rodents is, respectively, about 150 mW/g and 50 mW/g, it is less than 9 mW/g in toads and 3 mW/g in iguanas. Rapid reptile motion depends on short, anaerobically energized bursts that require subsequent long periods of recovery. In contrast, flying insects put out 0.12–0.58 W/g, or up to 100 times their RMRs (moths can go up to 150 times).

COT scales allometrically with body mass, with exponents for different modes of locomotion clustering around −0.3. Such large submerged marine swimmers as killer whales (2.5–5 t, 2–3 m/s) thus need less than 1 J/kg · m, and the best available rate for a 15-t grey whale is just 0.4 J/kg · m (if the exponent holds, a baleen whale would need <0.1 J/kg · m). In contrast, COT in mammalian surface swimmers (minks, muskrats, people) is at least 10 and up to 40 J/kg · m (fig. 4.5) (Williams 1999). Because the mechanical power of swimming goes up with the cube of velocity (v^3), the total energy demand for swimming is a sum of RMR (R) and v^3 (modified by a constant, k). The energy required per unit distance is $(R + kv^3)/v$, and its minimum value is when $v = (R/2k)^{0.33}$, which means that a high RMR goes with a high optimum speed and hence with high cost of transport (Alexander 1999b).

Surface swimming is so costly because of the elevated body drag (four to five times that of submerged swimming) and the necessity of pushing a bow wave; that is why more streamlined creatures (penguins) are more efficient surface swimmers than sea otters or people. Submerged swimming needs little or no energy to support the neutrally buoyant fish, and the effort is overwhelmingly channeled into overcoming the drag of the relatively dense medium (Videler 1993). There are three basic swimming modes: anguilliform propulsion of eels and blennies, which uses whole-body undulation; carangiiform motion of most species, which confines the undulation to the rear half or third of the body, often aided by lunate tails; and the mainly pectoral swimming of wrasses, parrot fish, and anglerfish.

Fish cruise by using alternate contractions of their aerobic muscles, which run the length of the body in a small red band just under the skin; massive white (anaerobic) muscles are used only for rapid, evasive bursts (fig. 4.5). But such powerful swimmers as tuna and lamnid sharks (makos and white) are distinguished not only by having

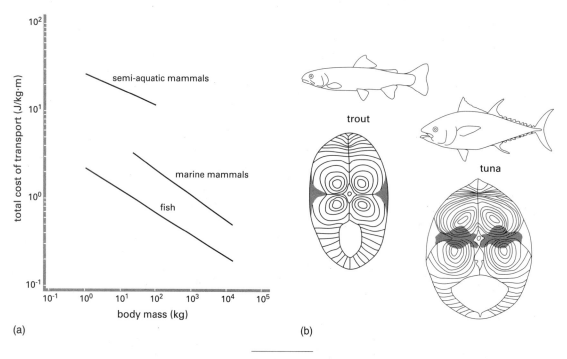

4.5 (*a*) Energy cost of surface and submerged swimming (based on T. M. Williams 1999). (*b*) Cross-sections of fish bodies showing two distinct placements of aerobic muscles: in most fish species they run laterally just under the skin, but in tuna and lamnid sharks they are deep inside the body (based on Shadwick 2005).

red (aerobic) muscles deep within their bodies, where other fish have only white anaerobic fibers, but also by their regional endothermy; those internal muscles can be 10°C–20°C warmer than the surrounding water (fig. 4.5) (Shadwick 2005). The aerobic red muscles are kept warm by constant contractions; if the fish stopped swimming, the muscles would stop working and might never bounce back (Bernal et al. 2005).

Flying animals can be found in each of the five classes of vertebrates. But it is the flight of birds, swift and ma-

neuverable, that has been so envied by humans and that has led to millennia of unfulfilled dreams and failed emulations (Hart 1985). Humans simply lack the radical avian adaptations: lengthening of forelimbs to produce wings delivering sufficient lift and thrust, muscles strong enough to sustain the flapping, a respiratory system superior to mammals' in its extraction of atmospheric oxygen. The minimum energy costs of flying can be estimated from the loss of potential energy in gliding. Birds and bats losing 1–2.5 m/s reduce their potential energy

(Mgh) by 10–25 W/kg of M. Level flight requires at least that much power, but because muscles do not sustain more than 20% efficiency, the metabolic cost of flapping flight should be most commonly 50–120 W/kg. The hovering flight of hummingbirds averages 98 W/kg, and it can reach 133 W/kg before aerodynamic failure (Chai and Dudley 1995).

For animals between 1 g and 1000 g, flying is always cheaper than running. COT in insects is more than 20 J/g · km, in largest birds just a few J/g · km. The aerodynamic power needed for flight is the sum of induced, profile, and parasite drag components. Induced power generates lift and declines with speed (it is highest while hovering). The standard understanding of bird and insect flight had to be revised with the discovery of leading-edge vortices: whereas the flow around the arm-wings remains attached (according to conventional aerodynamic principle), hand-wings can induce airflow separation resulting in leading-edge vortices and generating lift (Videler, Stamhuis, and Povel 2004). Profile power (friction drag on the wings) and parasite power (drag on the body) go up with speed (Norberg 1990).

Two dimensionless parameters inform about the performance of fliers and swimmers. The Strouhal number (St) divides the product of stroke frequency (f) and amplitude (A) by forward speed (v),

$$\text{St} = \frac{fA}{v},$$

and propulsive efficiency is high within a narrow range, $0.2 < \text{St} < 0.4$ (fig. 4.6). Taylor, Nudds, and Thomas (2003) showed that dolphins, sharks, and bony fish as well as cruising birds, bats, and insects converge on this narrow St range in order to achieve high power effi-

ciency. Reynolds (Re) numbers express the relation between viscous and inertial forces,

$$\text{Re} = \frac{\rho l v}{\mu},$$

where ρ is the density of the fluid (air or water), l is a characteristic body length (of wing or fin), and μ is the fluid's viscosity. Low Reynolds numbers are produced by small, slow fliers: for insects (with speeds 1–10 m/s) they are mostly 10^2–10^3, and for birds (with speeds 10–30 m/s) 10^4–10^6 (fig. 4.6).

Larger birds must fly faster to generate sufficient lifts: the expected exponent for wing loadings is $M^{0.33}$, actual mean is $M^{0.28}$. Because the power needed for flight goes up faster than the available power (a multiple of RMR), there must be a maximum size for flying animals. But power scaling shows how much we still do not know about bird flight. Metabolic power, a product of weight (M^1) and speed ($M^{1/6}$), should scale as $M^{1.17}$, but measurements of birds flying in wind tunnels show that it scales much less steeply, as $M^{0.78}$ in starlings (Ward et al. 2001) and only as $M^{0.35}$ in red knots (Kvist et al. 2001). The best explanation is that its conversion into mechanical output rises with fuel load. Even small shorebirds (red knots, weight 110–190 g) can thus make nonstop flights of 4000 km despite the high metabolic power input. Measurements also show that flight muscle efficiency is not, as has usually been assumed, 23% but rather 17%–19% in starlings and 8%–14% in red knots.

There is also a structural limit: flight muscles make up about 17% M (only in hummingbirds are they ~30% M), and hence birds heavier than about 13–16 kg (African kori, great bustards) cannot generate enough lift to support their weight, particularly during low takeoff speeds.

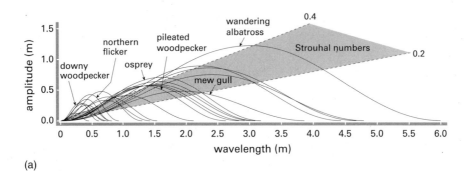

(a)

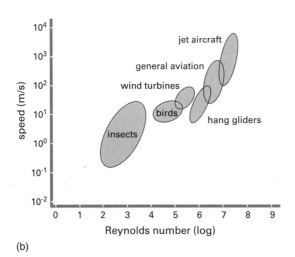

(b)

4.6 (*a*) Inverted waveforms of wing amplitude for 22 bird
species (plotted from data in Taylor, Nudds, and Thomas
2003). As expected, most of the Strouhal numbers fall between
0.2 and 0.4 (plot based on Corum 2003). (*b*) Speeds and
Reynolds numbers of various flyers (based, with permission, on
Filippone 2003).

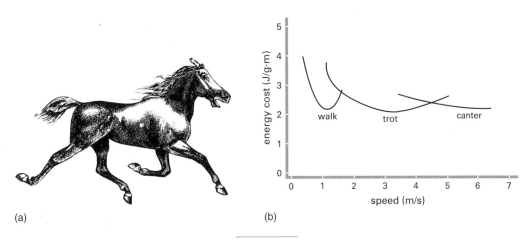

(a)

(b)

4.7 Engraving of a galloping horse based on Muybridge's (1887) classic photographs of animal locomotion. (b) Energy costs of different horse gaits (Carrier 1984).

But bustards rarely fly, and other large flyers, be it condors (8–10 kg) or albatrosses (7–9 kg), spend hours as gliders aided by wind currents rather than as active flappers. Satellite monitoring of magnificent frigate birds, the only species besides swifts to spend nights on the wing, shows them soaring in thermals up to 2.5 km and descending to the sea surface only once every 8 h or so (Weimerskirch et al. 2003). Some large birds save energy by using other birds' airstreams in V formations: measured values in great white pelicans show they save 11%–14% by extending their gliding time in a vortex wake (Weimerskirch, Martin et al. 2001).

Typical flight speeds are proportional to a wing loading to power ratio of 0.55, and hence to $M^{0.18}$. Speeds minimizing the cost of transport (in m/s) are equal to $14.6M^{0.2}$, resulting in usual velocities of about 9 m/s for starlings, 14 m/s for ducks, and close to 20 m/s for large geese. Whereas mammalian breathing is a succession of tidal flows, a complex arrangement of parabronchi and air sacs allows birds to use unidirectional flow through a large number of parallel tubes. Avian parabronchi remove almost 30% more oxygen than mammalian alveoli. Vertebrate runners become comatose at high altitudes, but even small songbirds fly for 10^3 km at altitudes of 3–6 km, and bar-headed geese were seen above Mount Everest. The energetics of long-distance migrations is explained by preflight accumulation of fat and by reliance on tailwinds. Lipid/lean dry weight ratio is 0.34–0.44 in sparrows, up to 3.42 in blackpoll warblers, and as high as 3.5 for ruby-throated hummingbirds ready to migrate (Blem 1980). In-flight use of 25% M is common; 50% loss appears to be the maximum.

Terrestrial gaits begin with walk and then speed up to running modes of trotting and galloping (fig. 4.7). Gaits are distinguished by their footfall patterns, duty factors (fractions of a cycle for which a given foot is contacting the ground), and Froude numbers (dimensionless $Fr = v^2/gh$, where h is a characteristic length, such as hip height). Run is a gait with duty factor <0.5, and changes from walk to trot take place when $Fr = 0.3$–0.5 and from trot to gallop when $Fr = 2$–3. There is no substantial difference between the COT of bipedal and quadrupedal running of animals who can do both (monkeys), nor between the bipedal running of birds and the quadrupedal motion of mammals. The number of steps a running animal must take per unit of distance will be inversely proportional to its length (roughly to $M^{0.33}$), and the work accomplished for each step must be proportional to its mass.

Net metabolic COT thus scales as $M^{-0.33}$; when running faster, the smaller creatures must take many more steps, each one needing power in direct proportion to its M (fig. 4.7). At the same time, all running animals have to deal with the same biomechanical constraints, and hence their mechanical COT (distinct from metabolic COT) is independent of mass. A 10-g lizard will have a metabolic COT 1 OM higher than a 100-kg mammal, but their mechanical COTs will be very similar, at about $1 \text{ J/kg} \cdot \text{m}$ (Alexander 2005). Maximum oxygen consumption increases with $M^{0.85}$, whereas the exponent for the total running costs is only 0.67. Because the available power goes up faster than the cost of running, large mammals (bears, hippos, buffaloes) are able to run much faster (up to ten times) than the smallest ones. Large mammals also save much energy in running because of the elastic structure of their legs. Kinetic or potential energy lost at one stage of the stride is temporarily stored as elastic strain in both muscles and tendons, to be used later as elastic recoil (Alexander 1984).

At high speeds some mammals, including humans, may save more than half the metabolic energy they would otherwise need in running. Similarly, penguin waddling conserves mechanical energy, and its high COT is due to the animal's short legs, which require rapid generation of muscular force (Griffin and Kram 2000). Cheetahs are the fastest runners, reliably measured at 105 km/h, or 29.1 m/s (Sharp 1997), but like other sprinters they cannot sustain speed for more than a few 100 m. Eaton's (1974) records show 60–90 m as the usual rushing distance, and during successful hunts the distance between cheetahs and the prey when it started to run was merely 45 m. The high energy cost of these rushes (up to 156 breaths/min after a dash and kill compared to 16 breaths/min at rest) requires subsequent recovery and limits the number of such intensive chases. In contrast, fast-moving elephants are not really running (Hutchinson et al. 2003). Although their duty factors are as low as 0.37 and their Fr values could surpass 3.0 (both far above quadrupedal walking gait), they always keep at least one foot in contact with the ground.

Jumping has a counterintuitive limit: energy output is directly proportional to the muscle mass, which is generally proportional to the total body mass; identical energy release per unit mass should then raise the animals to equal heights. This expectation is confirmed by measurements for animals whose body sizes differ more than 10^8-fold. A flea (0.49 mg) jumps 20 cm, a locust (3 g) 60 cm, and a human (70 kg) 60 cm (this refers to the lifting of the body center during a standing jump that does not utilize the kinetic energy of high jumping). Because air resistance is very important for tiny insects, the threefold difference between a flea and a person would virtually disappear in a vacuum. African bush babies (*Galago*) jump up to 2.25 m, but their jumping muscles are about twice as massive as in humans. Distance, not

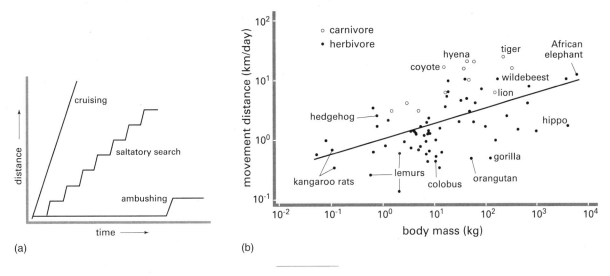

4.8 (a) Basic food-searching strategies (O'Brien, Bowman, and Evans 1990). (b) Scaling of the daily movement distance of mammals (Garland 1983).

height, is the objective of kangaroo jumping. Large Achilles and tail tendons store elastic energy for highly efficient locomotion. Whereas the energy cost of penta-pedal gait (four limbs plus tail) rises rapidly with speed, once kangaroos start hopping (~6 km/h), the cost levels off, even declines, and at about 17 km/h it becomes lower than the expenditure of equally massive running quadrupeds (T. J. Dawson 1977). Most of the energy invested in animal locomotion is in the search for food. The two extreme searching strategies are cruise (or widely ranging) and ambush (sit and wait) (fig. 4.8). Many animals fit neatly into these categories, hawks and tuna in the first one, rattlesnakes and herons in the second, but most foragers exhibit saltatory patterns of movement, alternating pauses and moves and fall-ing somewhere between true cruisers and ambushers

(O'Brien, Bowman, and Evans 1990). As they graze or hunt, many animals seem to be trying to maximize their total energy gain per unit of foraging time. In times of plenty, grazers select feed with higher energy density, and predators reduce pursuit of elusive prey.

Widely influential optimal foraging theory initially per-ceived such behavior as one of the great natural laws; its proponents believed that optimal foraging maxi-mizes animal fitness and plays a critical role in natural se-lection (Stephens and Krebs 1987). Critics have pointed out that foraging behavior is shaped by other factors than just searching for and consuming food; the necessity for constant antipredator vigilance is perhaps the most criti-cal concern for most grazers. The life of heterotrophs cannot be simply partitioned into independent activities. Optimal foraging is not one of the universal energetic imperatives of heterotrophic life, and it may be more rewarding to look at several important links between foraging and energy expenditure.

Garland (1983) has reviewed these critical relations—daily movement distance (DMD), incremental cost of locomotion (ICL), daily energy expenditure (DEE), and ecological cost of transport (ECT)—for mammalian species. DMD (in km, for animals ranging from 56 g, *Dipodomys*, to 6 t, elephant) scaled as $1.038M^{0.25}$ (M in kg), but it varied almost 2 OM for a given body mass (fig. 4.8). Carnivores move much more than herbivores (respective M multiples are 3.87 and 0.87) but the exponents are identical (0.22). ICL (in J/km, a constant independent of speed) scaled as $10,078M^{0.7}$, and DEE was a relatively uncertain variable that averaged about $800M^{0.71}$. ECT, expressed as a percentage of DEE (ECT = 100 [DM × ICL/DEE]) scaled as $5.17M^{0.21}$ for carnivores and as $1.17M^{0.21}$ for other mammals. ECT of small noncarnivorous animals may thus be less than 1% of their DEE, whereas large carnivores may spend 10%–15% or even a larger share of their DEE on locomotion.

4.4 Biomasses and Productivities

Global assessments of heterotrophic biomass can aim at nothing more than the right orders of magnitude. Large and unknown number of species, their highly variable densities, and their occupation of many extreme niches are the main factors militating against any reliable assessments of prokaryotic biomass. Enormous uncertainty regarding the overall diversity of arthropods, which account for most of the described heterotrophic species, and the variety, mobility, and variability of vertebrates pose additional difficulties for quantification of other heterotrophic biomass. For example, collection of biomass living within epiphytic tree ferns has doubled the estimate of invertebrate zoomass in a rain forest canopy (Ellwood and Foster 2004). Extrapolation and global aggregation of uncertain information carries large errors that are further increased when these totals are converted to energy equivalents. Carbon dominates the elemental composition of heterotrophs, but higher protein content accounts for much higher shares of N and S than in plants. Invertebrates are about 80% water, fish average about 75%, arthropods, birds, and mammals about 70%. Dry matter of metabolizing tissues is nearly pure protein, averaging about 22.5 kJ/g.

Differences in the energy density of heterotrophs are thus a function of mean body ash contents (about 17% of dry fat-free weight in mammals, 12% in birds) and accumulation of lipids. Both these shares change with age, and fat content in adult animals fluctuates frequently with the season and feeding opportunities or necessities. For most vertebrate species there is at least a twofold difference of energy densities depending on their growth stage; seasonal variations on the order of 20%–30% are common; and energy contents of different developmental stages or castes of insects differ by up to 40%. Approximate energy densities (all in kJ/g) are 18.5 for bacteria, 22 for fungi, 19 for protists, molluscs, and annelids, 23 for arthropods, 21 for fish, and 23 for mammals and birds. The only certainty is that endothermy limits the share of heterotrophic biomass; consumers must be mostly prokaryotes and invertebrates.

Available estimates do not distinguish between autotrophic and heterotrophic species, but the bulk of bacterial mass is clearly heterotrophic (decomposers, N fixers). Given the wide range of bacterial presence in soils (10^1–10^3 g/m^2) (Paul and Clark 1989; Coleman and Crossley 1996), global estimates differ substantially. Whitman,

Coleman, and Wiebe (1998), assuming an average of about 450 g/m^2, ended up with 57 Gt of soil bacteria. A more conservative assumption of 250 g/m^2 yields about 33 Gt. But these uncertainties are minor compared to those we face in estimating the mass of subterranean (and subsea) prokaryotes. Whitman, Coleman, and Wiebe (1998) offered a range of 48–477 Gt, but even this span may be much too narrow. Fungal biomass in soils also ranges widely, from 10^0 g/m^2 to 10^2 g/m^2 (Bowen 1966; Reagan and Waide 1996). A fairly conservative global aggregate would be 7 Gt, but a mass twice as large (about 100 g/m^2 of ice-free surface) may not be excessive.

Earthworms are the most conspicuous soil invertebrates, but their mass is usually just around 5 g/m^2, although in cultivated soils it may be well over 10 g/m^2 (Hartenstein 1986). Nematodes average about one-tenth of annelid biomass, and ants and termites each add typically no more than 0.1 g/m^2 (Brian 1978). Invertebrates (dominated by annelids, nematodes, and microarthropods) range between 7 g/m^2 and 10 g/m^2, and their global biomass may be close to 1 Gt, roughly 20 EJ. Reptiles and amphibians often dominate vertebrate zoomass in tropical forests, and their biomass density may rival that of invertebrates (Reagan and Waide 1996). Densities of rain forest mammals are generally low, 0.1–2 g/m^2, whereas in the ungulate-rich savannas they can surpass 3–4 g/m^2 (Plumptree and Harris 1995). Small mammals, mostly rodents, add generally less than 0.2 g/m^2, and zoomasses of insectivorous mammals are mostly below 0.05 g/m^2 (Golley and Medina 1975). And the total for all carnivores in the Ngorongoro Crater, one of the world's best places for large predators to hunt ungulates, is less than 0.03 g/m^2 (Schaller 1972).

Similarly, total avifaunas usually do not surpass 0.05 g/m^2 (Reagan and Waide 1996). In aggregate, vertebrate zoomass rarely sums to more than 1 g/m^2 over large areas, and uncertainties in its quantification thus hardly matter in the overall biospheric count. Errors inherent in estimating prokaryotic biomass are easily 1 OM larger. Published estimates of terrestrial invertebrate and vertebrate biomass (excluding domestic animals) range between 1 Gt and 2.1 Gt, with wild mammals contributing less than 10 Mt (Bowen 1966; Whittaker and Likens 1975; Smil 2001). Great depth of the inhabited medium, and extraordinary patchiness and mobility of many oceanic heterotrophs, makes the quantification of marine zoomass exceedingly difficult. Published estimates range from 0.7 Gt to 1.1 Gt, with invertebrates dominant and mammals contributing less than 5% of the total. Mann (1984) estimated total fish biomass at 300 Mt of fresh weight. The global grand total of close to 10 Gt, or roughly 200 EJ, of heterotrophic biomass equals less than 0.2% of all phytomass stores.

Although it is possible to construct similar sets of planetary totals for heterotrophic productivity (production/year), a high degree of uncertainty involved in these exercises, particularly because of the rapid turnovers of decomposers and invertebrates, may result in totals of a wrong order of magnitude. Sexual reproduction dominates the production in metazoa as genetic recombination provides a mechanism for purging deleterious mutations and improving chances of environmental adaptation. But sexual reproduction is not inherently superior; many successful organisms are asexual (fungi) or, as some parasitic protozoa, clonal (Ayala 1998). Sexual heterotrophs have a continuum of reproductive strategies ranging from a single prodigious reproductive bout (semelparity) to successive breeding (iteroparity), which produces fewer but larger neonates of greater competitive ability.

Borrowing the terminology of growth equations, MacArthur and Wilson (1967) labeled the first strategy r selection (r being the intrinsic rate of increase) and the other one K selection (K being the upper asymptote of population size). Clearly, r selectionists are great opportunists, pouring a much larger part of their metabolism into reproduction, and this makes them into obnoxious pests and efficient colonizers. But this prodigious channeling of energy into offspring severely limits the survival of parents and the chances for repeated reproduction. Endoparasites are notable exceptions to this rule; tapeworms reproduce copiously and survive for up to 15 years. In contrast, adaptation to limited resources makes the larger K selectionists long-term occupants in more stable settings. Most heterotrophs do not operate at these extremes; their reproductive strategies tend toward the r or the K end of the continuum.

Biochemical commonalities of the production of gametes and the growth of embryos and neonates put the maximum theoretical net efficiency of organizing food-derived monomers into zoomass polymers at about 96%, an impressive figure by any standard. This is only a theoretical maximum. Actual rates (considering digestive, molecular turnaround, molecular transport, and mechanical inefficiencies) would be just over 70%. Actual efficiencies can be measured in proliferating unicellular organisms; net values are 50%–65% for bacteria; 40%–50% for protozoan and yeast. Estimates for invertebrates show gross growth conversion efficiencies of 30%–65% for mollusca, 35%–55% for crustaceans, and 25%–60% for insects (Calow 1977).

Given the prominence of eggs in avian life and the frequent complexities of reproductive behavior among birds, ornithologists were among the earliest students of the energetics of reproduction and growth. Ricklefs (1974) put the energy requirements of testicular growth at no more than 0.4%–2% of BMR during the period of rapid gonadal gain. Female gonadal growth claims between 1.5%–6% of BMR, a small expenditure compared to the cost of eggs, a function of their energy density and their rate of formation. Egg energy density is related to the development of hatchlings. Precocial species, above all, waterfowl, have heavier yolks (30%–40% of egg weight) and shells, and their eggs have up to 7.6 kJ/g; in contrast, small altricial birds (born blind and featherless) have eggs containing as little as 4.1 kJ/g.

But the extreme case of high energy investment in an egg is New Zealand's kiwi, whose large eggs (at 400 g–435 g, roughly one-fifth of the female's weight) contain 10 kJ/g (yolk is 61% of total weight) and provide nutrition for one of the longest incubation periods (70–74 days), resulting in a fully feathered chick with adult-type plumage (Calder 1978). The energy cost of eggs among precocial species is commonly in excess of 110%, even 130%, of their BMR, whereas for passerines and raptors the markup may be as low as 30%–35%. Incubation of a chicken egg containing 360 kJ produces a chick (160 kJ) and leaves behind nearly 110 kJ of unused biomass, yielding a gross conversion rate of 60%. Incubating a clutch of eggs equal to body weight is less demanding for large birds, usually requiring markups of between 20% and 80% of BMR, whereas in the smallest species the energy investment is equivalent to 100%–300% of BMR.

The energy cost of mammalian gonadal growth is negligible compared to the investment in the gravid uterus, embryo, and enlarged mammary glands. Birth weights and gestation periods increase as power functions of maternal M; the larger species produce neonates of relatively smaller weights over much longer periods of time. But even for large ungulates the total cost of gravidity is proportional to maternal M in the same ratio as the mean for other mammals ($M^{0.7}$). The maintenance needs of the gravid uterus are the dominant gestation costs in all but the smallest rodents and insectivores, whose very short gestations and large litters make the growth more demanding. Among larger mammals at least 60%, among ungulates some 80%, of energy metabolized by the gravid uterus is for maintenance (Robbins 1983).

Mammalian milks have a wide range of dry matter levels, from 8.5% in asses to 64.4% in northern elephant seals (Oftedal 1984). Ungulates and primates have the most dilute milks, pinnipeds the most concentrated (to counteract conductive heat loss of newborn in cold waters and to deposit rapidly the insulating layer of subcutaneous fat). Most of the solids in dense milks are fats, whereas in thinner milks sugars dominate. Energy densities are up to 21 kJ/g for seal milk, as little as 1.45 kJ/g in ungulates. Cow's milk rates 2.95 kJ/g, human milk 2.87 kJ/g, a close match with obvious evolutionary consequences for domestication of cattle. Larger animals must have faster growth rates in order to attain their mature body mass. But as organisms grow, the number of cells that have to be supplied with energy scales faster than the capacity of the networks needed to supply them, and this results in generally valid S-shaped growth curves (West and Brown 2005). When growth data of all multicellular organisms are rescaled in dimensionless terms, they follow very closely a universal asymptotic growth curve (fig. 4.9).

Most of the gain comes obviously during a fairly linear stage, when the maximum daily gains can be well approximated as power functions of adult M. The general relation for placental mammals scales as

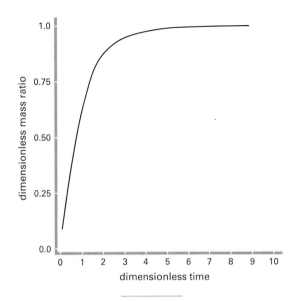

4.9 Universal asymptotic growth curve. From West and Brown (2005).

$0.0326M^{0.75}$ (in g/day). Notable departures from the trend are the fast-growing pinniped carnivores and the slow-growing primates. The energy content of these gains ranges widely as a function of different concentrations of body constituents. Mammalian neonates average about 12% of protein and 2% of fat, but gray seals have 9%, guinea pigs 10%, and humans 16% of fat and hence an extraordinarily high energy density of 8.75 kJ/g at birth, compared to typical mammalian values of 2.9–3.6 kJ/g. The conversion efficiency of these gains is remarkably similar at the earliest growth stage, averaging about 35%.

4.5 Heterotrophs in Ecosystems
Mass-related energy needs of heterotrophs must be reflected in the densities with which they are encountered

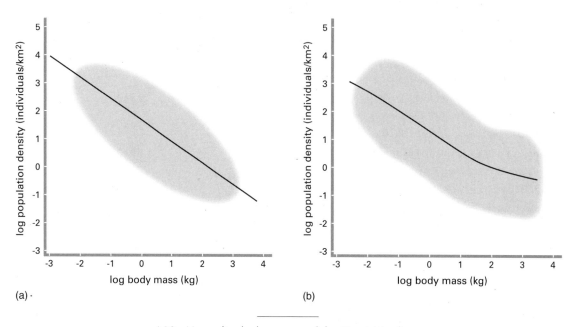

(a)·

(b)

4.10 Mammalian body masses and densities. (a) Log-linear
relation posited by Damuth (1993). (b) Nonlinear relation
found by Silva and Downing (1995).

in undisturbed ecosystems. Densities of the largest herbivores (number/area) are very low, but in some ecosystems their combined energy density rivals that of soil invertebrates. Up to the early 1970s the Ruwenzori National Park in Uganda supported as much as 145 kJ/m^2 of large herbivores, Serengeti about 60 kJ/m^2. These densities are not matched anywhere else on this planet. Such social insects as ants and termites usually do not add up to more than 20 kJ/m^2. Even in tropical forests total avifaunas barely surpass 1 kJ/m^2. Insectivorous mammals are frequently abundant, but their total zoomass is nearly always below 1 kJ/m^2 because their bodies are small. Carnivorous zoomass in the Ngorongoro Crater was less than 0.7 kJ/m^2, roughly 1% of the her-

bivorous prey (Schaller 1972). And a shift from strict herbivory to omnivory is well illustrated with primates: the densities of folivorous *Lemur fulvus* go up to 16 kJ/ m^2, those of herbivorous gorillas are about 1.5 kJ/m^2, and those of omnivorous chimpanzees are about 0.5 kJ/ m^2 (Bernstein and Smith 1979).

Damuth's (1981) plot of mammalian densities and body masses (fig. 4.10) showed an overall negative slope of −0.75 or −2.25 when related to characteristic body length (M is l^3, and hence $M^{-0.75}$ is equal to $l^{-2.25}$). More interestingly, if the BMR were perfectly proportional to $M^{0.75}$ (it is not; see section 4.1), this rule would imply that energy harvested daily per unit area is a constant independent of the unit mass. Larger species do not control larger amounts of trophic energy than smaller ones, and no herbivorous mammal could thus outstrip another energetically only because of its larger size. But Damuth's (1993) regressions of 39 mammalian dietary groups revealed slopes ranging from $+0.42$ (for temperate South American frugivores-herbivores) to -1.30 (for North American and Asian carnivores). A reexamination based on ecological density data for nearly 1,000 terrestrial mammal populations showed that density scales as $M^{-0.75}$ only for animals with body masses between 0.1 and 100 kg (Silva and Downing 1995). The overall relation is distinctly nonlinear, which means that previous log-linear models overestimated the densities of small species and underestimated those of large mammals (fig. 4.10).

More important, densities were well described by $M^{-0.75}$ for only about half of all studied populations (for the smallest and largest mammals, allometric exponents were significantly different from zero). The energy use of mammalian populations is not independent of M; it appears to increase linearly with M in small mammals, varies little for those up to 100 kg, and rises rapidly for the largest animals. The absence of an energy equivalence rule was confirmed by Schmid, Tokeshi, and Schmid-Araya (2000) in their analysis of population density and body size of invertebrates in two stream communities. But Carbone and Gittleman (2002) found that 10 t of prey biomass supports about 90 kg of carnivores regardless of body mass and that the ratio of carnivore number to prey biomass scales almost perfectly to the reciprocal of carnivore mass, $M^{-1.048}$.

The higher total energy needs of larger heterotrophs dictate that space used by animals must increase with M. An early analysis of home ranges for small mammals yielded the allometric ratio of $M^{0.63}$, and because this was very close to assumed universal BMR scaling of $M^{0.75}$, the ranges were seen to be simply a function of metabolic needs (McNab 1963). Reality is more complex. Examination of species belonging to 124 mammalian and avian genera showed the ranges of carnivorous passeriformes scaling as $M^{1.75}$, herbivorous galliformes as $M^{1.39}$, omnivorous rodents as $M^{0.97}$, and herbivorous rodents as $M^{0.81}$ (Mace, Harvey, and Clutton-Brock 1983). Range overlap is a major reason why space used by animals increases at steeper rate than would otherwise be expected. Small animals can maintain small, defensible, and hence overwhelmingly exclusive home ranges, but spatial constraints on effective defense mean that exclusivity of home range use declines with rising M and that large mammals may lose to neighbors over 90% of resources available within a home range (Jetz et al. 2004).

Size is much less predictably related to trophic levels. Elton (1927) was the first ecologist to recognize the declining numbers of heterotrophs in higher food web levels while pointing out their often increasing size. A decade later Hutchinson redefined the principle of the Eltonian pyramid of numbers in terms of productivity and opened the way for Lindeman's (1942) pioneering research on energy transfers in Wisconsin's lake Mendota in terms of progressive efficiencies (assimilation at level n/assimilation at level $n - 1$). Lindeman's results showed primary producers assimilating 0.4% of the incoming energy, while the primary consumers incorporated 8.7%, secondary consumers 5.5%, and tertiary consumers 13% of all energy that reached them from the previous trophic level. These results led to the formulation of the often-invoked 10% law for typical interlevel energy transfers, and Lindeman thought that the progressively higher efficiencies at higher trophic levels might represent a fundamental ecological principle. In reality, ecological efficiencies depart significantly from the 10% rate.

Ten years after the publication of Lindeman's work, Odum (1957) began his study of an aquatic ecosystem at Silver Springs in central Florida, which resulted in trophic efficiencies of about 16% for herbivores and, respectively, 4.5% and 9% for first- and second-level carnivores. Teal (1962), in his study of a Georgia salt marsh, obtained trophic efficiencies of nearly 40% for bacteria, 27% for insects, and only about 2% for crabs and nematodes. By that time Hairston, Smith, and Slobodkin (1960) elevated food chain dynamics to one of the key concerns of ecology. In contrast to the previous paradigm stating that heterotrophic numbers are primarily limited by available energy or habitats, they generalized this fundamental energetic concern by concluding that herbivores are limited most often by predators rather than by resources and hence are not likely to compete for common resources. In this three-link food chain only the predators are energy-limited, and their pressure on grazers does not allow these primary consumers to regulate energy intake.

Earlier studies of food webs found an average of 3 food chain lengths and maxima between 5 and 7, but later studies showed higher values, some with average length of 5 and maxima of 11–12 (S. J. Hall and D. Raffaelli 1991; Polis and Winemiller 1996). The earliest explanation ascribed the brevity of most food chains to energetic constraints, but the impacts of environmental instability limits the length long before such constraints become operative (fig. 4.11). There is no simple progression of animal sizes along these chains: for example, in a Caribbean tropical rain forest the fourth level of consumers includes not only snakes but also arboreal arachnids (Reagan and Waide 1996). In even-linked grasslands the dominant ungulates are larger than their predators; in odd-linked tropical rain forest small, folivorous grazers are controlled by larger predators; in many aquatic ecosystems abundant herbivores are much larger than autotrophs, and secondary consumers and predators are larger still. But in grasslands the presence of large mammals strongly influences the numbers and diversity of small mammals (Keesing 2000). With ungulates present, the densities of small mammals are lower and their diversity generally higher; the reverse is true if ungulates are absent.

Large differences in average lifespans and masses of producers and consumers result in broad-based terrestrial trophic pyramids. Phytomass density is commonly 20 times more abundant than the mass of primary consumers, and zoomass in the highest trophic level may be equal to a mere 0.001% of the phytomass. In lakes food chain length increases with ecosystem size, but it is not related to productivity (Post, Pace, and Hairston 2000). Short lifespans of marine phytoplankton and high energy throughputs among consumers usually reverse the layering in the ocean and produce inverted trophic pyramids: the standing heterotrophic biomass is at least twice, often even three or four times, as large as the oceanic phytomass.

All we can do in the absence of any grand patterns is to describe the substantial variability in energy transfer rates. As already detailed (see chapter 3), the share of PAR used in converting carbon in CO_2 to new phytomass is at best about 4%. After that rate is reduced by carbon losses due to respiration, the efficiency of the NPP, that is, the phytomass actually available for heterotrophic consumption, averages only about 0.33% for terrestrial ecosystems and (because of the low absorption of PAR by

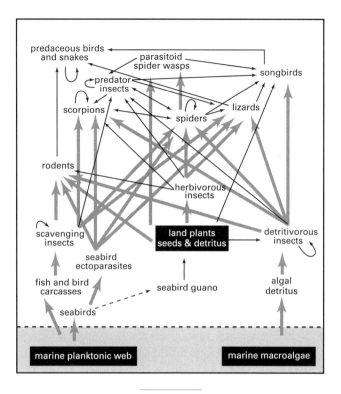

4.11   Trophic web for islands in the Gulf of California. From
McCann, Hastings, and Huxel (1998).

phytoplankton) less than 0.2% for the entire biosphere. Energy losses between subsequent trophic levels are never high; on average they are definitely lower than 10%.

The calculation begins by determining exploitation efficiencies, shares of the production at one trophic level that are actually ingested by organisms at the level above (the rest being accumulated and degraded by decomposers). Shares of the NPP actually consumed by herbivores are just 1%–2% in many temperate forests and fields, and 5%–10% in most forest ecosystems. They peak at 25%–40% in temperate meadows and wetlands

and, exceptionally, at 50%–60% in rich tropical grasslands; in the ocean they can even surpass 95% for some patches of phytoplankton (Crawley 1983; Chapman and Reiss 1999). When the calculation includes only the above-ground heterotrophs, the transfer is rarely above 10% in any temperate ecosystem, and when it is restricted to vertebrates, it is mostly about 1%.

The next calculation step considers ingested energy that is actually absorbed in the gut. Herbivores, subsisting on often digestion-resistant plant polymers, absorb feed with efficiencies that are often less than 30%, and carnivores absorb nutrients from their high-lipid, high-protein diets with efficiencies that may surpass 90%. But the most important determinant of zoomass abundance is production (growth) efficiency, the share of assimilated energy that is neither respired nor spent on reproduction. Here the thermoregulatory divide is obvious. The theoretical maximum is about 95%, actual peaks are 50%–65% for bacteria, and 40%–50% for fungi and protists. Ectotherms have high production efficiencies: invertebrates often channel more than 20% and nonsocial insects more than 40% of all assimilated energy into growth. Social insects, with much higher respiration, have production efficiencies around 10%, large mammals only about 3%, small mammals and birds 1%–2% (Humphreys 1979). Multiplying exploitation, assimilation, and production rates yields the result that trophic (ecological or Lindeman's) efficiency (energy available at one level that is actually converted to new biomass at the level above it) may be only a small fraction of 1% or well above 10%.

There are also no representative large-scale means of net energy transfer because there is no evidence of predictable taxonomic, ecosystemic, or spatial variation either for grazers or predators. The only permissible generalizations are that for the primary consumers' rates (excluding decomposers) fit overwhelmingly within 1%–10% and that the efficiencies of carnivores (invertebrate or vertebrate) are significantly higher than those of herbivores inhabiting the same environments. But a high trophic efficiency does not translate into abundance. Because of the maintenance of complex arrangements it is significantly lower among social insects, but this high energy price has been no obstacle to the abundance of ants and termites in all temperate and tropical ecosystems. And trophic efficiency is reduced even more by endothermy, but this price is offset by the competitive advantages of adaptability and mobility. Endotherms cannot dominate global heterotrophic biomass whose bulk is formed by prokaryotes, fungi, and invertebrates. They could thrive without vertebrates, but higher organisms could not survive without their irreplaceable decomposition and nutrient-cycling services.

Energetic imperatives also govern the ratios between energy fluxes and accumulated biomass. Large, long-lived animals have high capacities to accumulate zoomass, but they do so very slowly, and hence the energy flux through their populations is relatively low. For elephants the annual energy consumption factors (consumption/zoomass ratios) is less than 10, and for the smallest rodents as high as 800. Small mammals and birds have relatively large energy throughputs and high production/zoomass ratios (annual turnover rates), commonly 2–3 for wild rodents, no more than 0.2 for large Serengeti ungulates. For herbivorous mammals Calder (1983) found turnover time scaling with $M^{-0.33}$.

Arthropods are the most prolific producers, with annual turnover rates up to 10 for termites and some spiders. Their aggregate energy consumption densities may rival those of the largest herbivores; many ant species consume annually 200–350 kJ/m^2, or as much as elephants. Turnovers are generally higher for ectotherms (although only 0.3–0.9 for ants) with no obvious dependence on M, whereas endotherms show the expected decline with greater M, the values ranging from about 5 for shrews to 0.05 for elephants. McCullough (1973) pointed out that it is hardly accidental that the animals raised for meat are those combining the capacity to produce with the capacity to accumulate zoomass. The adult M of such compromise mammals is mostly 40–400 kg, and 1–10 kg for birds, for whom a large clutch size is also an asset.

5

HUMAN ENERGETICS

People as Simple Heterotrophs

Nutritional individuality is a characteristic of mankind, and this is as true of energy intakes and needs as of other attributes. Studies over the years have shown that individuals vary by a factor of two or more in their intakes of energy from the first year after birth to 75 years and over. The metabolic differences that must lie behind this are still not fully understood.

Elsie Widdowson, "How Much Food Does Man Require?" (1983)

Studies of human energy needs, conversions, and expenditures were in the mainstream of the new science of energy during the pioneering decades of both theoretical and applied research between 1840 and 1880. The contributions of that time were critical in formulating the canons of general energetics (see section 1.1). The subsequent invention and diffusion of small portable combustion engines and electric motors (see chapters 8 and 9) marked the beginning of the demise of humans as prime movers, a shift that has been largely responsible for removing human energetics from the core of energy studies and its relegation to specialized niches of physiological and nutritional interest. This reality has not changed the fact that no energy conversion is more immediately essential for human existence than the continuing oxidation of foodstuffs or lipid reserves (and if need be, protein reserves).

Specialized studies of human energetics have advanced in many fascinating ways, ranging from the minutiae of cellular biochemistry to appraisals of the biophysical limits of the human body as a machine. But many fundamentals concerning human nutrition remain uncertain and unknown. After more than a century of study we still cannot explain large observed differences in energy needs among individuals of the same body mass and build. Before looking at some of these uncertainties, particularly in relation to food energy requirements for basal metabolism, growth, activity, pregnancy, and lactation, I review the basic energy sources of human nutrition. Then I survey the basics of human thermoregulation, the energy needs of common work and leisure activities, and the physical limits of human performance. This chapter closes with a concise review of humans as simple heterotrophs, foraging societies whose only source of mechanical energy was human muscles.

5.1 Energy Sources and Basal Metabolism

Like all other heterotrophs, humans must consume three classes of nutrients available in plant and animal foods. These essential nutrients include preformed organic compounds of the three kinds of macronutrients as well as two distinct kinds of micronutrients, vitamins and mineral elements. Digested nutrients are absorbed in the small intestine and distributed by the blood in order to reproduce, grow, be active, and adapt. Energy-yielding macronutrients—carbohydrates, proteins, and lipids— are consumed at rates of 10^1 (proteins and lipids) to 10^2 (carbohydrates) g/day. The intakes of micronutrients range from 2.5 g/day for K and Na to a mere 3 μg/day for vitamin B-12, but their adequate presence is imperative for healthy growth and activity. Deficiencies of Ca, P, Mg, and Zn impede normal growth, and inadequate vitamin intakes disrupt the functioning of essential metabolic and maintenance processes, leading, for instance, to gastrointestinal disturbances with folacin shortages and epithelial hemorrhaging with low ascorbic acid (best known as scurvy, common on long sea voyages before the late eighteenth century).

Food energy requirements are not simply a matter of sufficient quantities of macronutrients. Most of the biosphere's carbohydrates (lignin, cellulose, and hemicellulose, including those in cell walls of ingested plant foods) cannot be digested by humans. Digestible carbohydrates are ingested as polysaccharides (starches, glucose polymers) and as simple sugars. These include two monosaccharides (fructose and glucose present in plants) and three disaccharides: lactose, natural milk sugar; sucrose in refined sugar; and maltose, produced by enzymatic degradation of starch and used as a sweetener. Proximate analysis of foods commonly finds the energy value of carbohydrates indirectly as a residual, after subtracting the analyzed amounts of proteins, lipids, water, and ash.

Dietary proteins have much more important qualitative roles than supplying energy. Human growth is impossible without concurrent digestion of nine essential amino acids in infants, eight in children and adults (Smil 2000b). Amino acids contain 15%–18% of N; 16% is commonly used as the average value. These acids are precursors of structural and functional proteins in muscles, bones, and internal organs; in enzymes, hormones, neurotransmitters, antibodies; and in metabolically active compounds. Additional N is also needed to replace small but constant protein losses caused by the breakdown and reutilization of the compounds, excretions of N (in urine, feces, and sweat), shedding of skin, and cutting of hair and nails (Pellett 1990).

Unavoidable (obligatory) N losses through excretion remain fairly constant in adulthood (41–69 mg/kg, average 53 mg/kg); other losses add about 8 mg/kg. Complete proteins, with more than adequate shares of all essential amino acids, are available only in foods of animal origin (meats, fish, eggs, dairy products) and mushrooms, whereas all plant foods have incomplete proteins, with one or more amino acids relatively deficient (cereal grains are deficient in lysine, leguminous seeds in methionine and cysteine). Lipids contain two essential fatty acids that must be present in all healthy diets, linoleic acid and α-linolenic acid, which must be digested preformed to become precursors of prostaglandins (to regulate gastric function and smooth-muscle activity and to release hormones) and parts of cell membranes. Food lipids also carry fat-soluble vitamins (A, D, E, and K).

The energy content of ingested food is measured accurately after complete combustion yielding CO_2 and H_2O. Part of the ingested (gross) energy is lost in feces and intestinal gases (H_2, CH_4) produced by microbial fermentation of unabsorbed carbohydrates in the colon. A small part of digestible energy (DE) is voided in urine and lost through the skin. A fraction of the remaining metabolizable energy (ME) is lost as the heat of microbial fermentation and as the heat due to dietary-induced thermogenesis (specific dynamic action of food, which elevates the BMR of individuals on a mixed diet by about 10%). A very small fraction of net metabolizable energy (NME) is lost to thermogenesis due to hormones, drugs, or effects of cold or stimulants, leaving the net energy for maintenance, which sustains basal metabolism and is available for physical activity (fig. 5.1) (FAO 2003).

The gross energy (GE) of carbohydrates is 17.3 kJ/g. These compounds have been the dominant energizers of human evolution. In the modern world their main sources are staple cereals (predominantly rice, wheat, corn, and millet, eaten as whole grains or milled into flours to make breads, pastas, and pastries), tubers (white and sweet potatoes, cassava), leguminous grains (beans, soybeans, peas, lentils), and sugar refined from cane and beets. Complex carbohydrates are preferable to the simpler varieties, especially when consumed as whole seeds (beans, peas, lentils) or whole-grain products containing indigestible dietary fiber whose regular intake (50–100 g/day) is a critical part of proper nutrition (FNB 2005).

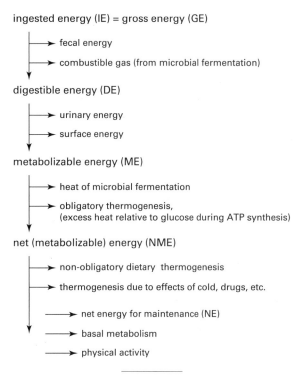

5.1 Energy cascade in human nutrition, from gross energy of foodstuffs to net metabolizable energy. From FAO (2003).

The gross energy of proteins is 23 kJ/g. The highest concentration in plant foods is found in legumes (particularly soybeans), cereals, and nuts; tubers, vegetables, and fruits have only trace amounts. Fats have a gross energy more than twice as high as carbohydrates and proteins (39 kJ/g) and are present in relatively high concentrations in plant seeds, fatty meat, fish, and dairy products. The average consumption of fats in preindustrial societies was very limited. Now they constitute as much as 40% of total food energy intake in rich nations, a dietary excess and a mark of unbalanced nutrition. Finally, alcohol (ethanol) is a peculiar energy source of relatively high density; its gross energy is 29.3 kJ/g. It can be utilized only by the liver at hourly rates usually not surpassing 0.1 g/kg body weight (∼190 kJ/h, or 53 W, for a 65-kg person).

Because ME or NME is appreciably lower than GE, it is necessary to use appropriate food energy conversion factors to express the actual energy available to humans (FAO 2003). The oldest solution (and still the standard) is the Atwater general factor system, introduced by W. O. Atwater and his collaborators at the U.S. Department of Agriculture in the late 1890s (Atwater and Woods 1896). Atwater factors correct for losses from the three macronutrients during digestion, absorption, and urinary excretion, and do not distinguish among foodstuffs. Their precise values are 16.7 kJ/g for carbohydrates and proteins, 37.3 kJ/g for lipids, and 28.9 kJ/g for ethanol, but these numbers are commonly rounded to, respectively, 17, 37, and 29 kJ/g. Food composition tables consider only actually available energy, which means that digestion in healthy people on balanced diets has a very high efficiency: 99% for carbohydrates and ethanol, 95% for fats, and 92% for proteins (more than 20% of dietary protein, about 5.2 kJ/g, is lost through urine).

The extensive general factor systems modifies and extends Atwater factors by adding separate values for monosaccharides (16 kJ/g) and for dietary fiber (8 kJ/g, assuming ∼70% of it is fermentable). The Atwater specific factor system refines the basic approach by assigning different, food-specific factors to macronutrients. The differences arise from different amino acid or fatty acid shares as well as from the way individual foodstuffs have been processed. Specific Atwater factors for protein range from 17 kJ/g for high-extraction wheat flour to 10.2 kJ/g for vegetables; for carbohydrates, from 17.4 kJ/g for white rice to 10.4 kJ/g for lemons; and for lipids,

from 37.7 kJ/g for eggs and meat to 35 kJ/g for grain products (Merrill and Watt 1973).

All these conversions refer to ME, but because not all of that is available for the production of ATP to energize metabolism, yet another conversion is made using NME factors. There is no difference between ME and NME factors for carbohydrates. The NME vs. ME values (in kJ/g) are, for ethanol, 26 vs. 29; for dietary fiber, 6 vs. 8; and of great practical significance, for protein, 13 vs. 17, a 24% reduction (FAO 2003). Using different methods to compare the energy content of actual diets shows that modified Atwater factors are reduced from standard ME valuations by no more than 2%–5% and that NME factors add another 2%–5%; consequently, the total difference between the classic Atwater values and the combined ME and NME factors can range from minor, about 7.5% for the UK urban population, to considerable, with the highest published value of nearly 20% for Australian Aborigines (FAO 2003).

Basal metabolism accounts for the largest share of daily energy needs in all but highly active individuals. It energizes the body's essential life processes, including cell functions, the synthesis of enzymes, the constant work of internal organs (heartbeat, breathing, peristalsis), and the maintenance of body temperature (homeothermy). The basal metabolic rate (BMR) in humans is measured in the supine position in a thermoneutral environment (no heat-generating or heat-dissipating responses) and after 10–12 h fasting (to eliminate metabolic response to food, which peaks ∼1 h after a meal and can boost BMR up to 10%). The share of metabolizing tissue is highly age-dependent, and gender is a key determinant of fat (adipose tissue), which accounts for most of human dimorphism (Bailey 1982). The differences in body composition are very small at birth but increase with age. Average percentages of fat in total body composition are, newborns, 14%; young Western men, 15%; men aged 60, 23%; young Western women, 27%; and women aged 60, 36%. Women store fat at a lifetime average rate of 0.3–0.4 kg/a, men at an average 0.15–0.25 kg/a. In 2005 the global anthropomass of 6.5 billion people equaled about 2 EJ.

The insidious storage of fat—at about 6–16 MJ/a (0.2–0.5 W) drawing away no more than 0.5% of daily food energy input—has its counterpart in the loss of lean body mass. Muscles compose only about 20% of a newborn's mass but average 52% of weight in young men and 40% in young women. After the third decade, the male's greater lean mass is lost more rapidly (2–3 kg/decade) than the female's (~1.5 kg/decade), and autopsies show people over 70 years with about 40% less muscle than they had as young adults. This loss represents a change 1 OM smaller than adipose tissue gain, as little as 600 kJ/a, or 20 mW, but it is an inexorable sign of aging even for those who avoided a significant fat increase.

Rapid buildup of the metabolizing lean body mass in childhood and adolescence and its later gradual decline are reflected in changed BMR. The extraordinarily high human encephalization quotient (actual/expected brain mass for body weight) is slightly over 6, compared to values between 2 and 3.5 for hominids and primates. At about 350 g, the neonate brain is twice as large as that of a newborn chimpanzee, and by age 5 it becomes more than three times as massive as the brain of our closest primate species (Foley and Lee 1991). As a result, the newborn brain (10% of total body weight) accounts for about 50% of BMR (fig. 5.2). The adult brain (2% of body mass) claims about 16% of BMR, and yet there is no sig-

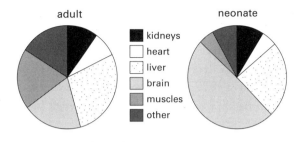

5.2 Partitioning of basal metabolic rates in adults and infants. Based on data in FAO (1985).

nificant correlation between relative BMR and relative brain size in humans and other encephalized mammals.

We do not have more of metabolically expensive tissues (internal organs and muscles) than would be expected for a primate of our size. Aiello and Wheeler (1995) suggested that the only way to support larger brains without raising the overall metabolic rate is to reduce the size of another major internal organ. With relatively little room left to reduce the mass of liver, heart, and kidneys, the gastrointestinal tract was the only metabolically expensive tissue whose size could be reduced with a better diet. The expensive-tissue hypothesis (well-supported by relative organ sizes) thus maintains that a higher-quality diet (richer in animal foods) allowed for the development of a relatively smaller gut and freed more energy for a larger brain.

As with animals, human BMRs have been measured mostly by converting the oxygen uptake into energy equivalents (average 20 kJ/L). Systematic larger-scale measurements of BMRs began in Western Europe and North America before World War I, and by the mid-1980s about 11,000 individual results were available for healthy individuals of both sexes and all ages (Schofield,

Schofield, and James 1985). The data set included adults of different stature and different weight-for-height, but nearly all its data came from Western nations, where unlike in poor countries, adequate nutrition led to full expression of physical growth potential. Moreover, nearly half of the data came from pre-1945 studies of Italian men. Naturally, doubts were raised about the set's global validity for deriving predictive regressive equations of BMR as was done in the 1985 report on food requirements (FAO 1985).

Henry and Rees (1991) found that Schofield's equations overpredicted the BMRs of people in the tropics, with the differences being largest for adults over 30 years of age. The average overprediction was 9% for males (ranging from 1.5% for Maya to 22.4% for Ceylonese) and 5.4% for females (up to 12.9% for Indian women). Explanations for these differences range from varying abilities in producing muscle relaxation to BMR-lowering effects of warm climate. The latter conclusion is supported by BMR changes in individuals moving from temperate to tropical climates. Piers and Shetty (1993) revisited these disparities by measuring BMRs of Indian women. They found that Schofield's equations overpredicted the basal metabolism by 9.2%; they also overpredicted the BMRs of American women and young Australians, and they appeared to be accurate only for predicting European BMRs (Piers et al. 1997).

New equations were generated by Ramirez-Zea (2002), but the latest expert consultation concluded that they are not robust enough to justify the abandonment of Schofield's set (FAO 2004). Inclusion of body height has no effect on improving the fits of these simple linear equations, but in order to avoid excessively high recommendations of average food energy intakes, desir-able body weights for given heights (rather than actual means or medians) should be used in these regressions in any country where obesity is widespread, notably in North America (Pellett 1990). The regression for men 30–60 years old,

$$BMR \ (MJ/day) = 0.048 \ kg + 3.653,$$

explains only 36% of variance; for women same age the equation

$$BMR \ (MJ/day) = 0.034 \ kg + 3.538$$

accounts for 49% of variance.

The regression fits are much better for children ($r = 0.97$ for those less than 3 years old) and adolescents ($r = 0.9$ for teenage boys). For males and females whose adult masses are, respectively, 65 kg and 55 kg, daily BMRs progress from about 700 kJ at birth to peaks of about 7.5 MJ (male) and 6.0 MJ (female) during the late teens and early twenties, then gradually decline to levels about 20% below the adult maxima after age 60. The total energy expenditure (TEE) of newborns (whose BMR is difficult to measure) is just over 3 W/kg, and the BMRs of 1–2-year-old girls and boys averages, respectively, 2.68 W/kg and 2.75 W/kg (FAO 2004). By the age of 10 the rate dips below 2.0 W/kg, and five years later it declines to <1.5 W/kg (fig. 5.3). The average decadal decreases are 2.9% for normal-weight males and 2.0% for normal-weight females (FAO 2004). By the late teens the rates are only slightly above the adult level, and after staying more or less stable for 40 years, they renew their decline after age 60. By age 70 the fire of life, at just 1 W/kg, burns at less than half the rate at birth.

CHAPTER 5

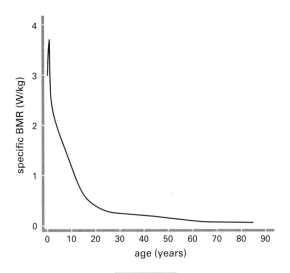

5.3 Decline of BMR with age. Plotted from data in FAO (2004).

5.2 Requirements and Uncertainties

Human energy requirements are defined as intakes that, in adults, maintain body weight and composition, and support a desirable level of physical activity and long-term health. In children they also have to cover the requisite growth needs, in pregnant women the deposition of new tissues, and in lactating women the production of milk. BMR usually dominates, and activity needs—energy spent at work and in leisure and varying with the nature, intensity, and duration of the tasks—account for most of the remainder. The requirements of growth (the sum of the energy values of the newly formed tissues and products, and the cost of their synthesis) claim a significant share in infants and children but are absent after adolescence. There are two other kinds of gender-specific requirements, the energy needs of pregnant and lactating women. The TEE of an individual is the sum of all these

requirements during a 24-h period. The physical activity level (PAL) for a 24-h period is a quotient of TEE/BMR (TEE is the product of BMR and PAL). PAL expresses the energy cost of a specific activity as a multiple of BMR, and the metabolic equivalent (MET), the quotient of work rate and RMR, is roughly equal to the cost of sitting.

Determination of energy costs of activities used to be a much greater challenge than measuring BMR. Rigging up and wearing the apparatus necessary for accurate respirometry (masks, hoods or mouthpieces, and hoses needed to measure gas exchange) is no problem when testing a cyclist on a stationary bicycle, but it is a challenge for measuring the costs of coal mining or mountain climbing. Three portable systems that can be used for continuous measurements of oxygen uptake for long periods are in widespread use (Patton 1997), and many values have been gathered by decades of dedicated research (Durnin and Passmore 1967; FAO 2003). Two convenient methods now allow nonintrusive measurement of actual energy expenditures: the doubly labeled water technique (see section 4.1) and heart rate monitoring (HRM). The DLW technique allows the calculation of total energy use over the course of several days; it has been accepted as the best way of measuring the TEE of free-living individuals. Heart rate monitoring measures daily energy expenditure by relating minute-by-minute pulse to the rate of oxygen consumption using calibrations previously determined by laboratory respirometry.

The energy needed for growth is about 40% of TEE during the first month, averages 35% for the first three months, is halved to 17.5% during the subsequent three months, and drops to 3% by the end of the first year and <2% a year later. The rate remains between 1% and 2% until middle adolescence and is absent by the age of 20.

(A small fraction of 1% of total energy is still needed for the renewal of adult tissues, principally sloughing skin and intestinal lining, and growing nails and hair.) The growth requirement is composed of the energy needed to synthesize new tissues (fat and protein) and the energy deposited in them. As the share of fat in newly deposited tissues declines, so does the energy accrued in normal growth, from about 25 kJ/g during the first three months to about 11 kJ/g by 12 months of age.

Adequate energy intake in early childhood is essential not only for transforming us from a state of immobile dependence to one of active exploration but to make us human. The neonate brain (at about 350 g, twice as large as that of a newborn chimpanzee) enlarges 3.5 times by the age of 5 to become more than three times as massive as the brain of our closest primate species (Foley and Lee 1991). Average childhood weights at particular ages differ appreciably among countries; better nutrition has been pushing them higher in most affluent nations over time. Long-term Japanese trends are an excellent illustration of this universal trend. Between 1900 and 2000, 11-year-old Japanese boys gained nearly 20 cm of height, with the hungry years of World War II representing the only interruption of a steady rise (fig. 5.4).

The increased energy needs of pregnancy are due not only to the growth of fetus, placenta, and maternal tissues (uterus, breasts, blood, fat) but also to the rise of BMR and the cost of increased cardiovascular and respiratory effort during activity. Large data sets show that desirable birth weights of 3.1–3.6 kg are associated with total maternal weight gain of 10–14 kg (FAO 2004). This means that the total average weight gain of 12.5 kg (baby's 3.4 kg, deposition of 925 g of protein and 3.8 kg of fat) costs about 335 MJ (Hytten 1980), or almost 27 kJ/g. FAO (2004) put it at 321 MJ, or about 0.35 MJ/

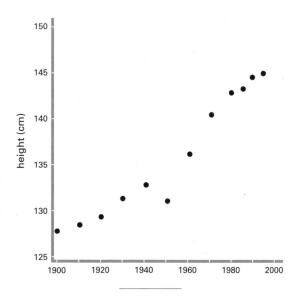

5.4 Average height of 11-year-old Japanese boys increased by nearly 20 cm during the course of the twentieth century. Plotted from survey data published regularly in *Japan's Statistical Yearbook.*

day, 1.2 MJ/day, and 2.0 MJ/day for respective trimesters. Average increases in BMR during the trimesters are on the order of 5%, 10%, and 25%, respectively, and the total BMR increment for the 12-kg weight gain is about 150 MJ.

The energy cost of lactation varies considerably with the volume and duration of milk production. The metabolizable energy of human milk is 2.59 kJ/ml (protein 8.9 g/L, 3.2% fat). In affluent countries the median milk production (with exclusive breastfeeding) rises from about 700 mL/day during the first month to 850 ml/day in the sixth month and declines afterward to around 500 mL/day by the end of the first year (FAO 2004). Conversion of food energy to milk has a high, 80%–85%, efficiency. With a daily mean of 800 mL for six months of lactation, the cost would average 2.8 MJ/day. A typical Western woman would start breastfeeding with some 150 MJ of fat reserves that will be converted to milk during the following six months, so the actual additional energy cost is only about 2.1 MJ/day. Still, this is nearly twice the extra energy needed in pregnancy, and the 370 MJ needed for half a year of breastfeeding represents a higher energy cost than carrying a baby to term after 280 days.

International consensus on how to combine these food energy needs and recommend desirable daily intakes has undergone a number of adjustments since the periodic standard-setting meetings of nutritional experts begun shortly after WW II. The first two international committees put the needs of a reference man (25 years old, weighing 65 kg, and living in a temperate zone) at 13.4 MJ/day and those of a reference woman (25 years old, weighing 55 kg) at 9.6 MJ/day (FAO 1957). The third committee redefined the reference adults: still at 65 and 55 kg, but 20–39 years old, healthy, working moderately for 8 h, sleeping 8 h, spending 4–6 h in very light activities, and moving or engaging in active recreation or household duties for the remaining time (FAO 1973). These moderately active adults needed, respectively, 12.5 MJ/day and 9.2 MJ/day. A multiplier of 0.9 was selected for light work, 1.17 for very active, and 1.34 for exceptionally active exertions. The 1985 consensus meeting adopted a different approach by expressing total energy requirements as multiples of the BMR (FAO 1985).

The latest consensus report (FAO 2004) continued that approach for adults but relied on DLW or HRM studies in setting the standards for infants, children, and adolescents. The advantage of the latter approach is particularly clear when estimating the energy requirements of infants because their BMRs are so highly variable, ranging from 180 kJ/g/day to 250 kJ/g/day. The new equation to calculate the TEE of infants from birth to 12 months of age is valid for both sexes:

$$\text{TEE (MJ/day)} = 0.416 + 0.371 \text{ kg}.$$

Measurements show that the TEE of breastfed infants is lower than that of babies fed a formula. Equations for the TEE of children and adolescents are sex-specific:

$$\text{TEE (MJ/day)} = 1.298 + 0.265 \text{ kg (boys)}$$

$$\text{TEE (MJ/day)} = 1.102 + 0.273 \text{ kg (girls)}.$$

As the children grow, their specific energy requirements decline linearly between ages 2 and 18, from about 350 kJ/kg/day to about 200 kJ/kg/day for boys, and from about 340 kJ/kg/day to 180 kJ/kg/day for girls. Supplementary food for catch-up growth of children with weight deficit requires about 21 kJ/g of newly added tissue.

Adult requirements are based on factorial estimates of habitual TEE. These, in turn, are derived from the average BMR and the PAL attributable to a particular population, and the 24-h PAL takes into account both occupational and leisure activities (FAO 2004). The minimal survival requirement for people who are not bedridden can be calculated by assuming 8 h of sleep and rest for the remainder of the day, except for token moving and standing briefly while washing and dressing; this gives PAL of 1.27. Survival minima for most adults would thus be between 6.2 MJ/day (50-kg woman) and 9.6 MJ/day (80-kg man), fluxes of about 70–110 W. Beyond these minimal requirements lies an endless variety of actual needs determined by the type and duration of activities.

Human energy costs can be measured directly and with great accuracy. The first human respiration chamber was built in the 1860s (Pettenkofer and Voit 1866), and the first ingenious calorimeter was used to determine both heat production and respiratory exchange just before the end of the twentieth century (Atwater and Benedict 1899). Since that time we have accumulated thousands of measurements for scores of activities, with energy cost now usually expressed in terms of physical activity ratio (Durnin and Passmore 1967; FAO 2003) or as MET (Ainsworth 2002). Thinking is at the bottom of that scale. The brain's constantly high BMR goes up only marginally during challenging mental tasks, which means that science is, energetically at least, very light work. Standing requires the deployment of large leg muscles, resulting in MET of 1.2–1.3. Light exertions are typical of numerous service jobs that now dominate modern economies. Typing, truck driving, food retail, or car repair have usual physical activity ratios of 1.8–2.0. Most modern manufacturing and construction work as well as mechanized farming belong to the same category.

Activities requiring moderate or heavy exertions are most common in traditional farming: plowing, manual planting, weeding, and harvesting have MET 4–6; digging and cleaning of irrigation canals, tree cutting, and some fishery tasks rank above 6. But even for the most demanding occupations average long-term expenditures may be only in the medium category, as the spells of taxing exertions are interspersed with periods of less demanding activity or rest. In traditional farming these disparities also had a seasonal flux. Averages for a group of North China peasants showed 11.5 h of moderate to heavy work during the summer harvest, with 6.5 h of

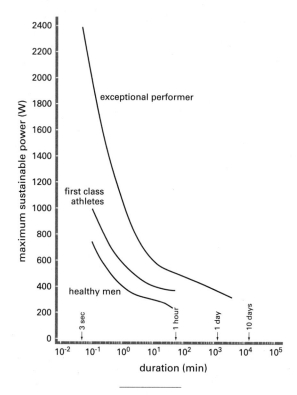

5.5 Energy costs of cycling. Based on D. G. Wilson (2004).

sleep, whereas during the mid-winter there were only 7 h of light work and 10 h of sleep (Chen 1981).

Energy costs of leisure and sport activities depend on their intensity. Cycling is a perfect example of this difference: its energy demands can range from very moderate in sedate pedaling to extremely heavy in record-breaking races (fig. 5.5). Because it deploys very efficiently the body's largest leg muscles, cycling is also the fastest mode of human locomotion. The fastest human-powered machines on land, water, and in the air have all been propelled by accomplished cyclists (D. G. Wilson 2004). Competitive cross-country skiing and medium-

and long-distance running have the highest energy cost among individual sports (MET 12–18), and basketball and soccer rank high in the group sports category (MET 8–10). Measurements of TEE during the 4000-km Tour de France showed daily means of about 25 MJ and individual rates up to about 32 MJ (Saris et al. 1989). Similarly, studies using doubly labeled water found actual energy expenditures during a climb of Mt. Everest at nearly 14 MJ/day (Westerterp et al. 1992); the addition of BMR brings the TEE of these climbers to over 20 MJ/day. These values are near the metabolic maxima for multiday efforts.

Integrating BMR with work and leisure requirements results in a PAL spectrum with extremes of 1.4–2.4. A sedentary or light-activity lifestyle has a 24-h PAL range of 1.40–1.69 (mean 1.53). The multiple rises to 1.70–1.99 (mean 1.76) for an active or moderately active lifestyle, and it goes to 2.00–2.40 (mean 2.25) for a vigorous lifestyle. PAL values in excess of 2.4 cannot normally be maintained over extended periods of time. Individual TEEs can rise or slide along this continuum depending on a particular day's activities. For example, a moderately active (PAL 1.75) 40-year-old 70-kg male (BMR 7.0 MJ/day) will have a usual TEE of 12.3 MJ/day (142 W; 175 J/g), but a day-long skiing trip (PAL 2.1) can boost it to 14.7 MJ/day (170 W; 210 J/g). Approximate daily energy requirements of adults whose work can be classified as light (most white-collar workers, many factory workers, and homemakers in the rich countries) correspond to PALs less than 1.75; in contrast, very few demanding jobs in affluent countries (lumberjacks, some miners) and traditional farmers have PALs exceeding 2.0.

The United States has the longest record among the countries that have issued their own national dietary recommendations. Recommended dietary allowances (RDAs) were first published in 1941, and their tenth edition came out in 1989 (FNB 1989). Recommended energy intakes for adults was calculated by using FAO (1985) end PALs of 1.6 for men and 1.55 for women. The approach changed in 1997 with the establishment of dietary reference intakes (DRIs). These are formulated not only in order to prevent nutrient deficiencies but also to lower the risk of chronic diseases. For food energy DRIs quantify total PAL-based daily requirements as well as recommended intakes of carbohydrates (130 g/day for adults), protein (56 g/day for males, 46 g/day for females), and two essential fatty acids.

DLW measurements have provided the most accurate check of the accuracy of recommended requirements (Black et al. 1996; FAO 2004). The highest daily PALs, 4.7 and 4.5, were found, respectively, in cyclists taking part in Tour de France and in men pulling a sled across the Arctic. Activity markups found by DLW studies for individuals leading normal lifestyles corresponded very well with expected multiples, ranging between 1.2 for sedentary subjects to 2.5 for exceptionally active individuals. PALs between 1.6 and 1.8, means for all adult subjects in the analyzed studies, would tend to encompass activity markups for most healthy, active individuals. At the same time, there are many uncertainties. For instance, the regression equations for adult BMRs (see section 5.1) account only for variances of 36% for men and 49% for women, leaving plenty of room for substantial individual departures from the means.

Limited databases are common. For example, the energy requirements of children and adolescents have been based on DLW or HRM studies of only 801 boys and 808 girls 1–18 years of age (FAO 2004). Moreover,

74% of boys and 86% of girls in these studies were from affluent Western countries, and all individuals from low-income countries were from just five countries in Latin America: the set had no entries from Asia or Africa, continents that contained nearly 75% of the world's population in 2005. But no uncertainty is more intriguing that the highly variable energy cost of pregnancy. The latest consensus advice for pregnant women is to add just 0.35 MJ/day during the first trimester, 1.5 MJ/day during the second, and 2.0 MJ/day during the third, and well-nourished lactating women should increase food intake by 2.1 MJ/day (FAO 2004). In low-income countries, where smaller women give birth to smaller babies, the costs should be lower but, astonishingly, several sets of detailed energy balance studies show that many women have, not as exceptional individuals but as groups, extremely low energy needs during pregnancy and lactation.

Prentice (1984) found that, compared to standard metabolic expectations, pregnant rural Gambian women appeared to have energy shortfalls of at least 600 kJ and as much as 2.15 MJ/day, even if they just slept, ate, and rested. But they not only worked hard but often engaged in tasks that would not be contemplated by most Western women, so their daily energy shortfalls would be up to 2.50–4.14 MJ/day when performing their duties. Adair and Pollitt (1982) found a similar situation in Taiwanese mothers giving birth to healthy children. And Norgan, Ferro-Luzzi, and Durnin (1974) found that among the Kauls of New Guinea there was no difference in energy intakes of nonpregnant and nonlactating women and pregnant and lactating women.

Not only pregnant women but whole populations have been shown to live with surprisingly low food energy intakes. Among the Senegalese, Ferlo proved that a large seasonal food deficit (nearly 1.25 MJ/day compared to standard recommendations) was not accompanied by any significant increase in malnutrition or any clinical signs of food deficiency (Benefice, Chevassus-Agnes, and Barral 1984). And in New Guinea the average energy intakes of the whole coastal Kaul tribe were just 27% (males) and 13% (females) higher than their expected basal metabolic rates, or about 2.4 MJ/day less for men and 2.9 MJ/day less for women than in a highland village, disparities unexplainable by differences in methodology (identical techniques were used), body weights, food availability, or work.

Given the large variability of adult BMRs (see section 5.1), recommended energy intakes based on BMR regressions have a high degree of uncertainty when applied to individuals. Looking back in 1983, Elise Widdowson, a pioneer of modern nutritional studies, noted that the fundamental question she had posed in the late 1940s—Why can one person live on half the calories of another and yet remain a perfectly efficient physical machine?—has never been satisfactorily answered (Widdowson 1983). This is still true a generation later; the difference must come down to metabolic efficiencies, and there is little doubt that not only some exceptional individuals but entire populations can use food energy much more efficiently than the standard expectations would have it. Such adaptations as reduced activity, weight loss, or lower milk production would be obviously undesirable, but evidence of harmless adaptations means that there is a range of long-term averages of energy intakes compatible with regulation of expenditures without exceeding the homeostatic limits.

Such conclusions are particularly noteworthy given the post-1980 epidemic of obesity in North America (and to lesser extent elsewhere), and growing indications that restricted energy intake promotes longevity in species ranging from microbes to chimpanzees and humans (Weindruch and Sohal 1997; Fontana et al. 2004). Humans are flexible converters of food energy, able to respond with altered metabolic efficiencies to different diets, environmental conditions, specific tasks, and health states. The questions about food requirements include not simply How much? but For what? and In what context? These questions remove the search for food requirements from the realm of quantifiable energetic considerations to the much larger and fundamentally unquantifiable setting of cultural preferences and social expectations. Human energetics is so contextual and so value-laden precisely because it concerns humans. Borrini and Margen (1985) summed up its challenge well: before defining specific food requirements, it is imperative to appreciate the perceived needs and wants of the people and their customs, the structure and dynamics of their societies, and the ecology of their environments.

5.3 Thermoregulation

Like all other homeotherms, humans are tachymetabolic (able to produce heat at the cellular level even in the absence of any muscular activity) and have an intricate neural regulatory system to maintain thermal homeostasis (Hardy, Stolwijk, and Gagge 1971; Blumberg 2002). As a result, the normal range of human temperatures is very narrow. The lowest mean is at 35.1°C in Andean Indians; 37°C is the standard physiological cove temperature. Infants have slightly higher normal temperatures, elderly people slightly lower, and a daily cycle (evening peak, morning trough) averages about 2°C. Humans have extensive sensors to guide their thermoregulation. Thermal receptors in the brain and blood vessels sense the body core temperature, and warm and cold fibers (maximum

firing rates at, respectively, about 40°C and 32°C) sense heat through the skin.

The surface area of adult bodies is 1.3–2.2 m^2, and the highest heat loss is in the neck ($>$180 W/m^2), the lowest in the feet ($\sim$2 W/m^2). Radiation is the largest heat loss route, accounting for 60% of the total from a naked body. According to the Stefan-Boltzmann law, the radiant loss or gain (F, in W) is proportional to the product of body surface area (A, in m^2) and the difference of fourth powers of ambient temperature (T_a) and body surface temperature (T):

$$F = \sigma A(dT_0^4 - eT^4),$$

where σ is the radiation constant and d is the skin's absorption coefficient (0.6–0.8 for light to dark skin). Conduction is the least important cause of heat loss, normally just about 3% of the total (heat loss in urine and feces is similar). Evaporation (about 15%) and warming and wetting of the air through breathing (10%) are other major forms of heat loss. Convection is important only when air or water move past the body.

Healthy people tolerate spikes up to 40°C–40.5°C that follow taxing work or exercise, and fever causes longer periods of such elevated temperatures without any lasting damage, but temperatures above 42°C are fatal. When the core temperature falls to $\sim$35°C, mental slowness becomes extreme, and at a core temperature below 33°C, life-threatening hypothermia ensues. Hypothermia if brief is survivable. In contrast, the Earth's inhabited regions have extremes of about 50°C in subtropical deserts and −60°C during Arctic winters. A hairless heterotroph has only four choices to live in temperatures below the zone of thermal comfort, which, for resting unclothed people is 24°C–29°C: to insulate by

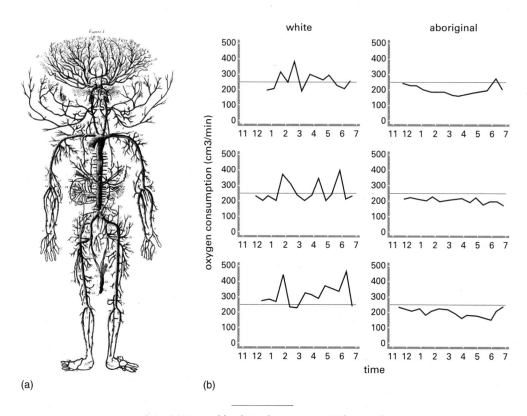

white   aboriginal

oxygen consumption (cm3/min)

time

(a)   (b)

5.6   (a) Human blood circulation system (Diderot and
D'Alembert 1751–1772). (b) Adaptive vasoconstriction of
Australian Aborigines (based on Scholander et al. 1958).

becoming sufficiently obese, to boost BMR in order to generate more heat, to cut conductance through vasoconstriction, or to create a portable microenvironment. The first adaptation is not an option for an originally tropical, long-limbed, upright runner even if the food to sustain it were available; pigs and seals have mastered the adaptation.

Different BMRs (highest in Northern Europe and among native North Americans) confirm the usefulness of the second approach. Complexity and adaptability of the human circulatory system offers some remarkable opportunities for heat-saving vasoconstriction (fig. 5.6). Its efficacy was first documented in the Australian Aborigines (Scholander et al. 1958). They, as well as white Australians, slept in the open in sleeping bags as the overnight lows dropped to 0°C. The whites Australians had recurrent erratic elevations of metabolic rate (up to more than twice BMR) associated with periods of restlessness and shivering, but the Aborigines slept undisturbed, their metabolism steady (fig. 5.6), in some cases even below

BMR, as vasoconstriction lowered skin temperature (feet minima down to 17°C) and reduced thermal conductance. Conserving heat the Aboriginal way is much more effective than shivering (thermogenesis), which only accelerates convectional heat loss (Taylor 2006).

The vasoconstriction response would be grossly inadequate to handle seasonal thermal deprivations, even in temperate latitudes where clothing and heating must be used to protect against the cold. Interestingly, the Inuit, whose harsh environment forced them to develop extraordinary multilayered clothing protection replicating the furry adaptations of Arctic animals, appear to have the poorest naked-body insulation (the highest thermal conductance) of any population studied. The microclimate of the Inuit's body was constantly almost tropical, and hence there was no need of metabolic adaptations. Other supposedly adaptive Inuit characteristics also have plausible energetic explanations (Rennie et al. 1962; So 1980). Their relatively high BMR is explained as being due to the specific dynamic action of diets rich in fat and protein.

Still, boreal populations have evolved a variety of minor but important adaptations: sweat glands more abundant on the face and greatly reduced on the body trunk and legs, and higher finger temperatures, advantageous when warm mitts have to be slipped off for outdoor activities. But intelligence (an enterprising and impressively effective mimicry of hunted mammals) was always the key to Arctic adaptation. Animal precursors can also be seen in the common heat adaptation of wearing dark-colored clothes in desert environments. As in the case of dark bird plumage, dark robes in the desert absorb a very large part of incoming radiation, which is subsequently lost by convective heat transfer without impinging on the body.

Adaptation to hot climates is not only fast but also very effective for all healthy people, and it is retained even by populations that have lived for generations in temperate or cold climates. Hanna and Brown (1983) include the ability to tolerate heat as a distinctly human characteristic alongside bipedalism, hairlessness, a large brain, and a symbolic linguistic ability. Both the rapidity and the effectiveness of the process can be perfectly illustrated by Wyndham's (1969) revealing experiments with South African white males with no recent exposure to heat during work. They had to perform moderate work in a hot (33°C) and humid environment, and the acclimatization period consisted merely of ten consecutive days of such a regime. Their response was remarkable: rapidly rising core temperatures reaching a danger zone, heartbeats approaching tolerable maxima, and low sweating rates were swiftly transformed to levels nearly identical to those of highly acclimatized native Bantu.

Other studies have demonstrated that such adaptive training requires just an hour or so per day regardless of the thermal environment for the rest of the time. A Brazilian study found increased sweating during a short-duration exercise in a temperate environment after only nine days of 1-h exercises on a cycle ergometer at 50% $VO_{2\,max}$ (Machado-Moreira et al. 2005). The initial response to heat is the dilation of peripheral skin blood vessels, compensated by vasoconstriction elsewhere, and the shifting of large volumes of blood into the hands and feet. The magnitude of skin vasodilation is striking: resting skin flow (in thermoneutral surroundings) is about 250 mL/min (dissipating up to 100 W), but vasodilation can increase skin blood flow to 6–8 L/min (or 60% of cardiac output) during severe hyperthermia (Charkoudian 2003). Perspiration begins, first on the trunk, then on the extremities, at 28°C–32°C of skin temperature.

HUMAN ENERGETICS

Sweat (99% H_2O, NaCl, KCl, traces of urea, lactic and fatty acids, and proteins) is produced by about 3 million ecrine glands (from $<100/cm^2$ on the thigh and leg to $>300/cm^2$ on the forehead), and studies have found no significant differences in their density among populations of different continents and climates (Taylor 2006).

Without active sweating, the average person loses about 12 W/m^2 of body surface, equally split between respiration and skin diffusion. Above 28°C, evaporative losses climb exponentially, so that even at rest in still air they remove about 130 W/m^2 at 48°C, or a total of up to 230 W. With work, the perspiration rates go up to surpass greatly those of sweating mammals: a horse can lose every hour 100 g/m^2 and a camel up to 250 g/m^2, but a person can perspire more than 500 g/m^2 (Folk 1976). Perspiration of 500 g/m^2 translates to heat loss of 550–625 W for most adults, sufficient to regulate temperatures even in extremely hard-working individuals. Designers of heating and ventilating plants use 586 W ($\sim$325 W/m^2) as the maximum heat output of a manual laborer, and under normal free-conversion conditions, with air moving at just 0.45 m/sec, virtually all of this load can be lost through evaporation of sweat. Similarly, pedaling (with adequate cooling) can sustain output of nearly 600 W without any noticeable increase in body temperature no matter how long the activity continues (D. G. Wilson 2004).

Acclimatized individuals can produce sweat up to 1.1 $L/m^2/h$, enough to remove 5 MJ of heat. At a rate of 1390 W, this exceeds all but the most strenuous athletic exertions. The highest reported short-term peak sweating rates are 4 L/h. At such levels rehydration is an acute necessity because humans can neither tolerate substantial dehydration nor store large volumes of water. However, limited voluntary dehydration (drinking less than

is perspired) is common during heavy exertions, with the deficit gradually replaced within a day. Most notably, elite-level marathon runners ingest only about 200 mL/h during the race, much less than the recommended volume of 1.2–2 L/h, and the latest guidelines for fluid replacement during marathon running urge slower runners to drink no more than 400–800 mL/h in order to avoid risks of hyponatraemia (IMMDA 2001).

5.4   Limits of Human Performance

There are three fundamentally different, time-dependent ways to energize physical performance. The first one is the anaerobic, alactic mode. ATP, the direct source of energy for muscular contractions, is stored in muscles at minuscule levels, averaging a mere 5 mmol/kg of wet tissue. With 20 kg of active muscles and 42 kJ/mmol of ATP, this is equivalent to 4.2 kJ, the total sufficient to energize contractions for 0.5–0.75 s of maximum effort. The most rapid recharge is through the breakdown creatine phosphate (CP). But CP's muscle stores are also limited (20–30 mmol/kg of wet tissue), and the recharge will last only 5–8 s. Maximum metabolic power achievable by this route is large, 3.5–8.5 kW for a 65–70-kg average man, and as much as 12.5 kW for a trained man, but the overall capacity averages just 20–40 kJ and reaches no more than 55 kJ for the best-adapted bodies.

Somewhat longer exertions are energized anaerobically by muscular glycogen, whose glycosyls break down into pyruvate and hydrogen. The pyruvate is converted to lactate (its accumulation in active muscles leads to the well-known weakness and pains), and a convenient indicator of anaerobic threshold is the breaking point when lactate blood levels start increasing exponentially. In sedentary people this is invariably when the workload is 50%–70% of the maximum aerobic power; in endurance athletes this threshold may rise to 85% of the maximum power. The maximum metabolic power of anaerobic glycolysis is 1.8–3.3 kW for average persons and up to 8.3 kW for trained persons; peak capacities average 75–100 kJ and can go up to 205 kJ, as much as three to four times the CP-derived total. Short (30–180 s) exertions are largely energized in this way. For example, models of 100-m runs (using speed curves of world champions) show anaerobic metabolism contributing 95% of all energy (Arsac and Locatelli 2002).

All prolonged efforts are powered primarily through aerobic (oxidative) recharge. Because the body's oxygen store of about 1 L can support moderate exertion for no more than 0.5 min, the subsequent demand requires linear increases in pulmonary ventilation. Sustained human power is largely a function of maximum oxygen intakes. Maximum aerobic power, sustainable for about 90 min, is only 10–20 mL/kg · min in people with chronic pulmonary diseases, and its range is 20–55 ml/kg · min, or 350–1350 W, in adults. A metabolic range of 600–900 W would include most mildly active people, but the rate goes over 90 ml/kg · min for elite endurance athletes. This is the equivalent of more than 2 kW, a flux 25 times BMR, an impressive metabolic scope in comparison with most mammals (see section 4.1). Aerobic capacities are slightly lower in women than in men of the same age, and after the adolescent peak the annual decline range is 0.4–1 mL/kg · min, so by the age of 65 aerobic are just capabilities 25–30 ml/kg · min. Peak aerobic capacities are 1.5–3.5 MJ for healthy adults and surpass 10 MJ for trained athletes, with maxima at 45 MJ.

Both glycogen and fatty acids are the substrates of the oxidative metabolism, with the acids' share rising to 70% during prolonged activities. But it is important to note that every sustained exertion also uses a small share of anaerobic glycogen breakdown, and conversely, as the duration of brief (predominantly anaerobic) exertions increases, aerobic recharge supplies rising shares of energy for efforts lasting 30–120 s. Individual limits of performance can change. Genetic endowment is a prerequisite of exceptional performance (Bouchard, Malina, and Perusse 1997), but the aerobic power of average individuals can go up by 20% with training. The concurrent increase of blood volume, roughly 1 mL/mL of aerobic capacity, raises cardiac filling pressure, output, and blood delivery to muscles and skin for cooling.

Walking and running are the two physical activities during which most individuals experience the limits of their performance, whether ascending stairs, carrying loads, or racing. In mechanical terms, walking can be modeled simply as a motion of an inverted pendulum, with the body mass center at its lowest point at heel strike and at highest level at midstance. Theoretically, no mechanical work is required to move a pendulum along an arc, but work is needed for the transition from one stance limb to the next, and experiments indicate that it increases with the fourth power of step length (Donelan, Kram, and Kuo 2002). This matters because a faster walk is normally an equal combination of increased step length and higher step frequency. In actual walking nearly half of the metabolic cost goes into generating horizontal propulsive force (Gottschall and Kram 2003).

Many studies have measured the energy cost of walking on the level, all indicating lower efficiencies at speeds both below and above the optimum range of 5–6 km/h. The minimum costs of walking on the level are about 1.5 ± 0.5 J/kg · m at the speed of 1.3 m/s. Higher speeds and walking uphill bring linear cost increases across a broad range of speeds, as much as 18 J/kg · m on a 45° slope (Minetti et al. 2002). The gross energy cost of walking also varies linearly with sex and age: as a group, adult slim women have the lowest, and heavier women the highest, requirements. Uneven surfaces (meadows, stubble field, plowed land), muddy roads, or deep snow will raise the costs of level walking by up to 25%–35%. Level walking with light loads is not too demanding, but heavier loads can be very taxing, especially when carried uphill.

Slow to moderate walking speed is 1.1–1.2 m/s (4–4.3 km/h), and 1.9–2.1 m/s (6.8–7.6 km/h) is the speed at which walkers voluntarily switch to running. Untrained runners manage speeds of 3–4 m/s. By 2005 the best men's marathon time was run at 5.6 m/s (compared to about 4 m/s in 1910); world record speeds in 10-km and 5-km races were, respectively, 6.2 and 6.4 m/s, and for 100-m races, just over 10 m/s (IAAF 2006). Running usually requires power outputs between 700 W and 1400 W (about 10–20 times BMR for adults). Compared to other mammals (see section 4.4), the energetic cost of human running is relatively high, but humans are unique in virtually uncoupling this cost from speed (Carrier 1984; Bramble and Lieberman 2004). Quadrupeds have optimum speeds and hence different COT for different gaits (e.g., horses walk, trot, and gallop). COT for human walking has a similarly U-shaped curve, but the cost of human running is essentially independent of speed between about 2 m/s and 6 m/s (fig. 5.7). Empirical data show that the metabolic COT roughly doubles from 1 J/kg·m to 2 J/kg·m as the walking speed goes from 1 m/s to 2 m/s (3.6–7.2 km/h), but the cost of running remains fairly stable, about 4 J/kg·m (Minetti and Alexander 1997).

Two factors explain this extraordinary capability: efficient heat dissipation (see section 5.3) and bipedalism.

In quadrupeds ventilation is limited to one breath per locomotor cycle (the thorax bones and muscles must absorb the impact on the front limbs as the dorso-ventral binding rhythmically compresses and expands the thorax space), whereas human breath frequency can vary relative to stride frequency. People thus have an option to run at a wide variety of speeds, but quadrupeds are largely restricted to structurally determined optima. Faster top running speeds are achieved by applying greater support forces to the ground, not by a more rapid repositioning of legs in the air. Experiments show that support forces to the ground rise with speed, whereas the time taken to swing the limb into position for the next step does not vary (Weyand et al. 2000). Peak running performance has a structural basis. The highest ground support forces exceed the body's weight fivefold, and hence more muscle (to generate the forces) and tendon and bone (to transmit them safely to the ground) are needed than in endurance running (Weyand and Davis 2005).

Expectedly, a study of the world's 45 fastest athletes showed that male body mass declines from more than 75 kg for sprinters to less than 60 kg for long-distance runners. Human excellence in running is also illustrated by steadily improving record speeds for every distance (fig. 5.8). Since 1910 the annual rate of improvements has averaged less than 1 m/min, a gain imperceptible in sprints now run at speeds over 10 m/s but a reduction of about 30 min in running the marathon. Actually, this 42,195-m endurance race is now run by many of the world's best athletes at a faster pace than the record 10 km run as recently as 1945. Ryder et al. (1976) believed that the historic rate of improvements could continue for decades to come. Whipp and Ward (1992) predicted that by 1998 women might run the marathon as fast as men.

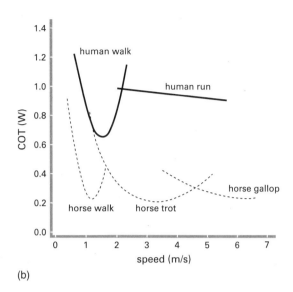

(a)　　　　　　　　　(b)

5.7 (a) Runner from Muybridge's (1887) classical studies of locomotion. (b) Energy cost of walking and running. Human walking, much like quadruped running, has different energy costs for different speeds, but human running is essentially independent of speed between about 2 m/s and 6 m/s (based on Carrier 1984 and Bramble and Lieberman 2004).

In reality, the marathon gap was still about 12 min in 2005, and the rates of improvement have been slowing down (McConnell 2000).

How efficient are these performances? How good a machine is the human body? Metabolism of each mole of glycogen-derived hexose (containing 2.81 MJ) conserves 39 mol of ATP (42 kJ/mol), a 50% efficiency; about 40% of this energy powers muscle contractions for a total gross efficiency of 23%. ATP conservation ranges between 40% and 70%, giving a range of 16%–20% for the final gross efficiencies. For comparison, a classic ergometric test of Benedict and Cathcart (1913) put the net mean at about 21% and the maximum gross effort for well-trained bicyclists at 16%–21%. Anaerobic conversion is much less efficient (about 10%–13%), and so the long-duration efficiencies will be highest for elite aerobic performers deploying their muscles at higher power rates in activities most conducive to peak kinetic power outputs.

Almost invariably this will be cycling, and an abundance of measurements illustrates both the top and average performances in pedaling (D. G. Wilson 2004). Elite athletes can produce bursts in excess of 2 kW lasting just a few seconds, can sustain 1 kW for more than 1 min, 500 W for 1 h, and 400 W for a day, whereas exertions lasting more than 1 h are limited to less than 200 W in untrained men. Regardless of the length of a race, cycling

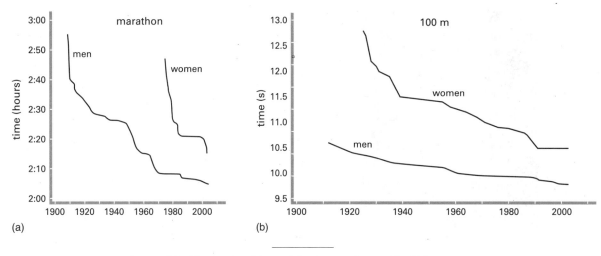

5.8 History of world records for the marathon and for 100 m.
Plotted from data in IAAF (2006).

is the fastest mode of human locomotion, and it has energized all record-setting machines not only on land but on water and in the air (D. G. Wilson 2004). *Monarch B* completed a triangular 1.5-km course at nearly 10 m/sec (Drela and Langford 1985). *Flying Fish II*, a watercraft with a hydrofoil ridden like a bicycle, outpaced a single rower at about 6.5 m/s.

A healthy adult should tolerate many hours of work at 40%–50% of maximum aerobic capacity, equal (conservatively) to gross metabolic power of 380–490 W for 60–70-kg men. With typical kinetic efficiencies around 20%, this would be 75–100 W of useful work. Measurements in heavy manual occupations confirmed that 350 W of total output are sustainable during an 8-h shift but rarely exceeded. At 70 W the total daily useful work capacity is just 2 MJ. That much kinetic energy can be liberated, even when the conversion efficiency of an engine is a mere 10%, by burning just 1 kg of ordinary steam coal or 500 g of fuel oil. Human effort, even at its best, is a most unimpressive source of mechanical energy.

Ultimately, the limits of human performance depend on adequate food supplies. The body's carbohydrate stores add up to a mere 400 g of glycogen in muscles, 100 g in the liver, and 3 g of glucose in the blood, a total of just 8.4 MJ, good for a day's minimal survival needs. Only 2–2.5 kg (33–42 MJ) of protein, or no more than 20%–25% of the body's total stores, can be metabolized without endangering life; 30–50 days of minimum needs would exhaust those reserves. Starving people must derive most of their energy from fat. Between 2 kg and 3 kg of lipids are a structural part of cells, but the rest (in adults 6–18 kg, or 226–678 MJ) can be converted to last at least 40–50 days. Obese people would have plenty of fat left for another month, but their protein reserves would be gone. The range of starvation survival is thus 4–8 weeks.

5.5 Gathering, Hunting, and Fishing

For more than 99% of its existence the genus *Homo*—beginning with *Homo ergaster* from Kenya's Turkana Basin 1.9–1.5 Ma ago (Wood and Collard 1999)—survived as a simple heterotroph by gathering, hunting, and fishing as an omnivorous user of basic tools without any permanent abodes. The genus differed from its predecessors in several key physical features, all with profound consequences for its energetics (Aiello and Wells 2002). They included larger bodies with relatively larger brains; smaller gut, teeth, and jaws; slower growth and delayed maturation; and outstanding capacity for running. Larger bodies required more energy but made thermoregulation easier, increased mobility, and expanded prey size. Larger brains were accommodated without higher energy consumption by a smaller gut (see section 5.1). But endurance running may have been more important for the evolution of human bodies than any other adaptation (Carrier 1984; Bramble and Lieberman 2004). Walking cannot explain such key physical changes as extensive springs in the leg and foot (long Achilles tendons), short toes, enlarged heel bone, hypertrophied gluteus maximus, short forearm, tall and narrow waist, and strong spinal extensor muscles and nuchal ligament that stabilize the trunk and balance the head.

Endurance running would have been a major advantage in both scavenging and hunting. Given their body size and lack of effective weapons, it is most likely that the earliest members of our genus were much better scavengers than hunters (Blumenschine and Cavallo 1992). Many large predators (lions, leopards, saber-toothed cats) left behind partially eaten carcasses. This meat, or at least the nutritious bone marrow, could be reached by endurance runners before it was devoured by vultures and hyenas or by other hominids. However, because these scavenging opportunities were often rare, and always unpredictable, hominids could never be strategic scavengers (as vultures are), merely opportunistic exploiters (Tappen 2001).

Weaponless runners could also chase animals to exhaustion, as some did even after 1900 (Heinrich 2001). Tarahumara and Navajo ran down deer; Paiute and Navajo, pronghorn antelopes; Kalahari Basarwa, duickers, gemsbok, and during the dry season even zebras. These fast animals could not match the human endurance, variable speed, and heat dissipation that opened up a new niche for *Homo* as a diurnal, hot-temperature predator. These humans ran barefoot, and studies show that running unshod not only reduces energy costs by 4% but causes fewer acute ankle and chronic lower leg injuries (Warburton 2001).

The energetics of hunting improved with the introduction of weapons. Throwing spears date back as far as 380,000–400,000 years ago (Thieme 1997), predating the emergence of early modern humans, now dated at 150,000–195,000 years ago (Trinkhaus 2005). Bows and arrows were used from about 25,000 years ago, line fishing from about 12,500 years ago, and nets made from twisted fiber, hair, or thongs from about 8,000 B.C.E. (Coles and Higgs 1969). The earliest date for the control of fire has receded to 790,000 years ago (Goren-Inbar et al. 2004). Cave and pit bones are used to identify a long list of animals eaten around prehistoric fires, and isotope ratios (^{13}C and ^{15}N) can separate plant remains into legumes, nonleguminous C_3 species, and C_4 varieties, and their determinations in human bone collagen can uncover relative amounts of terrestrial and aquatic foods in prehistoric diets (De Niro 1987). This evidence is insufficient for energy supply calculations. Plant remains have been preserved only infrequently, and we have no idea what fraction of killed animals was brought to camp sites or how many people it served. Moreover, extreme spatial variability of the fragmentary evidence permits no quantitative generalizations.

By the time anthropology was ready to study directly the energetics of foraging societies (only well after WW II), most such societies were either extinct or affected by contact with neighboring pastoralists or farmers. Some of the best records of relatively unchanged foraging subsistence are available for the !Kung, ≠Kade, and G/wi groups of the Basarwa (Bushmen, the San), gatherers-hunters of the late 1950s and early 1960s, just before the rapid disappearance of this traditional way of life (Lee 1979; Tanaka 1980; Silberbauer 1981), but they pertain to a marginal environment and tell nothing about the situation in more equable climates and fertile areas where gathering and hunting were abandoned millennia ago in favor of settled cultivation.

Systematic appraisals make clear that forager energetics is a matter of peculiarities and exceptions rather than of close similarities and general rules (Lee and De Vore 1968; Service 1979; Winterhalder and Smith 1981; Kelly 1983; Price and Brown 1985; Kelly 1995; Gowdy 1998; Lee and Daly 1999; Stanford and Bunn 2001; Panter-Brick, Layton, and Rowley-Conwy 2001; Barnard 2004; Frison 2004). Large differences in habitats and diets translated into population densities differing by up to 2 OM. Minimum population densities of groups that depended on mixtures of gathering and hunting activities were on the order of $1/100$ km^2 (tropical Aeta of the Philippines, the Semang of Malaya, the boreal Micmac of eastern Canada). The rates were 1 OM higher in seasonally dry tropical environments (Kalahari Basarwa at $7–10/100$ km^2). Groups heavily dependent on fishing had densities up to 1 OM higher (Pacific Northwest's

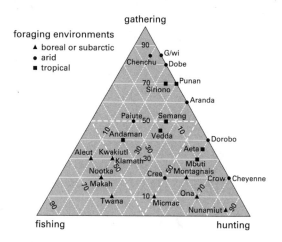

5.9 Approximate contributions of gathering, hunting, and fishing to the diets of some foraging societies that survived into the twentieth century. Plotted from data in Murdock (1967).

Nootka and Kwakiutl at about $60/100$ km^2, Makah at nearly $90/100$ km^2).

Regardless of the prevailing source of food energy (fig. 5.9), large seasonal variations of staple food abundance, as well as its annual fluctuations, resulted in highly irregular utilization patterns. Kelly (1983) suggested a coverage index (total exploited area/total residential mobility distance) to indicate the intensity of land utilization. Predictably, it would be highest for gathering societies and up to 1 OM lower for hunters. Primary and secondary productivity and the shares of accessible edible biomass are the key ecosystemic variables needed to evaluate forager energetics. Forest phytomass is mostly in indigestible lignin and cellulose of tree trunks; fruits and seeds are a very small portion of the total, are commonly inaccessible in high canopies and are often well protected by hard coats, requiring fairly energy-intensive processing before consumption.

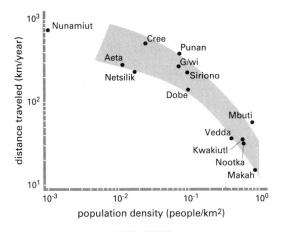

5.10 Annual mobility and population density of some foraging societies that survived into the twentieth century. Plotted from data in Kelly (1983) and Murdock (1967).

Tropical species diversity makes gathering energetically more taxing owing to often considerable distances between the plants of desirable species. In temperate and boreal forests seasonal cessation of primary productivity at and near the ground level and frequently heavy snow cover make for very low energy returns in winter. In contrast, grasslands and woodlands often produce an abundance of easily collectible fruits, seeds, and roots; and concentrated patches of tubers and rhizomes are a relatively important part of semidesert phytomass. Forest gatherers thus had to relocate their camps as many as 40–50 times a year, whereas foragers in grasslands and semideserts moved less than a dozen times a year but over longer distances. Ethnographic evidence confirms the expected inverse link between total annual mobility and population density (fig. 5.10).

Also as expected, average densities of foragers were always lower than those of similarly sized herbivorous mammals, which can digest a much wider range of available phytomass. Calder's (1983) allometric equation indicates five 50-kg individuals/km^2, but densities of large primates (orangutans, gorillas, chimpanzees) prorate in all cases to about two animals/km^2 (Bernstein and Smith 1979). The documented densities of tropical rain forest hunters were below 1 person/km^2, and those in boreal forests mostly 1 OM lower. Most tropical forest zoomass is folivorous, hence arboreal and relatively small and inaccessible, and thus less likely to be the subject of successful hunts than many larger grasslands grazers.

Sillitoe's (2002) detailed hunting energy budgets for the Wola of the Papua New Guinea Highlands are a perfect illustration of this counterintuitive reality as hunters in one of the world's richest ecosystems incur large energy losses in their pursuit of small birds, rats, and possums: on average they recover only 4% of the energy expended on hunting, and the returns (albeit somewhat higher) are very low for other studied New Guinea tribes. Bailey et al. (1989) concluded that there is no unambiguous ethnographic account of foragers who lived in tropical rain forest without some reliance on domesticated plants and animals. In contrast, grasslands and open woodlands offer excellent opportunities for both collecting and hunting. But this advantage has its obverse: large grazers are highly mobile, and hence the groups dependent on hunting them had to cover considerably larger (and increasing poleward) ranges.

Exploiting such resources becomes impractical without frequent camp shifts. As the distances of daily hunts grew, residential change was a must. The exploitation ranges of North America's buffalo hunters, woodlands Indians, and the Inuit extended over many thousands of square kilometers. Moreover, before a rewarding kill, a northern hunter had to invest a great deal of energy in search and pursuit and, in the case of large seasonal migrations, also in careful monitoring of herd movements in order not to miss a unique opportunity for a relatively large meat gain. Dependence on fishing was another key density-determining factor: in the Pacific Northwest, with its massive runs of salmon, fishing allowed the establishment of permanent settlements and sharp reduction of mobility needed for food acquisition.

Energetic imperatives also determined the minimum group sizes for successful hunting. An individual hunter's daily success rate in pursuing small mammals was rarely higher than 15%–30%, so at least three to six hunters were needed to ensure daily meat supply for their families. With an average of six or seven people per family, this translated to minimum group sizes of 18 to 40 people, an estimate well supported by archaeological and ethnographic evidence. The benefits of group hunting were clearly illustrated in Harako's (1981) studies of two Mbuti bands in the Ituri tropical rain forest. Solitary archers shooting monkeys had a daily success rate of a mere 10% and averaged daily just 110–170 g of meat per capita. Spear hunters averaged about 220 g, but net hunting yielded 370 g for every band member. This gain is even more impressive given that spear hunting was highly dependent on the superior experience of the oldest hunters (93% success rate for 40–50-year-olds, 2% for hunters under 20), and hence the yield was vulnerable to incapacity or death. A minimum number of adults was also required to butcher a large mammal and transport pieces or just stripped meat to a camp; this all added up to energy input beyond the capability of one or two families.

In common with chimpanzees, all foragers were omnivorous, eating dwarf willow leaves and the contents of caribou stomachs mixed with seal oil in the Arctic, and termites and roasted and ground ungulate hinds in Africa's savannas. Many gatherers ate parts of many plants. Gwembe Tonga (Zambezi Valley) collected 58 varieties of leaves and stalks, 53 kinds of seeds, nuts, fruits, and berries, and 17 bulbs, roots, or tubers (Scudder 1976). Basarwa foragers exploited 34–126 plant species. But only a few species were usually dominant. Nine out of 85 plants gathered by !Kung accounted for 75% of all edible phytomass, and *mongongo* nuts alone provided more than 50% of plant food energy. Australia's Anbarra collected 29 mollusc species, but 95% of their food energy came from just five bivalves and 60% of the gross weight was from a single species (Meehan 1977). A preference for energy-dense seeds is not surprising. Many groups on all continents collected grass seeds (typically around 15 MJ/kg); South Africa's *si* bean (*Bauhinia petersiana*) provided 17.7 MJ/kg and *mongongo* nuts (*Ricinodendron rautanenii*) as much as 25 MJ/kg, and piñon nuts gathered and stored by Californian foragers contained 26.5 MJ/kg.

The high value of meat in foraging societies is well documented by the hunters' willingness to spend much energy on its acquisition and on its sharing with other members of the group (Kelly 1995; Stanford and Bunn 2001). Mbuti women spent only a few hours a day gathering, but spear hunters pursuing larger game averaged about 11 h away from the camp, completing round-trips of over 20 km (Harako 1981). Some energy expenditures were extraordinarily high. A reliable report has the Kalahari hunters running at an easy trot for up to 30 km without pause and catching up with a wounded animal with an all-out spurt (van der Post and Taylor 1984). Basarwa hunting forays ranged between 13 km and

48 km. Almost invariably, only herbivores were hunted: hunting carnivores was energetically unprofitable given their low numbers and higher risks of injury to the hunters. Where available, larger species were preferred and Silberbauer (1981) noted that the G/wi generally selected the prey with the greatest food reward for the lowest expenditure of energy, and this approach was undoubtedly ubiquitous.

Wild herbivore meat is an excellent source of protein (20% of fresh weight) but a marginal source of fat. Providing only about 6 MJ/kg and less than 10% of lipids, it left people feeling hungry and craving fat. A richer ratio of fat to lean meat was particularly critical for survival in the Arctic (Cachel 1997). Hayden (1981) argued that the high regard for meat is actually a misconception promoted by ethnographers, that the meat itself was valued little but fat was the real prize. Studies of Basarwa groups showed a preference for fat underground mammals (porcupines, antbears) as well as for hippopotamus and eland, the fattiest of all antelopes (van der Post and Taylor 1984). But Wrangham et al. (1999) argued that in general it was the cooking of plant foods that enlarged diets and improved their quality much more than did additional intakes of meat. Cannibalism took place in many preagricultural societies, but the reasons for it were not solely energetic (starvation) and included also ancestor worship and terrorism (Brown and Tuzin 1983).

The best estimates based on a few studies of actual energy expenditure indicate that the PAL for adults in foraging societies were mostly in moderate category (Jenike 2001), but in general the energetic explanations of foraging have clear limits. For !Kung Basarwa abundant, energy-dense *mongongo* nuts provided the best energy return and much of the diet, but /Aise, another Basarwa group with access to the nuts, did not eat them because

they did not taste good to them (Hitchcock and Ebert 1984). Similarly, coastal groups in Southern Australia supported their high densities by fishing, but across the strait the Tasmanians did not eat fish at all. And among the Yanomami, Lizot (1977) found, a group surrounded by a particularly animal-rich forest consumed less than half as much animal food energy and protein as its neighbors did. His explanation: people of the first group were simply lazier and hunted infrequently, preferring to eat less well and to spend days taking hallucinogens.

In this and other instances energy supply may simply have been a function of the attitude toward work. And yet some foraging efforts were rather brief, just 2–5 h/day in many tropical and subtropical environments, and these relatively short workdays were perhaps the most important argument in attempts to portray foragers as the original affluent society, living comfortably in a kind of material plenty, filled with leisure and sleep, never in hurry, never worried (Sahlins 1972). Such theorizing ignored evidence of much less comfortable subsistence patterns as well as the frequency with which seasonal food shortages and famines ravaged the foraging societies (Rowley-Conwy 2001).

Our understanding, based on foraging groups that survived to the twentieth century, may not be representative of foragers during the time of their preagricultural dominance. Still, these studies make it possible to reconstruct the approximate energy returns of individual foraging activities. They are as high as 30–40-fold for gathering some energy-dense roots, 10–20-fold for all gathering, in the same range for hunting large ungulates, and minimal energy gain or even net energy loss for hunting smaller mammals. Much higher energy gains (up to 19-fold) in hunting a large (280-kg) eland rather than a smaller (70–80-kg) gemsbok (sixfold return) explain the preference for killing such large herbivores as bisons and woolly mammoths. Moreover, these animals yielded unusually high amount of lipids, a quality not encompassed by a simple energy ratio. Conversely, hunting certainly contributed to the emergence of bipedalism and running (due to the necessity of covering larger home ranges) as well as to the evolution of strategic thinking needed to search for and kill animals (Foley 2001).

Perhaps the most important outcome of foraging studies is the realization that many gatherer-hunter societies reached levels of complexity, including sedentism, high population densities, large-scale food preservation and long-term storage, social stratification, elaborate rituals, and plant cultivation, that are usually associated only with farming societies. This cultural complexity had its energetic foundations in exploitation of extraordinarily productive environments, substantial storage, and incipient agricultural practices. The image of ephemeral encampments and marginal existence in small groups describes most preagricultural societies, but it does not fit some foragers in the Upper Paleolithic, when the mammoth hunters in the Moravian loess region lived permanently in well-built semisubterranean dwellings, made a variety of tools, fired clay, and sculpted. Zvelebil (1986) argues that delayed acceptance of farming in Europe's northern and eastern forest zones (where farming was available in 5000–4500 B.C.E. but was not adopted until two or three millennia later) was the result of high foraging yields.

Similarly, the well-documented social complexity of the Upper Paleolithic groups of southwestern France was based on the high productivity of the continent's most southerly open tundra or steppe-like vegetation, which supported the largest herbivorous herds in the periglacial Europe (Mellars 1985). The highest population densities in foraging were associated with exploitation of marine resources. Maritime foraging was marked by high biomass and high resource diversity, reliance on migratory species, sedentism, technical complexity, cooperative resource exploitation, high per capita productivity, territoriality, resource competition, and warfare (Yesner 1980). Whereas foraging groups typically comprised 20–50 people, the Pacific Northwest's settlements commonly housed (in well-built wooden structures) several hundred people. The seasonal abundance of salmon and its preservation by smoking constituted the energetic basis of this extraordinary population density and resultant social complexity. Compared to cod (3.2 MJ/kg) or whitefish (6.3 MJ/kg), chinook salmon yields about 9.1 MJ/kg, largely owing to its high fat content (15%).

The effects of the high food energy value of migratory marine species are even better demonstrated in the case of Northwestern Alaskan Inuit. Despite an extremely harsh environment, and low and unpredictable density of large land mammals, these groups were able to secure a food energy surplus in less than four months of near-shore hunting of baleen whales during their migration. Sheehan's (1985) calculations for four precontact settlements with a total of about 2,600 people show that even a minimum estimate of the baleen whale harvest could—together with subsidiary exploitation of walruses, beluga whales, several seal species, and fish—result in food surpluses. Indeed, walrus and bearded seal were mostly taken for raw materials and dog food. The baleen whale's huge mass (even the most commonly landed immature 2-year-olds averaged nearly 12 t), its incomparable food energy density (~36 MJ/kg for blubber, 22 MJ/kg for *mukluk*, skin and blubber), and its easy storage in permafrost cellars offered a roughly 2000-fold net energy return (unmatchable by any other form of hunting) and supported large permanent settlements and impressive social complexity. This adaptation appeared to be self-amplifying: more people in settlements could field more whaling crews, resulting in more sightings and higher chances of hunting success. Eventual limits on population growth were imposed not by food but by the need to hunt the marine and land species for raw materials to make clothing, sinews, bedding, hunting equipment, bags, floats, and covers.

The reliance on seasonal food flows required extensive and often elaborate storage: caching in permafrost; drying and smoking of fish, fish eggs, clams, mussels, seaweed, berries, and various meats; storing seeds and roots; putting seal oil and blubber into large clay jars; preserving food in oil; making intestinal sausages and nut meal cakes and flours (Hayden 1981; Zvelebil 1986). Testart (1982) argued that large-scale, long-term food storage changed foragers' mentality, giving them new attitudes toward time, work, and nature. The need for planning and time budgeting was perhaps its key evolutionary contribution; tool making and maintenance and the preparation of storage items had to be concentrated in slack periods. Once this pattern was mastered, there was no turning back without a sharp reduction of prevailing population densities. Large-scale storage is not only incompatible with mobility but also highly conducive to the accumulation of other foods and goods. Many complex sedentary foragers found it natural to gradually incorporate incipient agricultural practices. The stage was slowly transformed to a fundamentally different way of subsistence and to widespread surplus accumulation. The process was evolutionary and multifocal, its onset spread over several millennia in different parts of the world, but

its outcome was universal. Humans ceased to be simple, opportunistic, omnivorous heterotrophs and became—through crop selection and cultivation, irrigation, and nutrient recycling—increasingly refined manipulators of solar energy flows, overwhelmingly herbivorous producers of a few staple crops, rapid learners of social and technical complexity.

6

TRADITIONAL FOOD PRODUCTION

Humans as Solar Farmers

Rain in the village must be plentiful.
I dream of fragrance with the rice-plants full.
Since Heaven's impartial in its overflow
Of grace, strong reeds and tares will likewise grow.
Men find such growths unwelcome from the harm
They always do those who work a farm.
Hence none of the good villagers can shirk
In seasonable tasks of weeding work,
Piling tares by the river in defense
Of cleaner crops. Grain is life's sustenance.

. . . .

Whatever grows will rise in mad confusion
And toil must guide the crop to its conclusion.
Du Fu (712–770), "Directing Farmers"

Du Fu's verse defines agriculture very well. On the most abstract level it can be seen as a prescient description of negentropic effort: agriculture as periodically strenuous energy investment combating the natural tendency toward weedy disorder and producing orderly harvests. An ecologist might note the emphasis on the separation of crops from other phytomass and on grains as key crops as a perfect description of farming as a manipulated ecosystem inimical to the maintenance of species diversity. An anthropologist might focus on the inevitability of collective participation in seasonally demanding labor alternating with periods of extended rest. Definitions of agriculture abound, and so do explanations of its origins (Reed 1977; Pryor 1983; Cowan and Watson 1992; Bellwood 2004). Arguing about the relative importance of individual drivers is counterproductive; such a complex process has no single cause, only a history with its many interdependent interactions (Rindos 1984).

Consequently, it is important to reject the idea of agriculture as a result of an invention (G. F. Carter 1977) as well as the notion of agricultural revolution (Childe 1951). The well-documented evolutionary nature of the process makes these ideas untenable. Nor were new tools, cereal processing, sedentism, or storage habits the necessary prerequisites. Simple tools (wooden sticks, stone-chip cutting blades) that were already in use for millennia sufficed. The earliest confirmation of pounding and grinding of some wild cereals and using the flour in baking comes from an Upper Paleolithic site (19,500 B.C.E.) in Israel, at least 12,000 years before the domestication of cereals (Piperno et al. 2004). Neolithic Çatalhöyük and Aşıklı on the Anatolian Plateau were settlements with complex arrangements of buildings housing thousands of people that subsisted mostly on hunting and gathering (Balter 1998), whereas the Tehuacán Valley had no permanent settlements but thousands of years of crop cultivation (Bray 1977). There are numerous examples of food storage among foragers and its absence among gardeners.

Many foraging societies coexisted side by side with agriculturalists; farming had no irresistible, automatic universal appeal. Why it arose—independently in at least seven locations on three continents, between 10,000 and 5,000 years ago—is perhaps the most challenging evolutionary, archaeological, and anthropological puzzle that may never have a definite solution (Megaw 1977; Mannion 1999; Armelagos and Harper 2005). Environmental explanations (well-documented climate changes) have a long tradition (Childe 1951; Byrne 1987). The latest contribution in this category concludes (despite the accumulated knowledge) that Paleolithic agriculture was impossible (climate too dry, CO_2 levels too low) whereas Neolithic agriculture was mandatory (Richerson, Boyd, and Bettinger 2001). In contrast, evolutionary explanations see diminishing returns in gathering and hunting brought by slow growth of foraging populations, gradual extension of incipient cultivation techniques (present in most foraging societies), and a slow adoption of a greater variety of cultivation practices (Cohen 1977).

But cultural factors should not be neglected. Crop cultivation fosters association, a desirable goal for our sociable species. At the same time, farming promotes individual ownership and accumulation of material possessions; it makes it easier to have larger families; and it facilitates warfare. Orme (1977) has gone too far when concluding that food production as an end in itself may have been unimportant, but there is no doubt that social co-factors are important, some having little or nothing to do with food. After all, net returns of many early field harvests were in no way superior to those of some foraging practices, and hence the adoption of farming cannot be seen as a quest for maximized energy returns. Quality considerations (ignored by common energy denominators) were important: domestication of cereal and oil plants can be readily explained because of their high protein or lipid content rather than because of any generalized quest to minimize energy expenditures.

Whereas the transition from foraging to farming was initiated and sustained by a complex of energetic, nutritional, and social impulses, the further evolution of agriculture can be seen as a matter of clear energy imperatives. Boserup (1965; 1976) elaborated this link with great clarity when looking at the evolution of peasant societies. As a particular food production system reaches its limits, the affected population can migrate, stay, and stabilize; stay and decline; or adopt a more productive subsistence. The last option may not initially be any more appealing or probable than the others. When it comes, the shift requires higher energy inputs, so even with higher food production density the net return ratio may decline, but the higher edible energy flux will support a larger population.

There will be a natural tendency to postpone the switch as long as a less intensive arrangement will do. This reality is illustrated by the lengthy coexistence of foraging and cultivation. For example, at Tell Abu Hureyra in northern Syria, hunting remained a critical source of food for 1,000 years after the beginning of plant domestication (Legge and Rowley-Conwy 1987). Another example is the contrast between intensive farming in plains and valleys and shifting cultivation in nearby mountains, a contrast commonly seen in Southeast Asia, even in the second half of the twentieth century. Intensification advances in stages, from long forest fallow (just one or two crops followed by a regeneration of 15–25 years), to bush fallow (with four to six crop years and a similar fallow), to short fallow (a crop or two followed by a year off), to regular annual cropping (fallow reduced to fall and winter) and finally, to multicropping, often irrigated (two or three grain or oil crops, or five to six vegetable crops, planted in rapid succession).

Each of these steps recovers more of the site's potential photosynthesis and supports more people per hectare of land (fig. 6.1), but it demands higher energy inputs, first for forest clearing, planting, and cultivation and eventually for repeated cultivation, terracing, and irrigation. These activities also require further energy investment for making tools and implements. And because a large part of the energy inputs is in the form of long-term investments (fields cleared of stones or terraced, irrigation systems, roads), and intensive cropping needs planning, storage, and trading, agricultural intensification has been a key ingredient of a civilization's complexification, promoting innovation, specialization, interdependence, and exchange of goods and techniques. And the process led inevitably to reliance on sources of energy other than human muscles. Plowing was either enor-

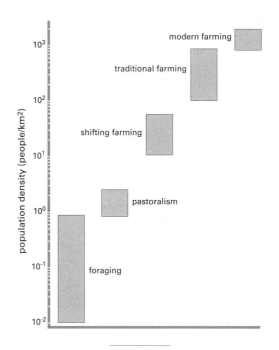

6.1   Ranges of population densities supportable by intensifying modes of food provision.

mously taxing or outright impossible without draft animals; manual threshing and milling of grains was so labor-intensive that inanimate (water and wind) power was necessary to process harvests for cities; long-distance distribution of grain also relied largely on animal power and wind; and iron for tools and implements was smelted with charcoal.

6.1   Extensive Practices

Low population densities and abundant land availability were the two key factors that favored extensive modes of food production. Nomadic pastoralism and shifting agriculture (slash-and-burn or fell-and-slash-and-burn farming) are two very dissimilar practices sharing the intermittent and extensive use of land, with orderly patterns (herds following seasonal rains, or planting the same sequences of crops in fixed cycles) imprinted on changing locations and over large areas. Shifting agriculture was a part of the evolutionary continuum from foraging to permanent cropping in forest and woodland environments. Pastoralism, a form of resource conservation, was an adaptation to exploit arid regions or a response to desiccation using domesticated animals to convert scarce, and seasonally nearly absent, grassy phytomass into food in the form of milk, blood, and meat.

Not surprisingly, the sustainable population densities of many nomadic pastoralists were not any higher than those of less widely roaming foragers. In suitable environments even the moderately productive forms of shifting cropping with long regeneration cycles could support populations 1 OM larger than the settled foraging communities, but commonly 1 OM smaller than that sustained by permanent field farming. For millennia these ways of life dominated huge areas of all continents (except Australia). Not infrequently, especially in Africa, they blended into mixtures of seminomadic agropastoralism, sometimes with a significant bit of foraging. Their rapid twentieth-century retreat is attributable to rising population densities, which lead, among pastoralists, to unsustainably large herds (helped by better water supply and control of animal disease vectors) and to overgrazing; and, among shifting farmers, to drastically shortened soil regeneration cycles, greater soil erosion, nutrient decline, and eventual loss of many sites.

Animal husbandry is a form of prey conservation, a strategy of deferred harvests whose opportunity costs are greater for larger animals (Alvard and Kuznar 2001). However, the smaller species (sheep, goats) were domesticated first because of their higher growth rates. There

have been many social studies of pastoral societies in transition (Helland 1980; Galaty and Salzman 1981; Khazanov 2001) as well as descriptions of the traditional pastoralist way of life (Irons and Dyson-Hudson 1972; Monod 1975; Salzman 1981; Khazanov 1984; Rigby 1985; Evangelou 1984; Salzman 2004), but accounts of pastoral energetics are rare, and only human energy inputs are easy to approximate. Traditional pastoralists did nothing to improve the pastures, and their labor was confined to herding the animals, guarding them against predators, watering them, helping with difficult deliveries, milking them regularly, butchering them infrequently, and building temporary enclosures.

These tasks usually required only light to moderate exertion for 2–6 h a day, and many of them were done by children. Evangelou (1984) showed that 92% of all herding and 42% of *boma* (enclosure) livestock work among Kenyan Maasai were done by children. A single East African herder managed up to 100 camels or 200 cattle and 400 sheep and goats (Helland 1980). Khazanov (1984) lists similarly high rates in Asia: two mounted shepherds for 2,000 sheep in Mongolia, an adult shepherd and a boy for 400–800 Turkmen cattle. Herding, even with additional labor (digging wells), was not labor-intensive, and this fact was one of the key reasons for the reluctance to convert to farming.

Given the enormous variety of natural settings, a numerical analysis of the relation of livestock to grazing has little significance unless one specifies a host of local factors determining the grazing potential (Monod 1975). Grazing intensity also depends on the kinds and mixtures of animals and on their metabolism, water requirements, and vulnerability to parasites (*Glossina*, tse-tse fly, has always excluded many areas from grazing). Animal resilience is a matter of large interspecific differences. In cool season with good pastures camels can go up to 90 days without water, sheep 30, cows only 3, and calves just 1. Similarly, camels can cover up to 80 km/day to reach new grazing ground, cows no more than 20, and calves 10. In East Africa, traditional diet (eating only old or diseased animals) required 7 lactating cows (or 4 camels) for a family of 6.5 adult equivalents, or 14–15 cows in total. With milch cows being about half of a typical herd, 30–40 animals per family, 5–6 heads of cattle, 2.5–3 camels, or 25–30 goats or sheep were a minimum standard per person.

This represents 2.5–3 livestock units (equivalent to one camel, two or three cattle, or ten heads of small stock). African experience shows that 6 ha of a good grassland may be a typical average to support one adult head of cattle, with a common range of 4–8 ha and extremes of less than 2 ha (where green grass is available all year) and 16 ha (Allan 1965). Some 36 ha of grazing land would be needed to support six cows per capita, prorating to an average population density of $2.8/km^2$. Actual densities (subject to long-term environmental fluctuations) have been considerably smaller, in East Africa largely between 0.8–2.2 people/km^2 and 0.03–0.14 heads/ha (Helland 1980; Evangelou 1984; Coughenour et al. 1985). The larger number of cattle among traditional Maasai (13–16 heads per capita) is explained by the status-seeking accumulation of animals or (in energetic terms) by the minimum requirements for the sustainable tapping of blood. Blood (harvested by piercing a tightened jugular vein with an arrow and letting out 2–4 L at intervals of five to six weeks) was nearly as important as milk. Two heads of cattle must be bled to feed a family of five or six people, a herd of 80 is needed during periods of low (drought-induced) lactation, or 13–16 animals per capita.

Shifting cultivation alternates variable but always short cropping periods (one to three years) with no less variable but fairly long periods of fallow (a decade or more). The practice was once ubiquitous on every continent except Australia, and it remained of major importance for tens of millions of families in Africa, Latin America, and Southeast Asia even in the second half of the twentieth century (Allan 1965; Spencer 1966; Watters 1971; Grigg 1974; Okigbo 1984). The cycle starts with clearing of natural vegetation, most often forest, climax, or secondary growth. The most labor-intensive tasks are felling of large trees, trimming and pollarding of smaller trees, and slashing of younger growth. After drying, the cut phytomass is burned. Fire clears away the litter, prepares the surface for planting, and reduces the invasion of weeds and pests. Most of the nitrogen is lost, but minerals are recycled.

A variety of edible, fiber, and medicinal species, dominated by grains (rice, corn, millet), roots (sweet potatoes, cassava, yams), and legumes (beans, peanuts) were grown in gardenlike arrangements with high degrees of interplanting and staggered harvesting. Just two to five staples provided most of the food energy, but there were rarely fewer than 12 crops, and often 30 to 50 species crowded a small area. Gardens were often fenced to keep domestic or wild animals away, and much time was spent in preventing predation and keeping the herbaceous and ligneous competitors in check by repeated weeding. Besides annual or semiannual harvests of major grain crops, there was continual digging of roots and picking of seeds, leaves, and stems.

The energetics of shifting cultures is not easy to study. Yields of small, scattered, continually harvested plots are seasonally variable and differ enormously depending on soils, fallow periods, and the quality of clearing, weeding, and protection; work expenditures vary greatly depending on the cleared vegetation, terrain, and crops. Rappaport's (1968) quantifications of Tsembaga horticulture in the highlands of New Guinea were the first solid account of the energetics of shifting farming. Growing nearly 40 species in their intercropped gardens, the Tsembaga harvested annually 23.7–26.9 GJ/ha of edible phytomass while expending 1.48–1.63 GJ of energy above basal metabolism, a 16-fold return on their efforts.

Not long after Rappaport's fieldwork a British team studied coastal (Kaul) and highland (Lufa) New Guinea tribes. It found that in spite of their low body weights, apparently high metabolic efficiencies (see section 5.2), and fairly sedentary lifestyle (walking and gardening occupied ~15% of their time), the tribes' food energy returns were very low. With labor investments of 3.20 MJ/day (Kaul) and 5.73 MJ/day (Lufa) and food intakes at about 33 MJ, the returns were just ten- and sixfold (Norgan, Ferro-Luzzi, and Durnin 1974). The corn harvest of the Guatemalan Kekchi Maya (16.4 GJ) yielded a minimum 30-fold return (W. E. Carter 1969). The Colombian Yukpa realized a 20-fold gain by growing corn, beans, manioc, and bananas (Ruddle 1974). The Hanunoo cultivated 68 food species, of which 39 were also grown for medicinal, ritual, cosmetic, or manufacturing uses in complex, staggered plantings, precluding a reliable quantification of total edible yield (Conklin 1957), but their rice cultivation had at least a 16-fold gain.

Other available figures support the conclusion that total labor inputs vary between 600 h/ha and 3200 h/ha and that the energy returns of shifting cultivation are 11-fold to 15-fold for small grains (whether Southeast Asian hill and swamp rices or African millets), 20-fold to 40-fold for most root crops, bananas, and good corn

yields, and maximally close to 70-fold for some roots and legumes in excellent locations. One person requires as much as 10 ha and as little as 2 ha of land in fallow and under the crops, with the actually cultivated area ranging from 0.1 ha to 1 ha. Compared to pastoralist population densities of 1–2/km², shifting cultivators' densities of 30–40/km² are 1 OM higher, and their energy returns and security of food supply are almost invariably superior.

Shifting agricultures shared a number of energetic commonalities. Clearings were chosen close to the settlements to minimize walking distance, but more remote locations saved labor on building fences against roaming domestic animals. Clearing the secondary growth, rather than a climax forest, was the preferred choice; only 1 out of 381 Tsembaga gardens was in virgin forest. Men did the heaviest tasks (tree felling, pollarding, fencing), and women carried a disproportionate share of repetitive chores (weeding and harvesting). But no form of food production is governed by maximization of energy returns. Nutritional imperatives are clear: legumes give superior energy return compared to cereals but are less palatable, and that is why wheat or rice are the staples. Rappaport (1968) showed that the energy to grow pig feed was often greater than the food energy of pork, but how else could starchy roots be turned into food with 27% fat and 11% protein? Simplistic comparisons of energy returns find little or no value in such practices, but the widespread human desire for fatty and meaty meals justifies them.

6.2 Permanent Cropping

All traditional Old World agricultures shared the same energetic foundations. They were powered by photosynthetic conversion of solar radiation that produced food

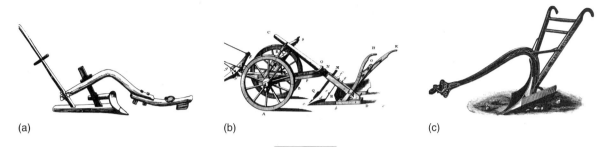

6.2   Evolution of curved moldboard plows. (*a*) Traditional
Chinese plow. (*b*) Eighteenth-century French plow. (*c*) American steel beam plow of the mid-nineteenth century. From Smil
(1994).

for people, feed for animals, organic wastes that were recycled to replenish soil fertility, and fuels that energized the smelting of metals needed to make simple farm tools. Consequently, traditional farming was, at least in theory, fully renewable because it relied on virtually immediate conversions of solar energy flows. But this renewability was no guarantee of sustainability. Conversion of forests to farmland, and the use of wood for fuel and charcoal, steadily depleted accumulated phytomass energy, and poor agronomic practices lowered soil fertility and often caused excessive erosion or desertification. Such environmental degradations lowered yields or even caused the abandonment of cultivation. These agricultures also shared the fundamental agronomic sequence of plowing, seeding, harvesting, and crop processing that was dictated by the nearly universal dominance of grains.

All the Old World's high cultures were creations of grain surplus, and regular plowing was their energetic hallmark (fig. 6.2). Plowing's antiquity is attested by the fact that both the Sumerian cuneiform characters and the Egyptian glyphs have pictograms for plows (Jensen 1969). Plowing opened the soil for planting of small

seeds on scales vastly surpassing those of hoe-dependent farming. The first primitive scratch plows (ards), commonly used after 4000 B.C.E. in Mesopotamia, were just pointed wooden sticks with a handle. Later most of them were tipped with metal, but for centuries they remained symmetrical (draft line in a vertical plane with the beam and share point) and light, able only to open a shallow furrow for seeds and leave the cut weeds on the surface. These plows were the mainstay of both Greek and Roman farming, and they were found in parts of the Middle East, Africa, and Asia well into the twentieth century.

Addition of a moldboard was a fundamental improvement. A moldboard guides the plowed-up soil to one side, turns it partly or totally over, buries the cut weeds, and cleans the furrow bottom. A moldboard also makes it possible to till a field in one operation rather than by the cross-plowing required with ards. The moldboard's draft line is displaced slightly toward the side of the turned-up soil, making the plow asymmetrical. The first moldboards were just straight pieces of wood, but before the first century B.C.E., the Han Chinese introduced curved metal plates joined to the plowshare. During the second half of the eighteenth century, Western plows still retained their heavy wooden wheels but carried well-curved iron moldboards. Steel replaced cast iron beginning in the 1830s.

In most soils, plowing leaves behind relatively large clods that must be broken up before seeding. Hoeing is too slow and too laborious, and hence various harrows were used by all old plow cultures. After plowing, harrowing, and leveling, the ground was ready to be seeded. Although seed drills were used in Mesopotamia as early as 1300 B.C.E., and sowing plows were used by the Han Chinese, broadcast seeding by hand (wasteful and resulting in uneven germination) remained common in Europe until the nineteenth century. Simple drills, dropping seeds through a tube from a bin attached to a plow, started to spread, first in northern Italy during the late sixteenth century, and many innovations turned them before too long into complex seeding machines. Sickles were the first metal harvesting tools to replace short sharp stonecutters. Serrated (the oldest designs) or with smooth edges, and with semicircular, straight, or slightly curved blades, they have been used in countless variations. More efficient scythes, equipped with cradles for grain reaping, were used for harvesting larger areas.

But sickle harvesting minimized grain losses, and it was retained throughout Asia in cutting easily shattered rice. Mechanical reapers came only after 1800. Harvests were brought home as sheaves carried on heads, in panniers hung on shoulder beams or sides of animals, and in wheelbarrows, carts, and wagons. A great deal of energy was spent on crop processing. Grain spread on a threshing floor was beaten with sticks or flails, sheaves were hit against grates or pulled across special combs. Animals treaded the spread grain or pulled heavy sleds or rollers.

Winnowing, the separation of chaff and dirt from grain, was done manually with baskets and sieves, later with crank-turned fans. Tedious manual labor was also needed for grain milling before animals, water, and windmills mechanized the task. Oil was extracted from seeds by manual or animal-operated presses, as was the juice from cane.

All traditional agricultures grew a variety of grain, oil, fiber, and feed crops, but the standard agronomic sequence was performed most often when cultivating cereals. The plowless Mesoamerican societies, which relied on corn, shared this practice, and even the Incas were only a partial exception: at high elevations they planted potatoes, but at lower altitudes they also cultivated corn, and on the high-lying Andean Altiplano, quinoa grain. There were many cereals of local or regional importance, but the main genera gradually diffused worldwide from their areas of origin. Wheat spread from the Near East; DNA testing pinpoints the domestication of einkorn wheat to the Karacadağ region in Turkey (Heun et al. 1997). Cereal gathering in the region dates to 19,000 years ago, and the earliest indehiscent (nonshattering) domestic wheat to about 9,250 years ago (Tanno and Willcox 2006). Rice came from Southeast Asia, corn from Mesoamerica, and millets from China (Cowan and Watson 1992). Grains became dominant because of a combination of evolutionary adjustments and energetic imperatives.

Foraging societies depended on a wide variety of plants, and either tubers or seeds provided most of their food energy. But the water content of fresh tubers is too high for long-term storage in the absence of effective temperature and humidity controls. The Incas solved the problem by making dehydrated chuño (produced by trampling and alternate freezing and drying), which was storable for years. Tubers are also a poor nutritional choice because of their low protein content and cannot be the only staple. Leguminous grains have commonly twice as much protein as cereals, but their yields are much lower. The dominance of cereals is thus an energetic as well as a nutritional imperative: they combine fairly high yields, good nutritional value (high in filling carbohydrates, moderately rich in proteins), relatively high energy density at maturity (roughly five times that of tubers), and low moisture content suitable for long-term storage.

The dominance of a particular species is largely a matter of environmental limits and taste preferences (all cereals have remarkably similar energy content, about 15 MJ/kg). Wheat spread to all continents because it does well in semideserts as well as in rainy temperate zones, at elevations ranging from sea level to 3 km, and in well-drained soils (Briggle 1980). In contrast, rice, originally a semiaquatic plant of tropical lowlands, grows in fields flooded with water until just before harvest. Its cultivation has also spread far beyond the original South Asian core, but the best yields have always been in rainy tropical and subtropical regions. Corn yields are best in regions with warm and rainy growing seasons, but it, too, prefers well-drained soils. Farmers had to set aside a portion of seed for the next year's planting; with low yields it was often as much as one-third or even one-half of medieval crops.

Manual grain harvesting (in double-cropping areas followed closely by the planting of a new crop) was the most time-consuming task in traditional farming. In South China, 94%–98% of all available labor had to be engaged between March and September (Buck 1937). In parts of India with a very short rainy season, the two peak summer months required more than 110% or even 120% of actually available labor, and a similar situation existed in other parts of monsoonal Asia (Clark and Haswell 1970). This need could be met only if all members of a farming family worked arduously long hours or by relying on migratory labor. These peak labor demands were among the most important energetic bottlenecks of traditional farming.

6.3 Muscles, Implements, Machines

People can pull primitive, surface-scratching ards, but pulling moldboard plows to open up new farmland or to deep-till heavy clay soils of temperate latitudes could be only an *in extremis* possibility (F. Bray 1984). Such plowing has always required traction and endurance much superior to those routinely deliverable even by a gang of strong men. Thousands of years of draft animal breeding resulted in a profusion of physiques and performance capacities (Rouse 1970; Cockrill 1974; Clutton-Brock 1992; Budiansky 1997). Indian bullocks weigh usually less than 300 kg, Italian Romagnola or Chianina draft cattle at least 650–700 kg. North China's Sanhe horses average only 350 kg, the heaviest European breeds (Percherons, Clydesdales) about 1 t (fig. 6.3). And water buffaloes range from 250 kg to 700 kg. With sustainable pulls at about 15% of body weight for horses and 10% for other species, typical drafts range between 20 kg and 80 kg.

Figure 6.3 superimposes the performance fields of common working species on power isolines, showing most horses below 1 hp and donkeys often no better than humans. Brief exertions are substantially higher. Maximum 2-h pulls during German drawbar tests were 260–290 kg for heavy horses and 170–190 kg for small

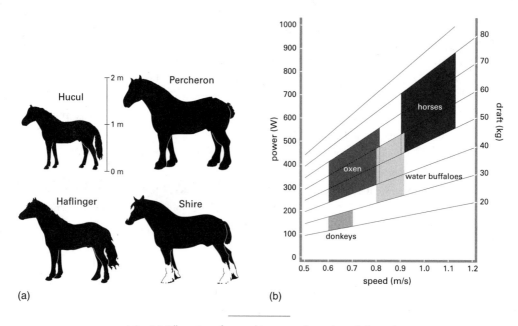

6.3 (a) Silhouettes of several important horse breeds (based on Silver 1976). (b) Useful power of the four common species of draft animals.

horses, 160–170 kg for mountain cows, and 140–150 kg for plains cattle (Hopfen 1969). Mechanical considerations favor smaller animals; with the identical type of harness their line of pull is lower, and the more parallel this line is with the direction of traction, the greater the efficiency of work. The lower pull line also lowers the uplift on drawn implements, resulting in much less strain on a plowman. The actual draft required for field work varies with the task and soil type. Deep plowing with a single share needs drafts between 120 kg and 170 kg; shallow plowing, heavy harrowing, and grass mowing require 80–120 kg; and cereal harvesting with a mechanical reaper and binder demands about 200 kg. This means that a pair of average horses can handle all these tasks but a pair of weak oxen would have difficulties with some of them.

Well-fed horses have also greater endurance than oxen and provide much higher power during maximum exertions. Brief pulls up to 35% of body weight are possible, equivalent to working rates of 2.2 kW or 3 hp, and during pulling trials horses developed up to 14.9 hp (11.1 kW) for a few seconds (Collins and Caine 1926). Horses, unlike cattle or humans, are also unique in not requiring any additional energy for standing (typically 10% above BMR). Their unusually powerful suspensory and check ligaments mean that they can rest (even sleep) in harness without any energy markup. Consequently, the traditional cropping of cereal fields in temperate climates was always done best with horses, and proper harnessing was a decisive factor in utilizing most efficiently their large draft capabilities (des Noëttes 1931; Haudricourt and Delamarre 1955; Needham 1965; Spruytte 1983). Mules, hybrids of male donkeys and mares, were favored in the southern United States because they could tolerate poor feeding and grooming, even neglect and abuse, better than horses (Kauffman 1993).

Throat-and-girth harness, documented in all ancient horse-using cultures, was not suitable for heavy draft. The breastband harness, a Chinese invention dating back at least to the early Han dynasty, increased the efficiency of draft, but the point of traction was too far back from the most powerful shoulder and breast muscles. The design spread across Euroasia and reached Italy as early as the fifth century, and northern Europe some 300 years later. The origins of the collar harness are also in China, perhaps in the first century B.C.E., as a soft support for the hard yoke, later transformed into a single component. The design reached Europe by the ninth century and was universally adopted by the end of the twelfth century. Usually made in one piece, with an oval wooden (later also metal) frame, softly lined to fit a horse's shoulders (often with a collar pad underneath) and with attachments just above the animal's shoulder blades to connect the draft traces, the collar harness offers the most comfort and the most efficient way to pull heavy loads (fig. 6.4).

Working bovines on all continents have been harnessed by neck or head yokes. The double neck yoke was the most widespread type used in Africa, the Middle East, and the Indian subcontinent. A beam, often shaped to fit the neck, is held in place by wooden sticks, chains, or ropes. Its throat fastenings tend to choke the animals, its point of traction is high, the animals must be of the

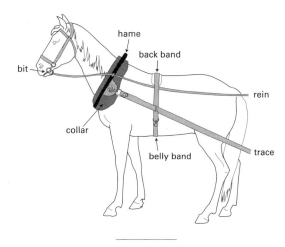

6.4 Major components of the modern horse collar harness. From Smil (1994).

same size, and a pair of animals must be used where one would otherwise suffice. The single neck yoke was dominant in East Asia as well as in Central Europe. The ancient Mesopotamian harness became common in Spain and Latin America; this double head yoke was fixed either at the front or the back of the head and was best suited for strong, short-necked animals. A more comfortable single head yoke was used in parts of Europe.

No draft animal offers a universal fit. Horses, powerful, fast, smart, and easy to handle, were often too light, underperformed in the tropics, and were expensive to harness. Oxen, slow and difficult to train, had the compensating advantages of stolidity, simple harnessing, and easy feeding. A working animal that best fits its environment is the water buffalo (Cockrill 1974). Male castrates used for work, as heavy as good-size horses, look clumsy. In reality, they are nimble and move easily on the steep, narrow earth bunds that divide rice fields as well as in the slippery, deep mud of the fields. Their large hoofs and flexible pastern and fetlock joints are of great help. They browse readily on hedgerow and oddland grasses, and they can even graze on aquatic plants while completely submerged. During slack periods, grasses and rice straw are enough to keep them fit. They mature fast and are superior converters of feed (43 MJ/kg of gain compared to cattle's 78 MJ/kg). They are docile (children are often in charge), are easily trained (a week may be enough), and can work for at least 10–15 years, although 25 years is not uncommon.

The working hours of draft animals were dictated by the pulse of seasonal cropping, and the annual totals were 130–140 days for buffalo in China's double-cropped fields but only 60–80 days for horses in single-cropped agricultures of North China or Central Europe (where the animals were used extensively for transport). Workdays lasted from 4–5 h to 8–12 h. In plowing, invariably the most difficult task, the daily performance ranged from as little as 0.15 ha for a single buffalo in wet fields to 0.5–0.8 ha for a good pair of horses during a long day of stubble plowing or grassland breaking. In the late-nineteenth-century United States gang plows pulled by a dozen horses could finish 1 ha in just 2.5 hours. In intensive traditional farming, animals usually worked no more than 1100–1400 h annually. Annual useful work, with average ratings about 500 W for oxen and buffalo and 700 W for horses, would be equivalent to 2–3.5 GJ for every healthy animal, or 10–20 MJ for every day of work.

The energetic value of working animals in comparison with manual labor is clear. Compared to maximum sustained human exertions at 50–100 W (mean 75 W), draft animals commonly used in fieldwork can deliver 400–800 W (average ∼600 W), typically an eightfold and usually not less than a sixfold difference. Moreover,

during numerous critical periods marking the course of every cropping year, when speed is essential for timely planting, harvesting, or storage, a well-fed animal can work long hours for a few days and accomplish close to 30 MJ/day of useful work, more than 13 times what a good human laborer could do. The question is, How much of an energy burden were the animals for a traditional farmer?

According to NAS (1978), the maintenance requirements of a mature 500-kg horse are about 70 MJ/day of digestible energy. Depending on the shares of highly digestible concentrates (corn, oats) and less digestible roughages (hays, straws), this may represent a very broad range of required gross feed energies, but values between 80 MJ/day and 100 MJ/day (90 MJ/day average) were most common. For actual working requirements we can rely on Brody's (1945) impeccable metabolic measurements: a 500-kg Percheron working at a rate of about 500 W metabolized about 10 MJ/h; a 700-kg horse delivering 750 W metabolized 14 MJ/h. With 8 h of work and 16 h of rest (at 3.75 MJ/h), this translates to 140–170 MJ/day. Half a century later Perez et al. (1996) confirmed this result: the daily energy expenditure of Chilean draft horses used for plowing was 2.24 times their maintenance requirements.

These values are confirmed by traditional feeding recommendations. Around 1900, U.S. farmers were advised to feed working horses 4.5 kg/day of oats and 4.5 kg/day of hay (Bailey 1908). During the 1910s an average horse was fed daily 1 kg of oats, 2 kg of corn and 4.5 kg of hay (USDA 1959). The Chinese norm issued in 1955 prescribed 7 kg/day of roughage (straw or grass) and 2.75 kg/day of unhusked grain (O. L. Dawson 1970). These amounts translate, respectively, to about 150, 120, and 125 MJ/day. The last two values are long-term means that agree closely with the working-day average: feeding 150 MJ during 140–160 working days and 90 MJ during the rest of the year gives a daily mean of 110–120 MJ. Calculating annual requirements is more complicated; usually only about two-thirds of animals worked, and there was a mixture of draft species, most of them of lesser working capacities and needs.

Rather than preparing hypothetical accounts of possible requirements, I offer two detailed calculations for the two extremes of traditional farming: the United States in 1910 (dependent on horse and mule farming energized by lavishly feeding grain and good hays), and China in the early 1950s (the world's largest traditional agriculture powered by more than 50 million oxen, horses, water buffalo, and donkeys).

In 1910 there were 24.2 million horses and mules on U.S. farms and only 1,000 small tractors; eight years later, horse numbers peaked at 26.7 million but there were already 85,000 tractors and 89,000 trucks. I assume two-thirds of all animals working and eating a standard ration of 110 MJ/day; for the other six months and for other animals, I assume maintenance feeding. The result: an annual requirement of about 50 Mt of feed, equivalent (with average yields of 1.5 t/ha) to some 35 Mha. USDA (1959) arrived at 29.1 Mha, or 22% of the nation's harvested area in that year, a very close agreement in approximations of this kind. Thus 20%–25% of farmland was needed to feed the country's working animals (the addition of horses and mules in cities and industries increased the area by some 15%).

The Chinese could not afford to devote 20% of their land to feeding their draft animals. In the early 1950s, before they started making small-sized and walking tractors, the average weight of their draft animals was only about 400 kg per adult, maintenance requirements were about 60 MJ/day of digestible energy, or (given a high shares of roughage) at least 80 MJ of gross feed energy. Typical workday needs were about 125 MJ. With 140 days worked on average by two-thirds of the animals and concentrate feeds accounting for just 25% for working animals and 10% for other animals (the rest being roughage from grazing, hay, and other crop residues), the country needed only about 20 Mt of grain feeds. That would have equaled a bit more than 10% of all unprocessed cereal harvests, but because the Chinese fed their animals a large portion of their grain-milling residues (for example, between 50% and 75% of wheat bran) and oil cakes whose total output at that time amounted to over 20 Mt/year, no more than about 8 Mt of unmilled grain was needed for all draft animals, claiming just about 7% of all available farmland.

A U.S. horse thus needed at least 1.2 ha, whereas a Chinese draft animal claimed only 0.13 ha, nearly a 1-OM difference, resulting from a combination of smaller sizes, less work, and poorer feeding in China. The gap was primarily a matter of farmland availability. In 1910, the United States had about 1.5 ha per capita and China in the early 1950s merely 0.16 ha per capita. India's draft animals were even less demanding, but M. Harris (1966) went too far when he concluded that India's people and cattle did not compete for crops. Fodder crops accounted traditionally for about 5% of India's farmland, with the ratio as high as 10%–20% in the north and northwest (Heston 1971). If only half of this had been fed to working bullocks, the nationwide average of farmland claimed by each animal would have been at least 0.06 ha in the 1970s, roughly half of China's rate.

With traditional dryland grain harvests not surpassing 1.3 t/ha, these land demands translate to preempting

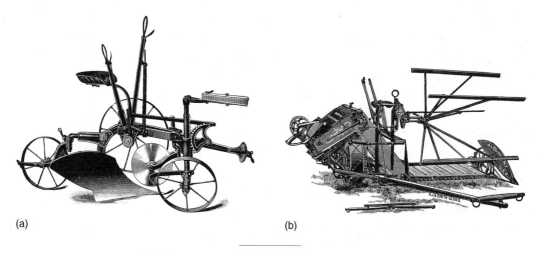

(a)   (b)

6.5   Two inventions that made large-scale grain cropping
possible. (*a*) Steel riding plow. (*b*) Harvester. Reproduced from
Ardrey (1894).

production of between 65 kg (India) and 1560 kg (U.S.) of grain, or roughly 1–25 GJ. In edible terms, this equals (with at least 10% milling rate) 0.9–22.5 GJ of food energy. The maximum was enough to support about six people on typical preindustrial grain-dominated diets, but the large, well-fed horses requiring so much feed were working at a rate at least tenfold higher (750–850 W) than an average human. Even when considering the necessity to feed the nonworking horses (about one-third of the total stock), the difference would scale to more than sixfold. Even a horse's claiming land capable of producing food for six adults made energetic sense. Moreover, heavy horses were able to provide traction for tasks that were both energetically and logistically beyond the practical means of human labor, tasks ranging from deep plowing of clay soil to pulling wheat combines, breaking up large expanses of natural grass lands, and performing critical tasks in a timely and efficient manner.

Animals harnessed to better implements and to new harvesting machines made it possible to cultivate grain fields on unprecedented scales, at the same time sharply reducing the demand for human labor (fig. 6.5). The nineteenth-century achievements of the United States are certainly the best example of this combined effect (Ardrey 1894; Rogin 1931). Charles Neubold's cast iron plow of 1797 was improved by Jethro Wood's interchangeable moldboard, patented in 1819, and superseded by steel walking plows first made by John Lane in 1833. The production was commercialized by John Deere in the 1840s and improved again by Lane with the introduction of a three-layer soft-center steel plow in 1868. Two- and three-wheel riding plows followed after 1864, and gang plows, with up to ten blades, drawn by 12 horses, were used by the late 1800s.

The first mechanical grain reapers were patented in England between 1799 and 1822, and two U.S. inventors built on this basis to develop practical mass-produced machines: Cyrus McCormick in 1831 and Obed Hussey in 1834. The first harvester was patented in 1858, the twine knotter was introduced by John Appleby in 1878, and it perfected the first fully mechanical harvester, a machine that was adopted with great rapidity, making possible large expansion of wheat-growing area before the end of the century. The peak of horse-drawn harvesting was reached when a combined header and thresher was marketed by the Stockton Works during the 1880s. Housers, the company's standard combines, cut two-thirds of California's wheat by 1900; the largest ones needed up to 40 horses and could harvest a hectare of wheat in less than 40 minutes. Such machines took animal-powered fieldwork to its practical limit: harnessing and guiding 16–40 horses was a logistic challenge. The time was ripe for a much more concentrated source of tractive power.

Horse power brought enormous time savings. By 1900 the most productive combination of horses, implements, and machines reduced required human labor by 95% compared with oxen-powered cropping in 1800. Energy costs are not so easily appraised. Net human energy—energy spent above the basic survival rate and equal to about 1.5 BMR—is fairly closely approximated by assuming an average exertion in traditional farming tasks equal to 4 BMR; 1 h of farming then costs 700 kJ. Since draft animals are kept solely for traction, their total annual feeding costs must be charged against the hours actually worked: a horse costs about 30 MJ/h, an ox 25 MJ/h. With these realistic assumptions the comparison shows little increase in overall energy cost. When deleting the hauling of grain to granary, to make the two data sets even more operationally comparable, the totals are virtually the same. This near-identity is clearly fortuitous, but

the similarity is not surprising. The useful work to be done was the same, harvesting about 1.3 t/ha of wheat, and the animate power, with an efficiency of 15%–20% for both humans and beasts, was the only direct energizer. But where in the early nineteenth century 1 h of labor was aided by draft work worth about 9 MJ, by the end of the century 1 h of labor in California fields controlled some 220 MJ of horse work. Farmers ceased to be the key energizers of the process and became controllers of larger energy flows.

6.4 Cropping Intensification

Agricultural intensification has three key ingredients: water, nutrients, and crop diversity. Two old Chinese peasant's sayings convey perfectly the dominance of these needs: "Whether there is a harvest depends on water; how big it is depends on fertilizer" and "Plant millet after millet and you will end up weeping." Water and nutrients open the photosynthetic work gates in intensive farming whose performance cannot be maintained at high levels with successions of monocultures: irrigation, fertilization, and crop rotations are thus the three principal roads to agricultural intensification. The relation between crop yields and water needs is complex, involving a host of environmental and genetic variables (Doorenbos and Kassam 1979; Rick 1990), but the total seasonal need is commonly about 1,000 times the mass of the harvested grain. Up to 1500 t of water are needed to grow 1 t of wheat; about 600 t of water suffices for 1 t of corn. With harvest indices no higher than 0.30–0.35, total water needs for the leading C_3 cereals are at least 300 mm and up to 500 mm.

Cultivation in arid and semiarid regions, whose annual precipitation may be less than 100 mm and rarely surpasses 250 mm, requires irrigation as soon as the cropping moves beyond the reach of seasonal floods that saturate alluvial soils and allow for the maturation of one crop, or as soon as the growing population requires the planting of a second alluvial crop during the low-water season. Such obligatory irrigation marked the gradual intensification of farming, first among the Sumerians, later in Egypt. A second type of traditional irrigation evolved in response to seasonal water deficits, especially in the northern parts of monsoonal Asia (Punjab, North China Plain). And, of course, rice cultivation required its own ingenious arrangements for flooding and draining the fields and for lifting irrigation water.

Gravity-fed irrigation is energetically most desirable, but in river valleys with minimal stream gradients and on cultivated plains, it is necessary to lift large volumes of surface or underground water, be it only 50 cm into the bunded fields or several meters from the steeply banked streams or wells. Even if only 50% of total need were supplied by lift irrigation, it would be necessary to raise at least 3,000 m^3/ha for a typical grain crop. Irrigation efficiencies (at best 50%, more likely 35%) at least doubled or tripled the theoretical need. Lifting 6,000 m^3 just 1 m needs roughly 30 MJ or (with 20% labor efficiency) some 150 MJ. A steadily working laborer (at 60 W) would need almost 700 h to accomplish the task. This is an extraordinary burden. A single laborer can hoe 1 ha of wheat field in 12–20 days of steady work, cut with a cradled scythe in 8 h, but supplying half of the field's evapotranspiration would take nearly three months of 8-h days.

Not surprisingly, traditional agricultures tried to do with as little irrigation as possible, or they employed a variety of ingenious mechanical devices (Ewbank 1870; Molenaar 1956; Forbes 1965; Needham 1965; Oleson 1984; Fraenkel 1986). The simplest ones were tightly woven or lined shovel-like scoops, baskets or buckets slung on ropes and handled by two people facing each other, dipping the device, and swinging it over a ridge never higher than 1 m. Worked by two pairs of laborers alternating in 2-h spells, the best performance with a typical 60-cm lift was about 5 m^3/h. A scoop or bucket suspended by a rope from a tripod was more effective, lifting about 8 m^3/h to a height of 1 m. But the oldest and simplest water-lifting mechanism that achieved widespread diffusion was the counterpoise lift (swape or well-sweep), perhaps best known as the Arabic *shādūf*. Recognizable first on a Babylonian cylinder seal of 2000 B.C.E. and widely used in ancient Egypt, it reached China by about 500 B.C.E. and eventually spread all over the Old World.

This simple machine was easily made and repaired, just a long wooden pole pivoted as a lever from a crossbar or a pole with a bucket dipper suspended from its longer arm and counterpoised by a large stone or a ball of dry mud to balance the weight of the full bucket. Its effective lift was 1–3 m, but serial deployment of the devices in two to four successive levels was common. With two workers spelling each other in 2-h shifts, hourly performance was 6 m^3 with a 2-m lift; in Egypt a single worker usually lifted 3 m^3/h a distance of 2.5 m. Whereas the *shādūf* required downward pulling on a rope to lift the counterweight, the Archimedean screw (Arabic *tanbūr*, Roman *cochlea*) needed tiresome cranking to rotate a wooden double helix inside a cylinder (150–250 cm long, 40–55 cm in diameter). Only low lifts were possible, and maximum capacities were about 30 m^3/h with a 25-cm rise when powered by two men, or 15 m^3/h with 75-cm lift.

Hand- or foot-operated paddle wheels were very inefficient until the lower parts of paddles were enclosed in a well-fitting box to reduce spillage and to increase the lift. Wheel diameters ranged between 1 m and 3.6 m, and the number of blades was from 8 to 24. With lifts of less than 50 cm, one man treading the wheel's perimeter could deliver up to 12.5 m³/h. These devices were commonly used in India, Korea, Vietnam, and Japan to irrigate small paddies. In China the same function was done largely by water ladders, commonly known as dragon backbone machines (*long gu che*). Square-pallet wooden chain pumps with a series of small boards passing over sprocket wheels formed an endless chain drawing water through a trough; the driving sprocket was inserted on a horizontal pole trodden by two or more men who supported themselves by leaning on a pole (fig. 6.6). Alternatives were a slower manual operation with cranks or animal traction transferred by the means of cogged wheels. With a typical lift of 0.9 m, two men could raise about 8 m³/h, and a recorded performance has four of them lifting 23 m³/h to the same height.

All of the following devices were always energized by animals or by running water (wind-powered machines were much less common). The rope and bucket lift, especially common in India (*monte* or *charsa*), used one or two pair of oxen walking down an incline while lifting a leather bag fastened to a long rope; instead of manual tipping a self-emptying bucket might be used. The arrangement worked well for lifts up to 8–9 m, with two-oxen drives and three workers delivering about 8 m³/h or four oxen and three workers, 16–17 m³/h. An ancient device best known by its Arabic name of *sāqīya* carried an endless chain of clay pots on two loops of rope upside down below a wooden drum to fill at the lower end and discharge into a flume at the top. The practical lift was limited by the power available, usually that of a single blindfolded animal walking in a circular 7-m-

6.6 China's *long gu che* (dragon backbone machine) was powered by people treading a spoked axle.

diameter path. Consequently a *sāqīya* was rarely used to lift water from wells deeper than 9 m and usually did not have discharges over 8 m³/h. An improved Egyptian version, *zawāfa*, delivered up to 12 m³/h from 6-m-deep wells.

Finally, the Arabic *noria* (from the more proper *naūra*) and the Chinese *hung che* (some with diameters up to 15 m) had clay pots, bamboo tubes, or metal buckets fastened to the rim of a single wheel and driven either through right-angle gears by animals or, if equipped with paddles in a stream, by water current. Discharges varied from 20–22 m^3/h with a lift of 1.5 cm to 8–10 m^3/h with high lifts around 9 m. The need to lift the buckets one wheel radius above the level of the receiving trough was highly inefficient and led to a better arrangement in the Egyptian *tablīya*. This improved device included a double-sided all-metal wheel scooping up water at the outer edge and discharging at the center into a side trough. Its best performance (when ox-driven) was about 20 m^3/h lifting water 1.5 m.

Energetic imperatives in traditional irrigation meant that a single laborer, working in 2–4-h spells at rates close to 100 W, could easily power all low-lift Archimedean screws, low-capacity water ladders, and counterpoise lifts. Two people were normally needed to energize high-capacity ladders and some Archimedean screws; 1-h spells by a single person (at close to 200 W) could cover the highest counterpoise performances. A single small ox could take care of a *tablīya* or a low-lift *sāqīya*, but lifts over 3 m required a pair of animals, as did all other high-lift methods. High-volume deep-well bucket lifts required three or four oxen (up to 1.6 kW). The energy costs of irrigation ranged from 100–250 kJ/m^3 of water for human-powered low lifts to as much as 4.5–6.5 MJ/m^3 for animal-powered medium and high lifts. Cost-benefit generalizations are precluded because of differences in crop sensitivities to water supply (interspecific variations and different responses to restricted water supply, with the flowering period the most vulnerable time).

A single specific calculation demonstrates the considerable energy returns of traditional irrigation. Spring wheat yields would drop 23% with a 20% water deficit spread over the entire growing period. Supplying this need of about 80–100 mm brings an additional yield of 300 kg/ha of grain. When irrigated with human-powered devices the cost, assuming 50% irrigation efficiency, would be 200–500 MJ/ha of additional food intake or only about 5%–10% of the yield gain. With animals, costs (4.5–6.5 GJ/ha) would be about equal to benefits (4.5–5 GJ/ha), but because oxen were fed solely crop- and grain-processing residues, energy return in terms of grain would be almost as favorable as with the human labor. Some irrigation systems required such enormous expenditures of human labor that their net energy return had to be very low or negative. Inca canals carved out of rocks (some main lines up to 10–20 m wide) carried water over astonishing distances. The main arterial canal between Parcoy and Picuy ran for 700 km to irrigate pastures and fields (Murra 1980).

Together with adequate water supply, nutrients are the critical inputs opening the photosynthetic work gates, and none is more important than nitrogen. Shortages of nitrogen have been encountered in all traditional agricultures. For example, a harvest of 1 t/ha of wheat would remove, in grain and straw, about 20 kg N, 4.5 kg P, 4 kg K, 2.5 kg sulfur, and 1 kg each of calcium and magnesium (Laloux, Falisse, and Poelaert 1980). Traditional farming resorted to three basic strategies to replenish nitrogen: returning a part of the phytomass to the soil by plowing in crop residues; recycling animal and human wastes and other organic materials; and planting green manures, mostly leguminous crops, to be incorporated into the soil to provide nitrogen for the subsequent grain crop. Of considerable antiquity, these practices also confer other important agroecosystemic benefits like improving the soil's moisture-holding capacity and tilth (Smil 1983).

Crop residues, above all cereal straws and stalks, are a large reservoir of recyclable nitrogen. Traditional cultivars stored 1.25–2.8 times as much mass in the residues as they did in grain. In a crop of medieval European wheat yielding 750 kg (N content 1.5%) and 2 t of straw (0.5% N) nitrogen was about equally divided among the grain and the residue, and the same was true for pre-1900 Japanese rice. Only a small fraction of crop residues were directly recycled; most of them were used as animal feed and bedding (then recycled in manures) as well as household fuel and raw materials for construction and manufacture. Often crop residues were simply burned in the field with virtually complete nitrogen loss.

Recycling of urine and excrement, practiced in Europe and in East Asia, can support high yields but only with much repetitive, heavy labor owing to the time-consuming handling and treatment of the wastes (provision of bedding material, cleaning of stalls and sties of confined animals, liquid fermentation or composting before applications). They also had low nutritional content (~0.5% N) and large preapplication and field losses (60%–80%) of the initially available nutrient (Smil 2001). Effective applications thus required a large mass of manure. In China in the early twentieth century the rates averaged about 7.5 t/ha for all farms and surpassed 10 t/ha in small holdings in the rice region (Buck 1937). Composting and regular applications of other wastes (from silkworm pupae to canal mud) further increased the burden of collecting, fermenting, and distributing, and at least 10% of all labor in Chinese farming was devoted to the management of fertilizers.

In the North China Plain heavy fertilization of wheat and barley was the single most time-consuming part of human labor (close to 20%) as well as animal labor (30%–40%) devoted to those crops. Even in China in the 1980s collection of urban night soil and organic wastes and their transport to the farm took up 2–4 h/day for those still willing or forced to continue that rapidly declining practice. The intensity of China's recycling was matched by the earlier European experience. In the eighteenth century in Flanders the annual application of manure, night soil, oil cakes, and ash averaged 10 t/ha, and rates up to 40 t/ha were not uncommon as organic wastes were brought from cities and towns by a large waste-handling and transportation industry (Slicher van Bath 1963). But the highest known applications of organic wastes, between 50 t/ha and 270 t/ha, of pig and human excrements took place in South China's Guangdong dike-and-pond region (Ruddle and Zhong 1988).

Again, a specific example illustrates the magnitude of rewards. During the 1920s wheat cultivation in a northern Chinese county yielded about 1.4 t/ha and required 307 h of human labor and 248.5 h of animal labor, of which fertilization took, respectively, 17% and 41% (Buck 1930). Assuming, quite conservatively, that the 10 t of fertilizer applied per hectare contained only 0.5% N, only half of which became actually available to the crop, and that each kilogram of N resulted in additional production of 10 kg of grain (Hanson et al. 1982), there is a yield increment of at least 250 kg (~4 GJ/ha) for an investment of 36 MJ of additional human energy inputs and animal feeding cost of 2.5 GJ. No more than about 5% of the latter, if any, would come from grains, resulting in a net gain of at least 3.8 GJ and a more than 20-fold return in edible cereal energy, clearly an excellent benefit-cost ratio.

Green manuring, used in Europe from ancient Greek and Roman times, and widely employed in East Asia from about the sixteenth century, relied mainly on the N-fixing legumes, above all, on vetches (*Astragalus,*

Vicia) and clovers (*Trifolium* and *Melilotus*). These plants fix 100–300 kg N/ha per year and, where climate allows, their winter growth of three to four months adds at least 30–60 kg N, enough to produce a good summer cereal crop. In the long run the provision of adequate N is of such importance that intensive agricultures cannot do without the N-fixing legumes and thus plant them in edible varieties. This desirable practice was perhaps the most admirable indirect energetic optimization in traditional farming, present in all intensive agricultural systems that relied on complex crop rotations.

The cereal-legume link was notable everywhere: China's soybeans, beans, peas, and peanuts alternating with millets, wheat, and rice; India's lentils, peas, and chick peas, with local grains, wheat, and rice; Europe's peas and beans, with wheat, barley, oats, and rye; and West Africa's peanuts and cowpeas, with millets. Food legumes left rather large quantities of N (10–40 kg/ha) for the subsequent grain crops, and while their absolute harvests were lower than those of grains, they yielded at least 50% more protein per hectare than cereals. Moreover, legumes, with lysine, complement the cereals, which are deficient in this amino acid. Soybeans and peanuts also provided edible oils to enhance the low energy density of vegetarian meals, and oilseed cakes and legume residues make excellent high-protein feeds or fertilizers. These quality considerations make any simple energy return calculations irrelevant, especially when the rotations are seen in wider agroecosystemic settings as excellent ways to reduce monocultural vulnerability to pests, limit soil erosion, and improve the tilth.

6.5 Traditional Agricultures

For most of recorded history, increases in agricultural production came from the expansion of cultivated lands. Cultivated areas grew, but yields showed hardly any upward trend because agronomic practices changed only very slowly. This is the key paradox, the principal advantage versus the fundamental weakness of traditional agricultures (subsistence peasant societies). Unlike the simpler survival cultures of foragers, traditional agricultures have been able to support much higher population densities by expanding cultivated land, gradually improving its productivity, and eventually bartering some products or services. Normal years provided a small food surplus, and these arrangements were, at least theoretically, renewable. Yet these peasant societies, with their high fertility rates and low per capita food production, lived barely above the existential minima and were always vulnerable to famines. This pattern of permanent misery, recurrent hunger, and massive death persevered in parts of Europe into the nineteenth century (for instance, the Irish famine of 1846–1851). It persisted throughout most of the poor world until the 1950s, and it was still in place in the poorest parts of sub-Saharan Africa at the beginning of the twenty-first century.

The most important difference between commercial agriculture, with its assured food surpluses, and vulnerable peasant farming is, not surprisingly, in their divergent energy conversion strategies. As Seavoy (1986) argued, this difference is perhaps best elucidated by posing a seldom asked question: Why do peasant societies increase their populations to the maximum carrying capacity during normal crop years and expose themselves periodically to seasonal hunger or famine during consecutive harvest failures? Moreover, why has this happened even in societies with low population densities, high soil fertilities, and fairly elaborate farming techniques? Despite enormous cultural differences, traditional peasant societies shared a strong preference for subsistence compromise, in which minimum levels of material welfare and food safety were acquired with the least expenditure of physical labor.

This predilection is confirmed by the persistence of shifting agriculture and by the reluctance to expand permanently farmed lands and adopt more intensive cultivation. As already mentioned (see section 6.1), shifting cultivation, with its absence of tillage, fertilization, and animals, requires relatively low and largely nonspecialized energy inputs, and it has been a preferred way of food production in all thinly populated forest regions. There it included even those populations that had long contacts with settled farmers, whether in Southeast Asia or Latin America or even in Europe, where the last recorded instances of the practice in Scandinavia and northern Russia date to the early decades of twentieth century. Permanent farming held little appeal until higher population densities forced more intensive use of smaller areas in order to maintain the accustomed nutritional levels. Increased energy expenditures were also needed to clear new lands for permanent arable land or to create new fields by terracing or to dig new irrigation canals. Again, these steps were taken reluctantly.

The villages of Carolingian Europe were overpopulated, and their grain supplies were constantly insufficient, but except in parts of Germany and Flanders few efforts were made to create new fields beyond the most easily cultivable soils (Duby 1968). Later European history is replete with waves of German migrations from densely populated western regions opening up farmlands in areas considered inferior by local peasants (Bohemia, Poland, Russia) and setting the stage for violent nationalist conflicts for centuries to come. A similar reluctance can be seen in Asia. In China the colonization of the fertile but cold northeast did not start until the

eighteenth century, and only post-1960 government-organized resettlement started the colonization of those Indonesian islands whose densities were extremely low compared to Java. Even in relatively densely populated regions of Asia and Europe, it took millennia to advance from extensive fallowing to annual cropping and multicropping.

The other important strategy for reducing labor inputs was to spread them as much as possible. Women, the low-status adults in all peasant societies, did a disproportionately large share of heavy work, and having a large family was the easiest way for parents to minimize their future labor exertions (Caldwell 1976). The energy cost of having an additional child is negligible (even pregnancy may have essentially no energy cost; see section 5.2) compared to the child's labor contributions, which start as early as four to five years of age and assure much less heavy work for parents in their old age. Seavoy (1986, 20) sums this up well: "Having many children (an average of four to six) and transferring labor to them at the earliest possible age is highly rational behavior in peasant societies, where the good life is equated with minimal labor expenditures, not with the possession of abundant material goods."

The process of very slow intensification of food production is best illustrated by focusing on three traditional agricultures of outstanding importance, in Egypt, China, and Europe (fig. 6.7). In Egyptian agriculture the limited cultivable area and the annual flood combined to produce an early shift toward intensification that, millennia later, resulted in some of the highest outputs achievable in solar farming. Predynastic agriculture coexisted with hunting (antelopes, pigs, crocodiles, elephants), fowling, fishing, and gathering. Emmer wheat and two-row barley were the first cereals, sheep the first domesticated

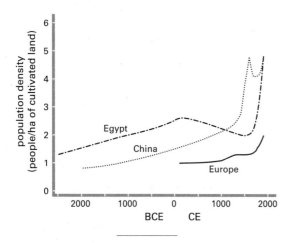

6.7   Long-term increases of population densities per hectare of arable land. From Smil (1994).

animals. October and November seeding followed the receding Nile waters, and harvests followed after 150–185 days.

Agricultural practices in place at the time of the first dynasty (3300 B.C.E.) changed little during the subsequent five millennia (Erman 1894). Harvesting was done with wooden sickles, and the straw was cut high above the ground, sometimes just below the heads. This practice made for easy harvesting, easier transport of the crop to the threshing floor, and cleaner threshing. Standing straw was cut later for weaving, brickmaking, or as kiln fuel. Inscriptions from Paheri's tomb express eloquently the energetic constraints and realities of the day (James 1984). An overseer prods the laborers: "Buck up, move your feet, the water is coming and reaches the bundles," but their reply, "The sun is hot! May the sun be given the price of barley in fish!" sums up perfectly both their weariness and their awareness that grain destroyed by water may be compensated by flood-borne

fish. And the boy whipping the oxen reminds them: "Thresh for yourselves, thresh for yourselves.... Chaff to eat for yourselves, and barley for your masters. Don't let your hearts grow weary! It is cool."

Besides chaff, oxen were fed the barley and wheat straw and grazed on wild grasses of the flood plain and on the cultivated *Vicia sativa*. Cattle were also seasonally driven to graze in the delta marshes. For plowing, oxen were harnessed by double head yokes to wooden plows, soil clods were broken by wooden hoes and mallets, and the scattered seed was trampled into the ground by sheep. Records from the Old Kingdom indicate not only large numbers of oxen but also substantial cow, donkey, sheep, and goat herds. Minimum populations supportable by predynastic farming were 1–1.3 people/ha of cultivated land (Hassan 1984). Subsequent millennia saw a substantial increase of these densities owing to greatly extended cultivation achieved by tree cutting, conversion of grasslands, diking, and drainage. There was no dynastic cultivation of summer crops because regular canal irrigation had only a marginal role.

The Nile's very small gradient (1:12,000) made radial canalization possible only in the Faiyum depression, and the absence of effective water-lifting devices limited dynastic irrigation techniques largely to regulation of high floodwaters, such as building higher and stronger levees, blocking off drainage channels, or subdividing flood basins (Butzer 1984). *Shādūf*, fit for just small-plot irrigation, is verified from the fourteenth century B.C.E.; animal waterwheels came only during Ptolemaic times (c. 300 B.C.E.). Butzer's (1976) reconstruction of ancient Egypt's demographic history has the Nile valley's population density rising from 1.3 people/ha of arable land in 2500 B.C.E. to 1.8 people/ha in 1250 B.C.E. and 2.4 people/ha in 150 B.C.E.

At that time Egypt's total cultivated land was about 2.7 Mha, and assuming 90% of it was planted with grains yielding no less than 900 kg/ha, the country produced (with 25% reserved for seed and 10% storage loss) about 1.5 times as much food as it needed for its nearly 5 million people and became the Roman empire's largest food surplus area. After centuries of decline and stagnation only perennial irrigation (introduced after 1800) boosted the multicropping index. By the mid-1920s it surpassed 1.5, and the still basically traditional farming, but one already helped by inorganic fertilizers, was feeding six people from every hectare of cultivated land (Waterbury 1979).

China, although far from immune to turmoil and stagnation, was a considerably more innovative civilization than Egypt. Chinese contributions to the art of irrigation, ranging from the invention of the square-pallet chain pump to extensive regional irrigation systems, have been outstanding. Other improvements included the horse collar harness and, about a millennium earlier than in Europe, an integrated dryland cultivation tool complex consisting of iron moldboard plow, multitube seed drill, and various horse-drawn hoes and ridgers for effective weeding. This traditional farming, able to support the world's largest culturally cohesive populations, was richly documented in classical Chinese writings. It survived largely intact until the 1950s, which afforded an outstanding opportunity for reliable quantification of its operations (Buck 1930; 1937; F. Bray 1984; Shen 1951; Perkins 1969; Ho 1975).

Intensive multicropping, extensive recycling of organic wastes, and widespread irrigation have been the hallmarks of traditional Chinese farming, but none of these practices is truly ancient. Clear evidence of manuring dates only after 400 B.C.E., and before the third cen-

tury B.C.E. there was no large-scale irrigation and little or no double-cropping and crop rotation. Dryland millet in the north and rice in the lower Yangtze (Chang) basin dominated the first millennia of Chinese farming; pigs were the most abundant and the oldest domestic animals. More than 2,000 years of subsequent intensification produced the world's most persistently self-sustaining farming system, admirably complex yet heavily labor-intensive and recurrently vulnerable. The typical size of fields was only about 0.4 ha, and they were on average just 600 m away from a farmhouse (Buck 1930; 1937). Nearly half (47%) of farmland was irrigated, 25% terraced, 95% cropped (buildings, ponds, roads, and graves covered the rest). Just over 90% of cropped land was in grains, about 4% in sweet potatoes, 2% in fibers, 1% in vegetables.

The second survey counted an astonishing 547 cropping systems in 168 localities. Recycling of organic wastes was the norm; dryland cereal yields were between 900 kg and 1400 kg, rice harvests up to 3.5 t/ha. The average farm family of six people put in 275 10-h days of labor measured in adult work units. Crops claimed the bulk of draft labor (90% for rice, 70% for wheat); manual tasks were more evenly divided among tillage, cultivation, and harvesting. Except for plowing and harrowing, Chinese fieldwork relied on human labor, and because oxen and water buffalo were fed hardly any grain, energy returns can be calculated by using only human labor budgets. Food energy returns (harvested energy/labor energy) were around 25 for staple grains and sweet potatoes, about 40 for corn, 15 for pulses, and 10 for plant oils. Virtually all edible crop harvests were consumed directly; the grains provided 90% of all food energy.

Repetitive diets dominated by wheat and millet flour, rice, cornmeal, beans, sweet potatoes, and cabbage, with only tiny amounts of animal foodstuffs on festive occasions, supported high population densities. The nationwide average was about 4.5 people per ha of sown land, or at least 5.5 people/ha of arable area. Rice-growing southern China could do even better, averaging 5.5 people/ha of arable land by the end of eighteenth century and surpassing 7 people/ha by the late 1920s. Double-cropping in the most intensively farmed areas (summer rice, winter wheat, rapeseed, or broad beans) could yield 36–48 GJ/ha, enough to feed 12–18 people. But like other peasant societies, China was vulnerable to recurrent famines caused by droughts and floods. During the 1920s peasants recalled an average of three crop failures serious enough to cause famines, which lasted on average about ten months, affected 65% of farmed area per county, forced 25% of people to eat bark and grasses, and forced nearly 15% to leave their villages in search of food. Even the most intensive traditional farming system could not rid itself of this vulnerability.

The European experience was less harsh owing to a more equable climate in the West and to generally lower population densities. But Europe experienced similar prolonged productivity stagnation, periodic deep declines, and centuries of gradual intensification punctuated by minor and major famines (Seebohm 1927; Abel 1962; Slicher van Bath 1963; Duby 1968; K. D. White 1970; Fussell 1972; Grigg 1992). Greek farming was not as impressive as its contemporary Middle Eastern counterparts, but the Roman experience, summarized by Cato, Varro, Columella, Virgil, and Palladius, was influential until the seventeenth century. Roman mixed farming (unlike China, Europe always had a strong animal husbandry component) included rotations of cereals and legumes, plowing-in of green manures, often intensive recycling of organic wastes, and repeated liming (using chalk or marl).

Oxen, often shod, were the principal draft animals, a good pair expected to plow 1 *jugerum* (2675 m^2)/day or 40 *jugera*/a of light soil (nearly 11 ha). Plows were wooden, sowing was by hand, harvesting was done with sickles (the Gallic reaper, described by Pliny and pictured on a few reliefs, was not widely used), threshing was done with flails, milling was manual, and the yields were low and highly variable. All these realities changed only very slowly during the millennium after the demise of the Western Roman Empire. Notable changes included use of the scythe instead of the sickle, shoulder collars for horses, and (in some regions) the emergence of horses as principal draft animals. Manuring was practiced with varying intensities, legumes were grown in diverse rotations, fallowing was common, and harvests remained extremely variable.

Times of relative prosperity, most notably 1150–1300 and 1450–1550, were marked by extensive conversions of wetlands and forests to fields and by a greater variety of *companagium* (accompaniments of the ubiquitous bread). Decades of decline were marked by famines, abandonment of villages, and soil erosion. Insecurity remained common right up to the eighteenth century, when fairly intensive cultivation became the norm with better field implements, progressive abolition of fallow, regular manuring, and diffusion of new cultivars. By far the most important step was the adoption of standard rotations, including legume cover crops, exemplified by Norfolk's four-year succession of wheat, turnips, barley, and clover, which at least tripled the rate of symbiotic nitrogen fixation (Campbell and Overton 1993). Chorley (1981) concluded that this step was of comparable significance to the concurrent diffusion of steam power. Industrialization would have been impossible without population growth, and the higher nitrogen supply allowed for higher population density per unit of arable land and for slow but steady improvement of the average diet.

The effects of these advances on human energy expenditures were noted in section 6.4. The effects on yields cannot be so easily demonstrated because many older figures are given as seed/yield ratios (these were sometimes negative) and the records show wide year-to-year fluctuations. Carolingian wheat ratios were no better than 2; in thirteenth-century England they rose to 3–4 (maxima 5.8). This translates to just above 500 kg/ha, and a doubling of this rate was irreversibly achieved only some 500 years later, a gain averaging just 2%/decade (Stanhill 1976). The yield surge came only between 1820 and 1860, a result of land drainage, crop rotations, and manuring. By the early 1850s the national mean surpassed 2 t/ha.

Although at that time British agriculture was still purely solar in terms of mechanical energy and nutrient provision, it was already benefiting from coal-based manufactures (better machinery). Indirect fossil energy subsidies also pushed Dutch yields above 1 t/ha by the year 1800. In contrast, pre-1900 French wheat yields rose only slowly, and there was no gain for the extensively grown U.S. crop. The subsistence capacities of European farming doubled from 2–2.5 people/ha of arable land in the Middle Ages to 4–5 people/ha by 1800, and in the most intensely cultivated regions they doubled again by 1900, when farming was changing into a new energetic hybrid where solar radiation was combined with substantial inputs of fossil energies (see sections 10.4 and 10.5).

TRADITIONAL FOOD PRODUCTION

7

PREINDUSTRIAL COMPLEXIFICATION

Prime Movers and Fuels in Traditional Societies

Nature is so subtil and so penetrating in her ways, that she cannot be used except by great craft; for she does not openly reveal that which may be completed within her; this completion must be accomplished by man.

Paracelsus (1493–1541), *Das Buch Paragranum*

Four kinds of energies powered the multifaceted evolution of preindustrial complexification: human and animal labor, flowing water, wind, and biomass. The capacities of animate prime movers were inherently limited, and higher performances could be achieved only through multiplication of small forces or technical innovation. These two processes were often combined. For example, the monumental architecture of antiquity required massed labor as well as labor-saving levers, inclined planes, and pulleys. Capacities and efficiencies of inanimate prime movers increased very slowly, but there were some remarkable advances by the time these machines began to energize the earliest stages of industrialization. The

spoon tilt hammer, the simplest machine (a lever) to use the energy of running water, did not involve continuous rotary motion, but large nineteenth-century vertical waterwheels commonly transmitted their power to heavy and complex forging hammers (fig. 7.1). Similar advances can be noted for other mechanical prime movers.

The horizontal wooden waterwheels of Mediterranean antiquity developed no more than 300 W, the seventeenth-century vertical machines had capacities ten times higher, and by 1854 the Lady Isabella, England's largest iron wheel, could deliver over 400 kW, a thousandfold power increment during two millennia. There are surprising contrasts even for draft animals. A massive (nearly 1 t), well-fed nineteenth-century draft horse (a Percheron or a Rhinelander), shod with new iron, harnessed with a collar, and hitched to a light flat-top wagon on a hard-top road, could move a load ten times as heavy as its much smaller (300 kg), poorly fed, unshod, breast-harnessed predecessor, which pulled a

7.1 Development of an idea. (*a*) Ancient Chinese spoon tilt hammer. (*b*) Medieval European forge hammers. (*c*) Nineteenth-century waterwheel-driven English hammer.

heavy cart on a muddy road of early medieval Europe. In contrast, the efficiencies of wood combustion (in simple fireplaces) or charcoal production (in earthen kilns) hardly changed for centuries. Similarly, human capacities remained basically unchanged. In the West the dominance of human and animal muscles, water and wind (harnessed by simple devices), and biomass fuels (burned with low efficiency) ended before 1900. Their combination was the not-so-distant foundation of our present affluence.

In many low-income countries animate labor remains an important prime mover, and hundreds of millions of people still rely on biomass fuels as their only or dominant source of heat (see chapter 9). Modern understanding of traditional prime movers and biomass fuels has benefited from works that focused on the history of individual cultures or particular periods as well as the studies that cut across historical eras. Notable contributions in the latter category include Neuburger (1930), Usher

(1954), Singer et al. (1954–1958), Klemm (1959), Burstall (1963), Forbes (1964–1972), Daumas (1969), Gille (1978), L. White (1978), Landels (1980), K. D. White (1984), T. I. Williams (1987), Basalla (1988), Pacey (1990), Cardwell (1991), Goudsblom (1992), McClellan and Dorn (1999), and Sieferle (2001). I appraise preindustrial achievements by looking first at animate power, then at water and wind conversions, and biomass fuels. The chapter closes with surveys of energy use in construction and transportation.

7.1 Animate Power: Human and Animal Muscles

For most of humankind, animate power remained the dominant prime mover until the middle of the twentieth century, and its limited fluxes, circumscribed by the metabolic and mechanical imperatives of animal and human bodies, marked the boundaries of human achievement and affluence. Societies that derived their kinetic energy solely from animate power (ancient Egypt, except for sailships) or mostly from animate power (medieval Europe, China until two generations ago) had little physical security and only a highly privileged affluence to offer. They could increase available power only through the mass concentration of labor and by deploying ingenious devices designed to overcome some limitations of human and animal bodies.

Despite the abundance of free slave labor, no ancient civilization took effective steps toward true mass manufacturing. What Christ (1984), writing about the Romans, calls atomization of production remained the norm. Ancient massed labor thus left its most impressive legacy in buildings of stunning dimensions and unsurpassed esthetic appeal. But their construction could not rely solely on massed animate power because its direct applications have obvious logistical limitations. Only a limited number of people can fit around the perimeter of a heavy object in order to grasp and lift it or simply push it. Similarly, only a limited number of animals can be harnessed together as a coherent team in order to perform a demanding task.

Without mechanical aids that help to overcome the effects of gravity and friction, individual human capacities to lift and carry loads are limited to modest burdens. Nepali Sherpas, the world's ablest load carriers, shoulder between 30 kg and 35 kg while walking to a base camp and less than 20 kg above it. Roman *saccarii* lifted and carried 28-kg sacks over short distances (Utley 1925). In the light version of the traditional Chinese sedan chair, two porters conveyed a customer, a load of 25–35 kg per porter (luxurious litters had as many as eight carriers). More efficient use of human power required devices that confer a significant mechanical advantage, usually by deploying a lesser force over a longer distance (Bloomfield 1997).

The Greeks and Romans were masters of five simple machines capable of such action (first enumerated by Philo in the third century B.C.E.): wheel and axle, lever, system of pulleys, wedge, and endless screw. The three simplest designs, levers, inclined planes, and pulleys, were used by all the Old World's high cultures (Lacey 1935; Needham 1965; Burstall 1968). Levers are rigid, usually slender objects whose pivoting around a fulcrum conveys mechanical advantage equal to the quotient of the length of their effort and resistance arms (measured from the pivot point). Archimedes' famous boast (quoted by Pappus of Alexandria more than 500 years after the physicist's death)—ΔΟΣ ΜΟΙ ΠΟΥ ΣΤΑ ΚΑΙ ΚΙΝΩ ΤΗΝ ΓΗΝ (Give me a place to stand, and I will move the Earth)—testifies to the ancient understanding of the lever's efficacy.

Three classes of levers are distinguished by the point at which the force is applied in relation to the object and the fulcrum (fig. 7.2). In levers of the first class, the applied force and load are at opposite ends with the fulcrum between them, and the force acts in a direction opposite to that of the displaced load. Familiar applications are seesaws, crowbars, scissors, and pliers (the latter are double levers). In levers of the second class, the fulcrum is at one end, and hence the force moves in the same direction as the load, the design shared by wheelbarrows and nutcrackers (a double lever). Levers of the third class do not offer any mechanical advantage because the applied force is greater than the force of the load, but the load moves faster because the force acts over a shorter distance than the displaced object. This is the case when one eats: the elbow is the fulcrum, force is applied to lift the forearm, and the food carried to mouth moves further (hence faster) than the actuated muscles. Catapults, hoes, and scythes are in this category, as

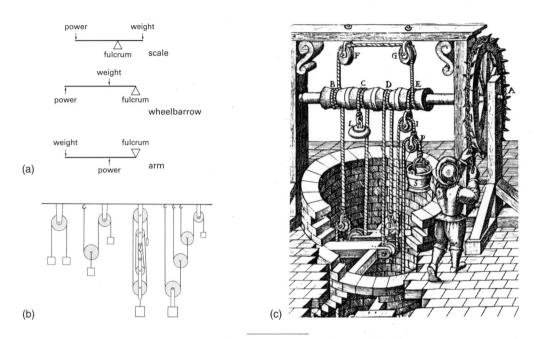

7.2 (a) Three classes of levers. (b) Simple pulley and compound pulleys. The third compound pulley could lift a 400-kg load with a force of only 50 kg, but a lift of 5 m would need pulling 40 m of the counterweight cord. (c) Compound pulley from Ramelli's (1588) book of machines.

are baseball bats, hockey sticks, and the arms of electric shovels used in surface mines.

Other common forms of ancient levers ranged from the water-raising *shādūf* (see section 6.4) to ship oars. The Chinese wheelbarrow, introduced during the Han dynasty, was a remarkably effective second-class lever. With its load above the axle of its central wheel, a worker could move 150–160 kg, sometimes lightening the task with a small sail. In contrast, European barrows, introduced during the high Middle Ages, were much harder to push because they had the fulcrum (wheel) at the end

rather than in the center; loads of 60–100 kg were maxima on good roads and with very slow speeds.

The wheel and axle form a peculiar lever whose long arm is the distance between the axle and the wheel's outer rim and whose short arm is the axial radius. A large difference between the two arms results in a huge mechanical advantage even with heavy wheels and high axial and surface friction. This invention, first used in Mesopotamian before 3000 B.C.E. and independently developed in China just a few centuries later, relied first on solid wooden wheels. Spoked wheels, introduced on chariots about 2000 B.C.E., lightened the device; iron rims reduced its friction and improved its durability (Piggott 1983). Besides serving in vehicles, wheels in the Old World were a part of countless mechanical arrangements. But the Americas and Australia had no native wheels, and extensive desert regions of the Muslim world relied on pack camels rather than on wheels (Bulliet 1975).

The mechanical advantage of an inclined plane (ramp) is expressed by multiplying the length of its slope by its vertical height, but because friction can greatly reduce this gain, the best efficiencies require very smooth or well-lubricated surfaces. According to Herodotus, an inclined plane was used to move heavy stones from the Nile's shore to the site of the great pyramids, and there has been much speculation on the use of ramps during the pyramids' construction (see section 7.4). Wedges are double-sided inclined planes. Wooden ones were used by ancient stonemasons to split large blocks of rock, and metallic forms are seen in knives, axes, and adzes. Screws are circular inclined planes, but in traditional societies, where metallic objects were scarce, they were commonly used as large wooden devices to lift water (see section 6.4) and extract plant oils and juices rather than as fasteners.

A simple pulley, a grooved wheel to hold a rope or cable, which was invented during the eighth century B.C.E., makes the handling of loads easier by redirecting the force in a rope, but it confers no mechanical advantage, and its use can result in accidental load falls. Ratchet and pawl take care of the latter problem, and multiple pulleys solve the first deficiency because the force required to lift an object is nearly inversely proportional to the deployed number of pulleys (fig. 7.2). Aristotle in *Mechanical Problems* clearly understands the device: a simple compound pulley halves the needed force, and the use of

block and tackle with several pulleys on every axle multiplies the mechanical advantage. The most famous testimony for the efficacy of these machines is Archimedes' demonstration to king Hieron (described by Plutarch in his life of Marcellus) of moving a large royal vessel loaded with passengers and freight from its dock as smoothly and evenly as if the ship were afloat by gently exerting traction with his hand through a system of pulleys. The Chinese were also masters of pulleys. Needham (1965) notes that even palace entertainments could not do without them: once a whole corps de ballet of 220 girls in boats was pulled up a slope from a lake.

Three classes of simple mechanical devices—windlasses, treadwheels, and gearwheels—were used to apply continuous animate power in lifting, grinding, crushing, and pounding (Ramelli 1588). Horizontal windlasses (winches) and vertical capstans enabled power transmission by ropes or chains through simple rotary motion (fig. 7.3). Handspikes in a windlass required shifting the grip, usually four times a revolution, a necessity that slowed down the work. These rotary tasks were made easier by cranks, a Chinese invention from the second century C.E. that was unknown in European antiquity. But a major disadvantage remained: the speed of a directly driven machine, such as a lathe, had to correspond to the speed of the hand-cranked or foot-treadle-operated wheel.

This inconvenience was eliminated by using an ingenious adaptation of a crank to power a great wheel, a medieval innovation that made it possible to operate lathes with a continuous rotary motion and accurately machine wooden or metallic parts (fig. 7.3). A large wooden or iron wheel was independently mounted on heavy shaft that rested on a sturdy trestle, and its rotation was transmitted to a lathe by a crossed belt. This arrangement

(a)

(b)

7.3 (a) Large vertical capstan powered by workers pulling a gold wire through a die. (b) Great wheel used to turn a metal-working lathe. From Diderot and D'Alembert (1751–1772).

made it possible to choose many gear ratios, and the momentum of a large flywheel smoothed the variation of human muscle power that was used to turn it and made it easier to maintain even revolutions during spells of higher or lower power need. Powering the wheel was a hard task when machining hard metals. In 1813, when George Stephenson's workers operated a great wheel while making parts for his first steam locomotive, they had to rest every five minutes (Burstall 1968).

Treadwheels provide more power than windlasses or capstans, and efficient transmission of their rotation via gearwheels brought a wide variety of final uses (fig. 7.4). The similarity of action with cycling, the most efficient form of human locomotion, is obvious as powerful leg, abdominal, and back muscles, supported by stationary arms, do nearly all the work. Landels (1980) described a well-designed treadwheel as a highly efficient mechanical device and, given the monotony of the work, one of the

most comfortable for the operator (many of whom were prisoners, often treading, side by side, a long cylindrical wheel). Internal vertical treadwheels were used widely in water pumping (fig. 7.4), and in preindustrial Europe they also powered large construction and dock cranes. Pieter Bruegel the Elder's painting of the Tower of Babel (1563) depicts a crane powered by six to eight men to lift large stones (Klein 1978). The same work was needed for one of the last treadwheel cranes (4.8 m diameter and 2.7 m wide) that operated on London docks during the nineteenth century.

Much less common, external vertical wheels maximized torque (when treading was on a level with the axle) and made it easy to adjust the pull on the hoisting ropes of cranes simply by shifting the operator's position higher on the wheel. Horizontal treadwheels were fairly popular in Europe, and inclined treadwheels were also in use (fig. 7.4). Drumlike devices could be rolled on

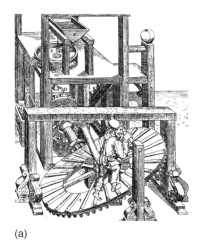  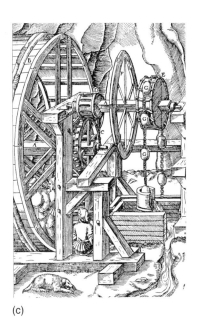

(a)   (b)   (c)

7.4   (*a*) Inclined, (*b*) horizontal, and (*c*) vertical treadwheels.
Reproduced from Agricola (1556) and Ramelli (1588).

a fairly flat surface to a new site, and until the introduction of steam-powered railway cranes, they remained a highly effective, practical way of tackling large lifting tasks. The maximum useful power driving these devices was inherently small. A single worker produced at best 100 W during hours of sustained effort, and the largest, eight-person treadwheels operated with brief inputs of about 1.5 kW and steady flux of up to 700–800 W.

The question of how much labor a man could be expected to do in a day became prominent during the late eighteenth century, when more labor was being hired for new manufacturing (Ferguson 1971). Measurements of work/day favored comparisons with horses, the output of 2.5–14 workers being equated to 1 hp. Two reliable rates include Guillaume Amontons' observations of glass polishers in 1699, whose exertions during a 10-h workday he equated with raising continuously a weight of 25 lb at 3 ft/s, and Charles-Augustin de Coulomb's 1798 figure of a fair day's work, 205,000 kgm, derived from climbing the 2923-m-high Tenerife Peak in the Canaries in a bit less than 8 h. Amontons' calculation is equivalent to 3.66 MJ of work at a rate of 102 W, and Coulomb's yields 2 MJ at about 72 W. We can do little to improve on these classic values.

Many vertical devices were also powered by animals walking inside a wheel or treading the outside rim, but donkeys, oxen, mules, and horses were used much more often to turn horizontal machines (Major 1980). This was done in three ways: animals (usually blindfolded) were coupled directly to the moving parts (beams attached to a central axle) to mill grain, crush seeds and fruits to extract oils and juices, or mill clay for tiles; animals turned a centrally mounted drum to wind up a rope and raise water, coal, ore, people, and animals from mines; and animals rotated whims that were attached to geared assemblies to enhance a mechanical advantage. Apuleius in *The Golden Ass* (second century C.E.) described the misery of these working animals, which went round and round day and night, their "chests ulcerated by the constant rubbing of the harness, flanks laid bare to the bone by incessant beatings, their hoofs splayed out to enormous dimensions by the constant turning." But these abused animals could rarely sustain more than 400 W. On commonly used tread-wheels it was impossible to fit more than two animals and hence to secure more than 1 kW.

Throughout the arid Old World, camels were used for some of the tasks that oxen and horses performed in Europe, and the Indian subcontinent and Southeast Asia were the only regions where working elephants were highly effective in doing heavy tasks like harvesting timber in tropical forests (Schmidt 1996). Unlike other domesticated animals, new working elephants (12–18-year-old adolescents) were taken from wild herds. A classic Indian text described how these beasts had to receive expensive feed (boiled rice and plantains mixed with milk and sugarcane) while in training (Choudhury 1734). Working elephants thus had an unusually high energy cost, but they more than made up for it by extraordinary power and longevity.

7.2 Water and Wind: Wheels and Mills

Antipater of Thessalonica, writing during the first century B.C.E., left the first extant literary reference to a primitive watermill, consisting of horizontal wheel with paddles or scoops turning a vertical shaft: "Demeter has reassigned to the water nymphs the chores your hands performed. They leap against the very edge of the wheel, making the axle spin, which...turns the heavy pair of porous millstones" (Humphrey, Oleson, and Sherwood 1998). And in 947 C.E., al-Masudi wrote the first extant description of the birthplace of simple horizontal windmills in a windy region of today's eastern Iran, where the machines were used to drive mills and raise water from streams (Harverson 1991). After their (highly uneven) diffusion throughout much of the Old World, both kinds of machines eventually assumed important roles in the economic life of preindustrial Europe, which helped to energize the beginnings of industrialization.

The origins of horizontal waterwheels (fig. 7.5), also called Greek or Norse wheels, remain untraceable. These wheels persisted for centuries in many regions of Europe and everywhere east of Syria; their most obvious appeal was that they could turn millstones directly without any gears. Small ancient horizontal wheels were inefficient, but some later designs—notably, the free-jet mills used in Persia and Spain, with heads up to 8 m and water ejected onto rotor blades through a wooden jet with a tapered bore (Wulff 1966)—were fairly efficient (>50%) as well surprisingly powerful (>3.5 kW). Vertical wheels (*hydraletae*), first described by Vitruvius in 27 B.C.E., turned millstones by right-angle gearing and were more efficient. They eventually became a common and persistent feature of many European landscapes.

In England in 1086 the Domesday survey counted at least 6,000 vertical waterwheels, one for every 350 people. By 1300 the total was about 12,000, and after an intervening decline and stagnation, about 30,000 waterwheels were in operation by 1850; Germany used 33,500

7.5 (a) Horizontal waterwheel (Ramelli 1588). (b) Overshot
wheel (Diderot and D'Alembert 1751–1772).

of them as recently as 1925 (Holt 1988; Muller and Kauppert 2004). Waterwheels reached their greatest prominence during the decades of early industrialization, and many systematic accounts of them exist, including Bresse (1876), Bennett and Elton (1898), Müller (1939), P. N. Wilson (1956), Moritz (1958), Forbes (1958; 1965), Hindle (1975), Fox, Brooks, and Tyrwhitt (1976), Meyer (1975), L. White (1978), T. S. Reynolds (1983), and Wölfel (1987).

The basic typology of vertical waterwheels is straightforward. The kinetic energy of low-gradient streams (with heads of 1–3 m) was exploited by undershot wheels, first built with simple radial boards, later with the backs preventing water from shooting over the floats. Their efficiency could be further improved by forming

the base below the waterwheel rim into a closely fitting breast over a 30° arc at the bottom center to increase water retention. This was particularly important in curved-blade wheels, which were introduced around 1800 by Jean-Victor Poncelet as the most efficient undershot machines. Wheel diameters were roughly three times as large as the water head for paddle wheels and two to four times for Poncelets. Locating them on swift currents was desirable because the theoretical power of undershots is proportional to the cube of the water speed (doubling the speed boosts the capacity eightfold).

Breast wheels combined the kinetic energy of the flowing water, which entered roughly at the level of the axis, with gravitational energy for heads between 1.5–5 m. Low-breast arrangements, with water entering below the elevation of the center shaft, were no more efficient than well-designed undershots, and high-breast machines approached the outputs of overshot wheels. Traditional overshots (fig. 7.5) exploited heads greater than 2–2.5 m (their diameters was equal to about 75% of the head) and hence were either confined to hilly or mountainous areas or needed a carefully regulated water supply, which often required construction of ponds and long races. Smooth downstream races were also needed in order to prevent any backing up of discharged water and excessive channel silting. Water was usually fed through troughs or flumes into buckets at rates between <100 kg/s and >1000 kg/s, resulting in 4–12 rpm. The kinetic energy of the water was relatively unimportant because the gravitational energy of the descending mass generated the bulk of the rotary motion.

Overshots became the favorite wheel for many applications besides efficient grain milling. Eventually they powered scores of previously manual tasks with relatively high efficiency, from sawing and wood turning to oil pressing, wire pulling, and majolica glazing. In England overshots were also used in mines for winding (Woodall 1982) and, particularly in the Northeast, for draining. Some were deployed in groups of three to draw water from a depth of nearly 80 m (Clavering 1995). Until the early decades of the eighteenth century it was widely believed that they were less efficient than undershots, but between 1752 and 1754 experiments by Antoine de Parcieux and John Smeaton, and calculations by Johann Albrecht Euler, proved the very opposite (T. S. Reynolds 1979). Smeaton then began promoting overshots, and within a few generations undershots had largely disappeared from England as the more efficient waterwheels delayed the diffusion of steam engines.

Smeaton's (1759) classic experiments indicate an average overshot efficiency of about 66% (52%–76%) and the best undershot performance at 32%. Smeaton also correctly concluded that a wheel's power is a function of the cubed velocity of the water. M. Denny's (2004) simplified theoretical analysis of the efficiency of waterwheels ended up with figures very close to Smeaton's experimental values: 71% for overshots, 30% for undershots, and about 50% for Poncelets. Properly designed twentieth-century overshot wheels operating within optimum parameters had efficiencies of 85% and shaft efficiencies close to 90% (Muller and Kauppert 2004), but 60%–70% was a common performance of the best preindustrial machines. In contrast, traditional, low-rpm undershots had efficiencies just around 20%, and the nineteenth-century wheels converted up to 35%–45% of water's kinetic energy into rotary power. However, the best German undershot designs of the 1930s reached efficiencies as high as 76% (Müller 1939).

Waterwheels were the most efficient preindustrial energy converters. They opened up new productive possibilities, particularly in mining and metallurgy. Animate energies could never generate kinetic energy at such high continuous rates. Waterwheels were a key ingredient of Europe's emerging technical superiority, and they were also a leading prime mover during the early stages of European and North American industrialization. Power of the simplest small machines was limited, but even they made a great difference. Two hard-working slaves in 1 h (useful input of 200 W) could grind 7 kg of flour with hand querns; a donkey-driven mill (300–400 W) produced about 12 kg; and a set of waterwheel-driven millstones (2.2–2.5 kW) typically produced 80–100 kg. If flour were to supply half of an average person's daily food energy intake, a single small watermill would have produced enough of it in a 10-h shift for about 3,500 people, a good-sized medieval town.

Unit capacities grew slowly. During the eighteenth century, European waterwheels still averaged less than 4 kW and rarely exceeded 7.5 kW; heavy parts and poor gearing led to low efficiencies. Larger outputs were achieved by multiple installations and later also by building massive wheels. An exceptional Roman mill-line at Barbégal near Arles had 16 wheels with at least 1.5 kW of useful capacity for a total of 24 kW (Sellin 1983). A water-pumping installation on the Seine at Marly, built between 1680 and 1688 to supply 1,400 fountains and cascades at Versailles, had 14 large (10-m diameter) wheels to drive over 200 pumps and raise about 150 m³/h of water 162 m high in three stages. The project exploited a site with the overall potential of nearly 750 kW, but because of the wasteful transmission of rotary motion via long reciprocating rods, the useful output was only about 52 kW, not enough to supply water for all the fountains (Brandstetter 2005). By the early 1830s, Shaw's water works at Greenock on the Clyde near Glasgow consisted of 30 units, situated in two rows on a steep slope and fed from a large reservoir to provide about 1.5 MW with possible extension to 2.2 MW.

The largest waterwheels had diameters around 20 m and capacities above 50 kW. The one at the lower end of the Greenock falls had a diameter of 21.4 and a width of 4 m. Burden wheel in Troy, New York, was 18.9 m across and 6 m wide. In Cornwall a wheel with 13.7 m diameter developed 52 kW (Woodall 1982). The largest wheel ever built was *Lady Isabella*, belonging to Great Laxey Mining Company on the Isle of Man (J. Reynolds 1970). This pitchback overshot machine had a diameter of 21.9 m and a width of 1.85 m; its 48 spokes (9.75 m

long) were wooden, the axle and diagonal drawing rods were wrought iron, the bearings were cast iron resting on two large oak beams, the total weight was about 80 t, and the wheel had 2.5 rpm. Streams on the slope above the wheel were channeled into the collecting tanks, and water was piped into the base of the masonry tower and ascended into a wooden flume. The power was transmitted to the pump rod at the bottom of a 451-m-deep lead-zinc mine shaft by the main-axle crank and 180 m of timber connecting rods. The theoretical peak was about 427 kW, and normal operation delivered about 200 kW of useful power. Built in 1854, the wheel worked until 1926, and it was restored after 1965.

There were also floating wheels set up on anchored vessels. In 537 C.E., Belisarius used them to grind grain in a Rome besieged by Goths, and many floating mills remained in Europe until the eighteenth century. Tidal mills were first documented in Basra of the tenth century, and during the Middle Ages they were built in England, the Netherlands, the Atlantic coast of the Iberian peninsula, and above all, in Brittany. Installations in North America and the Caribbean came later (Minchinton and Meigs 1980). Most of these mills worked only with the ebbing tide, but larger ones had reservoirs extending the operation to 16 h instead of the usual 8–10 h/day. Tide-driven undershots, including a large 9.75-m wheel, also powered the pumps that supplied London's water.

The first radical improvement of water-driven prime movers came only with the development of the water turbine. First, in 1832, came Benoit Fourneyron's reaction turbine (with radial outward flow), which was built to run forge hammers at Fraisans. Under the head of 1.3 m and with a rotor diameter of 2.4 m, it delivered 38 kW. By 1837 two of Fourneyron's machines for the Saint Blaisien spinning mill worked under high heads of

108 m and 114 m, and the power of 44.7 kW (with efficiency in excess of 80%) was large enough to run 30,000 spindles and 800 looms (N. Smith 1980). A better design (with inward water flow) was soon developed in the United States thanks to the experiments conducted in the 1840s by Uriah A. Boyden and James B. Francis. Although it was a product of several innovators and of what might be called prototypical industrial research (Layton 1979), this machine became known as the Francis turbine. Turbines became the dominant prime movers in regions where early steam power was more expensive. By 1875 they accounted for 80% of installed power in Massachusetts, where the Merrimack River provided about 60 MW, averaging some 66 kW per manufacturing establishment, mostly for textile mills (Hunter 1975).

The earliest history of windmills remains conjectural. The first horizontal machines, with sails mounted on a vertical axle (and thus able to turn the millstone without any gears), were housed in buildings with slots to lead the wind onto the cloth-covered sails, an arrangement that restricted their power and efficiency (Harverson 1991). Lewis (1993) believes that these mills spread during the eleventh century from eastern Iran to Byzantine territory, where they were transformed into vertical machines whose variety was brought to Europe during the Crusades. In any case, for more than seven centuries, beginning during the late twelfth century, windmills contributed appreciably to the gradual intensification of Western economic life.

Better vertical machines were readily adopted, above all, in arid regions with strong seasonal winds (throughout much of the Mediterranean) and in the lowlands of Atlantic Europe, where negligible water heads made them the only large inanimate prime mover. The early history of windmills, their evolution into highly complex

and fairly powerful machines, and their economic importance are well reviewed in Freese (1957), Needham (1965), J. Reynolds (1970), Beedell (1975), and Minchinton (1980). Details on British mills may be found in Skilton (1947) and Wailes (1975), on Dutch ones in Boonenburg (1952), Stockhuyzen (1963), and Husslage (1965), and on American ones in Wolff (1900) and Torrey (1976).

All early Western European machines, with the exception of low-power Iberian octagonal sail mills with triangular cloth, which were transferred from the Eastern Mediterranean, were post mills with vertically mounted sails and an ability to turn the driving shaft into the wind. Their gears and millstones were housed in a wooden structure pivoting on a sturdy central post, which was supported usually by four diagonal quarterbars ending in two foundation crosstrees (fig. 7.6). They could be turned to put the sails in the most efficient position, at right angles to the wind, but they could not realign themselves once the wind direction changed, and they were unstable in high winds. Being relatively low above the ground, they also had limited capacity because wind speeds rise exponentially with altitude and the power of mills goes up with the cube of speed. Post mills worked in parts of Eastern Europe until the twentieth century, but in the West they were gradually displaced by tower or smock mills.

Both of these structures had a fixed body, and only the top cap was turned into the wind, either from the ground or, with taller towers, increasingly from galleries. With the introduction of a fantail to power a winding gear (in 1745 in England) the sails could be kept automatically to the wind. Other pre-1800 innovations included improvements in sail mounting, a spring-sail, a centrifugal regulating governor that did away with difficult and

7.6   A French mid-eighteenth-century post mill. From Diderot
and D'Alembert (1751–1772).

often dangerous task of adjusting canvas to different wind speeds, and cast metal gearings. The Dutch, the world's leading windmill operators, adopted fantails only in the early nineteenth century, but they had the first efficient blade designs once they added a canted leading edge to previously flat blades (c. 1600). True airfoils, aerodynamically contoured blades with thick leading edges, were introduced in England only toward the end of the nineteenth century.

The first drainage mills date from Holland after 1300, but they became common only during the sixteenth century. By 1650, the United Provinces of Netherlands had at least 8,000 such machines, but the hollow-post *wip-molen* turning big wooden wheels with scoops, and the small mobile *tjasker* rotating Archimedean screws, had to be superseded by efficient smock mills before the country could begin large-scale reclamation of polders. European windmills were also used in grinding and crushing (chalk, sugarcane, mustard, cocoa), paper-making, sawing, and metal working (Hill 1984). New U.S. windmills appeared after the middle of the nine-teenth century, with the westward expansion across the Great Plains. They had many narrow blades or slats on solid or sectional wheels, equipped with a centrifugal or a side vane governor and independent rudders, placed on top of lattice towers, and used to pump water for households, cattle, or steam locomotives (fig. 7.7). These windmills, barbed wire, and railroads were the iconic artifacts that helped to open up the Great Plains (T. L. Baker 1985; A. M. Wilson 1999).

There is very little information on the energy output of early windmills. The first experimental measurements date from the 1750s, when John Smeaton matched the power of a common Dutch mill with nine-meter sails with the power produced by ten men or two horses (Smeaton 1759). This calculation, based on measurements with a small model, was corroborated by actual performance in oilseed pressing: while the wind-powered runners turned seven times a minute, two horses made scarcely 3.5 turns in the same time. Forbes (1958) estimated that a typical large eighteenth-century Dutch mill of 30-m span, when equipped with improved sails and turning in 8–9 m/s winds, could develop about 7.5 kW at the windshaft. The best modern sails and gearings could raise the output to as much as 15–22.5 kW. Measurements at a preserved 1648 marsh mill that lifted 35 m^3 of water at 2 m/min in 8–9 m/s winds showed a shaft power of about 30 kW but an actual output of just 11.6 kW, a transmission loss of 61%. This confirms Rankine's (1859) comparison, in which he assumed useful power of 2–8 hp (1.5–6 kW) for eighteenth-century post windmills and 6–14 hp (4.5–10.4 kW) for tower mills.

Wolff's (1900) measurements indicate just 30 W for U.S. wheels with a diameter of 2.5 m, and up to 1 kW for large 7.6-m devices. Representative ranges would be 0.1–1 kW of useful power for the nineteenth-century U.S. wheels, 1–2 kW for small post mills, 2–5 kW for large post mills, 4–8 kW for common smock and tower mills, and 8–12 kW for the largest nineteenth-century devices. Typical medieval windmills were thus as powerful as contemporaneous waterwheels, but by the early nineteenth century many hydraulic installations were four to five times more powerful than even the largest windmills. The importance of windmills peaked during the nineteenth century. As many as 10,000 of them worked after 1800 in England, 18,000 in 1895 in Germany, and some 30,000 (100 MW) by 1900 in countries around the North Sea (DeZeeuw 1978). Between 1860 and 1900 several million Halladays, Adamses, Buchanans,

(a)

(b)

7.7   Late nineteenth-century U.S. windmills. (*a*) Buchanan
machine. (*b*) Halladay windmill pumping water for steam
locomotives. From Wolff (1900).

and other brands (in 1889 there were 77 manufacturers) were sold in the United States, and many of these designs were also used in Australia, Latin America, and South Africa.

7.3 Phytomass Fuels and Metallurgy: Wood, Charcoal, Crop Residues

Woody matter (trunks, branches, twigs, bark, and roots), charcoal made from this phytomass, and crop residues (mainly cereal straws, legume stalks, and tuber vines) fueled the subsistence as well as the complexification of all preindustrial societies. Quantitative information on the use of these fuels in preindustrial households, metallurgy, and artisanal manufacturing is limited (Biringuccio 1540; Evelyn 1664; Buck 1930; Schott 1997; Sieferle 2001; Perlin 2004). However, the basic subsistence needs—for cooking and, in some societies, for space or water heating or food drying—have changed little over time, and valuable insights can be gained from modern studies of energy use in rural areas of low-income countries, where phytomass remains the dominant or even sole fuel (Earl 1973; NAS 1980; Hall, Barnard, and Moss 1982; Smil 1987; Vimal and Bhatt 1989; RWEDP 2000; Bailis 2004).

Despite the enormous diversity of woody phytomass, wood itself has a relatively uniform composition (Shelton and Shapiro 1976; Tillman 1978; Smil 1983). Typically, about 43% of it is cellulose, 28%–35% hemicelluloses, and the rest lignin; carbon makes up 49%–56%, oxygen 40%–42%. Ash (incombustible matter) varies from less than 0.5% to about 2%. Densities range from less than 0.4 g/cm^3 for poplar to nearly 0.7 g/cm^3 for white oak and 1.0 g/cm^3 for some eucalyptus. The energy content of wood goes up with the proportion of lignin, which contains 26.5 MJ/kg compared to 17.5 MJ/kg for cellulose

and hemicellulose; extractive resins have up to 35 MJ/kg. Differences in energy content among common species range between 17.5–20 MJ/kg for most hardwoods and 19–21 MJ/kg for softwoods. The energy density of typical wood (19 MJ/kg) slightly surpasses that of cereal straws (17–18 MJ/kg), and is about 80% of steam coal's (22–24 MJ/kg) and 45% of crude oils'.

All these energy densities refer to absolutely dry phytomass, but fresh mature wood averages 30% water for hardwoods and 46% for softwoods, and even air-dried wood (cut, stacked, sheltered, and dried for at least two months) still contains about 15% moisture, as do many straws. As a result, the net heating values of phytomass vary widely, from around 20 MJ/kg for very dry to about 15 MJ/kg for air-dried wood and ripe straws to as little as 5–6 MJ/kg for fresh (green) wood, crop stalks, and grasses. In contrast, charcoal is a high-quality fuel with an energy density (29.7 MJ/kg) equal to that of the best bituminous coal or twice the heat content of air-dried wood. But preindustrial production of charcoal, virtually sulfur-free and smokeless and hence suitable for indoor use, was often enormously wasteful.

The simplest process involved primitive earth or pit kilns in which the partial combustion of the heaped wood provided the heat necessary to initiate carbonization (fig. 7.8). Ancient charcoal makers had their preferences. According to Theophrastus (third century B.C.E.), very old trees were the worst, and smiths preferred charcoal from fir. Regardless of the charge, both the quality and the quantity of the final product were difficult to control in earthen kilns. Charcoal yields were just 15%–25% of dry wood charge by weight, and 1:5 was probably the best approximation of the typical charcoal:wood ratio during the preindustrial era. Assuming a net heat value of 14.5 MJ/kg of air-dried wood, the energy loss in tradi-

(a)

(b)

7.8 (*a*) Building a charcoal pile. (*b*) Charcoal production in
mid-eighteenth-century France. From Diderot and D'Alembert
(1751–1772).

tional charcoal making was about 60%. In volumetric terms, primitive earth kilns needed as much as 24 m³ of wood per tonne of charcoal, and even for good practices the average was between 9–10 m³.

Considering the generally low levels of final energy demand (for cooking, rudimentary space heating, and artisanal manufacturing), consumption of phytomass fuels in traditional societies was relatively high owing to often dismally low conversion efficiencies. Transition from uncontrolled and inherently inefficient open-air fires to enclosed, regulated, and efficient burners was very slow. Moving open fire inside made little difference. Fireplaces that dominated European cooking and heating for centuries performed quite poorly; as they drew the needed combustion air from the room, they warmed

the immediate vicinity of the hearth with radiated heat, but their operation amounted to an overall loss of interior heat.

There were some ancient ingenious ways of using phytomass efficiently, none more so than three space heating systems that provided an uncommon degree of comfort. The first two used combustion gases to heat raised room floors before leaving through a chimney. The Roman *hypocaust*—of Greek origin, with the oldest remains from the third and second centuries B.C.E., found both in Greece and in Magna Graecia (Ginouvès 1962)—was first used by the Romans in the *caldaria* of their public baths, then spread to stone houses in colder provinces of the empire. The Korean *ondol* (warm stone) led the hot combustion gases from the kitchen (or from additional fireplaces) through brick or stone flues under concrete-covered granite floors. The Chinese *kang* was just a large heated platform ($\sim$4–5 m^2) that served as a resting place during the day and a bed at night. The advantages of these arrangements are clear in comparison with brazier heaters that were common in both Asia and Europe (the British House of Commons was heated by large charcoal fire pots until 1791). The brazier heaters offered only a spot source of warmth and could produce dangerous levels of carbon monoxide.

Advances in residential heating and cooking were slow (W. Lawrence 1964). Draft chimneys appeared in Europe only during the late Middle Ages, when tile stoves were used in Germany and Scandinavia but smoky, inefficient fireplaces were still used in England (Edgerton 1961). In Europe iron plate stoves spread widely only during the seventeenth century, when they were brought by French and German settlers to North America. During the late eighteenth century Benjamin Franklin and Count Rumford (Benjamin Thompson) came up with major efficiency improvements (Brown 1999). The performance of these simple stoves varied widely with design and fuel, but they certainly did not surpass the levels measured in modern tests of their contemporary counterparts (mostly less than 25%). The best available studies of those societies that still rely almost solely on phytomass for their thermal energy put their annual consumption of woody matter and crop residues at less than 10 GJ per capita for the poorest communities in the warmest regions. The rates could have been 30–40 GJ per capita in relatively rich mid-latitude cities, where wood-based production of beer, bricks, tiles, metals, and glass added considerably to the residential demand.

This estimate receives excellent confirmation from a careful reconstruction of firewood demand in London, where per capita (dry matter) demand in 1300 averaged about 1.75 tons, or roughly 30 GJ (Galloway, Keene, and Murphy 1996). In societies on the verge of industrialization the needs were certainly much higher. Those nineteenth-century North American and European families who lived between 40° N and 50° N and in houses well heated with large iron stoves needed annually 50–200 GJ per household, or about 15–50 GJ per capita. Cooking, water heating, and industrial needs would have easily doubled these rates. Indeed, the best per capita fuel wood consumption estimate for the United States in 1850 is about 97 GJ (Schurr and Netschert 1960). Global estimates of preindustrial biomass consumption can aim only at the right order of magnitude. With 20 GJ per capita for the 1800 population of about 1 billion, it would be some 20 EJ, and up to 30 EJ is plausible. In gross energy terms, this would be at most one-tenth of the total energy content of fossil fuels that were burned annually in the year 2000, but considering the much higher efficiencies of modern fuel converters, the difference of actually available (useful) heat energy would be at least 30-fold.

The availability of phytomass was also a key determinant of metallurgical progress because charcoal was the only source of carbon used to reduce ores in preindustrial societies (Biringuccio 1540). Copper, the principal constituent of bronze (together with tin and smaller amounts of lead and silicon), was the first metal smelted in relatively large amounts. Forbes (1966) presented careful studies of Roman copper production that involved roasting and smelting of chalcopyrite followed by refining. Specific fuel needs were enormous, about 90 kg of wood per kilogram of copper, or 1.3 GJ/kg. With average wood growth of 900 kg per tree in 40 years, and 300 trees per hectare (about 6.75 t/ha · a), a smelter producing 1 t Cu/day would have consumed nearly 5000 ha/a. Demand for metallurgical charcoal rose sharply once iron began replacing bronze, starting c. 1000 B.C.E. Copper and tin melt, respectively, at 1083°C and 232°C, but iron melts at 1535°C. Unaided charcoal fire can easily reach 900°C, and even with simple forced air supply, its temperature can be increased to close to 2000°C.

Charcoal has only a trace of S and P, but its friability limited the mass of the smelting charge. This made no difference as long as the furnaces remained small. The simplest iron smelting furnaces were just partly enclosed hearths built on hilltops to maximize natural draft. They produced small masses (blooms; up to 50–70 kg) of iron heavily contaminated with slag. This low-carbon wrought iron had high tensile strength and was directly malleable; it was easily forged, but it could not be cast. Bloomery hearths were eventually succeeded by shaft furnaces, whose enlargement culminated in the medieval development of *Stückofen* and then *Blasofen*. Blast furnaces originated in the Rhine-Meuse region in the fourteenth century and produced cast (pig) iron, an alloy with up to 4% C that cannot be directly forged or rolled and is weak in tension but strong in compression.

Bloomery hearths and simple forges needed 3.6–8.8 times more fuel than the mass of charged ore (Johannsen 1953). With good ores averaging 60% Fe (75% of which ended up in the molten metal), this implies 8–20 kg of charcoal (240–600 MJ) per kilogram of hot metal. By 1900 a combination of technical advances brought the typical rate down to about 1.2 kg (36 MJ/kg), and the best rates in Swedish furnaces were as low as 0.77 kg (22.9 MJ/kg) (Campbell 1907; Greenwood 1907). In the year 2000, Brazil was the world's only consumer of metallurgical charcoal, using about 2.9 m^3 (725 kg)

of eucalyptus-derived fuel converted in masonry kilns for 1 t of pig iron, a rate of 21.5 MJ/kg of hot metal (Ferreira 2000). The high energy intensities of charcoal-fueled smelting caused extensive deforestation, first in copper- and lead-producing Mediterranean regions, then in the iron-producing Atlantic Europe. During the early eighteenth century a single English blast furnace, working from October to May, produced 300 t of pig iron (Hyde 1977). With as little as 8 kg of charcoal per kilogram of iron and 5 kg of wood per kilogram of charcoal, it needed some 12,000 t of wood.

By that time deforestation was so common that the wood was cut in 10–20-year rotations from coppiced hardwoods, whose annual increment would be 5–10 t/ha. In 1720, 60 British furnaces produced about 17,000 t of pig iron, requiring about 680,000 t of trees. Forging added another 150,000 t, for a total of some 830,000 t of charcoaling wood. At 7.5 t/ha, this represented about 1100 km^2 of forests and coppiced growth (equal to a 33-km square). Already in 1548 anguished inhabitants of Sussex wondered how many towns would decay if the iron mills and furnaces were allowed to continue (people would have no wood to build houses, watermills, wheels, barrels, and hundreds of other necessities), and they asked the king to close down many of the mills (Straker 1931).

Energetic constraints on preindustrial metal smelting were thus unmistakable, and widespread European deforestation was to a large degree a matter of horseshoes, nails, axes (and mail shirts and guns). Blast furnace locations were also constrained by the limited radius of animal-drawn transport used to bring in charcoal and the necessity for a continuous rapid water flow to power bellows. Access to ore was also essential, but ore made up only a fraction of charcoal's charge. Further, the maximum height of blast furnaces (about 7.5 m by 1750) was limited not only by charcoal's friability but also by the maximum air blast available from water-powered bellows. Both these limits were removed once coal was turned to coke (capable of supporting heavy charges) and converted to mechanical energy in steam engines.

Biomass fuels were also the principal provider of traditional lighting, beginning with fire glow and inefficient (and dangerous) resinous torches. The first oil lamps appeared in Europe during the Upper Paleolithic, nearly 40,000 years ago (de Beaune and White 1993); the first candles were used in the Middle East only after 800 B.C.E. Both were inefficient, weak, and smoky sources of light, but they were easily portable. Various plant oils and animal fats (olive, castor, rapeseed, linseed, whale oil, beef tallow, beeswax) provided the fuel, and wicks were made of papyrus, rush pith, flaxen, or hemp. Bright illumination was possible only through massive multiplication of these tiny and inefficient sources. Paraffin candles convert just 0.01% of energy in solid hydrocarbons to light and emit only about 0.1 lumen/W. The bright spot in their flame has irradiance only 20% higher than clear sky (1.0 Cd/cm^2 vs. 0.8 Cd/cm^2). The first eighteenth-century lighting innovations doubled and tripled the typical irradiance. In 1794, Aime Argand introduced lamps that could be regulated for maximum luminosity using wick holders with central air supply and chimneys to draw in the air (McCloy 1952).

Soon afterward came the first lighting gas made from coal; Paton (1890) calculated that gas jets converted less than 0.04% of that fuel to light. Outside major cities oil lamps were dominant during the entire nineteenth century. Until the 1860s they were fueled by an exotic biomass fuel, oil rendered onboard ships from the blubber of sperm whales. The poorly paid and dangerous pursuit of these giant mammals, portrayed so unforgettably in Melville's *Moby-Dick* (1851), reached its peak just before 1850 (Francis 1990). The U.S. whaling fleet, by far the largest in the world, had a record total of more than 700 vessels in 1846. During the first half of that decade about 160,000 barrels of sperm oil were brought each year to New England's ports (Starbuck 1878). The subsequent decline of sperm whale numbers, and competition from coal gas and kerosene, led to a rapid demise of the hunt.

7.4 Construction: Methods and Structures

Timber, stone, and bricks, either sun-dried or kiln-burned, were the dominant building materials of the preindustrial world, all suitable to construct the four components (walls, columns, beams and arches) that are needed to erect structures whose dimensions were surpassed only by the advent of inexpensive structural steel and modern concrete during the late nineteenth century. Only animate labor aided by a few simple tools was needed to extract, transport, shape, and emplace these materials. Sun-dried mud bricks, made of compacted mixtures of clays, water and chaff, or chopped straw and shaped in wooden molds, were the least energy-intensive building material. Fired bricks were used first in ancient Mesopotamia, and later they were common in both the Roman Empire and Han China. Their firing in open piles or pits was extremely wasteful, and only enclosed kilns, with regularly spaced flues, eventually resulted in more even baking and lower fuel consumption. The dimensions of fired bricks ranged from chunky square Babylonian pieces ($40 \times 40 \times 10$ cm) to slim oblong Roman shapes ($45 \times 30 \times 3.75$ cm).

Preindustrial construction involved a skillful integration of large numbers of workers in order to accomplish tasks that appear extraordinarily demanding even by modern standards, and yet many of these structures were built in a very short time, the Parthenon in just fifteen years (447–432 B.C.E.), the Pantheon in about eight (118–125 C.E.). There is no mystery about how these stunning structures were built (W. J. Anderson 1902; Coulton 1977; Adam 1994; Erlande-Brandenburg 1995). Massive stone components (Parthenon architraves weighed almost 10 t) and large timbers were lifted by compound pulleys or by cranes that were powered by workers or animals turning capstans, windlasses, and treadwheels. The Romans pioneered the use of concrete. The Pantheon's bold dome (no preindustrial builders ever topped its 43.2-m span) consists of five rows of square coffers of diminishing size that converge on the central oculus (MacDonald 1976). Its most intriguing property is the vertical decrease of specific mass achieved with progressively thinner layers of a lighter concrete (Mark 1987). Builders of medieval cathedrals and castles included many craftsmen using specialized tools (Fitchen 1961; C. Wilson 1990; Courtenay 1997).

In contrast to the well-known building techniques that were used in antique temples or medieval cathedrals, the logistics that made it possible to erect massive structures such as prehistoric megaliths and stone pyramids are speculative. There is no doubt that wooden sledges or rollers, ramps, and massed labor were used to move heavy stones or statues. Two Assyrian bas-reliefs from Kuyunjik (c. 700 B.C.E.) show giant statues moved on wooden sledges with about 50 men pulling, 10 operating a large lever at the back, and another 10 or so laying and raising logs and wedging a large lever. In brief spells such groups could develop about 14 kW of useful power. A better-known Egyptian painting of transporting a 50-t colossus from a cave at el-Bersheh (1880 B.C.E.) depicts 127 men (peak useful power of over 30 kW) pulling a sledge whose path is being lubricated by a worker pouring water from a vessel (fig. 7.9).

Protzen (1993) determined that the heaviest of enormous Inca stone blocks in the southern Peru, a 140 t stone at Ollantaytambo, required the force of 120.4 t to be pulled up the ramp, an effort calling for the concerted exertion of some 2,400 workers. The peak power of this group would have been around 600 kW, but we know nothing of the logistics of such an enterprise: How were more than 2,000 workers harnessed to pull in concert? How were they arranged to fit into the confines of the narrow (6–8 m) Inca ramps? Most intriguingly, construction of the three great pyramids at Giza remains a matter of conjecture. There are no contemporary descriptions or depictions of these projects. We know that the core stones were quarried near the pyramid site, that limestone for the facing was brought from Tura quarries across the Nile, that massive granite blocks inside the structure were shipped from southern Egypt, and that all this material had to be moved up on a ramp to the building plateau (Lepre 1990; Lehner 1997).

We also know that very large stones were transported on boats. A unique image from Deir el-Bahari shows two 30.7-m-long Karnak obelisks carried on a 63-m-long barge pulled by about 900 oarsmen in 30 boats (Naville 1908). And Spence's (2000) demonstration of how the pyramids were so near-perfectly oriented—using the alignment of Mizar and Kochab, two circumpolar stars, a method that also explains slight deviations in the

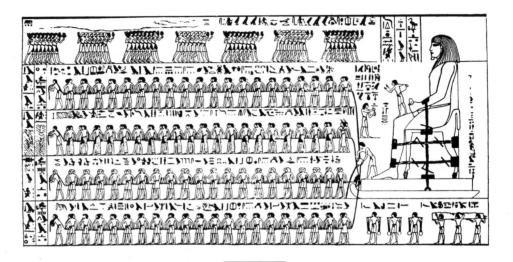

7.9   Moving a 50-t statue at el-Bersheh on a wooden sledge,
1800 B.C.E. From Rühlmann (1962).

orientation of all major pyramids—offers the most plausible explanation of the challenge while also narrowing the
beginning of the Great Pyramid's construction to 2485–
2475 B.C.E.). But arguments continue about the actual
ways of building the core of over 2 million blocks, and
emplacing and precisely finishing thousands of smooth
casing stones (Tompkins 1971; Mendelssohn 1974; Grimal 1992; Wier 1996; Edwards 2003).

Three principal explanations have been offered to explain the construction. Many Egyptologists have believed
that pyramid builders used gigantic clay, brick, and stone
ramps, but this explanation is highly unlikely. A single
suitably inclined plane (slope 10:1) would have had to
be extensively enlarged after finishing every layer of stone-
work, and its volume would have far surpassed that of
the pyramid itself. Winding ramps encircling the pyramid
would have had to be very narrow (hence difficult to

negotiate) and supported by very steep walls. Most
tellingly, there are no signs of a vast volume of ramp-
building rubble anywhere on the Giza Plateau. In contrast, the proponents of lifting have offered many
solutions of how the work could have been done with
the help of levers or simple machines. Hodges (1989)
argued for just levers and wooden packing pieces to lift
stone blocks, and rollers to emplace them.

Objections to this explanation include the large
number of horizontal transfers that would have had to
be performed for every block as the pyramid grew higher,
and the need for extraordinary caution and precision to
avoid repeated accidents in manipulating 2–2.5-t stones.
Edwards (2003), assuming an average sustained pull of
68 kg per worker (the value derived from moving the el-
Bersheh colossus) and a 0.2 coefficient of friction, argued
in favor of gangs of about 50 workers using the pyra-

mid's steep (51°52') smooth slopes, wetted for easier sliding, as inclined planes to pull the stones. These assumptions led him to calculate the total labor force needed to build the Great Pyramid at about 10,000 people. In contrast, the oldest surviving account of pyramid construction by Herodotus, written after the historian's visit to Egypt two millennia after the structures were completed, relates that the blocks were lifted from the ground level by contrivances made of short timbers and that a total labor force of 64,000 men was employed for 80 days a year for 20 years. Pliny put the total at 366,000 laborers.

Mendelssohn (1974), using basic physical considerations including the average force in pulling a laden sledge (10–15 kg/worker) and the labor needed to build approach ramps, came up with 70,000 seasonal laborers and up to 10,000 permanent masons. Wier (1996) approached the problem of labor force by considering the aggregate effort needed to lift all those stones above the plateau. His calculation of the Great Pyramid's potential energy is correct (2.52 TJ), but his assumption of average daily work accomplished per worker (240 kJ) is too conservative. After he factored in inefficiencies, he ended up with the maximum labor force of less than 13,000 people. Assuming an average daily rate of just 0.25 m^3 of stone per worker, only about 1,500 quarrymen using copper chisels and dolerite mallets and working 300 days a year were needed to cut 2.5 million m^3 of stone in 20 years (the length of Khufu's reign). If three times as many masons were needed to square and dress the stones and move them to the pyramid site, the total labor force would be about 5,000 people.

With net inputs of useful mechanical energy at 360 kJ/day only about 1,000 workers were needed to lift the stone blocks. Doubling that number for the workers

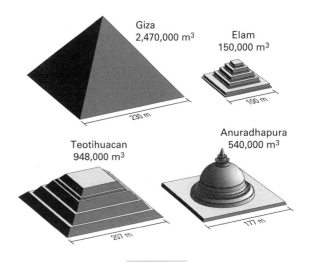

7.10 Scaled oblique views, dimensions, and volumes of Khufu's pyramid at Giza, Choga Zanbil ziggurat at Elam, the Sun Pyramid at Teotihuacan, and the Jetavana stupa at Anuradhapura. From Smil (1994).

needed to emplace the stones, and adding the labor needed to build and repair sleds, ropes, and timbers and to cook meals, adds to the total about 10,000 men, clearly a manageable body. The gross inputs of useful labor embodied in Khufu's pyramid (about 2.5 MJ/worker·day) add up to some 150 TJ (fig. 7.10). For comparison, Falkenstein (1939) calculated that the construction of Anu ziggurat at Warqa required at least 1,500 laborers working 10 h/day for five years (embodied energy <6 TJ), and E. R. Leach (1959) estimated that the largest Anuradhapura stupa (fig. 7.10), built of about 200 million of mostly rough-laid bricks, needed only about 600 laborers working 100 days/a for 50 years (∼7.5 TJ).

Some European cathedrals embodied a comparable total of energy (5–10 TJ), others required considerably more because their construction, employing only hundreds of lumbermen, quarrymen, carpenters, stonemasons, glassworkers, and haulers at a time, lasted scores of years, even centuries. In contrast, Teotihuacan pyramids, the largest such structures in the Americas (fig. 7.10), needed much less energy because their cores were made up of earth, rubble, and adobe bricks, and only the exterior was faced with cut stone (Baldwin 1977). No other single preindustrial structure embodied as much animate energy as did Khufu's pyramid, but that total was easily surpassed by two of the classical world's greatest engineering networks, Rome's aqueducts (Van Deman 1934; H. B. Evans 1994; Hodge 2002) and the Roman Empire's system of roads (Pekáry 1968; Chevalier 1976; Sitwell 1981; Kolb 2001).

My calculations, based partly on assumptions detailed in Blackman and Hodge (2001), indicate that the channels for Rome's 11 aqueducts (515 km), built between 312 B.C.E. and 226 C.E. and using nearly 6 m³ of stone per meter of the conduit, required as much stone as did Khufu's pyramid. To this must be added stones for about 58 km of elevated arches, and the volume of soil and rock that had to be removed for the sloping aqueduct ditches clearly surpassed the total volume of stone used to build water channels. Understandably, Pliny in his *Historia Naturalis* (book 36), wrote that "there was never any design in the whole world ... more admirable than this," and Frontinus, in his detailed treatise on the city's aqueducts, displays the pride of a practical engineer (and a Roman) by imploring his readers to compare these structures with "the idle pyramids, or else the indolent but famous works of the Greeks." Some Roman aqueducts had a much higher energy cost because of large amounts of lead needed for high-pressure siphon pipes that were used to cross river valleys; siphons in the

Lyon water supply needed about 15,000 t of the metal (Hodge 1985).

Principal Roman highways were 12 m wide, country roads less than 3 m (Forbes 1965). If one assumes, conservatively, an average width of 6 m and a depth of 1 m for all hard-surface Roman trunk roads (85,000 km by 300 C.E.), these roads would have required the emplacement of no less than 500 million m³ of stone for the base and the top *summa crusta* as well as gravel, sand, and lime after first moving some 750 million m³ of earth and rock for the roadbed and *agger*. An earthen embankment, the *agger* often measured up to 1.8 m above its surroundings and up to 12 m wide. With about 0.5 m³ of stone and gravel and 5 m³ of soil handled per capita per day, the tasks of quarrying, cutting, crushing, and moving stones and gravel, fashioning embankments, preparing concrete and mortar, and laying the road required on the order of 1 billion labor days, or 2.5 PJ of embodied energy.

7.5 Transportation: Roads and Ships

Preindustrial land transport was dominated by draft animals, and its capabilities depended not only on the kind, size, and health of commonly used species but also on the ability to reduce friction, that is, on the quality of roads and vehicles. Wheels varied from heavy, segmented solid disks fixed to a rotating axle (still in use in parts of Asia until the twentieth century) to light, multispoked arrangements that rotated on a fixed axle. The front axle itself was either pivoted (in ancient Persia, Celtic Europe) or fixed (during the Roman era). Roads ranged from muddy ruts and sandy trails to hard-top *viae*, and in many parts of the Old World their quality had greatly declined during the Middle Ages and improved only during the nineteenth century. On a smooth, hard, dry road only about 30 kg of force could wheel a 1-t load, 150 kg might be needed on a loose gravel surface, and more than 200 kg on sandy or muddy roads. In addition, smaller and inadequately fed animals were often quite unimpressive performers.

Ancient societies reserved most of their horses for fast action: war chariots and cavalry, races—Homer's *Iliad* gives the earliest description—and mounted messengers (Clutton-Brock 1992; Hyland 2003). Horses use a remarkable energy-saving mechanism. The muscular work of galloping is halved because they store and return elastic energy strain in muscle-tendon units (about 1 kJ/stride for a 500-kg animal) (Wilson et al. 2001). The horse-drawn transport of goods became more common only during the last centuries of the Roman Empire, particularly using the *cursus publicus*. This public transportation system, set up by Augustus primarily to relay important official dispatches, was also used to a much lesser degree to haul imperial freight and military supplies (Hyland 1990; Kolb 2001). Both draft and pack horses were widely used during the Middle Ages, but they remained fairly small; the stallions of medieval knights were typically just 140–145 cm at the shoulders. More powerful breeds were introduced only during the eighteenth century.

War horses represented a major energy drain on many preindustrial societies. They consumed feed and required much human labor for their breeding, care, and trading). This was particularly so in dynastic China, where the necessity to import and maintain horses in the large numbers needed for recurrent military campaigns against the encroaching nomads represented a major strain on the state's resources (Creel 1965). But superior animals and skillful horsemanship decided many battles even during the nineteenth century. Perhaps most notably, on June 18, 1815, at Waterloo the heavier English horses and the charge of the Household Brigade led the British to victory (Vesey-Fitzgerald 1946). Cavalry remained an important military component during WW I, and both the Red Army and the Wehrmacht relied on horses during WW II (Edgerton 2007).

A pair of early twentieth-century European draft horses could pull a 3–4-t rubber-tired cart on hard-top roads, but their smaller Chinese counterparts could pull no more than about 800 kg/head with wooden- or iron-rimmed carts on good dirt roads. Forbes (1965) estimated that the tractive effort was no more than 680 kgm/s for a typical preindustrial European horse cart, and the Theodosian code (438 C.E.) prescribed 490 kg as the maximum load for goods wagons pulled by oxen. The low speeds of walking animals restricted the daily range of freight carts and wagons to just 15–20 km for oxen and no more than 30 km for horses. People with wheelbarrows could do just 10–15 km, but where the loads had to be carried, they had the advantage of being able to walk on very narrow paths. Similarly, mules and donkeys with panniers were preferred to pack horses because of their better mobility on narrow paths, their higher resilience (harder hooves, lower water requirements), and their endurance. Typical loads were about 30% of an animal's weight (50–120 kg) on the level and 25% in the hills, and speeds did not exceed 5 km/h.

Preindustrial land transport was unfit for the large-scale movement of goods, a reality perfectly illustrated by Diocletian's *edictum de pretiis rerum venalium* (301 C.E.): it cost more to move grain 120 km on roads than to ship it across the Mediterranean. And even after the Egyptian grain arrived at Ostia, only 20 km from Rome, it was reloaded by *saccarii* to *naves caudicarii* (barges) that were drawn by oxen upstream to the capital rather

7.11 High density of London horse traffic. From the *Illustrated London News*, November 16, 1872.

than being hauled by wagons (Utley 1925). Land transport was thus the most ubiquitous energy-limited hindrance to the complexification of preindustrial societies; only railways changed that. But even as railways took over the intercity traffic, horse-drawn transport reached an unprecedented, albeit short-lived, intensity in the rapidly growing cities of late-nineteenth-century Europe and North America (fig. 7.11). New affluence brought more private coaches and cabs; omnibuses, first seen in London in 1829; and delivery wagons (Dent 1974). In 1901, at the end of Queen Victoria's reign London had some 300,000 horses, and some New Yorkers were thinking about creating a large suburban belt of pastures to accommodate the swelling numbers of urban horses.

In contrast to slow land transport of goods, many preindustrial societies perfected the art of fast horse riding,

which was used in most remarkable ways in efficient networks of long-distance communication. Horseback riding began most likely around 4000 B.C.E. among the people of the Sredni Stog culture in today's Ukraine (Anthony, Telegin, and Brown 1991). The still inconclusive evidence is based on distinctive fractures and beveling of premolars of animals that had been bitted. Riding a horse has always been a physical challenge. Because the horse's fore-end contains about three-fifths of its body weight, the only way for the vertical planes intersecting a rider's and an animal's center of gravity to coincide is for the rider to sit forward.

But an upright forward position leaves the rider's center of gravity much higher than that of the horse. This can produce a rapid lever action by the rider's back when the horse moves forward, jumps, or stops fast.

Consequently, the most efficient position requires the rider to put his center of gravity not only forward but low (K. S. Thomson 1987). The jockey's crouch is the best way, but skillful riders have used other postures that also allowed them to discharge weapons in gallop (Clutton-Brock 1992). Classical riders were even more disadvantaged because they did not have the stabilizing stirrups that spread westward through Eurasia after the third century. Good riders had no difficulty doing 50–60 km/day on a fit animal on a passable road, and with horse changes they could cover well over 100 km/day in emergencies.

Much higher record claims were made for the Mongolian *yam* (message delivery) service (Marshall 1993), and in 1860 William F. Cody's legendary ride covered 515 km in 21 hours and 40 minutes (R. A. Carter 2000). But typical performances were carefully optimized. Minetti (2003) concluded that relay postal systems based their selection of average speed (13–16 km/h) and daily distance (18–25 km) covered by an animal on minimizing the risk of damage to the horses, avoiding the emptying of spleen and anaerobic metabolism and allowing for adequate cooling. Remarkably, this optimal choice applied equally well to the ancient Persian service set up between Susa and Sardis by Cyrus after 550 B.C.E., which Xenophon called "undeniably the fastest travel by land possible by humans"; to messengers in the far-flung Mongol Empire of the thirteenth century; and to the famous Overland Pony Express of 1860, which reached California before the telegraph and the railway.

Compared to the flows of 100–1000 W that dominated the energy inputs in animate land transport, human-powered waterborne movement operated at much higher rates. Oared vessels offered some spectacular examples of clever design and mass labor integration

albeit often terrible suffering of crews chained to the oars of large galleys (Morrison and Williams 1968; Hocker and Ward 2004). Assuming 80% rowing efficiency and useful power between 100 W (sustainable for hours) and 200 W (shorter bursts), the 50-rower *penteconteres* that took the Greek troops to Troy were propelled with 3.5–7 kW. Triple-tiered Greek and Roman triremes were probably the best-performing classical warships. Their seating of 170 rowers has been a perpetual puzzle for naval historians. Their 24-kW power was enough to propel the ship over 20 km/h into a devastating ramming attack. Equipped with bronze rams, these ships made it possible for a smaller Greek force to defeat a large Persian fleet at Salamis in 480 B.C.E., and later they became the most important warships of republican Rome. Their successful full-scale reconstruction was finally accomplished during the 1980s (Morrison, Coates, and Rankov 2000).

The sequence of larger oared ships ended with a failure: a 128-m-long *tessarakonter*, built in the Ptolemaic Egypt (210 B.C.E.). This vessel, designed to carry more than 4,000 oarsmen and nearly 3,000 troops and to be propelled with over 5 MW of muscle power, proved too heavy to use. Large European oared vessels remained in use until the seventeenth century, when some Venetian galleys had 56 oars, each crewed by five men (Lane 1934). Large Maori dugout canoes were rowed by almost as many warriors (up to 200), indicating a universal limit of 12–20 kW for aggregate human power in sustained rowing applications. In addition to oared ships there were vessels powered by pedaling or stepping on treadmills. The Sung dynasty Chinese built paddle-wheel warships powered by up to 200 men, and in Europe tugs powered by 40 men turning capstans or treadmills appeared after 1550 (Needham 1965).

PREINDUSTRIAL COMPLEXIFICATION

Animals were used in waterborne transport from antiquity. As already noted, oxen hauled barges from Ostia to Rome, and Horace in his *Satires* describes a journey on a canal boat pulled by a mule. Their use increased with the construction of major canals, first in China, where the Grand Canal, begun during the Han dynasty, promoted economic development in the country's core economic areas (Chi 1936; Needham 1971), and much later in Europe and North America. Chinese canal boats were pulled by gangs of laborers or by oxen or buffalo. In Europe, where canals reached their greatest importance during the eighteenth and nineteenth centuries, horses were the main prime movers. They walked on tow paths, pulling the barges at speeds of about 3 km/h when loaded, and on a well-designed canal a single heavy horse could pull a load of 30–50 t, 1 OM more than could be managed on the best hard road. Steam engines gradually replaced barge-towing animals, but many horses still worked on smaller canals during the 1890s (Hadfield 1986).

All preindustrial seaborne transport of goods was dominated by sailing ships, machines to convert the wind's kinetic energy into forward motion through the use fabric airfoils (sails) and stable and steerable hulls (Van Loon 1935; Torr 1964; Chatterson 1977; Block 2003). The pressure differences generate lift and drag forces acting on the sail (Marchaj 2000). With wind astern, lift far surpasses drag and propels the ship; with wind on the beam or slightly ahead of it, the force pushing the vessel sideways is stronger than the force propelling it forward. If the ship were to steer even closer to the wind, the drag would pushed it backward. The oldest vessels, with square sails set at right angles across the ship's long axis, proceeded fast only with wind directly astern or less than 30° off course. The only way to cope

with this limitation was to keep changing the course by resorting to wearing (making a complete downwind turn). Consequently, Roman ships pushed by the north-westerlies could sail from Messina to Alexandria in just six days, but the return could take ten times as long. Duncan-Jones (1990) argues that it is almost impossible to say what speeds were typical.

Medieval square-rigged ships could proceed slowly with the wind on their beam (90°), and their post-Renaissance successors could move at an angle of about 80° into the wind. The canted square sail appeared in the Indian Ocean during the third century B.C.E. This was a clear precursor of triangular (lateen) sails of the Arab world after the seventh century. In Europe only the late medieval combination of square rigging and triangular sails made sailing close to the wind possible. Gradually, these ships were rigged with a larger number of loftier and better adjustable sails. Ships with fore-and-aft sails could tack, turning their bows into the wind and catching it on the opposite side of the sail. Ships that combined square sails with triangular mizzens could manage 62°, and fore-and-aft rigs (including triangular, lug, sprit, and gaff sails) could come as close as 45° to the wind (modern yachts come close to the aerodynamic limit of 30°). Capabilities for sailing into the wind have thus advanced by more than 100°.

Better sails, deeper hulls, the stern-post rudder (a Chinese invention), and the magnetic compass (in China after 850, in Europe around 1200) combined to make ships into more efficient wind converters. The simplest vessels of this improved design began the great European voyages. Portuguese sailors advanced first along the western coast of Africa (Senegal in 1444, the equator in 1472, to Angola in 1486). In 1492, Columbus crossed the Atlantic; in 1497, Vasco da Gama rounded the Cape

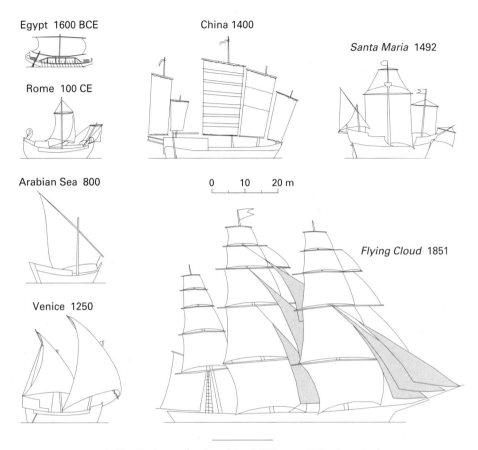

Egypt 1600 BCE

Rome 100 CE

China 1400

Santa Maria 1492

Arabian Sea 800

0   10   20 m

Venice 1250

Flying Cloud 1851

7.12 Evolution of sailing ships, 1600 B.C.E–1851. From Smil (1994).

of Good Hope and crossed the Indian Ocean to India; and in 1519, Magellan traversed the Pacific and the *Victoria* (captained by Sebastian Elkano) completed the first circumnavigation of the world. European expansion and the rising maritime trade followed as accurate guns turned larger ships into powerful tools of long-distance conquest (McNeill 1989). These compact devices allowed relatively small crews to deploy unprecedented amounts of

inanimate energy to roam and to destroy, and in doing so advanced Europe's global ascendance (Cipolla 1966).

But comparisons of record tonnages do not show any huge gains between antiquity and the late medieval and early modern eras (fig. 7.12). Ships with capacities in excess of 1000 t had been built already by the Romans, but a common Roman cargo vessel carried less than 100 t, voyages of great European discoveries started with ships nearly as small (*Santa Maria*, 165 t), Great Armada (1588) vessels averaged 515 t, and typical late eighteenth-century ships of the India fleet rated 1200 t. But speeds have roughly quadrupled. Roman cargo ships usually sailed at 2–2.5 m/s, whereas in 1853, the Boston-built and British-crewed *Lightning* logged the longest daily run, 803 km at an average speed of 9.3 m/s (Wood 1922). And in 1890, *Cutty Sark* (a famous tea clipper), ran 6000 km in 13 consecutive days, averaging 5.3 m/s (Armstrong 1969).

R. Unger (1984) tried to quantify the contributions of sailing ships by making a set of assumptions to calculate their energy use during the Dutch Golden Age at about 200 TJ (6.2 MW) a year, roughly equal to the total energy output from all Dutch windmills, as estimated by DeZeeuw (1978), but only a small fraction (less than 5%) of the country's huge peat consumption. But such comparisons mean little. On the one hand, no amount of peat would have made the seaborne trips to the East Indies possible; on the other, the useful energy gained from the peat was almost certainly less than one-quarter of its gross heat value. Moreover, it is questionable to compare a limited and rapidly depleting store of fossil energy with an abundant and renewable resource.

8

FOSSIL FUELS

Heat, Light, and Prime Movers

Nature, in providing us with combustibles on all sides, has given us the power to produce, at all times and in all places, heat and the impelling power which is the result of it.

Sadi Carnot, *Réflexions sur la puissance motrice du feu* (1824)

All preindustrial societies derived their energy from sources that were almost immediate transformations of solar radiation (flowing water and wind) or that took relatively short periods of time to become available in a convenient form: just a few months of photosynthetic conversion to produce food and feed crops, a few years of metabolism before domestic animals and children reached working age, or a few decades to accumulate phytomass in mature trees to be harvested for fuel wood and charcoal. Solar radiation remains as important as ever because it drives the Earth's climate, energizes its biosphere, powers all photosynthesis, and makes every build-

ing a solar one. But modern civilization has added two kinds of energy sources to set itself apart from all preindustrial societies: fossilized stores of solar energy extracted in the form of coals and hydrocarbons (crude oils and natural gases), and electricity generated mostly by burning these fuels as well as by water and nuclear fission and, to a much lesser extent, by wind and the Earth's heat.

Unlike the nearly instantaneous solar flows and their modestly aged phytomass and animate energy transformations, fossil fuels were formed through slow heat and pressure transformation of accumulated biomass, both terrestrial and marine, which lasted typically 10^7–10^8 years. Dukes (2003) estimated that typical preservation factors (fractions of phytomass carbon that remained in a fossil fuel) for coal are just above 10%, but for crude oil in marine reservoirs less than 0.5%. Consequently, during the later 1990s, annual global consumption of fossil fuels burned organic carbon that required about

400 years of current global NPP. In a fiscal analogy, preindustrial societies relied on instantaneous or minimally delayed and constantly replenished solar income, whereas modern civilization is withdrawing accumulated solar capital at rates that will exhaust it in a tiny fraction of the time needed to create it. Preindustrial societies relied on constantly renewable energies (on a time scale of many millennia) and were thus truly sustainable, at least in theory, because some of their practices caused excessive deforestation and soil erosion, which reduced or even prevented regeneration of phytomass energy.

In contrast, modern civilization rests on the indubitably unsustainable harnessing of a unique solar inheritance that cannot be replenished on a civilizational time scale. We are living in an energetic interlude because the stores powering our way of life are finite, and even the best conversion efficiencies and the utmost conservation measures cannot extend their life beyond several hundred years. Yet we will never exhaust all the recoverable reserves of fossil fuels, the share of overall resources that can be produced with available techniques at a known cost. Long before reaching that point we will either go back to immediate solar flows harnessed in ways vastly superior to preindustrial practices, or we will put in place new arrangements dependent on another, more durable class of stores, such as advanced nuclear options or entirely new, as yet unknown conversions. In any case, this means that modern economies are, from the fundamental energetic point of view, unsustainable.

But tapping the available solar capital of amassed fossil fuels made it possible to temporarily harness sources of extraordinarily high energy that are also relatively easy to store and transport. Our ingenuity has been converting them into a variety of final energies (including thermal, kinetic, chemical, and electric) that can be used with unprecedented efficiency and flexibility. Both past achievements and the immediate prospects of modern civilization are defined by this consumption of fossil fuels. In this chapter I first describe their attributes and resources, and then survey the progress of their extraction, transportation, combustion, and conversion to electricity. Fossil-fueled civilization as a high-energy system is the subject of chapter 9.

Although Thomas Gold and some Russian and Ukrainian geologists argued the case for the abiogenic origin of some hydrocarbons (see section 8.2), most fossil fuels are clearly organic mineraloids with minor admixtures of inorganic compounds and alkaline and metallic elements. Their physical state at ambient temperature divides them naturally into solids (coals), liquids (crude oils), and natural gases. Plurals are necessary in order to convey the considerable heterogeneity of these fossil fuels. Coal qualities range widely, crude oils are, as far as heating value is concerned, much more uniform, and the energy density of natural gases varies within a very narrow range. But even fuels with identical energy content command different prices because of the presence or absence of undesirable constituents (sulfur, ash, heavy metals).

The properties of fossil fuels have been studied systematically since the first half of the nineteenth century, and some classical accounts still appear useful (Sexton 1897; Poole 1910). Modern knowledge of fossil fuels is reviewed comprehensively in Francis and Peters (1980), Bartok (1991), Berkowitz (1997), and Odell (2004), and in detailed treatments focusing on coals (C. L. Wilson 1980; Ward 1984; Berkowitz 1985; Yang 1997; L. P. Thomas 2002; WCI 2005), and crude oils and natural gases (Hunt 1996; Royal Dutch-Shell 1983; Ikoku 1984; Tiratsoo 1986; Brooks 1990; Rojey et al. 1996; Gluyas 2003). News about resources, production, and technical advances in fossil fuel industries are reported in key periodicals like *Fuel, Fuel and Energy Abstracts, Coal Age, Coal Mining Journal, Coal Outlook, Natural Gas, Natural Gas Transportation and Distribution, Oil & Gas Journal, Petroleum Economist, Petroleum Engineer International, Petroleum Science and Technology*, and *World Oil*.

All fossil fuels fit into a broad category of nonrenewable mineral resources. Their exploitation has been seen as an inexorable march toward physical exhaustion. Strictly speaking, this is an incorrect view. Fuels are forming steadily, but extraction vastly surpasses the rate of replenishment, so unsustainable resources would be a more fitting term. More important, these resources should not be seen within a rigid stock exploitation frame because estimates of their magnitudes change with prices, input costs, and advances in exploratory, production, distribution, and conversion techniques. The traditional classification of resources spans the extremes of an enormous resource base, the total quantity of the minerals in the Earth's crust, unknown or only very roughly estimable; and the relatively small reserves, a tiny part of the base, whose spatial distribution and recovery costs at current prices and with existing techniques are known in detail. McKelvey's (1973) box, shown in figure 8.1, is perhaps the best-known graphic summary of this approach, whose inadequacy is in its static definition of availability under a given set of specifications.

As Tilton and Skinner (1987) argued, this may be useful but it is not enough. What happens as we move toward and past the prescribed conditions? A superior measure of resource availability is the cost of producing

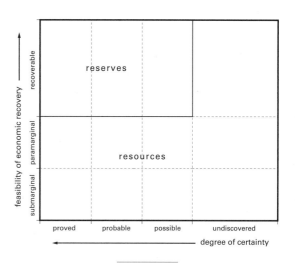

8.1   McKelvey's box, the standard U.S. resource classification system. From McKelvey (1973).

additional or marginal units of a resource. This dynamic approach takes into account improvements in techniques and the ability to pay the price of recovery. Exhaustion is then not a matter of actual physical depletion but rather a burden of persistent and eventually insupportable real cost increases resulting in declining availability of a resource. This is a gradual process allowing for adjustments and countermeasures, including innovative techniques, various conservation efforts, and often surprisingly sweeping resource substitutions. Consequently, there are no sudden ends, just variable and gradual shifts onto new supply planes. This understanding is critical in appraising the rise and prospects of fossil-fueled civilization. Although its energetic base is the recovery of unsustainable stocks, this reality does not imply any fixed dates for the physical exhaustion of fossil fuels. Nor does this mean an early onset of unbearable costs to tap these resources.

8.1 Coals: The Earliest Modern Fuels

Coals are sedimentary rocks whose main constituents are partially decomposed organic matter, various minerals, and water. Their genesis begins with anaerobic decomposition of accumulated plant debris in wetlands. Peats are the youngest, only partly fossilized products of this process. The Carboniferous period (360–286 Ma ago) left behind the largest coal resources, followed by the Jurassic (213–145 Ma ago), the Cretaceous (145–65 Ma ago), and the Paleocene (65–55.5 Ma ago). Poor-quality lignites and peats are the products of the Quaternary period (starting 1.8 Ma ago). The bulk of coal mass is composed of macerals, highly heterogeneous organic compounds derived from woody phytomass (Stopes 1935; Carpenter 1988). This complex mixture is mostly composed of vitrinite produced by the transformation of cellulose and lignin; H-rich liptinite from the waxy and resinous biomass; and C-rich inertinite from charred phytomass. Their specific density is 1.3–1.4 g/cm^3, and commonly present minerals except for sulfides are about twice as heavy. Ultimate elementary analysis on a moisture-free basis reveals the expected dominance of carbon: over 90% in anthracites, 80%–85% for good bituminous varieties, less than 75% for lignites.

Oxygen's share is 3%–20%, hydrogen's 4%–6%, and nitrogen's 0.5%–2% in U.S. bituminous coals. Sulfur in sulfides (pyrites), sulfates, and organic matter ranges from fractions of a percent to over 5%, and some coals have elevated content of trace elements, including As and various heavy metals (Mo, Cd, Hg, Pb) and trace amounts of U, Th, and Ra. More customary proximate and empirical analysis reports the shares of carbon, volatile matter, moisture, ash, sulfur, and energy content of coals. The World Coal Quality Inventory is the most comprehensive repository of both proximate and ulti-

mate analyses of samples from all coal-producing countries (USGS 2001). Coal qualities span a very wide continuum, with substantial variations common even in a given seam. While the best anthracites contain less than 5% moisture and 95% C, the mass of some lignites may have up to 60% water and 15% C. Volatile matter may be totally absent or as high as 85%. Ash is a collective term for incombustible minerals whose presence varies widely, from little to more than 40% of the fuel's mass in the poorest lignites.

The heating value of coals ranges about fourfold on fresh-weight basis, between 8.3 MJ/kg and 35.0 MJ/kg. Poor lignites are greatly inferior to air-dried wood (~15 MJ/kg), and the heat content of the best anthracites approaches that of some crude oils. This variability makes classification of coals difficult and explains the absence of a uniform global standard (Lett and Ruppel 2004). U.S. classification recognizes anthracite, bituminous (at least 26.7 MJ/kg), and subbituminous (at least 22.1 MJ/kg) coals and lignites (down to 14.7 MJ/kg). European classification has nine classes of hard coals (at least 23.86 MJ/kg) and six classes of lignites (brown coals). For decades international energy statistics defined the standard coal equivalent as the fuel with 29.3 MJ/kg or 7 Mcal/kg. Only the best bituminous coals rate that high, with most steam coals having 20–25 MJ/kg. Reliable conversions to energy equivalents require a detailed knowledge of specific coal qualities.

Coal seams outcrop in many parts of the world, and hence the fuel was known and locally used from antiquity, when its most important application was in iron making during the Han dynasty (Needham 1964). European extraction is documented in Belgium in 1113, the first shipments of coal to London date to 1228, and the first exports from the Tynemouth region to France date to 1325. But coal burning was never any more than a marginal source of energy in ancient or medieval societies, and it had no influence on determining the technical development of those societies (Nef 1932). England was not the first country to shift from phytomass to a fossil fuel. That primacy belongs to the seventeenth-century Netherlands, when the Golden Age of the country's economy and creativity was energized largely by peat (DeZeeuw 1978).

But England was the first country to accomplish the shift from phytomass fuels to coal, and this conversion preceded the often described onset of the late-eighteenth-century industrial revolution by some 200 years (Nef 1932; J. R. Harris 1974). During the sixteenth century demand for coal was rising steadily, and by the beginning of seventeenth century the fuel was commonly used in households as well as to make bricks, tiles, earthenware, glass, starch, soap, and sugar, and to extract salt. Between 1540 and 1640 mining began at all major English coalfields, and between 1580 and 1660 coal shipments to London rose about 25 times. The ensuing emissions gave the city a deserved reputation as being unfit for human habitation. By 1650, Britain's annual coal output passed 2 Mt; 3 Mt were extracted by the early eighteenth century, and over 10 Mt by its end (Hatcher et al. 1984–1993). Coal did not simply supplant wood. Its large-scale use required solving many technical and organizational problems connected with its mining, transport, and industrial uses.

Rising coal demand led to larger and deeper mines as the small surface pits or shallow shafts producing just 1–2 t/day were replaced by larger collieries. The deepest shafts surpassed 100 m in the early eighteenth century and reached 200 m by 1765 and 300 m by the 1830s. Extraction was initially energized solely by human labor.

Hewers, working with picks, wedges, and mallets, were assisted by putters filling baskets, loading them on wooden sledges, and dragging them to pit bottom. There onsetters hung the baskets on ropes, windsmen hauled them up, and banksmen carried them to storage heaps. Boys as young as 6 years did lighter tasks, and the heaviest work often fell on women or teenage girls as they carried coal to the surface by ascending a series of ladders with heavy baskets tied to forehead straps (Ashton and Sykes 1929).

The severity of these exertions was incredible. Ascending 35 m to the surface with up to 75 kg of coal translates, when assuming body weight of 60 kg and speeds of about 0.2 m/s, to exertions of close to 300 W, clearly near the limit of human performance, especially when considering the precarious nature of the task. This is a painful illustration of the inevitability of subsidizing the introduction of a new energy source by liberal use of the dominant energy, in this case the exertion of female muscles. Where human muscles were inadequate, horses became indispensable for powering the whims or treadmills used for pumping water from deep shafts and for hoisting coal from larger pits.

After 1667 horses and donkeys were also used underground for hauling. Waterwheels and windmills did some of the pumping and hoisting, and they and the horses were only slowly replaced by Newcomen's very inefficient steam engines, which converted just 0.5% of coal into useful motion. Until the widespread employment of steam engines, both coal mining and transport were overwhelmingly energized by animate labor. Transportation of coal in heavy horse-drawn wagons was feasible only over short distances. The use of ships for longer distances led to extensive development of canals, whose era ended only with the expansion of railways. Coal substituted for wood or charcoal in many manufactures, but its direct use, imparting undesirable impurities to the final product, was impossible in glass making, malt drying, and, most important, iron smelting.

Glassmakers solved the problem by introducing reverberating (heat-reflecting) furnaces, where the raw materials were heated in closed vessels. Coke was first used in malt drying during the 1640s. In 1709, Abraham Darby succeeded in smelting pig iron with it, but charcoal use persisted to the end of the eighteenth century (Harris 1988). Blast furnaces charged with coke could be built taller and more voluminous, raising productivity and creating more demand for the fuel. But by far the greatest boost for coal extraction came with James Watt's improved steam engine, patented in 1769. At that time Britain was still producing more than 90% of world coal output, and it dominated global extraction until the late 1870s. Its share of global extraction was about 53% in 1870 and 36% in 1890, when it was still about 30% ahead of the rising U.S. output, which became the world's largest just at the century's turn. Greater availability of wood and water power in the United States and in parts of Europe made the process slower than in the United Kingdom. For example, in Germany the switch required state subsidies to favor coal (Sieferle 2001), and U.S. energy use was dominated by wood until the 1880s, Russian energy use until the early twentieth century (Smil 1994).

The first detailed global survey of coal deposits was prepared for the Thirteenth International Congress of Geologists (McInnes 1913); it listed 6.402 Tt of resources and 671 Gt of recoverable reserves. By the year 2004 the global total of proved recoverable coal reserves stood at 909 Gt, with 479 Gt (about 53%) in bi-tuminous coals, nearly 30% in subbituminous deposits, and the remainder in lignites (WEC 2004). The energy density of recoverable reserves varies with the thickness of seams and the heat content of the fuel. A 1-m-thick seam of poor lignite (Germany's *Braunkohle* with 8.5 MJ/kg and specific density of 1.4) stores no more than 12 GJ/m^2; the same thickness of excellent bituminous coal (29.3 MJ/kg) contains up to 40 GJ/m^2; and anthracite has about 50 GJ/m^2.

Seams range from less than 30 cm (the usual minimum for reserve accounts) to over 100 m, with modal values between 0.5 m and 2 m (for example, Appalachia's average is 1.5 m). Typical energy densities of large coal basins are 10^1 GJ/m^2. The Pittsburgh bituminous bed and the huge North Plains lignites and subbituminous deposits (nearly 100,000 km^2 in the Dakotas, Montana, and Wyoming, with some 40% of the U.S. coal reserves) average around 50 GJ/m^2. The richest mines have densities 1–2 OM higher. Those in Wyoming's Powder Basin (Wyodak seam) rate about 450 GJ/m^2. Fortuna/Garsdorf, the Rhineland's largest brown coal mine during the 1980s, came close, with 430 GJ/m^2, and Garzweiler and Hambach, the largest German opencast lignite mines of the early 2000s, have reserves of around 200 GJ/m^2. Arizona's Black Mesa rates about 200 GJ/m^2, Montana's Ashland 415 GJ/m^2, and parts of Victoria's Gippsland Basin (up to 100-m-thick lignite seams of the Latrobe Valley) about 1 TJ/m^2. So does the Number 3 seam of Queensland's Blair Athol, which averages 29 m; in parts it is 32 m thick, which means that with high-quality bituminous coal of 24.5 GJ/t, its energy density is up to 1.1 TJ/m^2.

Coal reserves are actually distributed more unevenly than the reserves of crude oil. Just five nations account for about 76% of the world's total coal and for 73% of all bituminous coal. The United States leads, with 247 Gt of the total and 111 Gt of bituminous coal, followed by Russia (157 Gt and 49 Gt), China (~115 Gt and 62 Gt), India (92 Gt and 90 Gt), and Australia (79 Gt and 39 Gt). Asia, Europe, including Asian Russia, and North America have very similar shares (roughly one-third of global reserves each), and South America, with only about 2%, is coal-poor. In 2005 the reserves to production ratio (R/P) was 165 years globally, and the ratios for the countries with the largest reserves and high extraction were more than 500 years for Russia, 245 years for the United States, about 230 years for India, 215 for Australia, and 60 years for China.

Differences in reporting criteria (there is no uniformity in defining maximum depth and minimum thickness of potentially recoverable seams) and in accuracy of national estimates make the resource totals unreliable (Fettweis 1979). The United States and Russia, thanks to its huge but low-quality Siberian deposits, have the world's largest coal resources. The WEC (2001) survey lists the following quantities of additional U.S. coal resources in place: 445 Gt for U.S. bituminous coal (vs. 250 Gt of reserves), 274 Gt of subbituminous fuel (vs. 167 Gt of reserves), and 394 Gt of lignite (vs. 40 Gt of reserves). Australia and Germany have proportionately similar or even larger increments. Global coal resources are on the order of 3 Gt, but knowing their precise total is irrelevant because most of them will always remain undisturbed, either because they would be too expensive to extract or because the environmental consequences of coal combustion will continue to be much more important in determining the extent of production than concerns about its physical availability.

The major products of coal combustion include CO_2, H_2O, SO_2, and nitrogen oxides (NO and NO_2, commonly labeled NO_x, mostly from the splitting and subsequent oxidation of atmospheric N_2). Most coals have always been used directly to produce heat, but they can be also used as feedstocks for conversions producing secondary fuels in all three states. Metallurgical coke, produced by carbonization of low-ash, low-S bituminous coals in the absence of oxygen at temperatures up to 1400°C, is a highly porous solid fusion of carbon and residual ash strong enough to support ore and limestone charges. Its high heating value (29.6 MJ/kg) provides energy for reducing the ore. Briquetting, making solid shapes from coal dust or crushed and dried lignites by pressure molding, uses otherwise unwanted waste or inferior fuel.

The first coal-derived gaseous fuel was the low-energy town (coal) gas, first produced in 1812 in London by gasification of the fuel in closed retorts (net heating values 16–19 MJ/m³). This gas was used for urban lighting until the early twentieth century before it was displaced by electricity. High-energy synthetic gas (net heating values equal to that of natural gas, 30–38 MJ/m³) is made by a variety of gasification processes; the leading ones are Lurgi, Koppers-Totzek, and Winkler. Less than a stoichiometric supply of O_2 and combustion temperatures above 700°C produce CO- and H-rich gas that can be used directly as an energy source or as a feedstock for the catalytic Fischer-Tropsch process to produce liquid hydrocarbons, synthetic substitutes for crude oil-derived fuels. This method, as well as direct hydrogenation (Bergius process, the reaction with H_2 under elevated pressure and temperature), was deployed on a large scale by Germany during WW II.

Since 1945 its only major users are South African SASOL plants.

8.2 Hydrocarbons: Crude Oils and Natural Gases

Chemically, hydrocarbons are the simplest organic compounds composed of straight-chained, branched, or cyclic molecules that contain only carbon and hydrogen. In energetics, hydrocarbons are liquid and gaseous fuels that contain mixtures of these compounds. Crude oils exist as liquids both in underground reservoirs and after they have been brought to the surface. They were formed by heat and pressure transformation of marine biomass in sedimentary (mostly marine and also lacustrine) basins and are found either in continuous basin-centered accumulations or in structural or stratigraphic traps. This generally accepted genesis of crude oils has been challenged by an abiogenic theory advanced by some Russian and Ukrainian geologists (Kudryavtsev 1959; Simakov 1986) and promoted by Thomas Gold (1999). Instead of considering highly oxidized low-energy organic molecules as the precursors of all highly reduced high-energy hydrocarbons, as does the standard account of the origins of oil and gas, this theory attributes the formation of such highly reduced molecules to high pressures encountered in the Earth's mantle.

Accordingly, hydrocarbons should be found, and some were, in such nonsedimentary regions as crystalline basements, volcanic and impact structures, and rifts as well as in strata deeper below the already producing reservoirs. Moreover, if the origin of hydrocarbons is abyssal and abiotic, then there is a possibility that existing producing reservoirs can be gradually replenished, a phenomenon that would remove many oil and gas deposits from the category of nonrenewable energy resources (Mahfoud and Beck 1995; Gurney 1997). The theory has been dismissed by all but a few Western geologists. The isotopic composition of hydrocarbons—their content of ^{13}C matches that of terrestrial and marine plants and is lower than the isotope's presence in abiotically generated CH_4 (Hunt 1996)—is the best proof of their biogenic origin, and crude oil production from weathered or fractured crystalline rocks can be explained by migration from a flanking or overlying source rock.

Crude oils range from light mobile liquids of reddish-brown color to highly viscous black materials. Their composition is dominated by three homologous series of hydrocarbons: cycloalkanes (naphtenes or cycloparaffins in the oil industry, C_nH_{2n}), alkanes (commonly called paraffins, C_nH_{2n+2}), and arenes (aromatics, C_nH_{2n-6}, starting with benzene). Cycloalkanes are the most abundant compounds in crude oil (typically half of the weight), and methylcyclopentane (C_5H_{11}) and methylcyclohexane (C_6H_{13}) are the two most common cycloalkanes. The lightest alkanes, methane (CH_4) and ethane (C_2H_6), are gases at atmospheric pressure; propane (C_3H_8) and butane (C_4H_{10}) are also gases but are easily compressible to liquids (hence, known as liquid petroleum gases, LPG). Chains with five (pentane) to sixteen carbons are liquids, and the remainder are solids. Aromatics are unsaturated, highly reactive liquids named after the members with pleasant odors that share at least one benzene ring (C_6H_5) to which are attached long, straight side chains. Carbon accounts for 83%–87% of crude oil and hydrogen for 11%–15%, and the H/C ratio is 1.4–1.8, compared to about 0.8 for bituminous coals. Sulfur is the most common contaminant; sweet crude oils contain less than 0.5%, very sour crudes more than 4%. Other elements present are N and O (both <0.5%), and there may be traces of metals (Al, Cu, Cr, Pb). Oil's energy content is fairly uniform at 42–44 MJ/kg.

Different shares of constituent compounds result in specific densities, mostly between 0.78 and 0.89, but this attribute is commonly measured in °API (American Petroleum Institute) gravity, where °API = (141.5/specific density) − 131.5 (oil with specific gravity 1.0 has °API 10). Oils with gravities above 31.1° are classified as light, heavy oils have °API below 22.3. Many of the world's important crude oils are medium or only moderately light (Platt's 2005). Saudi crudes rate only °API 28–33, other Persian Gulf export streams range mostly between 30 and 35, and crude from Alaska's North Slope has °API 29. In contrast, some North African (Libyan and Algerian) and Nigerian oils are very light, with °API 37–44 and with very low pour points (−21°C to −36°C).

Natural gases are largely mixtures of the three simplest alkanes: methane (CH_4), ethane (C_2H_6), and propane (C_3H_8). Data available for distribution systems in the United States show CH_4 ranging from 73% to 95%, C_2H_6 from 3%–13%, and C_3H_8 from 0.1% to 1.3%. Butane, pentane, and sometimes a few higher homologues are also present in the extracted gas and are separated as natural gas liquids. CO_2, H_2S, N, He, and water vapor are found in many gases. The heat content of natural gases obviously declines with the presence of these impurities and rises with the share of higher alkanes. Extreme values for raw natural gases are approximately 30–45 MJ/m^3, with pure CH_4 at 35.5 MJ/m^3. They are the least-polluting fossil fuels, but because their energy density under normal pressure is only 1/1000 that of crude oil, their use as portable fuel is limited. Natural gases are commonly associated with or dissolved in crude oils, but they also exist as free (dry) gas that is not in any contact with crude oil in a reservoir or that occurs in entirely separate formations. A share of reservoir gas that is liquefied once it reaches the surface is known as natural gas liquid.

Use of hydrocarbons as fuels is of very recent origin. Because of seepages of oils "burning pillars" of natural gas, and bitumen pools, hydrocarbons were known from antiquity but were used only infrequently in building materials or as protective coatings (Forbes 1964). Notable exceptions include the burning of bitumens in Constantinople's *thermae* during the late Roman Empire and the Chinese use of natural gas in Sichuan, mainly to evaporate brines (Adshead 1992). This extraction began during the Han dynasty (200 B.C.E.), and it was made possible by the invention of percussion drilling, with teams of laborers raising and letting fall (by jumping on a lever) heavy iron bits attached to long bamboo cables from bamboo derricks (Needham 1964). In 1835 the deepest well reached the depth of 1 km (Vogel 1993).

The place with the longest tradition (since the early Middle Ages) of local crude oil use is the Absheron peninsula of the Baku region on the Caspian Sea in Azerbaijan. By the late eighteenth century there were scores of shallow wells from which oil was extracted for the production of kerosene (by primitive thermal distillation) for local lighting as well as for export (in skins) by camels and ships. In 1837 Russians set up the first commercial oil-distilling factory in Balakhani, and in 1846 the world's first exploratory oil well was drilled to the depth of 21 m in Bibi-Heybat. During the 1850s the higher cost of whale oil used for illumination led a number of entrepreneurs toward the small-scale beginnings of alternative oil-based industries.

In 1853 Abraham Gesner from Nova Scotia started producing kerosene from coal in his North American Kerosene Gas and Lighting Company on Long Island.

Ignacy Łukasiewicz, a Polish pharmacist, followed Gesner's work and became the first chemist to distill kerosene from oil. America's first oil well was hand-dug near Black Creek hamlet in a swamp in Lambton County in southwestern Ontario in 1858 by Charles Tripp and James Miller Williams. Percussion drilling, powered by a steam engine, was used for the first commercial Western oil well by Colonel Edwin L. Drake at Oil Creek in Pennsylvania on August 27, 1859, generally considered the beginning of modern oil era. The well penetrated 10 m of rock to strike oil at a depth of 21 m (Brantly 1971).

Kerosene shipments expanded rapidly during the 1860s, but the total consumption of refined oil products for lighting, heating, and lubrication grew slowly during the decades of rapid coal-mining expansion before 1900. This pioneering era bequeathed us a unit whose universal use has resisted replacement by a metric measure: a barrel of oil. Abbreviated as bbl (originally for blue barrel, the standard container of 42 U.S. gallons), this volume measure was adopted by the U.S. Bureau of the Census in 1872 and is used worldwide both for crude oil reserves and production. The different densities of crude oils preclude a single conversion rate to mass. Crude oils densities range from 740 kg/m^3 to 1040 kg/m^3, and 1 t thus equals 6.04–8.5 bbl. Most oils fall between 7 bbl/t and 7.5 bbl/t; 7.33 bbl/t is a common average, and BP uses 7.35 bbl/t (BP 2005).

Systematic estimates of the world's oil resources and reserves became possible once some exploratory drilling had been done in most of the world's promising oil provinces. *World Oil* and *Oil & Gas Journal* publish annual surveys of world oil reserves, and these, together with various primary sources and OPEC data, have been used to prepare BP's annual worldwide summary (BP 2006). The increase in global crude oil reserves during the third quarter of the twentieth century was particularly spectacular, from 85 Gbbl in 1950 to 715 Gbbl in 1974. By the late 1990s the total had surpassed 1 Tbbl, and in 2005 it was at least 1.19 Tbbl (BP 2005). Just five Persian Gulf countries, Iran, Iraq, Kuwait, Saudi Arabia, and United Arab Emirates, had about 60% of this total. The richest deposits outside the Middle East were in Russia and Venezuela (each about 6% of the total) and in Libya and Kazakhstan (each about 3%).

The worldwide R/P ratio has undergone major shifts since the end of WW II, when it stood at just 20 years (fig. 8.2). New discoveries pushed the ratio to nearly 40 years by the late 1950s, and after a period of decline and fluctuation, it reached a low of 26 years in 1979. Then it recovered and rose to more than 40 years by the end of the 1980s. At the beginning of 2005 it stood at 44 years, according to *Oil & Gas Journal*, or at 40.5 years, according to BP. This is a remarkable achievement. The difficult process of discovering and developing new oil reserves was not only able to maintain the global R/P ratio above 25 years but actually to increase it above 40 years despite the nearly 12-fold rise of global oil extraction between 1945 and 2005.

But this is not a universally shared appraisal. Unfortunately, absence of rigorous and uniform international standards in reporting oil reserves makes it certain that many national totals reported by these surveys are suspect, and this makes the global R/P ratios questionable. Reserve assessments lump figures for proved reserves with 50% probability estimates (P50 amounts that are as likely as not to be produced by a well or an oil field over its lifetime) and with estimates of possible reserves (probabilities exceeding these amounts are only 10% or 5%).

After sifting through the world's most extensive oil field database, Campbell and Laherrère (1998) concluded

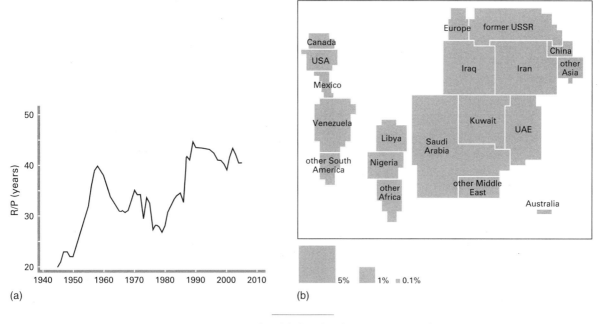

8.2 (a) R/P ratio for global crude oil, 1945–2005 (Smil 2003; BP 2006). (b) Map showing the areas of nations and continents proportional to the size of their conventional crude oil reserves (excluding oil sands). Constructed from data in BP (2006).

that the 1996 P50 reserves of conventional oil were only 850 Gbbl, almost 17% lower than the *Oil & Gas Journal* total of 1019 Gbbl and nearly 27% below the *World Oil* figure. The key reason for this discrepancy was a large jump in reported world oil reserves that came about not because of new discoveries but as a result of an accounting upgrade performed in 1987 by six OPEC members, reported in *Oil & Gas Journal* ("New data" 1987). Iraq and Iran were at war in 1987 and reported huge reserve additions over the previous year: Iraq's estimate was up by 112%, Iran's by 90%. Upward revaluations of existing reserves are to be expected, but Campbell and Laherrère

(1998) concluded that the magnitude of this increase (a 27% jump compared to 1986) and its timing in the wake of weakened world oil prices made it a prime example of politically generated reserve numbers, and that the totals' validity could be verified only if the relevant records of state oil companies were open for inspection.

Global estimates of undiscovered conventional crude oil resources have been even more contentious (Masters, Attanasi, and Root 1994; C. J. Campbell 1997; Odell 1999; USGS 2000; Deffeyes 2001; Odell 2004). Scores of estimates of ultimately recoverable oil have been published since 1940, and all but four totals published since

1980 are above 2.2 Tbbl, or 300 Gtoe, with the midpoint of recent assessments at about 3 Tbbl, or just over 400 Gtoe.

But because almost two-thirds of the world's oil reserves are in more than 300 giant fields, whose discoveries peaked in the early 1960s, Campbell and Laherrère (1998) saw no possibility of large-scale additions to the known oil pool and concluded that some 92% of all oil we will eventually recover has already been discovered. They also questioned the common expectation that additions to reserves in existing oil fields can appreciably delay the onset of declining output. In contrast, Odell (1984) thought 2 Tbbl of undiscovered oil to be an ultrapessimistic estimate, concluded that "the world is running into oil, not out of it" (Odell 1992, 285), and put the total oil in place at 3 Tbbl (Odell 1999). The USGS world assessment (Ahlbrandt et al. 2006) concluded that the mean value of the grand total of undiscovered conventional oil, reserve growth in discovered fields, and cumulative production up to the year 2000 was 3.021 Tbbl, 72% above the Campbell and Laherrère value.

Laherrère (2001) dismissed the precursor of this assessment, but he previously conceded that adding median estimates of undiscovered natural gas liquids (200 Gbbl) and nonconventional oil (700 Gbbl) would yield up to 1.9 Tbbl of oil yet to be produced (Laherrère 1996). Long-term prospects for the duration of an oil era must also consider the reserves in nonconventional deposits, which include very heavy crude oils, oil shales, and tar sands. Heavy crudes require special methods of recovery, and shales and sands have typically only small sharcs of oil (<10%) in the parental rock. The Athabasca oil sands in northern Alberta contain about 2.5 Tbbl of bitumen in place, of which some 315 Gbbl are potentially recoverable. In 2004, Canada's conventional crude oil and condensate reserves were about 5.2 Gbbl, but the Canadian Association of Petroleum Producers lists 6.9 Gbbl in oil sands under production, and the Alberta Energy and Utilities Board puts the remaining oil sands reserve under active development at 174.4 Gbbl of bitumen (CAPP 2005). When the *Oil & Gas Journal* added the latter total to its summary of crude oil reserves (Radler 2002), it boosted the global aggregate by about 17% and also diminished OPEC's share by more than 10%.

Long-term global assessments of natural gas supplies must also consider the existence of nonconventional resources. Conventional natural gas reserves added up to about 180 Tm^3 by the end of 2000, an equivalent of about 1.1 Tbbl (or roughly 150 Gt) of crude oil. They more than doubled between 1980 and 2005, and this substantial recent increase means that despite the nearly doubled extraction since 1980, the global R/P ratio in 2005 was about 67 years, compared to 56 years in 1980 and to just over 40 years in the early 1970s (BP 2005). Conventional reserves are concentrated in Russia (~27% of the total), Iran (~15%), Qatar (~14%), and Saudi Arabia and the United Arab Emirates (4% each). The Middle East claims about 40% of all reserves, much less than its share of crude oil. National R/P ratios range from less than 10 years for the United States to just over 80 years for Russia and more than a century for every major Middle Eastern natural gas producer.

As for the fuel's future, there are two opposing schools of thought. Pessimists argue that the ultimate resources of natural gas are actually smaller than those of crude oil, whereas those who believe in a gas-rich future say that the Earth's crust holds much more gas, albeit at depth and in formations currently categorized as nonconventional resources. Laherrère (2000) estimates ultimately recoverable gas at 1.68 Tbbl of oil equivalent, or about 96% of his total for ultimately recoverable crude oil. But because only about 25% of all gas has been already produced, compared to almost 45% of oil, the remaining reserves and undiscovered deposits of gas add up, according to this accounting, to about 1.28 Tbbl of oil equivalent, or roughly 30% more than all remaining conventional oil. Limited resources would thus make it impossible to contemplate natural gas as a long-term substitute for oil.

Again, as with crude oil, the latest USGS (2005) assessment of global resources of natural gas (Ahlbrandt et al. 2006) is much more optimistic, putting the total at about 415 Tm^3, or an equivalent of roughly 2.6 Tbbl (330 Gt) of crude oil and about 50% above the Campbell and Laherrère figure. Breakdown of the USGS estimates shows that only about 11% of ultimately recoverable gas has been produced, that remaining reserves account for some 31%, that their eventual growth should amount to nearly 24%, and that the undiscovered potential is almost exactly one-third of the total. Odell (1999) forecasts global natural gas extraction peaking at about 5.5 Gtoe by the year 2050 and declining to the year 2000 level only at the very beginning of the twenty-first century.

All these figures refer to conventional natural gas, the fuel that escaped from parental rocks and accumulated in nonpermeable reservoirs. Nonconventional gas comprises resources that are already being recovered, above all, methane in coalbeds, as well as much larger deposits in tight reservoirs, high pressure aquifers, and methane hydrates whose eventual recovery still awaits needed technical advances (Rogner 2000). Gas in coalbeds is absorbed into coal's structure. Gas in tight reservoirs is held in impermeable rocks whose leakage rate is slower than their filling rate; these would have to be fractured inexpensively in order to allow economic extraction. Global resources of geopressured gas were estimated to be more than 100 times the reserves of the fuel (Rogner 2000).

Methane hydrates (clathrates) were formed by the gas released from anoxic decomposition of organic sediments by methanogenic bacteria and trapped inside rigid lattice cages formed by frozen water molecules (Kvenvolden 1993; Lowrie and Max 1999). The upper depth limit for their existence is close to 100 m in continental polar regions and about 300 m in oceanic sediments, and the lower limit in warmer oceans is about 2000 m. Fully saturated gas hydrates have one CH_4 molecule for every 5.75 molecules of H_2O, which means that 1 m^3 of hydrate can contain as much as 164 m^3 of methane (Kvenvolden 1993). There are no reliable estimates of the total amount of methane in hydrates, but coastal U.S. waters may contain as much as 1,000 times the volume of U.S. conventional gas reserves (Lowrie and Max, 1999). The USGS estimates the global mass of organic carbon locked in gas hydrates at 10 Tt, or roughly twice as much as the element's total in all fossil fuels (Dillon 1992).

Nearly 20,000 hydrocarbon fields have been discovered worldwide, but more than 70% of recoverable oil and gas is in just 500 giant formations, each containing at least 80 Mm^3 of oil or 85 Gm^3 of gas or any combined energy equivalent and located mostly in five (of 260) producing basins: Persian Gulf–Zagros, West Siberia, Gulf of Mexico, Volga-Ural, and Maracaibo (Nehring 1978; Perrodon 1985; Tiratsoo 1986; Brooks 1990; Downey 2001). Of the world's 15 largest fields, containing more than half of all recoverable hydrocarbons, 12 are in the Persian Gulf–Zagros basin. In 2005 the region contained just over 60% of the world's crude oil reserves, and Saudi Arabia alone more than 20%. Discoveries of giant fields have become rare, and new reserves will have to come from smaller continental finds and from offshore areas that may contain as much as 40% of the world's undiscovered oil.

The Persian Gulf–Zagros Basin will remain an astounding singularity. Two supergiant oil fields, the Saudi al-Ghawār with reserves of at least 550 EJ and the Kuwaiti al-Burkān with at least 470 EJ, will retain their position as the world's largest oil reservoirs (fig. 8.3). The Cantarell complex in Mexico and Venezuela's Bolivar rank third and fourth, followed by Safānīya-Khafjī near the Saudi-Kuwait border. Of the world's ten largest natural gas fields, eight are in Russia. The Urengoy field in Western Siberia, discovered in 1966, producing since 1978, and holding some 270 EJ of reserves, is unlikely to be surpassed as the world's largest natural gas deposit. Its neighboring Yamburg field and Orenburg field in the Volga Basin rank second and third.

Resources in place in any small hydrocarbon fields prorate to less than 1 GJ/m^2, and extensive giants such as Hugoton-Panhandle in Texas or Alberta's Pembina have less than 10 GJ/m^2. The richest fields contain 10^1–10^2 GJ/m^2. Prudhoe Bay rates about 25 GJ/m^2; the Algerian Hassi Messaoud, 35 GJ/m^2; the Saudi al-Ghawār, originally 100 GJ/m^2; California's Wilmington, 200 GJ/m^2; California's Ventura-Rincon, 300 GJ/m^2, and the Kuwaiti al-Burkān, more than 1 TJ/m^2. The Green River formation, the world's largest concentration of oil, interspersed in the shales of Colorado, Utah, and Wyoming, has a total energy content rivaling al-Burkān's density, but deposits of richer shales yielding 100–400 L/t of rock prorate to no more than 185 GJ/m^2 (Dinneen and Cook 1974). The extraordinarily high energy cost

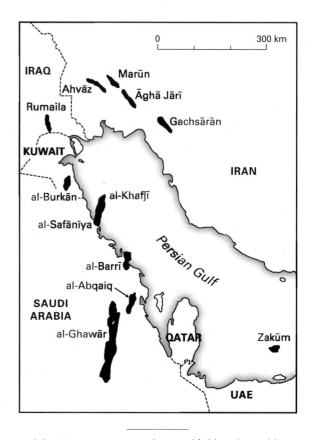

8.3 Major supergiant and giant oil fields in the world's richest oil province, the Persian Gulf–Zagros Basin. From Smil (2003).

of extracting liquid fuel from these rocks makes it highly unlikely that this will be a commercial source of oil.

The total bitumen content of Canada's oil sands prorates to about 100 GJ/m^2, and the ultimately recoverable share has an energy density of nearly 13 GJ/m^2. Only about 20% of the recoverable share could be reached by surface mining; the rest would have to be extracted *in situ*. The first operation of this kind, Imperial Oil's Cold Lake Project, uses injections of hot, pressurized steam (300°C, 11 MPa) to recover about 25%, and with follow-up processes up to 35%, of the bitumen present in sands (Lui 2004). Another large-scale nonconventional operation is the extraction of bitumen from the enormous deposits in the Orinoco Belt. The final product, 70% natural bitumen, 30% water, and a small amount of an additive that stabilizes the emulsion, is sold as liquid Orimulsion by Petróleos de Venezuela (PDVSA 2001).

At the beginning of 2005 total conventional reserves of the three principal fossil fuels added up to about 31 ZJ, with coals accounting for about 60% of the total and hydrocarbons 40%. A gradual shift of some nonconventional oil and gas resources into the recoverable category (already under way with heavy oils, oil sands, and coalbed methane) could substantially increase the hydrocarbon share. The continental division of fossil fuel reserves shows Asia's lead, with nearly 50% of the global total, and Africa's less than 10%. In per capita terms, Russia has the largest reserves among the major economies, and China, because of its large population, the smallest.

8.3 From Extraction to Combustion: Modern Fossil Fuel Industries

Modern extraction, transportation, and processing of fossil fuels are characterized by an extraordinarily high degree of mechanization, which makes it possible to produce and distribute enormous masses of solids, liquids, and gases at affordable prices. Fossil fuel industries are among the least labor-intensive sectors of modern economies; their activities and markets are truly global; and their ability to deliver energy at a low, often declining real cost underpins the current high-energy civilization. Coal dominated globally until the early 1960s, and as oil

and gas began to contribute more than one-half of the world's fossil fuels output, their extraction began to shift increasingly offshore.

Coal extraction remained muscle-powered and very risky until the early twentieth century. Traditional methods were first replaced by handheld pneumatic hammers, then by mechanized cutters and loaders, and finally by continuous miners, machines that grind coal from the face and dump it on an adjoining belt. In 1920 all U.S. coal mined underground was manually loaded into mine cars, but by the 1960s nearly 90% was machine-loaded (Gold et al. 1984). The most economical and safest method of underground extraction is longwall mining, introduced during the 1960s (Ward 1984; Barczak 1992). This technique uses large drum-shaped shearers to cut panels up to 2.5 m high, 80–200 m wide, and up to 1.5 km long that are delimited by two side tunnels. Miners remain always under a protective canopy of hydraulic steel supports, which are advanced and reset as a face is completed, leaving the unsupported roof behind to cave in (fig. 8.4). This technique recovers more than 90% of coal compared to just 50% in room-and-pillar operation, and it now accounts for just over half of U.S. underground extraction (EIA 2006b).

The largest coal production increase came from new large-scale opencast operations in the United States, Russia, Australia, and China. By the year 2000 the largest of these mines produced annually more coal that did the entire formerly prominent coal-mining nations from their underground seams. With nearly 75 Mt a year mined in 2004, the Powder River Coal Company's North Antelope/Rochelle mine in Wyoming extracted more solid fuel than did the Ukraine, the world's tenth largest bituminous coal producer (EIA 2005c). Much better overall safety and the virtual elimination of traditional

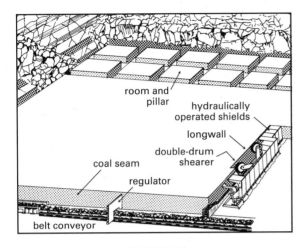

8.4 Comparison of traditional room-and-pillar extraction and modern longwall mining. From Smil (2003).

mining health risks are additional factors responsible for the growing adoption of surface mining. By 2000 it produced 65% of U.S. output, about 40% of Russian output, but less than 10% of Chinese extraction.

Remarkable economies of scale were achieved by using larger shovels and walking draglines to remove thick overburden layers (Hollingsworth 1966). In the United States, Bucyrus-Erie and Marion were the dominant makers of these machines (G. B. Anderson 1980). Capacities of the largest shovel dippers rose from 2 m^3 in 1904 to 138 m^3 in 1965. Similarly, bucket volumes of the largest walking draglines increased from less than 3 m^3 before WW I to 168 m^3. These machines had to be erected on site, and they consumed as much electricity as a city of 15,000–20,000 people (the world's largest one, as much as 100,000 people). With 12,700 t, the Marion 6360 shovel, built in 1965 for Captain Mine in Illinois, was the world's heaviest excavator. With 12,247

t, the Bucyrus-Erie 4250W walking dragline at Ohio Power Company's Muskingum mine was not far behind. Such machines allowed the maximum overburden to coal ratio to rise from 1–2 before WW II to 5–6.

Mining mechanization resulted in the consolidation of coal extraction into a smaller number of larger operations, and the rising productivity of labor led to falling labor force totals and plummeting occupational deaths and injuries. In Germany's Ruhr region the share of mechanical coal extraction rose from less than 30% in the early 1950s to more than 95% by 1975 while the total number of operations fell by about 90% (Erasmus 1975). The average productivity of U.S. underground mining rose from less than 1 t per work shift in 1900 to more than 3.5 t per labor hour by 2003. In surface operations it approached 10 t per labor hour, with productivities as high as 6 t per labor hour in longwalls and 35 t per labor hour in the surface mines in Wyoming (EIA 2006b). By the year 2000 U.S. coal mining produced almost four times more fuel than it had in 1900, with less than 20% of the former labor force. At the same time, accidental deaths declined by 90% since the early 1930s (MSHA 2000).

Extraction productivities span a broad range, from no more than a few hundred kilograms (2–6 GJ) in primitive small rural Chinese mines to more than 2,000 t per workday (>40 TJ) in the largest U.S. surface mine. Differences in seams and extraction techniques result in underground power densities as high as 1–2 kW/m^2 in longwall mining but only 100–200 W/m^2 in smaller pits with thin seams. Surface mining needs more land than just the area overlying the worked seam, for temporary displacement of overburden before eventual reclamation and for on-site transportation and buildings. The power densities of extraction in opencast mines are commonly 2–8 kW/m^2, with the most productive operations having 10–30 kW/m^2. During the early 2000s, Wyoming's large surface mines averaged 8–10 kW/m^2, Victoria's Latrobe Valley extraction rated up to 28 kW/m^2, and Queensland's Blair Athol approached 33 kW/m^2 (Australian Government 2005).

Among the world's remaining large coal producers, only China followed a singular path as the slow progress of mechanization in its large, state-owned enterprise was accompanied by two waves of indiscriminate openings of nonmechanized small coal mines (Smil 2004a). The first brief pulse began in 1958, a part of Mao Zedong's delusionary Great Leap Forward, when some 20 million peasants opened up more than 100,000 small mines to produce coal mostly for local smelting of inferior pig iron. The second expansion was a part of post-1980 economic reforms. By 1997 half of China's 1.3 Gt of raw coal was coming from some 82,000 small mines known for their low productivity and dangerous working conditions. Their number was cut to 36,000 by the year 2000. China is also the only major producer that uses mostly raw fuel. Elsewhere the fuel is processed before marketing by washing (based on the difference of specific gravities between lighter coal and heavier incombustible waste, including pyritic sulfur), screening, and crushing to sort the fuel to uniform sizes. Coal preparation facilities have power densities of 8–10 kW/m^2.

Seaborne trade of hard coal—Australia, China, and Indonesia are leading exporters—uses large bulk carriers with capacities of up to 200,000 dwt. Long-distance land transport of coal remains dominated by railways. Unit trains are the best solution for moving bulky and dirty solids. They are permanently coupled assemblies of powerful diesel locomotives and about 100 lightweight aluminum cars, each with capacity of up to 90 t, that shuttle between a mine and an electricity-generating plant or a port on runs of 10^2–10^3 km (Glover, Hinkle, and Riley 1970). At arrival, a rotary car dumper turns the cars 140°–160° to unload coal, or mechanical trippers open the hatches for bottom-dump unloading. Typical corridors claimed by railways are 20–30 m wide, but cuts and fills may easily double that width. Dedicated unit train railroads that move coal 500–1500 km to large power plants have annual throughput densities between 100 W/m^2 and 400 W/m^2, depending on the length of the run, width of the right-of-ways, and capacity. Moving coal at the plant from bunkers, silos, or outdoor stockpiles to coal mills or directly to boilers is done by belt conveyors, stackers, reclaimers, in-ground hoppers, or plow feeders (McGraw 1982).

All these advances have been essential in order to multiply the world's coal output (fig. 8.5). The 1900 total of less than 800 Mt of hard coals and lignites was doubled by 1949 to nearly 1.3 Gt of hard coals and about 350 Mt of lignites. Yet another doubling of the total coal tonnage (in terms of hard coal equivalent) took place by 1988, and global extraction peaked the next year at nearly 4.9 Gt, with about 3.6 Gt contributed by hard coals and 1.3 Gt by lignites (UNO 1990). About 40% of that year's lignite production was coming from the now defunct Communist states of East Germany and the USSR, whose lignites were of particularly low quality, averaging just 8.8 and 14.7 GJ/t, respectively (UNO 2000). The 1988 total was matched in 2003 with 4.04 Gt of hard coals and 886 Mt of brown coals and lignites, and by 2005 the output grew to 4.97 Gt of hard coal and 905 Mt of brown coal and lignites (WCI 2006). Growing output has been accompanied by declining relative importance, but there has been a recent gain.

FOSSIL FUELS

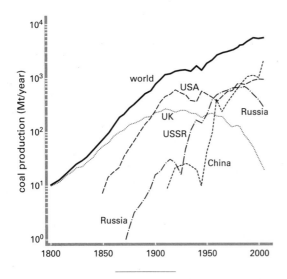

8.5 Coal production, 1800–2005 (Smil 2003; BP 2006).

In 1900 coals accounted for about 95% of the world's total primary energy supply (TPES). Coal's share then declined during every year of the twentieth century, falling to below 75% by 1939 and to less than 50% by 1962. OPEC's oil price rises of the 1970s engendered unrealistic (albeit widespread) hopes of coal's comeback (Wilson 1980). Coal's share slipped to just below 30% by 1990 and to no more than 23% of the world's TPES by 2000, but it rose to nearly 28% by 2005. While the ranks of hydrocarbon producers have been steadily expanding, those of coal-mining countries have been declining. In 2000 only 16 countries extracted annually more than 25 Mt, and the six largest producers (in the order of energy content, the United States, China, Australia, India, Russia, and South Africa) accounted for slightly over three-quarters of the world's output. The United Kingdom, the world's second largest coal producer in 1900,

extracted less than 20 Mt in 2000, and its peak labor force of 1.25 million miners in 1920 was reduced to fewer than 10,000 (Hicks and Allen 1999). Many African and Asian countries use no coal, but in 2005 the fuel still provided 76% of South Africa's, nearly 70% of China's, and nearly three-fifths of India's TPES, but only 24% of the U.S. and 16% of the Russian (BP 2006).

Another perspective on coal's absolute rise and relative decline is offered by the fact that during the twentieth century the energy content of extracted coals increased less than 4.5 times although the world's total fossil fuel consumption rose 15-fold. But coal mined during the twentieth century contained more energy than any other fuel, about 5.5 ZJ. In contrast, the energy content of crude oil extracted between 1901 and 2000 was about 5.3 ZJ, but during the century's second half it surpassed that of coal by roughly one-third. China and India are the only major producers where coal is still an important household fuel; elsewhere it is used mostly to generate electricity and to produce metallurgical coke and cement.

More efficient iron smelting lowered the specific use of coke from 1.3t/t of hot metal in 1900 to less than 0.5 t/t in 2000 (Smil 1994; de Beer, Worrell, and Blok 1998). Steel recycling (some 350 Mt of the scrap metal, an equivalent of ~40% of global steel output, is now reused annually) and direct iron reduction made blast furnaces less important, and injection of 1 t of pulverized coal directly into a furnace displaces about 1.4 t of coking coal (WCI 2005). Coke now needs only about 17% of the world's extracted hard coals, or just over 600 Mt in the year 2000 (WCI 2005). In contrast, coal use by cement production has been increasing (more than 1.8 Gt annually since 2002, with the processing requiring 3–9 GJ/t). In 2000 it amounted to about 150 Mt, mostly in China, now the world's leading producer, with Japan, the United States, and India each using less than 10 Mt/year (WCI 2005).

The first major technical advance in oil and gas extraction was the replacement of percussion (cable-tool) rigs. The first rotary rig drilled the Corsicana field in Texas in 1895, but the technique began to spread only after WW I (Brantly 1971). Its main components are a heavy rotating circular table, drill pipes inserted in its middle, and a bottom hole assembly at the drill string's end (Devereux 1999). Tables were originally driven through gearings by steam engines, later by diesels or diesel-powered electric motors. As the drilling progresses, sections of threaded drill pipes (9 m long) are added, but as a drilling bit wears out, the pipes have to be withdrawn from the well, stacked, and reattached after a new bit is mounted (the process called tripping). The weight of the drill string is increased by heavy drill collars placed just above the bit.

Drilling mud (water-, oil-, or synthetics-based fluid) is pumped at high pressure down the drill string and through the bit to cool it, to remove cuttings, and to exert pressure on the well sides to prevent the hole from caving in (Van Dyke 2000). Casing is installed and cemented in place to stabilize the well. Rotary drilling was much improved by replacing fishtail and circular-toothed drills (only good for penetrating soft formations) by a new bit invented by Howard R. Hughes in Texas in 1907. His design speeded up the drilling tenfold, and his company's engineers subsequently introduced a number of improvements, including a three-cone bit in 1934 that provided much better support on the well bottom, reduced vibration, and resulted in faster yet smoother drilling. Other innovations included the first diamond drill in 1919 (modern bits are covered with a layer of fine-grained synthetic diamonds), heavy drill collars to add weight and rigidity, and various well control devices to cope with high pressures in the well and to prevent catastrophic blowouts.

Efficient cementing of wells and automatic well logging spread after WW I. Erle P. Halliburton patented his cement jet mixer in 1922 (Allaud and Martin 1976; Haley 1959). In 1912, Conrad Schlumberger proposed the use of electrical measurements to map subsurface rock bodies, and in 1927 his son-in-law, Henri Doll, produced the first electrical resistivity well log in the Pechelbronn field in France. In 1931 the company introduced electrical well logging, the simultaneous recording of resistivity and spontaneous potential produced between the drilling mud and formation water present in permeable beds. A multivalve stack to control well flow was introduced in 1922, and by the 1990s the best control-flow devices could hold pressures up to 103 MPa. Myron Kinley and Red Adair pioneered the risky controls of well blowouts and fires (Singerman 1991).

Faster operation and increased drilling depths—less than 2000 m before WW I, 4500 m by the late 1930s, and 6000 m a decade later—led to a spate of new discoveries. In the United States 64 giant oil fields were discovered between 1900 and 1924, and 147 were added during the subsequent 25 years (Brantly 1971). The largest pre-1950 finds included supergiant Kirkūk in Iraq (1927), Abqaiq (1940) and al-Ghawār (1948) in Saudi Arabia, al-Burkān in Kuwait (1938) and oil fields in Venezuela. During the 1950s the addition of Safānīya-Khafjī in Saudi Arabia (1951), Rumaila in Iraq (1953), and Ahvāz in Iran (1958) turned the Middle East into the world's largest oil province (Nehring 1978). By the 1970s production from wells deeper than 5000 m became common in some hydrocarbon basins, particularly in Oklahoma's Anadarko Basin.

FOSSIL FUELS

While reserve densities of commercially exploited coals and hydrocarbons overlap, ranging from 10^0 GJ to 1 TJ/m^2, power densities of oil and gas extraction are necessarily much lower when prorated over the total reservoir area. Coal mining will remove at least half and often all the fuel in accessible seams while oil and gas production exhausts the reservoir gradually, over a period of many decades, and the best unaided recovery rates do not surpass 35% of the original oil content. Even in the richest fields, extraction densities rarely surpass 200 W/m^2 of total area. Al-Burkān, as delineated from satellite images, rated about 250 W/m^2 in the early 2000s, Ventura-Rincon some 130 W/m^2 since its beginnings in 1916, Hassi Messaoud just over 30 W/m^2, and al-Ghawār about 10 W/m^2 since 1948, all values much below the typical coal extraction densities of 1–20 kW/m^2.

Much higher power densities result when counting only the land that is actually claimed by extraction. In the world's richest oil fields, individual wells produce annually at least 1 PJ of crude, the peak Middle Eastern output was nearly 12 PJ/well, and al-Ghawār used to flow at up to 40 PJ/well. Only a small percentage of these oil field areas are actually taken up by surface structures or reserved for the right-of-ways of gathering pipelines. Actual extraction power densities are thus at least 10–20 kW/m^2. In contrast, the annual production average for more than 500,000 U.S. wells operating in 2005 was just 34 TJ/well. Dominant (>95%) lift wells averaged less than 25 TJ/well, and the best estimates available for U.S. oil fields fall within 1–3 kW/m^2. The typical power densities of natural gas extraction are 10–15 kW/m^2 but are reduced by up to 1 OM by adding the right-of-ways for gathering pipelines and space for field gas processing.

Rapid post–WW II expansion of the oil industry, from less than 500 Mt in 1950 to about 3.5 Gt by the year 2000 when there were almost 1 million wells in more than 100 countries, and to about 3.9 Gt in 2005 (fig. 8.6), drove the scaling-up of all of its infrastructural elements. This was a particular challenge for offshore production, which now supplies nearly one-third of the world's crude oil. The first drilling from wharves was done in California as early as 1897, and the first platforms were extended into Venezuela's Lake Maracaibo in 1924. The first well drilled out of the sight of land (nearly 70 km from the Louisiana shore) was completed by Kerr-McGee Corporation in the Gulf of Mexico in just 6 m of water in 1947 (Brantly 1971). Offshore drilling then progressed from small jackup rigs for shallow near-shore waters (the first one, *Offshore Rig 51*, in 1954) to drill ships capable of working in up to 3000 m of water and semisubmersible rigs.

The Transocean company pioneered self-propelled jackups, dynamically positioned semisubmersibles, and rigs capable of year-round drilling in extreme environments (Transocean 2003). During the 1970s the company introduced the *Discoverer* class of drill ships; by the end of the twentieth century their fifth generation was capable of working in waters up to 3000 m deep. But Shell Oil was the first company to deploy a semisubmersible rig, *Bluewater I*, in 1961 in the Gulf of Mexico. By the century's end the industry's worldwide surveys listed nearly 380 jackups (mostly for work in waters of 30–90 m), about 170 semisubmersibles (for water depths up to 300 m), and about 80 drill ships and barges for a total of nearly 640 marine drilling rigs.

There was a concurrent increase in the size of offshore production platforms. In November 1982, *Statfjord B*, a concrete platform weighing more than 800,000 t in the

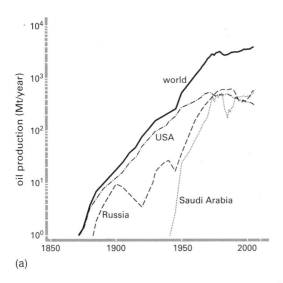

(a) (b)

8.6 (*a*) Global crude oil production, 1900–2005 (Smil 2003; BP 2006). (*b*) *Ursa* tension leg platform operated by SEPCo 120 km southeast of New Orleans (photograph courtesy of SEPCo, Houston).

Norwegian sector of the North Sea, became the heaviest object ever moved by people. By 1989, Shell's *Bullwinkle*, sited in 406 m of water and weighing about 70,000 t, became the world's tallest pile-supported fixed steel platform (SEPCo 2004). By 1999, Shell's *Ursa* tension-leg platform was the largest structure of its kind, displacing about 88,000 t, rising 146 m above water, and requiring 16 steel tendons to anchor it to 340-t piles placed 1140 m below the surface (fig. 8.6) (SEPCo 2004). SPAR, another innovative design, is also moored by steel lines, but its deck is supported by a single large-diameter cylinder.

Reservoir yields increased as extended reach, horizontal drilling, and improved methods of secondary and tertiary oil recovery extracted more oil from parental rocks.

Directional drilling uses steerable downhole motors that are powered by the pressurized mud flowing in cavities between a spiral-fluted steel rotor and a rubber-lined stator (SPE 1991; Cooper 1994). Steerable downhole motors are also used by a new drilling technique that was introduced during the late 1990s and that unreels narrow (5–7 cm diameter) steel tubing, wrapped on a large drum mounted on a heavy trailer, into a well (Williams et al. 2001). The benefits of directional and horizontal drilling are clear; it can reach much larger volume of hydrocarbon-bearing strata from a single drill site. Even for the lightest oils, natural reservoir pressure releases no more than 40% of the fuel originally present in the parental rock. For heavy oils the share is below 10%, and means in most oilfields are 25%–35%. Various

FOSSIL FUELS

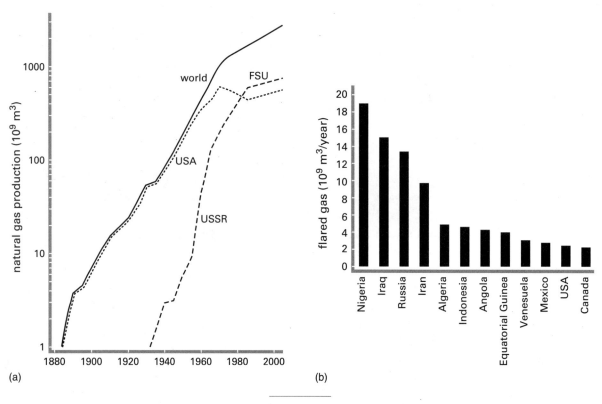

(a)

(b)

8.7 (a) Global natural gas production, 1900–2005 (Smil 1994; data from BP 2006). (b) Major locations of natural gas flaring (data from Mouton 2005).

techniques enhance these rates (Baviere 1991; Carcoana 1992; Lakatos 2001). Water flooding or gas injection lift 5%–10% more oil, steam drive can recover 30%–60% more, CO_2 flooding 20%–30%, and surfactant flooding 15%–40%.

Before WW II associated natural gas was mostly flared. The postwar expansion of petrochemical industries and the rising household and industrial demand for clean fuel pushed natural gas consumption from about 200 Gm^3 in 1950 to 1.2 Tm^3 by 1975 and 2.7 Tm^3 in 2005, but this rise was accompanied by further absolute increases in gas flaring (fig. 8.7). By 1975 about 170 Gm^3/year were flared, equal to 14% of total production. Subsequent reduction brought the total to just over 100 Gm^3, equal to about 4% of global extraction in 2005 or to half of Russia's gas exports (Mouton 2005). The leading wasters are Nigeria (nearly 20 Gm^3/year), Iraq, Russia, Iran, and Algeria, and the major sites of these flarings stand out on the night-time satellite images as the brightest spots on the Earth's surface.

Large tankers made crude oil shipping so inexpensive that the distance from market became almost irrelevant (Ratcliffe 1985). Between 1884, the year of the first such vessel, and 1921 record tanker capacities rose from just over 2000 dwt to more than 20,000 dwt, and then they stagnated for over a generation. In 1959 came *Universe Apollo*, the first 100,000-dwt ship; in 1966 the record-sized 150,000-dwt *Tokyo Maru* of Ishikawajima-Harima Heavy Industries; and later in the same year the 210,000-dwt *Idemitsu Maru*. By 1973 there were 366 very large or ultralarge crude oil carriers, with the largest vessels in excess of 300,000 t (Kumar 2004). The long-term trend clearly pointed to ships of 1 Mdwt. But the growth peaked with the launching of *Seawise Giant* in 1975 and its enlargement three years later.

The world's largest ship was hit in 1988 during the Iran-Iraq war, it was subsequently relaunched as the 564,650-dwt (nearly 459 m long). *Jahre Viking*—and now, renamed yet again (*Knock Nevis*)—is moored in Qatar as a floating storage and offloading unit. Supertankers reached their limit because of operational considerations. Very large ships have to reckon with the depths of sea routes and the long distances needed to stop; accidental oil spills affect marine life and pollute beaches and shores for years to come; and the penalties sought, as illustrated by the *Exxon Valdez* Alaska spill on March 24, 1989, can reach billions of dollars and tanker insurance becomes prohibitively expensive. Also, megaships could find only a handful of ports in which to moor, losing the flexibility with which multinational oil companies use their tankers.

Shipments of natural gas in liquefied natural gas (LNG) tankers (fig. 12.6) are much more expensive. These vessels carry CH_4 at $-162°C$ in four or five insulated spherical containers. They were used for the first time in 1958 to ship Algerian gas to the Britain (*Methane Pioneer*) and France, and in 1960 (*Bridgestone Maru*) to take Indonesian gas to Japan (Corkhill 1975). By the year 2005 there were 29 liquefaction facilities in 17 locations in Africa (Algeria, Libya, Egypt, Nigeria), the Middle East (Oman, Qatar, United Arab Emirates), Asia (Brunei, Indonesia, Malaysia), Australia, Alaska, and Trinidad. Scores of LNG tankers, most with capacities in excess of 100,000 m^3, carried about one-quarter of globally traded natural gas to 40 regasification sites in Japan, South Korea, the United States, Europe, and the Caribbean (IEA 2005). LNG terminals are relatively compact but require extensive security zones; their throughput densities range between 2 kW/m^2 and 60 kW/m^2; Qatar's LNG terminal, the world's largest, rates 2.8 kW/m^2. Regasification terminals are equally compact, with throughputs between 6 kW/m^2 and 50 kW/m^2. Increasing LNG trade has led to many new designs for offshore regasification terminals known as floating storage and regeneration units.

Pipelines are the least expensive choice for moving hydrocarbons on land, thanks to their compactness (a 1-m-diameter line can carry 50 Mt oil/a), cleanliness, reliability, safety, and hence excellent environmental acceptability. The first long-distance oil lines were laid in the United States during the 1870s, but worldwide expansion began only after WW II. U.S. lines from the Gulf to the East Coast were eclipsed by the world's longest crude oil pipelines laid during the 1970s to move Western Siberian crude oil to Europe. The Ust'-Balik-Kurgan-Almetievsk line, 2120 km long and with a diameter of 120 cm, can carry annually up to 90 Mt of crude oil from a supergiant Samotlor oil field to European Russia, and then almost 2500 km of large-diameter lines are needed to move this oil to Western European markets. The TransAlaska line, also 120 cm in diameter, is 1280 km long and annually moves about 100 Mt of crude oil from the North Slope to Anchorage.

Natural gas is not as easily transported as crude oil (Poten & Partners 1993; IEA 1994). The specific energy needed for its pumping is about three times that of energy required to move crude oil. The longest and widest natural gas pipelines (6500 km long and up to 142 cm in diameter) carry fuel from the supergiant fields of Medvezh'ye, Urengoy, Yamburg, and Zapolyarnyi in the Nadym-Pur-Taz gas production complex in Western Siberia to European Russia and then all the way to Western Europe, with the southern branch going to northern Italy and the northern branch to Germany and France. Undersea links are practical only over short and shallow stretches, as in the North Sea (bringing the gas to Scotland and the Continent) and across the Sicilian Channel and the Messina Strait to Italy (bringing Algerian gas to Europe).

Natural gas is stripped of any undesirable ingredients, most often H_2S and moisture, before it is compressed and piped, and this preparation can be accomplished with minimal land requirements. The throughput power densities of gas-processing plants may be as high as 70 kW/m^2 and rarely are below 50 kW/m^2. Pipelines usually need a 25–30 m corridor for construction; afterwards only access strips of up to 10 m may be necessary. Compressor stations take up to 20,000 m^2 at 80–120-km intervals; pumping stations claim up to ten times as much area every 100–160 km (the TransAlaska line has 11 of them, one every 116 km). Aggregate U.S. data and assumptions of 7–10-m right-of-ways for operating lines prorate to average throughputs of 200–300 W/m^2 for natural gas and 350–480 W/m^2 for crude oil. Major lines do much better. Even with a 30-m right-of-way the TransAlaska line has maximum design throughput power density of 3.7 kW/m^2, and its actual maximum in the year 2000 prorated to 1.7 kW/m^2 (Alyeska 2003).

While modern coal preparation is just a set of simple physical procedures, crude oils undergo a complex process of refining, a combination of physical and chemical treatments. Refining separates the complex mixture of hydrocarbons into more homogeneous categories and adds value to final products. During straight thermal distillation light naphtha (pentanes to heptanes) boils away at 27–93°C, heavy naphtha (compounds with up to 10 C atoms) at 93–177°C; then comes kerosene (jet fuel) at up to 325°C, followed by light (diesel) gas oil and heavy gas oil (up to 565°C). Residual solids (coke, asphalt, tar, waxes) boil only above 600°C. Early refining relied strictly on heat (delivered as high-pressure steam at 600°C) to separate major fractions. Given the quality of dominant flows, this produced largely medium and heavy products, and without an effective technical solution the extent of driving and flying would have remained restricted because of crude oil quality.

In 1913, William Burton patented thermal cracking of crude oil, which relied simply on the combination of heat and high pressure to break heavier hydrocarbons into lighter mixtures. A year later Almer M. McAfee patented the first catalytic cracking process, which used aluminum chloride to break long-chained molecules into shorter, more volatile chains. But because the relatively expensive catalyst could not be recovered, thermal cracking remained dominant until 1936, when Sun Oil put on line the first catalytic cracking unit, designed by Eugène Houdry, to produce high-octane gasoline (Houdry

1931). This method required shutting down the operation to regenerate the aluminosilicate catalyst. Warren K. Lewis and Edwin R. Gililand introduced a moving-bed arrangement, with the catalyst circulating between the reaction and the regeneration vessels. By 1942, 90% of all U.S. aviation fuel came from catalytic cracking.

An even greater yield of high-octane gasoline was achieved in 1942 with the commercialization of airborne powdered catalyst (Campbell et al. 1948). This fluid catalytic cracking, invented by a group of four Standard Oil chemists in 1940, was improved in 1960 with the addition of synthetic zeolite, a crystalline aluminosilicate with uniform pores, to act as an exceptionally active and stable catalyst facilitating the cracking of heavy hydrocarbons (Plank and Rosinski 1964). Zeolite Y improved the gasoline yield by as much as 15%. During the 1950s, Union Oil Company developed hydrocracking, which combines catalysis at temperatures above 350°C with hydrogenation at relatively high pressures (10–17 MPa). Large-pore zeolites loaded with a heavy metal (Pt, W, or Ni) serve as dual function catalysts to produce high yields of gasoline and low yields of less desirable CH_4 and C_2H_6.

Large refineries process in excess of 100,000 bbl/day, or at least 5 Mt/a of crude (input rating 6.6 GW). The average for Texas refineries is twice as large, and the state's largest refinery processes annually nearly 28 Mt of crude oil (input rating 37 GW). In the early 2000s the world's largest refineries were in South Korea, where Ulsan's capacity was nearly 44 Mt/a and Yosu's 32 Mt/a, and in Singapore, where Exxon's Jurong Island rated 30 Mt/a. Europe's largest facility was Pernis in the Netherlands (nearly 21 Mt/a) and the Middle East's was Saudi Rās Tanūra (26 Mt/a). Crude oil refining, producing a wide variety of highly flammable gases and liquids, requires safety precautions in locating the processing and storage facilities, preventing spills, and facilitating fire fighting.

Minimum spacing of 60–75 m is mandatory to separate many parts within a refinery; large storage units cannot be placed closer than one-sixth to one-third of the sum of adjacent tank diameters. The standard planning requirement for modern refineries prorated to throughputs of about 3300 W/m^2 during the last quarter of the twentieth century (Gary and Handwerk 1984). Actual footprints depend on the extent of buffer zones and on-site storage, on the mode of crude oil delivery and processing, and on areas reserved for possible expansion. As a result, actual throughput power densities range from as low as 60 W/m^2 at the Los Angeles refinery complex to more than 7 kW/m^2 at Exxon's huge Baton Rouge refinery in Louisiana. Values of 1–4 kW/m^2 are perhaps most common for large operations; the largest U.S. refinery, Exxon's Baytown in Texas, rates nearly 4 kW/m^2.

The chemical energy of fossil fuels is converted to heat by combustion, the rapid oxidation of carbon and hydrogen (reaction times are 0.1 ms for gases, 1 s for pulverized coals), which is the single most important anthropogenic energy conversion in industrial civilization. Combustion temperatures range from the minimum needed to support a stable flame (just short of 1000°C) to maxima posed by the practical difficulties of containing the hot flame within solid walls. The hottest firebox flame in large boilers is about 1600°C. Complete combustion of 1 g of carbon requires 2.66 g of O_2 (11.53 g of air), producing 3.66 g of CO_2 and releasing 33 kJ/g. The corresponding figures for complete combustion of hydrogen are 7.94 g of O_2 (34.34 g of air), producing 8.94 g of H_2O and releasing 121 kJ/g. Oxidation of sulfur or H_2S represents a negligible contribution.

FOSSIL FUELS

8.4 Mechanical Prime Movers: Engines and Turbines

An enormous variety of combustion devices have been designed to use released thermal energy directly in space heating and industrial and agricultural processes or to convert thermal energy to kinetic energy in mechanical prime movers. The trend of these conversions has been toward higher combustion temperatures and higher power densities under more controllable conditions. For example, small hand-stoked eighteenth-century units burned lumps of coal on simple grates with densities of about 100 kW/m^2 to heat unpressurized water, whereas modern electricity-generating plant boilers consume pulverized coal at rates up to 10 MW/m^2 to heat pressurized water to more than 600°C. The steam engine was the first practical converter of coal into kinetic energy. Its slow introduction took up the entire eighteenth century, and its subsequent rapid rise shaped the advance of industrializing societies during the nineteenth century. The steam engine has been the subject of many studies. Dickinson (1939) and von Tunzelmann (1978) give perhaps the best overviews; Jones (1973) offers a good collection of illustrations. The evolution of the first practical inanimate prime mover started with Denis Papin's 1690 experiments with a toy atmospheric. Papin's tiny device was followed in 1698 by Thomas Savery's only partly successful steam-operated pump, with a maximum rating of about 750 W, and in 1712 by Thomas Newcomen's engine, working at atmospheric pressure with a maximum effective output of 3.75 kW. Condensing steam on the underside of the piston made Newcomen's engine hugely inefficient (0.5%–0.7%), but John Smeaton's improvements around 1770 doubled this performance.

James Watt's revolutionary contribution in his appropriately titled 1769 patent, "A New Method of Lessening the Consumption of Steam and Fuel in Fire Engines," was the introduction of a separate condenser (fig. 8.8). An insulated steam jacket around the cylinder and stuffing box, an air pump to maintain vacuum in the condenser, sun-and-planet gearing, and later a double-acting engine with steam moving the piston also on the down stroke and a centrifugal governor to maintain constant speeds with varying loads were other notable innovations making Watt's engine a rapid commercial success. By 1800, when the 25-year extension of the original patent expired, Watt and Matthew Boulton, his financing partner, built about 500 engines (60% rotary, the rest pumping), rated mostly at 8–16 kW (the largest ones were just over 100 kW).

The steam engine's average rating of 20 kW was much higher than the means of eighteenth-century watermills (3.7 kW) or windmills (7.5 kW) but lower than the power of many large waterwheels built to serve expanding manufactures. But unlike windmills or waterwheels, the engine allowed unprecedented freedom of location, and its diffusion led to the emergence of new industrial centers not only in or near major coalfields but also in locations easily accessible by cheap water transport. An intense period of innovation during the first half of the nineteenth century, including Trevithick and Evans's introduction of a high-pressure boiler in 1802 and Corliss's invention of a valve mechanism in 1849, made the steam engine more efficient and much more versatile. Its portability, adaptability, dependability, and durability assured its place as the most common prime mover of nineteenth-century industrialization. There was little the engines could not do; their original uses for pumping in mines and for rotary power in textile mills were soon augmented by a host of stationary and mobile applications.

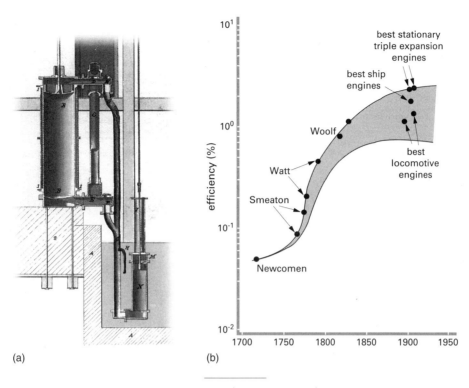

8.8 Watt's steam engine with its separate condenser (Farey 1827). (b) Efficiencies of the best steam engines, 1700–1930 (Smil 1994).

Stationary uses ranged from sawing wood and stones to powering the belt drives in factories, from compressing air to driving the first electricity generators. Its kinetic uses revolutionized both land and waterborne transport with railways and steamships. Other mobile applications included heavy self-propelled engines, cranes, pile drives, hammers, fire engines, tractors, trucks, cable ploughs, and steamrollers. The lightest steam engine was built by Sir Hiram Maxim in 1894 to power his never-flown airplane (Gunston 1986); it weighed a mere 1.04 g/W, and with a boiler 4.09 g/W, lighter than contemporary internal combustion engines. Large-scale manufacture of steam engines led to improved machining and integrated design practices that served as foundations of twentieth-century engineering advances. The availability of such concentrated power changed industrial production and transportation. Both transformations were soon reflected in extensive urbanization, migration, growth of trade, and shifts in international relations. However, British data show that to relate nineteenth-century economic growth primarily to steam is misconceived (Crafts and Mills 2004).

In a way, the steam engine became a victim of its own success. As its performance grew, more was demanded of it than it could deliver. Its improvements and adaptations during the nineteenth century were admirable. Its maximum ratings went up from about 100 kW to 3 MW, largely owing to a 100-fold increase of operating pressures (from 14 kPa to 1.4 MPa) and resulting in the best efficiencies climbing from just 2.5% to 25% (fig. 8.8). But the engines had their inherent weaknesses. They were massive and hence impractical for very large stationary applications and for light mobile use, and they were relatively inefficient. They could not fill two important emerging needs: to supply unprecedented capacities for efficient and convenient generation of electricity; and to provide a convenient energizer for mechanized road and airborne transport, which required lightweight, compact power plants. Steam turbines, internal combustion engines, and gas turbines filled these needs; only in railway transport did steam engines retain their global indispensability until after WW II.

The superiority of steam turbines for delivery of rotary power is clear. Steam engine rotations rarely surpassed 100 rpm, whereas modern turbines have up to 3600 rpm and work under pressures of 14–34 MPa and temperatures up to 600°C. They can be built in capacities ranging from 10^4 W to over 1 GW; their mass to power ratio (1–3 g/W) is only a fraction of that of steam engines (250–500 g/W); and their top efficiencies are 40%–42%. In sum, steam turbines are an excellent source of power for electric generators, compressors, centrifugal pumps, and ship propellers. Charles Parsons followed his first patented 1884 reaction turbine design with a 75-kW public power station in Newcastle in 1888, the first condensing turbine of 100 kW in 1891, and the first 1-MW unit for the Elberfeld station in 1900. It took less than

two decades to demonstrate the superiority of the steam turbine over the steam engine (fig. 8.9). The exponential rise of turbine ratings was interrupted only during the late 1920s, resumed in the mid-1950s, and reached a plateau in the early 1970s.

Many unsuccessful attempts preceded the first commercially successful design of an internal combustion engine, a horizontal double-acting machine with slide valves to admit a mixture of illuminating gas and air and to release the burned and expanded gas. This was patented in 1860 by Jean Joseph Étienne Lenoir and subsequently sold as a 2-kW machine for workshops. This slow (200 rpm) engine ran on an uncompressed explosive mixture of gas and air ignited with an electric spark, and its efficiency was a mere 4% (Smil 2005a). In 1862, Alphonse Eugène Beau (or Beau de Rochas) outlined the operation of a four-stroke cycle but did nothing to translate the concept into a working machine. That pioneering step was taken by Nicolaus August Otto who, after years of building improved Lenoir engines, patented a horizontal four-stroke design powered by coal gas in 1877. This was a very successful machine. Some 50,000 units, in sizes from 375 W to 12 kW, were eventually sold, but because it was slow (160 rpm) and heavy (250 g/W), it was suitable only for stationary uses.

Decisive breakthroughs into the transportation market came only with the introduction of gasoline engines mounted on road vehicles, steps taken independently during the mid-1880s by Gottlieb Daimler and Wilhelm Maybach in a suburb of Stuttgart and by Karl Benz in Mannheim. With about 33 MJ/L gasoline has about 1,600 times the energy density of the illuminating gas (usable in transportation only if highly compressed), and its low flash point (−40°C) makes it ideal for easy starting. At the same time, this low flash point makes it haz-

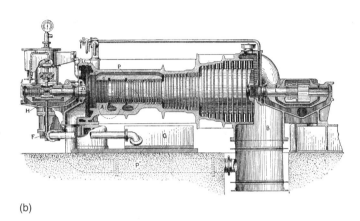

(a)   (b)

8.9 (a) Charles A. Parsons (1854–1931) and (b) his first
1-MW steam turbine. From Smil (2005b).

ardous to use. Daimler and Maybach built a prototype of a high-revolution (~600 rpm) gasoline-fueled engine with a surface carburetor and hot-tube ignition in 1883. Two years later they fastened this version onto a bicycle, creating the prototype of the motorcycle, and in 1886 they mounted a larger (0.462 L, 820 W), water-cooled version on a coach chassis.

Working independently, Karl Benz was granted a patent for a three-wheeled vehicle powered by a four-stroke, single-cylinder gasoline engine in January 1886 (fig. 8.10). This motorized carriage was driven publicly for the first time on July 3, 1886, along Mannheim's Friedrichsring. The light vehicle—total weight was just 263 kg, including the 96-kg single-cylinder, four-stroke engine—had a less powerful (0.954 L, 500 W) and slower-running (250 rpm) engine than did Daimler

and Maybach's carriage, and it could go no faster than about 14.5 km/h. The car had to be started by turning a heavy horizontal flywheel behind the driver's seat, but water cooling, electric coil ignition (highly unreliable), spark plugs (two lengths of insulated platinum wire protruding into the combustion chamber), and differential gears were key components of Benz's design that are still standard features in modern vehicles.

These concurrent innovations introduced the key ingredients of the modern car engine and launched the still continuing expansion of the automobile industry. After a relatively slow period of advances during the 1890s, the new century brought a flood of innovations ranging from electrical starters to antilock brakes. Many impressive gains in efficiency and reliability followed, but the development of automotive Otto cycle engines

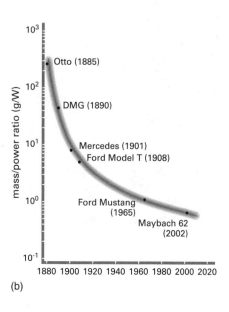

(a)

(b)

8.10 (a) Karl Benz's pioneering vehicle (courtesy of Daimler-Chrysler Classic Konzernarchiv, Stuttgart). (b) Decline of weight/power ratios of automotive and aero engines (Smil 2005a).

has been surprisingly conservative. Major historical shifts have included a steady rise in compression rates, an increase in average power, and a decline in weight/power ratios. In the United States typical compression ratios rose from about 4 in the early 1920s to about 10 by the 1960s, then declined slightly, averaging 8–9 by the mid-1980s. The first mass-produced car, Ransom Olds' Curved Dash, had a single-cylinder 5.2 kW engine; Ford's model T (16 million produced between 1908–1927) had a four-cylinder 15 kW engine; and most passenger cars built at the beginning of the twenty-first century were powered by engines rated between 85 kW (Honda Civic) and about 450 kW (most powerful Mercedes).

Otto's engine needed nearly 270 g/W in 1880, Daimler and Maybach's radical redesign brought the ratio down to 40 g/W by 1890, the best ratios approached 5 g/W before the end of the century, and since then the ratio has declined to as low as 3 g/W for trucks and about 1 g/W for passenger cars. The decline of weight/power ratios was much faster for aircraft engines (Gunston 1986). A four-cylinder engine powering the Wright brothers' first flight on December 17, 1903, rated 9.1 g/W; the U.S. V-12 Liberty engine mass-produced during WW I had 1.34 g/W; and the Wright Cyclone R-3350 air-cooled 28-cylinder engine powering B 29 bombers during WW II had 0.67 g/W, which was only slightly surpassed in the 1960s by another Wright (R-1820-82A) with 0.59 g/W.

These excellent achievements did little to boost the relatively low efficiency of gasoline engines. Although the ideal efficiency of Otto cycle engines is 60%, actual peak performance is no higher than about 25%, inferior to engines patented by Rudolf Diesel in 1892 and commercialized after 1900. These are distinguished by high compression ratios (15–24) designed to produce spontaneous ignition of the injected fuel and hence no need for a carburetor or a sparking device. Their weight/power ratio was 40–60 g/W, maxima up to 120 g/W, and their speed only about 300 rpm. The disadvantages of a heavier, low-speed engine were offset by superior thermal efficiency and the possibility of using a cheaper, more energy-dense, less volatile, less flammable liquid fuel. In his classic tests Clerk (1911) calculated the following efficiencies for a 445-kW diesel engine: mechanical efficiency 77%, indicated thermal efficiency 41%, brake thermal efficiency 31.7%. In contrast, typical efficiencies of Otto engines were 14%–17%. Engines with low volatility and flammability are particularly suitable for vessels as well as for the use in tropics to minimize evaporation from truck and bus fuel tanks.

The first niche conquered by the engine was in marine propulsion, where its weight was of little consequence. Submarine engines were followed by increasingly larger ship plants. By the year 2000 some 90% of the world's largest cargo ships, including supertankers, were powered by diesel engines. MAN, Wärtsilä, Mitsui, and Hyundai are their leading makers. The maximum sizes of these machines are still increasing; in 1996 the world's largest marine diesel engine rated about 56 MW, a decade later the largest Wärtsilä engines had capacities in excess of 80 MW. On land the replacement of steam and gasoline engines by diesels began after WW I, the first diesel trucks in 1924 and the first heavy passenger cars in 1936. By the late 1930s most new European trucks and buses had diesel engines, and this dominance was extended after WW II to the rest of the world. Bus engines with up to 350 kW, weighing 3–9 g/W, can log up to 600,000 km without overhaul. Similarly, diesels dominate all nonelectrified railway traction with engines up to 3.5 MW weighing 5–10 g/W.

The weight/power ratios of automotive diesels eventually declined to 2–5 g/W, and today's lightest passenger car diesels are only slightly heavier than gasoline-fueled engines. Rare in North America, diesels had about 45% of the new car market in Western Europe by 2005. Very large stationary diesels (up to ~50 MW) have been used for electricity generation in remote locations as well as for standby capacity. Ratings of the largest diesel-powered stations reached 200 MW by the late 1990s. Efficiencies have surpassed 42%, and performance above 35% should be attainable with most well-maintained engines. Diesels also dominate in heavy construction machinery, tractors, self-propelled harvesters, and locomotives on nonelectrified railways.

The third type of internal combustion engine that has revolutionized many industries appeared during the late 1930s, when Frank Whittle in England and Pabst von Ohain in Germany began testing their independently invented prototypes of gas turbines to power military jet aircraft (Constant 1981; Smil 2006). The post–WW II perfection and diffusion of this powerful prime mover brought total displacement of reciprocating engines in fighters and bombers and, starting in 1958, on transoceanic passenger flights. (The British Comet, introduced in 1952, did not succeed owing to structural defects rather than to the inadequacy of jet engines.) This transformation reached a temporary plateau with the introduction of hundreds of jumbo airplanes during the 1970s (the first flight of a Boeing 747 was in 1969). These airplanes were powered by large turbofan engines capable of nearly 250 kN of thrust at the sea level and delivering about 65 MW during the flight at the cruising altitude. A low weight/power ratio and a high ratio of thrust per frontal area characterized the evolution of these increasingly powerful aircraft gas turbines.

At the beginning of the twenty-first century the best commercial turbofan engines were rated above 500 kN (0.06–0.07 g/W), had a thrust/weight ratio above 6 (8.5 for military engines), and had high bypass ratios (90% of the compressed air bypasses the combustion chamber, thereby lowering specific fuel consumption and reducing engine noise.) Their high reliability made it possible to deploy two-engine aircraft not only on trans-Atlantic crossings but even on trans-Pacific ones. Other notable niches conquered by gas turbines have been in driving centrifugal compressors for natural gas pipelines, starting in the late 1940s, with current capacities up to 15–30 MW; oil fields and oil refineries; chemical syntheses, most notably in ammonia production since the early 1960s; steel mills; powering fast trains, hydrofoils, and military and cargo ships with engines of up to about 80 MW; and driving electric generators (15–80 MW to 150 MW) used for emergencies, peaking service, and base load (Islas 1999). The best efficiencies in these applications, just above 40%, match the performance of the best steam turbines, and the weight/power

ratios of the largest stationary gas turbines are around 2 g/W.

The only prime movers that outperform gas turbines in terms of weight/power ratio are liquid- or solid-propellant rocket engines for missiles and space vehicles. These large-thrust jet propulsion engines are used to accelerate loads to high velocities, often in stages, in short periods of time. Intensive engineering development of this ancient idea started only during the first decades of the twentieth century, but the founders of modern rocket science, Konstantin Tsiolkovsky and Hermann Oberth, correctly envisaged rapid advances (von Braun and Ordway 1975). In 1942 the 13.8-m-long ethanol-powered German V-2 missile had a range of 340 km and a destructive payload of 1 t. Its 931-kg engine had a sea-level thrust of 249 kN, imparting the maximum speed of 1.7 km/s during its 68-s burn. This translates to a maximum power rating of about 6.2 MW and an engine weight/power ratio of 0.15 g/W.

In contrast, during the 150-s firing, the liquid-fuel (kerosene and hydrogen) engines of the 109-m-tall Saturn C 5 rocket, which sent Apollo 11 on its journey to the Moon on July 16, 1969, had to impart an escape velocity of 11.1 km/s to a mass of 43 t by providing a combined thrust of nearly 36 MN, or an equivalent of about 2.6 GW. Even when including all the fuel in the weight of the three booster rockets with 11 engines, their weight/power ratio would be just 0.001 g/W. The thrust/weight ratio of individual rocket engines is as high as 150, 1 OM above the best military jets. This rapid transformation of chemical energy in fuels to kinetic energy of rising rockets is the most dramatic demonstration of powerful conversions mastered since the end of WW II. But the most important, although certainly much less spectacular, energy transformation that matured during the twentieth century is the large-scale fossil-fueled generation of electricity.

8.5 Fossil-Fueled Electricity: Generation and Transmission

The advent of electricity represents an unprecedented technical revolution (Smil 2005a). Previous inventions filled specific needs and could be readily inserted into existing systems. In the energy industry Newcomen's engine displaced horse-powered ones; in manufacturing, Watt's invention severed the dependence on water-wheels; and Fourneyron's turbine prevented the demise of water power by allowing for better efficiencies and capacities of water-powered machinery. Only the subsequent diffusion of these innovations, their greater reliability, lower cost, and higher performance led to a gradual overall transformation of systems into which they were introduced. Not so with electricity. Here the whole system had to be put in place *before* the idea became viable. The current had to be generated on a scale sufficiently large to allow for transmission and distribution to numerous, distant consumers, who first had to acquire electricity-powered converters that were superior to existing devices such as oil lamps and gas lights for illumination or steam engine-powered shaft-and-belt transmission in factories. One man's vision was central to this endeavor:

Edison was a holistic conceptualizer and determined solver of the problems associated with the growth of systems.... Edison's genius lay in his ability to direct a process involving problem identification, solution as idea, research and development, and introduction into use.... Edison is most widely known for his invention of the incandescent lamp, but it was only one component in his electric lighting system and was no more critical to its effective functioning than the Edison Jumbo generator, the Edison main and feeder, or the parallel-distribution systems. (Hughes 1983, 18–21)

There were other inventors of light bulbs and large generators, but only Thomas Edison had the vision of a complete system as well as the determination and organizational talent to make the whole work. The rapidity with which he and his many dedicated co-workers put the system into place is astonishing (Israel 1998). On October 21, 1879, after months of repeated failures, a cotton sewing thread was carefully carbonized, inserted into a glass globe evacuated by suction pump, and connected to an electricity supply from a dynamo. In Edison's words, the light bulb "burned like an evening star." Less than three years later, on January 12, 1882, the first power plant built by Edison Electric Light Company of London at Holborn Viaduct started to generate electricity. The first U.S. station followed in September 4, 1882. Located in New York's financial district at 257 Pearl Street, it lit 5,000 lamps by the end of 1882. The diffusion of electricity generation was accelerated by the victory of alternating current (AC) over direct current (DC) (in what became known as the battle of the systems) as well as by rapid development of reliable transformers and displacement of steam engines by steam turbines.

Edison, a staunch defender of direct current, was on the losing side of the transmission battle (as was Lord Kelvin). Because electric power is the product of current (amperes, A) and voltage (volts, V), and because voltage equals current multiplied by resistance (ohms, Ω), power is the product of $A^2\Omega$. This means that transmitting the same amount of power with 100 times higher voltage will result in cutting the current by 99% and reducing the resistance losses by the same amount. Edison's DC had voltages to match the load (a light or a motor) or had to be reduced to that level by another converter placed in series or by a resistor that wasted the difference. Raising voltage and reducing current in order to limit DC transmission losses would have resulted in dangerously high voltages in houses and factories. In contrast, high-voltage alternating current (HVAC) is transmitted over long distances with minimized losses and than is reduced to acceptably low voltages by transformers.

AC transmission, supported by George Westinghouse and Sebastian Ferranti, prevailed fairly rapidly; the conflict was basically over by 1893 with the Niagara Falls project designed to deliver high-voltage AC. Generation of the current at low voltages, its transmission as HVAC, and its distribution to customers at low voltages were made possible by the introduction of transformers whose fundamental designs emerged in the United States (William Stanley and George Westinghouse) and in Hungary (Ottó Bláthy, Károly Zipernowsky, and Miksa Déri) between 1883 and 1886. These simple but flexible and durable devices, converting high current and low voltage into low current and high voltage, and vice versa, opened the way for centralized electricity generation and hence for the enormous economies of scale associated with the larger sizes of turbines and plants.

All early generating stations produced rotary power for their dynamos by steam engines. Edison's first New York station had four Babcock & Wilcox boilers (about 180 kW each) on the ground floor and six Porter-Allen engines (each 94 kW and 320 g/W) and six large direct-connected Jumbo dynamos on the reinforced second floor of 257 Pearl Street (T. C. Martin 1922). Parson's reaction turbine, patented in 1884, soon displaced

these bulky and expensive machines. Seven years later, his pioneering 100-kW machine rated 40 g/W, more than 80% below the best comparable steam engines. That ratio fell below 10 g/W by 1914, and it was eventually brought down to just over 1 W/g for the largest machines built after the mid-1960s (Hossli 1969).

The quest for cheaper and better illumination was Edison's great incentive for designing a new energy production and distribution system, but the largest demand for electricity lay in the substitution of clumsy, noisy, and inefficient steam-powered shaft-and-belt drives used to run machines by electric motors in scores of industrial enterprises. This huge market was unlocked by Nikola Tesla's designs of a practical AC induction motor, patented in 1888. What Edison, Westinghouse, Parsons, Stanley, and Tesla accomplished during the 1880s put in place such solid foundations of a new industry that within two generations the electric industry became a mature branch of engineering. Its pioneers would find little incomprehensible about today's arrangements, although they would certainly appreciate all the major quantitative and qualitative improvements. Principles remain but, as the following contrasts indicate, particulars have changed (Smil 2005a).

The largest turbogenerator sizes increased by 5 OM, rating as much as 1.2–1.5 GW before reaching a plateau during the 1980s (fig. 8.11). Working steam pressure rose from about 1 MN/m^2 for the first commercial units to 35 MN/m^2 for supercritical turbines introduced in the 1960s; however, 31 MN/m^2 appears to be the optimum for greatest efficiency gains (fig. 8.11). Steam temperatures rose from 180°C for the first units to 650°C by 1960, with more recent optima at between 560–600°C. Coal-fired units (boiler-turbine-generator) account for most of the largest sizes, and in 2005 coal generated

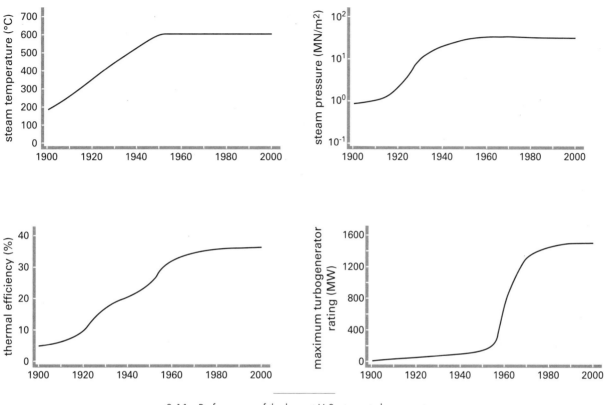

8.11 Performance of the largest U.S. steam turbogenerators, 1900–2000. From Smil (2005a).

about 40% of the world's electricity. The average efficiency of U.S. thermal generation rose from 4% in 1900 to 13.6% by 1925, then almost doubled by 1950, to 23.9% (Schurr and Netschert 1960). The nationwide mean surpassed 30% by 1960, but it has stagnated since, never exceeding 33% (fig. 8.11).

Mechanical stokers were used to feed small lumps of coal (no larger than 0.5–1.5 cm) onto a grate. Although the first experiments with finely milled coal took place before WW I, pulverized coal combustion was widely commercialized only after 1920. Pulverized coal combustion involves blowing a mixture of air and finely milled coal—70% of particles with diameter less than 75 μm, similar in size to baking flour—into a boiler and burning it in a swirling vortex. Unfortunately, this efficient mode of combustion is a major source of air pollution. Pulverized coal burning proceeds with excess air at flame temperatures between 1600°C–1800°C, ideal conditions for producing large volumes of NO_x and for oxidizing virtually all of the sulfur present to SO_2. The

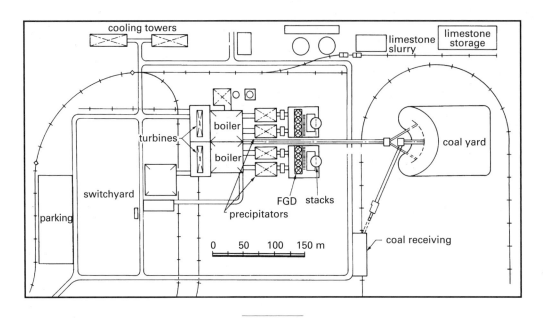

8.12 Layout of a typical late-twentieth-century coal-fired
electricity-generating plant.

incombustibles generate hard, slagging fly ash and heavy metal emissions.

Fly ash emissions have gradually been eliminated by installation of electrostatic precipitators (patented in 1909 but universally used only after 1950), and various flue gas desulfurization (FGD) techniques, commercialized during the 1970s, can remove most of the SO_2. By the year 2000 about 40% of the world's coal-generating capacity (excluding China) had some FGD. These controls made coal-fired generation environmentally more acceptable, but disposal problems with the abrasive fly ash and wet sulfate sludge from FGD have considerably lowered overall power densities. Combustion of fossil fuels in large boilers proceeds at very high power densities (2–13 MW_e/m^2), and modern turbogenerators are also quite compact. The power density of Parson's first 100-kW turbine was about 26 kW_e/m^2; turbogenerators over 1 GW rate up to 10 MW_e/m^2. The heart of any fossil-fueled power plant, boilers and machine halls, is thus only a small part of the site (fig. 8.12).

Once-through cooling, which draws water from streams or the ocean, makes minimal spatial claims, but spray ponds need about 400 m^2/MW_e (handling 2.5 kW_e/m^2), and ordinary cooling ponds 1 OM higher (commonly 5000 m^2/MW_e). Cooling towers cut this huge demand while allowing for a greater density of water reuse on major streams. Counterflow wet cooling towers with natural draft were the dominant choice until the 1970s; they required 25–50 m^2/MW_e, which means they serviced 20–40 kW_e/m^2. These power densities are calculated per square meter of tower base, but large towers require considerable spacing in order to prevent high wind loads induced by the Venturi effect. Wet mechanical draft units of the cross-flow type need only one-half or one-third of the area for the same cooling load and can easily handle 80–100 kW_e/m^2, or as little as 10 m^2/MW_e. Switchyards typically take up 50–75 kW_e/m^2, but overall densities are determined by the plant's size, its fuel storage, air pollution control, and water cooling facilities. Economies of scale are substantial.

On-site storages of plants supplied by gas pipelines or fuel oil barges may be quite small, but coal piles required for 60–90 days of operation are necessarily extensive. With heights between 5 m and 12 m, they have storage densities of 25–100 kW_e/m^2, similar to those of switch-yards. Because of large volumes of hot gas, electrostatic precipitators take up about as much space as boilers, and FGD can be even more compact (up to 400 kW_e/m^2), but the disposal of captured fly ash and sulfate sludge requires much land. Combustion of steam coal with 22 GJ/t and 10% of ash will generate annually about 250 t/MW_e. The ash is deposited in fills or ponds to a depth of 5–10 m, requiring annually 20–40 m^2/MW_e, assuming specific density of 1.3 g/cm^3. During its lifetime of 35–50 years, a coal-fired power plant will thus need 700–2000 m^2 of ash disposal space per MW_e. Lifetime disposal of FGD sludge requires 200–600 m^2/MW_e.

Stations with once-through cooling, no fuel storage (mine-mouth plants), no FGD (burning low-S coal), and off-site ash disposal or commercial ash sales can have power densities of 5–10 kW_e/m^2. Rates for large plants with coal storage and cooling towers can be up to 4 kW_e/m^2, and with on-site disposal of ash and FGD sludge they decline to mostly between 1–2.5 kW/m^2. In contrast, the power densities of plants burning fuel oil are often higher than 3 kW/m^2, and those burning natural gas in stationary gas turbines can be twice as high. Nuclear plants require no extensive fuel storage and no air pollution controls, and their land claims are dominated by exclusion and buffer zones. Their densities of about 1–1.5 kW/m^2 are less than those of hydrocarbon-fired plants but more than those of many coal-fired stations.

Perhaps the most promising technique to solve the triple problem of particulates and sulfur and nitrogen oxide emissions is fluidized bed combustion (FBC). First patented for gas generators by Fritz Winkler in 1921 and used extensively by the Germans during WW II to produce feedstock for gasoline synthesis, the technique has been used commercially in power plants only since the early 1980s (Valk 1995). A fluidized bed is a layer of small noncombustible particles, usually limestone, kept aloft by air forced through perforations in a base plate. Pulverized fuel is introduced in quantities of less than 5% of the total load, and superior mixing rates make it possible to burn the fuel at 760°C–930°C, well below 1370°C, the temperature at which NO_x begin to form.

In addition, finely ground limestone ($CaCO_3$) or dolomite ($CaMg(CO_3)_2$) could be mixed with combustion gases in order to remove as much as 95% (and commonly no less than 70%–90%) of all sulfur present in the fuel by forming sulfates (Henzel et al. 1982; Lunt and Cunic 2000). Atmospheric FBC is available in sizes up to 300 MW, able to burn coal competitively compared to new natural gas-fired combined-cycle systems (Schimmoller 2000). More than 600 AFBC boilers, with a total capacity of 30 GW, operated in North America in the year 2000, and a similar capacity was installed in Europe. Pressurized FBC produces gas that can drive turbines; about 1 GW of such capacity existed worldwide by the year 2000 (USDOE 2001). FBC stations can have power densities in excess of 5 kW_e/m^2.

During the twentieth century the power of the largest transformers increased 500 times, and their voltage about 15 times, while their weight/power ratio declined by 1 OM, and their efficiency reached practical limits at over 99% (Coltman 1988; Smil 2005a). U.S. transmission started with wood poles, cross arms, and solid copper wires at a mere 4.6 kV and rose in a series of leaps: 6.9 kV, 23 kV, 69 kV, 115 kV, 230 kV, 345 kV, 500 kV, 765 kV. Transmission corridors widened from just over 10 m for 115 kV to 40 m for 765 kV, but the exponential increase in power loading of three-phase, 60-Hz lines means that the latter link will have more than 50 times higher capacity. Typical rates for the land taken over by tower pads and access roads, and hence unavailable even for cropping or grazing, range between 10,000 m^2/km and 15,000 m^2/km.

DC transmission came back for undersea cables and long-distance connections between large, remote hydrogenerating stations and major load centers. In 1954 the pioneering Sweden-Gotland cable carried 20 MW at 100 kV over 96 km; the first English Channel crossing in 1961 had a capacity of 160 MW at $\pm$100 kV; and New Zealand's two islands were connected in 1965 by a $\pm$250 kV tie carrying 600 MW. The West Coast Pacific Intertie in 1970 carried 1440 MW at $\pm$400 kV over 1330 km, followed in 1972–1977 by Manitoba's Nelson River-Winnipeg link (1620 MW at $\pm$450 kV over 890 km) and in 1976 by Zaire's Inga-Shaba line (560 MW at $\pm$500 kV over 1700 km). These ties were surpassed in the late 1980s by the link between Itaipú hydrostation and São Paulo ($\pm$600 kV over 800 km).

9

FOSSIL–FUELED CIVILIZATION

Patterns and Trends

The oil rigs in Bahrain imply a buyer
who counts no cost, when all is said and done.
The logs give back, in burning, solar fire

but Good Gulf gives it faster; every tire
is by the fiery heavens lightly spun.
Nothing is lost but, still, the cost grows higher.

So guzzle gas, the leaden night draws nigher
when cinders mark where stood the blazing sun.
The logs give back, in burning, solar fire;
nothing is lost but, still, the cost grows higher.
John Updike, From "Energy: A Villanelle" (1985)

The principal attributes distinguishing a mature fossil-fueled civilization from its predecessors are easy to describe. Most notably, there is a secure, abundant, and varied food supply, ensuring average per capita availability levels far in excess of any imaginable needs. (The persistence of significant intake disparities and even of hunger in rich societies is a matter of distributional, that is, social inequalities.) The abundant food supply reflects the high output density of modern agriculture, which is subsidized by numerous direct and indirect inputs of fossil energies (see chapter 10).

There is a high level of material affluence, reflecting large amounts of energy invested in producing basic commodities like metals, building materials, and modern synthetic compounds as well as in providing energy-intensive services. Mechanization of agricultural and industrial labor transferred most of the daily work to the light exertion category (see section 5.2) as humans became primarily controllers and managers of fossil fuel and electricity flows and gained more time for leisure.

A notable component of this affluence is the high mobility of products via trucks, railways, ships, and planes, and of people via private and public carriers. This mobility was made possible by the introduction of progressively more powerful, lighter, and more efficient prime movers (see section 8.4), and it became a mass reality with the increasing affluence accompanied by greater amounts of leisure time.

The complexity of the systems that have evolved to produce, convert, distribute, and consume energies in an affluent civilization is revealed by looking at the essential attributes of fossil energy industries, the contribution of nonfossil conversions, the grand patterns of energy utilization, the energy costs of energy, materials, products, and services, and the considerable, but ultimately futile, opportunities for energy conservation.

9.1 Fuels and Fossil-Fueled Electricity: Energy Production and Trade

Perhaps the clearest long-term exponential trend that can be traced since the onset of the fossil-fueled era is the growth of the total primary energy supply (TPES), encompassing fuel extraction and primary electricity generation. Its smooth progression has been interrupted only by wars or severe economic setbacks. This aggregate growth has been driven by, and has in turn stimulated, increased unit sizes for individual techniques and complete production systems. It has resulted in the gradual concentration of outputs and in extensive transportation and distribution systems creating strong national and international dependencies and a global market. Qualitative changes have been no less obvious. Production of hydrocarbons has grown much more rapidly than the mining of coals, and a higher proportion of fuels has been converted to electricity, the most convenient and flexible of all energies. Improvements in performance extend to entire systems. All these trends and accomplishments are assessed in this chapter.

The modern commercial production of fossil fuels can be charted with acceptable accuracy almost from its very beginnings. Important coal-mining nations in Europe and North America have fairly good output statistics going back to at least 1850, and nineteenth-century hydrocarbon production is recorded even more accurately (USBC 1975; Etemad et al. 1991; Mitchell 1992). The inevitable uncertainties and omissions are probably less important than the cumulative errors inherent in converting the raw fuel data to energy equivalents. An often cited reconstruction of worldwide fossil fuel production spanning the years 1860–1953 (UNO 1956) probably overestimates the total pre–WW II output, but the error is no greater than 5%–10%. Better figures are available for post-1945 production in the United Nations *World Energy Supplies* series until 1979 and in the *Yearbook of World Energy Statistics* from 1980 as well as in annual publications by the Energy Information Agency and BP (2006).

A semilogarithmic plot of global fossil fuel production shows a rise from about 2.5 EJ (80 GW) in 1850 to 22 EJ (700 GW) in 1900 and nearly 330 EJ (10.44 TW) in 2000, when hydro and nuclear electricity (converted using thermal equivalents) added about 50 EJ to TPES of roughly 380 EJ (fig. 9.1). This spectacular rise can be broken into four distinct periods. There were three generations of exponential rise averaging 4.3% a until the beginning of WW I. Then a spell of slow increases and temporary declines before WW II was followed by a resumption of vigorous growth (nearly 4.5% a year) that lasted until OPEC's first round of crude oil price increases in 1974. During the following decade average growth slowed down to about 2%, and between 1984 and 2004 it declined further, to about 1.25%. The British

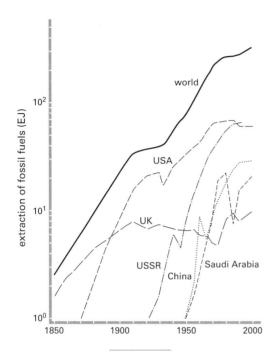

9.1 Exponential rise of global fossil fuel production, 1850–2000. Plotted from data in UNO (1956), Etemad et al. (1991), and BP (2006).

production lead ended in the early 1890s, and the United States has been dominant ever since. The USSR was in second place from the late 1940s until its demise in 1990, and Russia retained the second place until 2003, when its primary commercial energy production was surpassed by that of China.

The recent origins of global TPES have been somewhat less uneven than in the past, but the skewness is still considerable. In 1975 the two superpowers, with nearly 12% of the world's population, produced about 47% of all fossil fuels. In 2005 the top two producers, the United States and China, with 25% of the world's population, extracted nearly 30% of all fossil fuels; the five largest producers (United States, China, Russia, Saudi Arabia, and Canada) accounted for 51% of the world's output of coals, oils, and gases. The aggregate growth of global energy output has been accompanied by a universal transition toward higher-quality fuels, from coal to hydrocarbons. Oils and gases were a marginal part of the total output until the first decade of twentieth century. Their heat content surpassed coal's contribution by 1960, and they have been providing more than two-thirds of all fossil energies since 1970 (the 2005 share was about 70%). The extraction of natural gas has been rising much faster than the production of crude oil, and by 2005 gases accounted for just over one-quarter of all fossil energies.

For many decades these energy transitions appeared to follow a remarkably orderly, predictable course. Marchetti (1977) found that full cycles of shares of TPES delivered by individual energy sources approximate normal Gauss-Laplace curves. The pattern looks particularly impressive when market shares (f) of primary commercial energy consumption are plotted as logarithms of $f/(1 - f)$; the ascents and descents then appear as straight lines (fig. 9.2). The process is very slow, with every new source taking about a century to penetrate half the market share, and for decades it was surprisingly regular. Despite wars and periods of economic stagnation and rapid growth, penetration rates remained constant during the first three-quarters of the twentieth century. That is why Marchetti and Nakićenović (1979, 15) concluded, "It is as though *the system had a schedule, a will, and a clock*" and "All perturbations are reabsorbed elastically without influencing the trend."

As coal displaced wood, and oil displaced coal, so would economic and technical imperatives ensure the displacement of oil by natural gas, nuclear, and solar

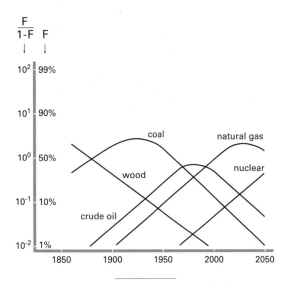

9.2   Marchetti model of global energy substitutions and major departures from the expected trends. From Smil (2003).

energies on a highly predictable global schedule. According to Marchetti, trying to change the course of these developments is futile: the system makes the decisions, and we are at best optimizers. However, this was an incorrect conclusion. After the oil price increases of 1973–1974 many forces began reshaping the global energy system on a massive scale. The result has been a shift from a regime of seemingly preordained energy substitutions to one of surprisingly stable energy shares with relatively little structural change. In just a decade the actual share of oil in global energy consumption was well ahead of the forecast value. By the year 2005, reality and Marchetti's model were far apart. Crude oil supplied 36% of the world's primary commercial energy needs, nearly 50% above the prediction of 25%; coal supplied 28%, nearly three times as much as the prediction of 10%; and natural

gas delivered about 24% of all primary energy, much less than the predicted share of just over 50%.

Oil's share is not shrinking rapidly, the share of natural gas is increasing much more slowly than predicted, and coal is still claiming a larger share than expected. The shares are given on the basis of commercial energy consumption. Marchetti's substitution scheme had wood falling to just a fraction of 1% after the mid-1990s. That is another error. Phytomass fuels still provided about 10% of the world's TPES in the year 2000. This was more than all nuclear electricity in gross energy terms, although not in net terms because of much lower conversion efficiencies; 40–45 EJ of biomass fuels produced much less useful energy than 25 EJ of fission-generated electricity used by industries and households. In sum, the world still uses much more biomass energy than projected by Marchetti's model. Not surprisingly, IIASA forecasts stopped referring to inevitable substitutions and projecting the rapid end of the fossil fuel era (WEC/IIASA 1998).

Another key shift toward higher quality of the final energy supply is a move away from the direct use of fuels, especially coals, through the rapidly expanding generation of electricity. Early strategies of this expansion, especially evident during the 1920s and 1930s, included several universal components (Hughes 1983). Pursuit of economies of scale, concentration of fossil-fueled capacities in or near large load centers, development of high voltage links to transmit electricity from remote hydro stations, promotion of mass consumption, charging of differential rates, and interconnections of smaller systems resulted in greater supply security, lower installed and reserve capacities, and automated central controls. After WW II most of these thrusts intensified as turbogenerators and plants grew bigger, higher voltages spanned longer distances, and interconnections mushroomed to embrace centrally controlled systems with capacities of 10^1–10^2 GW.

The only major departure from earlier trends was the location of fossil-fueled power plants away from cities, a shift spurred by stronger air pollution regulations and scale economies of huge mine-mouth stations. In fact, post–WW II growth has been so rapid that some components of power generation clearly overshot the operational optima. Global generation has grown smoothly, with just a short break in the early 1930s. Between 1900, when worldwide output was about 8 TWh, and 1930, when it stood at 180 TWh, the average annual expansion rate was approximately 10.5%, and between 1935 and 1970 just over 9%. Later the growth was roughly halved, to 4.5% for the years 1970–1985, a reflection of declining demand throughout the rich world.

The roles of primary electricity in this growth (hydrogeneration, geothermal, nuclear, solar) are appraised in section 9.2. Although many countries have always produced most of their electricity from falling water, fossil-fueled generation has been globally dominant, accounting for just short of 60% in the 1920s, 67% in 1950, 70% in 1975, and just over 60% in 2000 (the relative decline was caused largely by rising nuclear generation). Assuming an average e_1 of 33% global fossil-fueled electricity generation would have consumed about 30% of worldwide production of fossil fuels in 2000. Fossil-fueled electricity generation is even more unequally distributed than fuel extraction. In 2000 the U.S. share was nearly 30% of the world's fossil-fueled generation, and the five largest producers (United States, China, Japan, Russia, and India) claimed about 57% of the world total.

During the first century of fossil-fueled electricity generation, installed capacities of the largest power plants rose from a few hundred kW to several GW, and individual utilities grew from enterprises serving a handful of city blocks to regional and national systems with capacities up to 10^{10}–10^{11} W serving areas of 10^4–10^6 km^2. Moreover, international interconnections in Europe and North America make it possible to trade electricity at multi-GW levels on semicontinental scales. But electricity trade amounts to less than 5% of global generation, and a large part of it derives from one-way sales of hydroelectricity (see section 9.2). In contrast, international fuel sales have steadily increased. During the early years of the new century, some 15% of coal, nearly 60% of crude oil, and 25% of natural gas, amounting to some 120 EJ (or one-third of global fuel extraction), were traded internationally.

Coal exports are dominated by fuel for electricity generation (steam coal) while coking coal sales have steadily declined; by 2005 they were only about one-quarter of the traded total. The fuel is now the world's most important seaborne dry-bulk commodity (after surpassing iron ore), and hence its exports set the freight market trends. Australia, with more than 200 Mt of steam and coking coal shipped annually, is the world's largest exporter, followed by Indonesia, Russia, South Africa, and China (WCI 2006). Japan is the largest importer of steam and coking coal (over 150 Mt by 2005), followed by the South Korea, Taiwan, and in a shift unforeseeable a generation ago, by the two former leading coal producers, Germany and Britain, using cheaper foreign coal for electricity generation.

Crude oil leads the global commodity trade both in terms of mass and value. In 2005, 47% of total crude oil production (just over 1.8 Gt) was sold abroad, and exports of refined products added about 600 Mt. These exports were valued at about $780 billion, or roughly 55% of all international fuel sales, a total was higher than the world's imports of food (about $680 billion) but lower than the value of automotive products (about $910 billion). About 45 oil-producing countries are exporters, and more than 130 countries importers, of crude oil and refined oil products. The global dominance of Middle East exports is obvious. The six largest exporters (Saudi Arabia, Iran, Russia, Norway, Kuwait, and United Arab Emirates) sell just over 50% of the traded total, and the six largest importers (United States, Japan, Germany, South Korea, Italy, and France) buy 70% of all shipments (BP 2006).

As noted earlier (see section 8.3), pipelines are superior to any other form of oil transportation over land. Only large riverboats and ocean tankers are cheaper carriers of energy. The United States has had long-distance pipelines for domestic distribution of crude oil since the 1870s. These were eclipsed only during the 1970s by the world's longest crude oil pipelines laid to move Western Siberian crude oil to Europe.

Russia, Canada, Norway, the Netherlands, and Algeria are the largest exporters of piped natural gas, accounting for just over 90% of the world total. Europe has a particularly complex network of trunk lines bringing gas from Russia, the North Sea, and Africa (fig. 9.3). Shipments from Indonesia, Algeria, and Malaysia dominate the

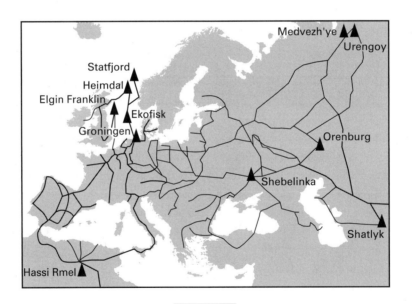

9.3 Europe's complex network of major natural gas pipelines (only the largest natural gas fields are identified). Western Siberia, the North Sea, and North Africa are major sources of supply.

CHAPTER 9

LNG trade. The largest importers of piped gas are the United States, from Canadian fields in Alberta and British Columbia; Germany, from Siberian Russia, the Netherlands, and Norway; and Italy, mostly from Algeria and Russia. Japan buys more than half the world's LNG, mainly from Indonesia and Malaysia. Other major LNG importers are the South Korea and Taiwan, both from Indonesia and Malaysia; and France and Spain, from Algeria.

Most of the techniques underpinning these far-flung, complex systems have either ceased growing or are increasing at much slower rates than they did until the 1970s. As a result, we may never see a larger overburden-removing shovel, refinery, or supertanker. The reasons for this end of growth are not primarily technical but rather environmental, economic, and social (see chapters 11 and 12). Concentration of fuel extraction and processing and electricity generation in progressively larger facilities has further increased the inherently high power densities of fossil-fuel-based energetics. Most of the world's fuel extraction goes on with densities surpassing 1 kW/m^2, and fuel-processing facilities have similarly high throughput densities. But the overall power densities of production systems are considerably lowered by the extensive transportation and transmission networks needed to deliver fuels and electricity as well as by the requirements of pollution and heat control. Some power production densities may be well below 50 W/m^2.

9.2 Nonfossil Contributions: Biomass and Primary Electricity

Nonfossil contributions range from critical but inefficient and environmentally ruinous inputs (fuel wood and crop residues in the poor world) to mere curiosities (tidal electricity generation), from universally adopted large-scale conversions (hydroelectric generation) to promising but still marginal techniques (photovoltaics), and from systems that could conceivably displace fossil fuels (nuclear fission) to options that will always remain restricted (geothermal electricity generation). Although biomass is by far the most important source of nonfossil energy, it is impossible to offer an accurate account of its consumption because most of these fuels are gathered directly by more than 500 million poor world families rather than being traded.

Uncertain conversion factors used to express these uses in common energy units introduce additional errors, and hence it is not surprising that plausible estimates of the global use of biomass energies differ by more than 10%. FAO (1999) estimated that about 63% of 4.4 Gm^3 of harvested wood were burned as fuel during the late 1990s. With about 0.65 t/m^3 and 15 GJ/t of air-dried wood, this would be an equivalent of about 27 EJ. In some countries a major part, and even more than half, of all woody matter for household consumption is gathered outside forests and groves from bushes, tree plantations (rubber, coconut), and roadside and backyard trees. Rural surveys show that nonforest fuel wood accounted for more than 80% of wood burned in Bangladesh, Pakistan, and Sri Lanka (RWEDP 1997). A conservative estimate of this nonforest woody biomass could raise the total to anywhere between 30 EJ and 35 EJ.

Crop residues produced annually in poor countries added up to about 2.2 Gt of dry matter during the late 1990s (Smil 1999a). Burning in the field, recycling, and feeding to animals account for most of their disposal, and if about 25% of all crop wastes (mostly cereal straws) were used by rural households, this would add about 8 EJ. Collected dried dung amounts to less than 1 EJ. A minimum estimate for the year 2000 was close to 40 EJ, and more liberal assumptions, higher conversion factors, and addition of minor biomass fuels (grasses, dung) would raise the total closer to 45 EJ. For comparison, Hall (1997) estimated that biomass supplied as much as 55 EJ during the early 1990s, the WEC (1998) assessment of global energy resources used 33 EJ, and Turkenburg (2000) bracketed the total at 45 ± 10 EJ.

China and India are the largest consumers of wood and crop residues in absolute terms. During the mid-1990s, China's annual consumption was at least 6.4 EJ and India's about 6 EJ (Fridley et al. 2001; RWEDP 2000). Brazil and Indonesia rank next, but in relative terms sub-Saharan Africa, with biomass supplying in excess of 80% of fuel in most of its countries, comes first (UNDP 2001). Among the most populous modernizing nations the shares are still close to 15% in China, roughly 30% in India and Indonesia, and about 25% in Brazil (IEA 2001). Fuel availability, climatic differences, and cooking and heating habits explain large consumption variation; most surveys indicate needs of 0.5–2.5 m^3 of air-dried wood per capita per year, or about 5–25 GJ.

Published rates for countries whose rural population is still almost totally dependent on wood, including charcoal, range from 8–10 GJ for Zambia, Zimbabwe, Madagascar, Kenya, and Ethiopia to about 15 GJ for Angola, Ghana, Cameroon, Sudan, Nigeria, and Thailand, and up to 34 GJ for equatorial Gabon, still with extensive tropical rain forest (Smil 1983; RWEDP 2000). For heavily deforested, densely populated nations whose people burn any available phytomass and dried dung, the rates are much lower, a mere 1.9–3 GJ in Bangladesh, 7–8 GJ in China, and 6–8 GJ in India. Yet even this low consumption causes extensive environmental degradation because the loss of vegetation accelerates erosion and reduces the soil's nutrients, organic matter, and moisture-holding capacity. The most acutely affected areas are Africa's Sahel and Namibia, Swaziland, Lesotho, and Botswana, the Nepali hills, large parts of India and interior China, Bangladesh, Pakistan, Afghanistan, Thailand, and much of Central America.

The best conclusion would be that those traditional rural societies still dependent on biomass fuels for all their household heat consume annually as little as 5–9 GJ per capita of crop residues and dung just for simple cooking (and even then experience seasonal shortages), and they are reasonably well off with 15–30 GJ in tropical and subtropical climates even with inefficient stoves. Potential savings from introduction of better stoves are enormous, but actual achievements have been mixed. Traditional open or partly enclosed fires convert less than 10% (even less than 5%) of fuel's energy to useful heat for cooking, and primitive stoves have efficiencies between 5% and 15%. Unfortunately, many projects to diffuse improved stoves, built mostly with locally available materials, were largely disappointing (Kammen 1995).

Many stove designs were still too expensive to be easily affordable or not sufficiently durable or easy to repair. Moreover, a good design is only a part of a much broader effort that must also include training of local craftsmen to build and repair such efficient units, active promotion of new stoves, and where needed, financial help in their purchase. The first encouraging success came with China's National Improved Stove Programme, launched in 1982 and initially aimed at 25 million units within five years. This effort was gradually transformed into a commercial venture, and by the end of 1997 about 180 million stoves were disseminated to some 75% of China's rural households (Smil 1987; K.

Smith 1993; Wang and Ding 1998). These stoves, operating with thermal efficiencies of 25%–30%, are credited with annual savings of close to 2 EJ of fuel wood and coal (Luo 1998).

Biomass provides only between 1% and 4% of the TPES in the world's richest countries, with Sweden (at more than 15%) being a major exception. Woody biomass consumed in the rich countries, mostly by lumber and pulp and paper industries and only secondarily for household heating (in old-fashioned fireplaces or high-efficiency wood stoves), amounted to no more than about 5 EJ in the year 2000. Nearly all this energy is used directly as space, cooking, and processing heat (particularly by the pulp and paper industries), and only a minuscule part is converted to electricity and liquid fuels. The world's largest effort to produce fuel ethanol, Brazil's PROALCOOL, began in 1975, and it remains the most productive solar alternative for converting phytomass to liquid fuels. Cultivation of sugarcane under optimal tropical conditions, with bagasse used to fuel the distillation of ethanol, results in power production density of about 0.45 W/m^2 (Macedo, Leal, and da Silva 2004). The U.S. corn-based system has much higher land requirements because of its inherently lower yields; during the early 2000s it produced about 7 MJ of ethanol, or 0.22 W/m^2.

The power densities of other biomass energy uses are similarly low, reflecting the inherently low efficiency of photosynthesis. Woody phytomass burned in rural areas is gathered with densities ranging from just 0.02 W/m^2 for twigs, branches, and leaves in arid environments to about 0.15 W/m^2 for selective stem cutting in moist forests. Plantations of fast-growing willows, poplars, eucalypti, leucaenas, or pines yield only 0.1 W/m^2 in dry northern climates and 1 W/m^2 in the best stands in

humid regions or with irrigation; values around 0.5 W/m^2 are more typical upper rates. Logging residues from clear-cutting can provide a one-time yield of 1–4 W/m^2, but most of them may not be usable (too remote, too dirty) and will be burned on site. Crop residues used for household combustion are harvested with densities ranging from a mere 0.01 W/m^2 from low-yielding cereals (assuming straw yield of 1 t/ha, and one-third of it used for fuel) to 0.4 W/m^2 for sugarcane bagasse (cane yield of 100 t/ha, all bagasse burned).

Conversion of water's potential energy to electricity is the second most important nonfossil energy input. Potential energy of global runoff is about 367 EJ, almost exactly the world's commercial TPES in the year 2000. If this natural flow were to be used with 100% efficiency, the gross theoretical capability of the world's rivers would be about 11.6 TW. Competing water uses, unsuitability of many sites, seasonal fluctuations of flow, and the impossibility of converting water's kinetic energy with perfect efficiency at full capacity mean that although the exploitable capability (share of the theoretical potential that can be tapped with existing techniques) can be well in excess of 50%, it is commonly just around 30% for projects rated above 1 MW. The worldwide total of technically feasible capacity must be thus constructed by adding specific national assessments. ICOLD (1998) puts it at about 52 EJ (>14 PWh/a) of electricity, or roughly 14% of the theoretical total. Asia claimed 47% and Latin America nearly 20% of the total, and China has the largest national total (about 15%).

Not everything that is technically possible is economically feasible, and projects in the latter category added up to about 30 EJ worldwide (just over 8 PWh), or roughly three times the total of 2.6 PWh/year (700 GW, with another 100 GW under construction) that was actually exploited by 2000 (IHA 2000). Europe has exploited 75% of all technically feasible capacity, North America nearly 70%, but Asia has tapped less than 25% and Africa less than 10% of this potential. The major advantages of hydroelectricity are mature dam construction and turbine manufacturing techniques, high reliability of generation, ability to run at synchronized zero load (spinning reserve), rapid start-up during peak demand periods, lowest operating cost, and longest plant life compared to all other options and, of course, the renewability of water flows. In addition, most hydro projects also provide water for drinking or irrigation, control floods, and support fisheries and recreation uses.

Hydroelectric generation began on a very small scale in 1882 (the same year as thermal production), and before 1900 new turbine designs and newly discovered dynamite brought rapid advances in construction of increasingly higher dams in Alpine countries, Scandinavia, and the United States. After WW I came the start of state-supported development of large hydro projects in the United States (most notably the Tennessee Valley Authority), and in the USSR (part of Lenin's drive for electrification). The Grand Coulee dam, the large U.S. project on the Columbia River, was completed in 1942. Post–WW II worldwide expansion turned hydro power from a globally marginal contributor to a source of nearly 20% of all electricity. The thermal equivalent used to convert primary electricity to a common denominator made global hydroelectricity generation equal to less than 3% of global primary energy in 1950 and to just over 6% in the year 2005.

The total number of newly built dams peaked in Europe and North America during the 1960s, in Asia and Latin America during the 1970s, and in Africa during the 1980s (WCD 2000). In 2000 the United States had the largest number of plants with capacities in excess of 1 GW as well as the largest installed capacity (98.5 GW), followed by Canada (66.8 GW), China (65.1 GW), and Brazil (56.8 GW). The next year China took second place, and by 2002 these four countries accounted for 43% of all installed hydro capacity and for 44% of all hydrogeneration, which supplied 17% of global electricity. Canada, the world's largest producer (just ahead of China), derived nearly 60% of its electricity from water. Hydrogeneration is even more important for many modernizing countries; the shares are over 90% in many African countries and 80%–90% in South America. Most of these countries also have considerable potential to develop small-scale hydro resources, the sites with capacities below 2 MW; China has led these efforts.

Peak technical achievements in large dam construction are displayed in the earth-and-rock-fill Rogun Dam on the Vakhsh River in Tajikistan (335 m high), completed in 1985; the Nurek Dam on the same river (330 m high); the Bratsk Dam on the Yenisey River in Russia (reservoir capacity ~170 Gm^3); Yacyretâ embankment dams (~65 km) on the Paraná River between Paraguay and Argentina; and dam volumes in excess of 200 Mm^3. The largest turbines, invariably Francis machines, have rated capacities of 700 MW; there are 26 of them in the world's largest hydro station, China's 18.2-GW Sanxia (Three Gorges) on the Yangtze (Chang) River. Other large projects are hydroelectric plants at Itaipú on the Paraná River between Brazil and Paraguay (12.6 GW), Guri on the Caroni River in Venezuela (10.3 GW), Grand Coulee on the Columbia River in the United States (7.079 GW), and Sayano-Shushensk on the Yenisey River in Russia (6.4 GW).

Hydrogeneration demands much space. The reservoirs behind the world's large dams (higher than 30 m) now cover almost 600,000 km², an area nearly twice as large as Italy, and those used solely or mostly for electricity generation cover about 175,000 km². This prorates to about 4 W/m² in terms of installed capacity and to about 1.7 W/m² in terms of actual generation. The power densities of individual projects span 3 OM. There is a predictable rise of power density with increasing installed capacity. Projects with 2–99 MW$_{ei}$ averaged just 0.4 W/m², those with 500–999 MW$_{ei}$ rated 1.35 W/m², and mean power density for the world's largest dams (3–18.2 GW$_{ei}$) surpassed 3 W/m² (Goodland 1995).

Dams on the lower courses of large rivers impound huge volumes of water in relatively shallow reservoirs. The combined area of the world's seven largest reservoirs is as large as the Netherlands and the top two, Akosombo on the Volta River in Ghana (8730 km²) and Kuybyshev on the Volga River in Russia (6500 km²), approach the size of small countries like Lebanon or Cyprus. The power densities of these projects are below 1 W/m² (fig. 9.4). In contrast, dams on the middle and upper courses of rivers have power densities on the order of 10¹ W/m². Itaipú has power density of 9.3 W/m², Grand Coulee rates nearly 20 W/m², China's Sanxia (the world's largest at 17.68 GW when finished in 2008) will have about 28 W/m², and some Alpine stations surpass 100 W/m². The world's record holder will be the Nepali Arun project, whose 42-ha reservoir and 210 MW installed capacity translates to about 500 W/m².

All these cited density values are calculated by using installed capacities; actual generation, determined by water availability, translates into substantially lower rates. Load factors of most hydro stations are highly variable, but the average rates are commonly below 50% (Itaipú was

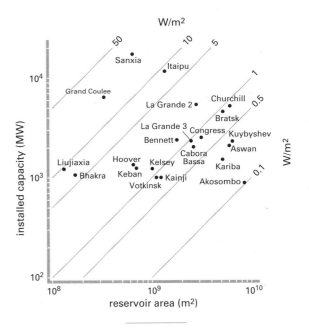

9.4 Power densities of the world's largest hydro projects, calculated using total installed capacities. Actual electricity generation densities depend on variable load factors.

designed for 68%, Sanxia for 53%). Many projects are built primarily as peaking power facilities. Consequently, effective power densities of most hydro stations are merely one-third to one-half of the theoretical rates. Pumped storage has become an increasingly important component of hydroelectric generation. It uses nighttime or weekend low-demand power to pump water to an elevated reservoir to be released during peak-demand hours through turbines into a lower reservoir. Generation can start in as little as 10 s.

With heads up to 1400 m (commonly 200–500 m) and limited load factors, pumped storage generation has quite high power densities, mostly 1–3 kW/m². These relatively expensive plants were built first during the

1890s in Italy and Switzerland, and by the year 2000 their global capacity surpassed 90 GW (about 13% of all installed hydro power), with Europe having one-third of the total. Italy, Japan, and the United States have the most high-capacity pumped storage facilities. The two largest stations are 2.88-GW Lewiston in Niagara and 2.7-GW Bath County in Virginia (ESA 2001). Other large projects are 2.4-GW Guangdong, part of the Day-awan nuclear station designed to cover peak demand for Hong Kong and Guangzhou, and 2.3-GW Dniester in southern Russia. Reversible pump turbines serving these projects rate 300–400 MW.

Adaptation of nuclear fission for electricity generation progressed very rapidly. The first proof of fission was published in February 1939 (Meitner and Frisch 1939); the first sustained chain reaction took place at the University of Chicago on December 2, 1942 (Atkins 2000); and Hyman Rickover's leadership to apply the reactor drive to submarines led to the launch of the first nuclear-powered vessel, *Nautilus*, in January 1955 (Rockwell 1992). Rickover was put immediately in charge of reconfiguring General Electric's pressurized water reactor (PWR), used on submarines to build the first U.S. civilian electricity-generating station in Shippingport, Pennsylvania. It went on line in December 1957, more than a year after the world's first large-scale nuclear station, British Calder Hall (4×23 MW), was connected to the grid on October 17, 1956. Regardless of the reactor type, nuclear generation uses the heat released by the fission of heavy elements (mainly ^{235}U) to generate steam for electricity generation. The PWR, with the coolant circulating through the core in a closed loop and transferring heat into a steam generator, became the dominant choice of a rapidly growing industry. This arrangement results in e_1 of 32%.

A boiling water reactor (BWR) generates steam by passing water directly between the elements of nuclear fuel. The efficiency of this simpler arrangement is just over 30%. Both PWR and BWR use ordinary water as a coolant and as a moderator to slow down the neutrons. In pressurized heavy water reactors (PHWRs), used by Canada's national program, D_2O is a moderator and organic compounds, gas, H_2O, and D_2O are working fluids. Gas-cooled reactors (GCRs), the foundation of British nuclear development, use graphite as moderator and He or CO_2 as coolant. The largest PWRs rate nearly 1.5 GW (the two largest are the Soviet-built Ignalina in Lithuania at 1.45 GW and the French Chooz at 1.457 GW). The largest nuclear power plant, Japan's Kashiwazaki-Kariwa, with 8,206 MW in seven units, was the world's largest thermal electricity-generating facility by the end of the twentieth century.

The ten years between 1965 and 1975 saw the greatest number of new nuclear power plant orders. Europe, including the former USSR, eventually ordered about twice as many power reactors as did the United States. Expert consensus of the early 1970s envisaged worldwide electricity generation in the year 2000 dominated by inexpensive fission. Instead, the industry has experienced stagnation and retreat. In retrospect, it is clear that commercial development of nuclear generation was far too rushed and that too little weight was given to public opinion on commercial fission (Cowan 1990). The economics of fission generation was always arguable because its costs did not include either the enormous government subsidies for nuclear R&D or the costs of decommissioning the plants and safely storing highly radioactive waste for millennia.

Weinberg (1994, 21) conceded that "had safety been the primary design criterion [rather than compactness and simplicity, which guided the design of the submarine PWR], I suspect we might have hit upon what we now call inherently safe reactors at the beginning of the first nuclear era." Moreover, promoters of nuclear energy ignored Enrico Fermi's warning that the public might not accept an energy source that generates large amounts of radioactivity as well as fissile materials that might fall into the hands of terrorists. By the early 1980s other un-expected factors—declining growth rates of demand for electricity, escalating costs in an era of high inflation and slipping construction schedules, and changing safety reg-ulations that had to be accommodated by new designs—helped to dim fission's prospects. Many U.S. nuclear power plants eventually took twice as long to build and cost twice as much as initially anticipated.

Safety concerns and public perceptions of intolerable risks were strengthened by an accident at the Three Mile Island plant in Pennsylvania in 1979 (Denning 1985). By the mid-1980s the shortlived fission era appeared to be over everywhere in the Western world with the exception of France. Accidental core meltdown and the release of radioactivity during the Chernobyl disaster in Ukraine in May 1986 made matters even worse (Hohenemser 1988). Although the Western PWRs with their contain-ment vessels and tighter operating procedures could not experience such a massive release of radiation as did the unshielded Soviet reactor, that accident only reinforced the widely shared public perception that all nuclear power was inherently unsafe.

Still, by the year 2005 nuclear generation was produc-ing nearly as much electricity as did hydro stations: 441 nuclear reactors with a total net installed capacity of 369 GW generated just over 2.6 PWh (fig. 9.5). Reactors accounted for about 11% of all installed electricity-generating capacity, but because of their high availability

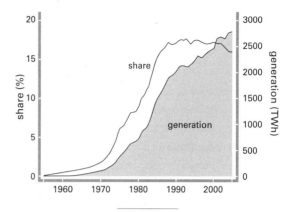

9.5 Nuclear electricity, 1956–2005: share of total electricity production and worldwide generation. Plotted from data in UNO (1990) and WNA (2006b).

factors (global average of about 80% during the late 1990s) they generated about 16% of all electricity in 2005 (IAEA 2006). The highest national contributions were in France, where 79% of electricity was generated by PWRs. Lithuania, with its large Soviet-built station in Ingalina, came second (70%) and Belgium third (56%). Japan's share was 29%, the U.S. share 19%, Russia's 16%, India's 3%, and China's 2% (IAEA 2006).

The future of commercial fission reactors is inherently limited by the fact that they produce heat by splitting ^{235}U, an isotope that makes up just 0.7% of the element in the Earth's crust. Any large-scale substitution of fis-sion electricity for fossil-fueled generation would soon encounter shortages of inexpensive uranium (Mobbs 2005). Fission cannot become a more important long-term contributor to global TPES without the com-mercialization of fast breeder reactors (MIT 2003). Breeders use fast neutrons to convert the dominant ^{238}U into ^{239}Pu, and this plutonium (after reprocessing) is substituted for ^{235}U. But technical problems led to shutdowns of all fast breeder prototypes, in the United States in 1983, in France in 1990, and in Japan in 1995, and no new viable commercial design is ready for construction.

The power densities of nuclear fission are high. In nuclear engineering the measure denotes the core generation densities (power per volume). They can go up to 110 kW/dm^3 for PWRs and are highest in fast breeder reactors (the French Phenix operated with 646 kW/dm^3). In terms of power densities as used throughout this book (power per unit area), the rates for reactors are between 50–300 MW/m^2, 1 OM higher than those of fossil-fueled boilers. Nuclear power plants do not need any extensive fuel-receiving and storage facilities, and because the low-level wastes held temporarily at the site occupy very small areas, the overall generation densities are high, typically 2–4 kW/m^2. Land claims for the complete fuel cycle (mining and processing of ores, uranium enrichment, production of fuel elements, fuel reprocessing, and storage of radioactive wastes) lower these values significantly. Gagnon, Bélanger, and Uchiyama (2002) offer an average of about 230 W/m^2.

Compared to hydro and nuclear generation other modes of nonfossil electricity production are globally much less significant. Rapid post-1995 growth elevated wind-powered generation to third place, but by 2005 its share was still less than 1% of the world's electricity production. Combined geothermal and photovoltaic generation do not add up to 0.5%. Modern wind energy projects began with U.S. tax credits in the early 1980s (Braun and Smith 1992). By 1985 the United States had capacity of just over 1 GW, and the world's largest wind facility was at the Altamont Pass in California. Expiration of the tax credits in 1985 ended this episode. The early 1990s brought better turbine designs and larger machines, whose average size rose to 500–750 kW by the late 1990s and surpassed 1.2 MW by 2003, with 5-MW machines in development (Øhlenschlaeger 1997; DWIA 2005).

The second wave of wind projects was pioneered by Western Europe, above all, by Germany, Denmark, and Spain, where new laws guaranteed a fixed price for wind-generated electricity. The Danish government has been a particularly active promoter. That country now has the world's highest per capita installed capacity and dominates the world export market in efficient wind turbines (Vestas 2001). The exponential growth of new capacities lifted the global total from less than 2 GW in 1990 to 17.3 GW by 2000 and 74.2 GW by 2006, with Germany accounting for more than one-quarter of the total (20.6 GW), and the United States and Spain each having nearly 12 GW (AWEA 2007). With roughly 25% load factor this implies still no more than about 85 TWh/year, only about 0.5% of all electricity generated worldwide. The highest national shares were in Denmark (about 20%), Germany (6%), and Spain (5%); the U.S. share remained below 1%.

Very large amounts of wind power could eventually produce some nonnegligible climatic change at continental scales (Keith et al. 2004), but more practical limits on global wind capture are imposed by average annual wind speeds (at least 6.9 m/s are needed for low-cost generation) and by the height of durable structures. Archer and Jacobson (2005) put the global wind power potential at 72 TW, but only a very small share of that could be captured, and even at the best sites wind can be converted to electricity with only very low power densities. For example, Grubb and Meyer (1993) put the practical potential at 6 TW. Windy sites with annual mean speeds of 7–7.5 m/s produce power densities of 400–500 W/m^2 of vertical area swept by rotating blades 50 m above ground. Spacing equal to five rotor diameters is enough to avoid excessive wake interference, but at least twice that distance is needed for wind energy replenishment in large installations.

Moreover, no more than 16/27 (59.3%) of the wind's kinetic energy can be extracted by a rotating horizontal-axis generator (the Betz limit), and the actual capture is about 80% of that. For example, in locations with wind power density averaging 450 W/m^2 (mean value of power class 4 common in the Dakotas, northern Texas, Western Oklahoma, and coastal Oregon), machines with a 50-m high hub would intercept about 7 W/m^2, but average 25% conversion efficiency and 25% power loss caused by wakes and blade soiling would reduce the actual power output to about 1.3 W/m^2 (Elliott and Schwartz 1993). Vertical power density of 700 W/m^2, turbine efficiency of 35%, and power loss of 10% would more than double that rate to about 3.5 W/m^2. Actual rates are highly site-specific. Altamont Pass averaged about 8.4 W/m^2 (D. R. Smith 1987). The most densely packed wind farms rate up to 15 W/m^2; more spread-out sites 5–7 W/m^2 and as low as 1.5 W/m^2; and the best offshore sites 10–22 W/m^2 (McGowan and Connors 2000).

Geothermal electricity generation can tap only a tiny part of the immense flux of the Earth's heat (see section 2.5). Drilling to depths of more than 7 km to reach rock temperatures in excess of 200°C is now possible, but it would be economically prohibitive to do that in order to inject water for steam generation. Consequently, today's conversion techniques could harness at best about 72 GW of electricity-generating capacity; enhanced recovery and drilling improvements currently under development could enlarge this total to about 138 GW (Gawell, Reed,

and Wright 1999). Geothermal electricity generation began at Italy's Larderello field in 1902. New Zealand's Wairakei was added in 1958, California's Geysers in 1960, and Mexico's Cerro Prieto in 1970. Only later did geothermal generation expand beyond these four pioneering high-temperature vapor fields.

By the year 2000 the United States had installed nearly 2.2 GW, the Philippines 1.9 GW, and Italy 785 MW. The global total was 7.7 GW, and annual generation reached nearly 52 TWh, implying an average load factor of 7% (IGA 2005). Power densities of 20–50 W/m^2 are comparable to those of Alpine hydro stations. Global capacity of 8 GW is only about 11% of the total that could be harnessed with existing techniques, and the prospective potential of 138 GW is less than 5% of the world's electricity-generating capacity in the year 2000. Geothermal energy will remain a globally marginal source of electricity, but it can be a locally important supplier of industrial and household heat. U.S. capacity was about 3.7 GW_t in 2000, followed by China (2.3 GW_t) and Iceland (1.5 GW_t). Global capacity was about 15.1 GW_t in the year 2000 (IGA 2005).

The photovoltaic (PV) effect, whereby electricity generation of an electrolytic cell made up of two metal electrodes increased when exposed to light, was discovered by Edmund Becquerel in 1839. In 1873 came the discovery of the photoconductivity of selenium, which made it possible for W. G. Adams and R. E. Day to make the first PV cell just four years later; the conversion efficiencies of such cells were a mere 1%–2%. The decisive breakthrough came only in 1954, when a team of Bell Laboratories researchers produced silicon solar cells that were 4.5% efficient and raised that performance to 6% just a few months later. By March 1958, when Vanguard I became the first PV-powered satellite (0.1 W from

(a)

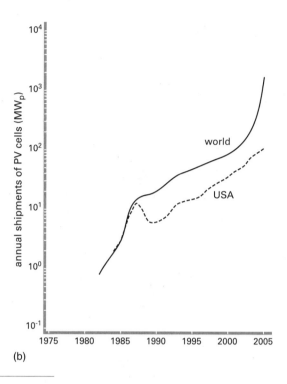

(b)

9.6 (a) Telstar I, the world's first telecommunication satellite
powered by PV cells, was launched in 1962 (photograph
courtesy of Lucent Technologies). (b) Worldwide and U.S.
shipments of PV cells, 1980–2005 (plotted from data
published by the U.S. Energy Information Administration).

about 100 cm^2), Hoffman Electronics had cells that were 9% efficient; it began selling 10% efficient cells a year later (PVPRS 2001).

In 1962, Telstar, the first commercial telecommunications satellite, had 14 W of PV power (fig. 9.6), and two years later Nimbus rated 470 W. PV cells became an indispensable ingredient of the burgeoning satellite industry, but land-based applications remained uncommon even after David Carlson and Christopher Wronski at RCA Laboratories fabricated the first amorphous silicon PV cell in 1976. After years of exaggerated expectations and missed targets, global PV electricity generation finally began to expand exponentially during the late 1990s. In 1990 the worldwide shipments of PV cells and modules were less than 50 MWp/a; by 2005 the total was just over 1.7 GWp/a (fig. 9.6). Solarex, Kyocera, Sharp, and Siemens were the leading producers. Installed capacity reached nearly 5 GWp worldwide, still only a tiny fraction of more than 2.1 TW in fossil-fueled generators (BP 2006).

The maximum theoretical efficiency of PV cells is limited by the range of photon energies, and large efficiency gaps remain between theoretical, laboratory, and field performances (Perlin 1999; Markvart 2000; Archer and Hill 2001). Thin PV films are made of amorphous Si or of GaAs, CdTe, and CuInSe$_2$. Theoretical single-crystal efficiencies are 25%–30%. Lenses and reflectors can boost these efficiencies, and stacking cells sensitive to different parts of the spectrum could push the theoretical peak to 50%. Actual efficiencies of new commercial single-crystal modules are 12%–14%. Thin-film cells can convert 11%–17% in laboratories but as little as 3%–7% after several months in field operation. Multijunction amorphous Si cells convert as much as 11% in large modules. With an eventual 17% field efficiency PV cells would be averaging about 30 W/m^2.

Actual field performance of PV cells has been, so far, much lower. Generation at the world's largest PV site—Bavaria's Solarpark, with installed capacity of 10 MW, peak power of 6.3 MW, and area of 250,000 m^2 divided among three sites (PowerLight Corporation 2005)—prorates to only about 3 W/m^2. The site's high cloudiness is the obvious reason for this poor performance; the location was chosen because of generous subsidies and not because of its suitability for PV conversion. These subsidies also explain why by the end of 2005 eight out of ten of the world's largest PV projects were in cloudy Germany and only two in the United States (in Tucson and in Rancho Seco, California).

The use of mirrors to concentrate solar radiation in order to heat water for electricity generation has seen only limited trials. The world's largest solar tower project, Solar Two in Barstow, California, with a solar field of 81,000 m^2 and annual generation of 17.5 GWh, had overall power density of less than 25 W$_e$/m^2 (SolarPA-

CES 2005). So far, direct solar conversions using flat plate collectors for space and water heating have been much more important than PV cells. Flat plates operate at up to 100°C and transfer up to 60% of insolation to water at 40°C–50°C. Well-designed systems rate 30–100 W$_t$/m^2 and in peak noon-time as high as 500 W$_t$/m^2. Globally, these installations, ranging from home rooftop heaters to industrial arrays, count in the millions, but it is impossible to offer an accurate estimate of their aggregate power. Finally, the world's sole tidal plant, the 240-MW Rance plant near St. Malo, Brittany, which was completed in 1966, has power density of about 14 W$_e$/m^2. Plans to harness the world's highest tides in Nova Scotia's Bay of Fundy have been repeatedly laid aside. The Bay of Fundy tides would deliver 15–16 W/m^2.

9.3 Global Consumption Patterns: Growth and Inequality

Perhaps the single most prominent characteristic of fossil-fueled civilization is the exponential increase in per capita energy consumption. Typical annual wood and charcoal consumption in richer preindustrial societies ranged between 20 GJ and 40 GJ per capita. In forest-rich, thinly populated North America of the eighteenth and nineteenth centuries, it peaked at atypically high rates of 70–100 GJ per capita (the latter was the U.S. rate in 1860). If the e_1 of fireplaces, stoves, and simple furnaces averages 10%, such consumption rates would prorate to no more than 2–4 GJ, and exceptionally to 7–10 GJ, of useful energy per capita. During the nineteenth century countries with rich resources of coal increased their rates of gross fossil energy consumption rapidly, from just 20 GJ in 1800 to 116 GJ in 1900 in Britain, from a negligible amount to 105 GJ during the same period in the United States. Per capita consumption of coal in countries that had to import most of that fuel, like France or Sweden, remained relatively low; their energy use surged only after WW II with large imports of crude oil. Japan followed a similar pattern.

In 2003, Canadian TPES, with primary electricity converted at the prevailing rate for thermal generation, averaged 450 GJ per capita, and the U.S. mean was about 360 GJ per capita. In contrast, Western Europe averaged 160 GJ, ranging from 117 GJ in Portugal to 197 GJ in France, and the Japanese mean was 185 GJ per capita (EIA 2005b). With overall mean conversion efficiency of about 40%, the North American mean prorated to about 145 GJ of useful energy, a flux roughly 15-fold higher than during the last decades of the biomass era. The rates of the world's most populous modernizing countries remain far behind the affluent rates. In 2003, China averaged less than 40 GJ, India less than 15 GJ, and Nigeria less than 10 GJ per capita. Brazil surpassed 50 GJ, still below the global mean of 70 GJ per capita. But because of a highly skewed consumption pattern this global mean is largely irrelevant. Only three countries, Argentina, Croatia, and Portugal, have consumption rates close to it, whereas the modal (most frequent) national mean is below 20 GJ per capita, and affluent countries average above 120 GJ per capita (fig. 9.7).

In 2003 the absolute range was from less than 1 GJ per capita for the poorest African countries to 450 GJ for Canada. This comparison leaves aside anomalous rates for such countries as Qatar or the United Arab Emirates, or such territories as Gibraltar or the U.S. Virgin Islands, where extraordinarily high consumption reflects, respectively, the cost of hydrocarbon processing for global export and ship refueling. Needless to say, all these measures are just statistical abstractions dividing a country's total commercial primary energy use by its population.

Because the poorest people in the poorest countries do not directly consume any fossil energies or primary electricity (their only links to modern energies are indirect, via purchases or donations of tools, clothes, or food imported during famines), the difference in modern energy consumption between a subsistence pastoralist in the Sahel and an average Canadian may easily be larger than 1,000-fold.

Another way to convey this great global disparity is by contrasting the shares of populations with those of TPES. Although post-1950 economic development reduced the relative gap between haves and have-nots, the disparities are unacceptably large. In 1950 industrialized countries consumed about 93% of the world's commercial TPES; by 1985 the rich world containing one-quarter of global population consumed about 82% of all primary energy; and in 2005 its share fell to just over 60% versus about 15% for population. The United States alone, with 5% of world population, used about 22% of total energy, and the G8 countries, with 12% of the world's population, consumed about 46% of the world's energy. In contrast, the poorest one-quarter of humanity (sub-Saharan Africa, Nepal, Bangladesh, most of rural India) consumes less than 3% of the world's TPES. Few other comparisons illuminate more starkly the existential chasm separating the two worlds (fig. 9.7).

This gap is even greater with respect to the most flexible form of energy. At the turn of the twentieth century, average per capita electricity consumption in affluent countries was about 10 MWh, the global mean (once again, a rather unrevealing figure) was nearly 2.5 MWh; the world's poor countries averaged just 1 MWh; the least developed nations not even 100 kWh; most sub-Saharan countries below 50 kWh; and some two billion people (nearly one-third of humanity) had no access to

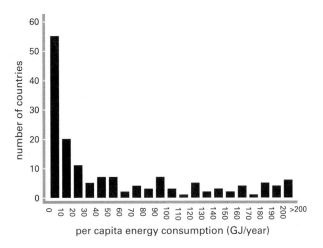

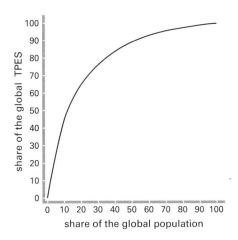

9.7   Highly skewed distribution of average national per capita energy consumption in the year 2000, shown as a frequency graph (*above*) and as a Lorenz curve (*below*). From
Smil (2003).

electricity. Global electricity consumption is thus even more unevenly distributed than is the supply of fossil fuels while serving as an effective indicator of affluence and economic success. Norway's extraordinarily high rate approaches 30 MWh per capita, the Canadian mean is nearing 20 MWh, the United States consumes in excess of 13 MWh, and most of Western European means are about half of the U.S. value.

While aggregate national consumption figures for individual fuels and electricity have been available for decades, systematic analytical studies of energy flows through national economies are of surprisingly recent origin. In the United States it was only three decades ago that SRI (1971) outlined the movements of fuels and electricity, and allowed a closer look at the patterns of consumption. After 1973 it became obvious that no sensible management of energy use can be done without first appreciating the complexities of actual flows, and this realization led to a proliferation of sectoral use studies. National energy flow graphs (often referred to as spaghetti charts) came into widespread use, tracing the inputs of primary energies, their multiple transformations (with different degrees of detail), and their final uses (fig. 9.8). Longitudinal comparisons of sectoral uses show four gradual trends that reflect the changing economic structure and rising affluence of modern societies: declining shares of industrial consumption and slowly increasing shares of residential, commercial, and transportation demand.

At the end of the twentieth century, industrial energy use was below 50% of the total in all rich Western countries, about 45% in Japan compared to 65% in the early 1960s, and 35% in the United States compared to nearly 50% in 1950. In contrast, Chinese industries consumed 70% of the country's TPES in the year 2000. Breakdown by final uses shows the largest share of U.S. industrial energy going for process heating (35%), the second largest for machine drives like electric motors (15%), and smaller shares for heating, ventilation and air conditioning, electrochemical use, and lighting. As for major industrial categories, the dominant uses are extraction of fossil fuels and mineral ores, crude oil refining, chemical syntheses, ferrous and color metallurgy, and food processing.

Increasing specialization and rapid globalization of industrial production has made comparisons of national industrial energy uses much less revealing than during the earlier era of considerable economic autarky. Some nations have emerged as leading producers of energy-intensive metals (Canada, South Africa), others have become the workshops of the world (Japan, China, South Korea), and these idiosyncracies determine the extent and intensity of national industrial energy use. For example, during the late 1990s the three leading U.S. sectors (fuels, chemicals, and paper) accounted for 40% of the country's industrial energy use, whereas China's three largest consumers (smelting and rolling of ferrous metals, chemicals, and nonmetal mineral products) claimed just over 50% of the country's large industrial PES.

Collectively, buildings are either the largest or second largest consumers of energy (behind industrial conversions) in all rich societies. In the United States they took about 40% of all fuels and 75% of electricity in the year 2000. Space heating is the dominant demand in every rich nation. Depending on the climate it is between 50% and 80% of all residential consumption; the U.S. rate is just over 50%. Next come appliances (~25% in U.S. households) and water heating (~20% in U.S. households). Air conditioning use has been rising but is still rather limited. It accounts for only about 5% of U.S. final household energy use, but it claims one-quarter of all

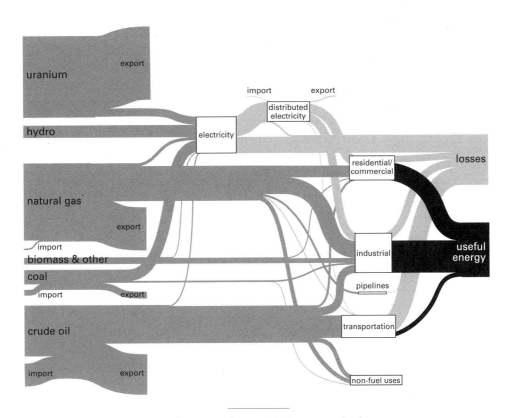

9.8 Canada's energy flows in 2002: an example of a comprehensive national energy flow graph. Simplified from a final version of a graph issued by Natural Resources Canada in March 2006.

household electricity and as such it plays a key role in boosting peak summer demand.

The pre-1973 performance of residential and commercial buildings with respect to energy conservation was astonishingly poor in North America and better, although far from satisfactory, in Europe. Since then many improvements have taken place, but further large savings remain possible (Baird 1984; Rosenfeld and Hafemeister 1988; Schipper and Meyers 1992; Roaf and Hancock 1992; Gonzalo 2006). For private houses standardized comparisons of energy requirements are usually done per unit area per heating degree-day (in °C to the base of 18°C). The average in kJ/m^2 per degree-day for U.S. housing stock in 1990 was about 150; new buildings, 100–120; superinsulated houses, 30–50; and the most efficient designs, 15–20. In terms of actual consumption, national means of total residential energy consumption at the end of the twentieth century prorated to less than 10

W/m^2 of floor area in China, about 14 W/m^2 in Japan, 15–20 W/m^2 for all types of U.S. housing (mean of ∼17 W/m^2 for single-family detached houses), and 25 W/m^2 in Canada (EIA 2001a; Natural Resources Canada 2000; Zhang 2004).

These power densities reflect more than just differences in climate. Japan averages annually 1,800 degree-days, the United States about 2,600, and Winnipeg 5,900. But my superinsulated house in Winnipeg consumes annually just 35–40 W/m^2 of its foundations, a rate nearly identical to a typical two-story detached U.S. house in a climate with less than 50% of Winnipeg's heating degree-days. In addition, different lifestyles and affluence levels set different consumption baselines. The Japanese and British tolerate much lower indoor temperatures (below 15°C) than North Americans and Canadians (typically above 18°C) and commonly heat only some rooms in a house or, in the Japanese case, merely parts of some rooms (traditionally with charcoal-burning *kotatsu*, now with electric heaters). Traditionally there has been no winter heating in China's provinces south of the Yangtze despite the fact that the room temperatures repeatedly dip below a comfortable level. Canadians and Swedes heat on average 2.5–3 times more water than do most Europeans or Japanese. And in low-income countries electricity use is dominated by meager lighting, whereas in affluent countries appliances consume the most.

Commercial buildings have a higher specific energy use than households. Most multistoried glass structures put up in North America between the early 1950s and the early 1970s averaged 110–140 W/m^2 of floor area. A typical 20-story glass building of those years required 2–2.5 kW/m^2 of its foundation, and New York's 110-story World Trade Center twin towers used about 12 kW/m^2

of their foundations, or a total of 80 MW per building. By the mid-1980s the primary energy required by new office buildings was below 50 W/m^2, and many all-electric buildings have been designed for 10 W/m^2 or below, less than 30 W of primary energy per square meter of occupied area (or, for a 50-story structure, no more than 1.5 kW/m^2 of the foundations). An extensive survey of U.S. commercial buildings shows a nationwide mean of about 33 W/m^2 for all structures, with the rates ranging from less than 15 W/m^2 for storage to more than 90 W/m^2 for food service and health care buildings (EIA 2005a). On average one-third of this flux went for heating, one-fifth for lighting, and only about 6% for office equipment.

The rising share of energy use for transportation is a universal marker of economic development. In 2000 that sector claimed nearly 30% of U.S. TPES, 25% in Japan, but less than 10% in China. Waterborne transport powered by diesels and gas turbines is relatively efficient and incontestably the cheapest way to move goods or people. Pipelines are best for the bulk movement of liquids and gases, but trains, powered by diesels or electric locomotives, are the best general freight performers on land (Smil 2006). Electric motors also power all high-speed passenger trains, most notably Japan's *shinkansen* and the French TGV, *train à grand vitesse* (fig. 9.9). In road transport diesels are the preferred prime movers for trucks and buses, and they now also account for nearly half of new European passenger cars; elsewhere gasoline-powered vehicles dominate. In 2005 about 25% of global refinery output was motor gasoline, 33% middle distillates, and about 6% aviation kerosene.

Gasoline and diesel fuel consumed by vehicles on busy U.S. urban interstate highways (up to 14,000 vehicles a day per lane, lane width 3.6 m) translates to power den-

(a)

(b)

9.9 (a) Japanese *nozomi shinkansen*. (b) French TGV Sud.
From Smil (2006).

sities of 150–200 W/m^2 of road, assuming an average rate of 13 L/100 km. Short-term rates limited to smaller areas during rush-hour periods or extensive traffic jams can surpass 250–300 W/m^2. These rates are more revealing (because of their contribution to the urban heat island) than any long-term, large-scale averages that are composed of extended periods of no or minimal traffic as well as daily and seasonal peaks of extraordinary intensity. Climate, building heights, concentration, and specific performance of commercial buildings and the typical density of traffic combine to produce an order of magnitude range for consumption power densities of urban central business districts (CBDs). Low-profile and low-density downtowns in smaller cities in mild climates, with buildings only a few stories high and lightly traveled roads, average no more than 100 W/m^2, and extensive

parking and greenery can bring this down to below 50 W/m^2 of the entire area. In contrast, high-rise (average 20 stories), high-density (75% built-up area) CBDs in large cities in cold mid-latitudes will rate, with heavy traffic, over 500 W/m^2 with peaks in excess of 700 W/m^2.

Sectoral changes of final energy uses have shifted the dominant demand decisions from industries and governments to individuals. At the end of the twentieth century the combined share of energy use by the residential sector and by private transportation was less than 15% of China's TPES, surpassed 20% in Japan, and approached 45% in the United States. Consequently, in such low-income, rapidly industrializing countries as China, where household energy use is frugal and car ownership still very limited, decisions about using some 85% of TPES are in the hands of the state and businesses. In contrast, in the United States, where some 80% of all transportation fuel is used by passenger cars, SUVs, and light trucks, the combined share of TPES that is determined by individual decisions is now approaching half the total, and this decision-making decentralization has made it much harder to anticipate future consumption trends. This situation would be reversed only with a new program of aggressively set corporate automobile fuel efficiency (CAFE) standards.

The energy requirements of buildings and transportation combine to reach their highest densities in cities and industrial conurbations. Surprisingly, the power densities of built-up areas, including roads, in such poor hot-weather cities as Kolkata or Mumbai (Bombay) are 10–15 W/m^2, the same as in warm but affluent Los Angeles or Tokyo. In the former instance low per capita energy use and relatively low rate of car ownership are counterbalanced by high dwelling densities and the presence of energy-intensive industries in residential areas. In the latter case the extraordinary urban sprawl that characterizes Los Angeles and the mostly low-rise housing of Tokyo dilute the effect of very high demand for transportation fuels and air conditioning. Some of the highest power densities of total energy use are found in Asian megacities. Shanghai's density is about 80 W/m^2 (in the city's built-up area, not the entire municipality, which includes a great deal of farm land); Seoul's close to 50 W/m^2; and Hong Kong's (with only 130 km^2 of built-up land) has surpassed 110 W/m^2 (Warren-Rhodes and Koenig 2001).

9.4 Qualitative Changes: Transitions and Efficiencies

A key qualitative shift in modern energy consumption was the substitution of coal by hydrocarbons. Higher heat content, easier and safer extraction, inexpensive seaborne and virtually invisible continental transportation, cleaner and much more convenient combustion, and incomparable flexibility of utilization account for the ascent of hydrocarbons in general and refined liquid fuels in particular, and for the relative decline of solid fuels. Shifting patterns of final uses mark distinct eras of fossil-fueled civilization. When coal overtook wood in the United States in 1885, 40% of it was used by railways, about 15% was converted to coke, and the rest was split among industrial boilers and residential heating. By 1945 railways used just over 20% and power plant consumption and coking were about equal. The railway market disappeared by the early 1960s, and by 2005 the coking share fell to just over 2% of the total, whereas electricity generation took more than 92% all coal (EIA 2006b). The breakdown of final uses is similar for affluent countries as a group. By 2003 they used nearly 75% of their coal supply for electricity generation (EIA 2005b).

Crude oil's first market was to replace whale oil in lighting, and only the invention of Otto and diesel engines made its lighter fractions dominant in transportation. Replacement of coal-fired steam locomotives by diesel engines boosted the typical conversion efficiency by nearly 1 OM. Even the best post–WW II steam locomotives were no more than 10% efficient, and large railway diesels have e_1 of at least 35%. Particularly after engine knocking was solved by the addition of tetraethyl lead, first introduced in the United States in 1924, gasoline helped to sustain successive waves of automobile use, first in the United States, from 1950 in Europe and Japan, and then in China. The automobile was a European invention, but U.S. affluence and mass production combined to give that country more than 90% of the world's cars, trucks, and buses before WW II and still 60% by 1960.

But Europe matched the U.S. total by 1983 and is now the world's largest market for new vehicles; China became the fastest growing new car market during the 1990s. In 2005 there were more than 700 million passenger cars on the roads, and their typical performance remains unimpressive. The best practical e_1 of the Otto cycle engine is about 32%; frictional losses bring this down to 26%, and partial load factors, inevitable during the urban driving that constitutes most car travel, reduce it to 19%–20%. Accessory loss and automatic transmission may nearly halve the total, so the effective e_1 is no more than 10%–12% and often as low as 7%–8%. Consequently, the greatest reductions of energy use in transportation can come from the long overdue radical redesign of passenger cars.

National differences are clearly discernible in the shares of final product uses. Gasoline makes up 50% of U.S. demand for liquid fuels, whereas Japan's share of the lightest fraction is less than 20%, but residual fuel oil accounts for nearly one-third of Japanese use, whereas the U.S. share is less than 10% of the total. The U.S. preference for large cars, decades of declining oil prices, and a tradition of heavy Detroit design also meant that the performance of the U.S. car fleet was actually getting worse. By 1974, U.S. cars needed roughly 15% more energy per kilometer than their counterparts of the 1930s. A sharp turnaround was brought about by OPEC's oil price increases and by the growing importance of European and Japanese car imports. Between 1974 and 1988 the fuel consumption mean for the U.S. car fleet fell by almost 50%, to 3.1 MJ/km.

Regrettably, the return of low oil prices and a decade of economic vigor reversed this trend as pickup trucks, vans, and the ridiculously named sport utility vehicles—all used mostly as passenger cars but exempt from passenger-car strict CAFE standards (minimum of 27.5 mpg or 8.6 L/100 km)—gained nearly half the U.S. car market. The average performance of these outsize vehicles in 2000 was only 17.5 mpg. Some of them weigh more than 4 t and need at least 15 L/100 km in highway driving. Among 2005 models, General Motors' Yukon and Chevrolet's Suburban and Tahoe were in this category. As a result, there has been no improvement in specific fuel consumption of the U.S. road vehicle fleet. The average for all passenger cars at the beginning of the 2000s was just 22 mpg (EIA 2006a). This stagnating performance was accompanied by a steady increase in average distance traveled annually. That rate hardly changed between 1950 and 1975, rising about 3% to 15,400 km/vehicle, but between 1975 and 2000, it rose 26% to about 19,500 km/vehicle. Yet there is no shortage of efficient cars on the market. The best-selling Honda Accord is highway-rated at less than 6.5 L/100 km, the Honda Civic at 5.7 L/100 km, and the hybrid Honda Insight at 3.3 L/100 km.

The high density and portability of refined fuels have also made them the only practical choice for aviation. Reciprocating gasoline-powered engines were displaced in nearly all military uses and all long-distance commercial flying by gas turbines. Kerosene is a much better fuel for jet engines than gasoline. It has a higher specific density (0.81 g/L vs. 0.71 g/L) and hence a higher energy density (35.1 MJ/L vs. 31.0 MJ/L), so more of it can be stored in tanks. It is also cheaper, has lower evaporation losses at high altitudes and a lower risk of fire during ground handling, and it produces more survivable crash fires. Jet A fuel, used in the United States, has a maximum freezing point at −40°C. Jet A-1 fuel, with freezing point at $-47°C$, is used on most long international flights and on northern and polar routes during the winter. Jetliners store this fuel in their wings; some also have a central (fuselage) tank and a horizontal stabilizer tank.

The aviation market witnessed by far the highest growth rates among all transportation modes as intercontinental flights became an unremarkable experience for tens of millions of business and leisure travelers. The annual total of passenger-kilometers flown globally by scheduled airlines surpassed 40 billion in the early 1950s and reached nearly 3 trillion passenger-kilometers in 2000 and 3.7 trillion in 2005. Growing even faster than passenger traffic, worldwide air cargo services rose from 750 million t-km in 1950 to about 108 billion t-km in 2000, and to more than 140 billion t-km in 2005 of which 90% was logged on international routes (ICAO 2006). But airplanes carry only a tiny fraction of commercial cargo. Even in the United States the share is less than 0.5%, compared to just over 40% for railways and nearly 30% for trucks (USBC 2006).

Another key qualitative shift marking modern energy use has been the rising importance of electricity. Rising demand for electricity spurred the expansion of hydrogeneration, whose contribution was negligible in 1900, and the introduction of nuclear fission, which became commercially available in 1956 (see section 9.2). But, above all, modern energy use is characterized by an increasing share of fossil fuels being converted to thermally generated electricity. In 1900 less than 2% of all U.S. fuel was converted to electricity; by 1950 that share had risen to 10%, and by 2000 it had reached 34% (EIA 2001b). The universal nature of this process is best illustrated by the rapid rate at which China has been catching up. That country converted about 10% of its coal (at that time virtually its only fossil fuel) to electricity in 1950. By 1980 the share surpassed 20%, and by 2000 it was about 30%, not far behind the U.S. share (Smil 1976; Fridley et al. 2001). Because of this strong worldwide trend, the global share of fossil fuels converted to electricity is now above 30%, compared to 10% in 1950 and just over 1% in 1900.

Given electricity's many advantages, this shift has been unavoidable. Only electricity offers instant, effortless consumer access; the ability to fill every consuming niche and be converted into motion, heat, light, and chemical potential; serving as the sole energizer of electronically transmitted information with unmatchable control, precision, and speed; silent and clean conversion; extremely reliable individualized delivery; and easily accommodating growing or changing uses. And electrical energy can be produced from a wide variety of (often inferior) fuels; its conversion to heat can be accomplished with nearly perfect efficiency; it can provide temperatures higher than can be attained by combustion of any fossil fuel; and its utilization requires no inventory.

Not surprisingly, electricity became the energy of choice for residential use, and with increasing consumer affluence its consumption became more diversified. At the beginning of the twentieth century, low-level lighting consumed nearly all of its use, but in 2001 the largest U.S. share, for air conditioning, was about 16%, refrigeration 14%, heating 10%, and lighting 9%, and water heating 9% (EIA 2001a). Refrigerator ownership is now nearly universal not only in rich countries but also in urban Asia. Other appliances with high ownership rates include color TVs (99% in the United States), microwave ovens (88%), and clothes washers (79%). Clothes dryers, dishwashers, and freezers are less common outside the United States. And even when everything is switched off, an average U.S. household now leaks constantly about 50 W, or about 4%, of all residential electricity, mostly owing to remote-ready TVs, VCRs, audio equipment, and communication devices (Meier and Huber 1997; Thorne and Suozzo 1997). Fortunately, these standby losses can be cut to less than 1 W per device.

That lighting takes such a small share of the U.S. household electricity demand is due not only to great disparities in typical power ratings of lights and common appliances (standard incandescent light bulb 100 W, standard fluorescent tube 40 W, and compact fluorescent 23 W, compared to 1 kW for a small toaster) but also to increasing efficacy (the quotient of the total luminous flux emitted by the rated power, measured in lm/W) of light (Smil 2003). In 1882 carbonized fibers in Edison's first lamps produced a mere 1.4 lm/W, and by 1900 their performance improved to 3–3.55 lm/W. Osmium, introduced in 1898, delivered 4 lm/W, tungsten filaments in vacuum 10 lm/W and in inert gas-filled bulbs (in 1912) 12 lm/W. Today's standard incandescent lights rate about 15 lm/W, which means (using 1.47 mW/lm as the standard mechanical equivalent of light) that they convert about 2.2% of electricity into light (fig. 9.10). New techniques boosted this performance. Fluorescent lights go as high as 100 lm/W (15% efficiency), and the best performers are low-pressure sodium lamps used for outdoor illumination (up to 175 lm/W, or nearly 26%). The next wave will see spreading indoor applications of light-emitting diodes (LEDs), which are already widely useed in brake lights and flashlights. By 2006 the best efficacy of white-light LEDs surpassed 100 lm/W.

Before post–WW II affluence spread the residential use of electricity beyond basic lighting and simple refrigeration to space heating, air conditioning, freezing, wash-

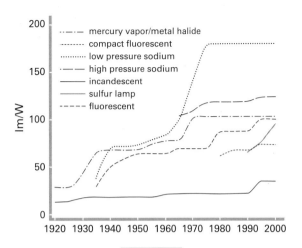

9.10   Increasing efficacy (lm/W) of electric lighting, 1920–2000. From Smil (2003).

ing, and scores of other uses served by a still-growing assortment of electronic devices, electricity's ascent first revolutionized industrial production when it became the dominant source of motive power. As revolutionary as was the substitution of waterwheels by steam engines in nineteenth-century factories, that step did not change the mode of distributing the mechanical energy needed for countless processing, machining, and assembling tasks. Factory ceilings were clogged by complex arrangements of iron or steel line shafts connected by pulleys and belts to parallel countershafts, which were belted to individual machines. Disabled, the prime mover (waterwheel, steam engine), a cracked line shaft, or a slipped belt shut down the whole assembly; conversely, if most of the machines did not need to work, the whole system still kept on running.

Electric motors changed this rapidly, first driving relatively short shafts for groups of machines and since the first decade of the twentieth century increasingly as unit drives. Devine (1983) and Schurr (1984) document the rapidity of this critical transition in the United States. While total installed mechanical power in manufacturing roughly quadrupled between 1899 and 1929, electric power capacities grew nearly 60-fold, from less than 5% to over 82% of total power. Since then, the share of electric power has changed little; the substitution was practically complete in just three decades. Its benefits translated into superior productive efficiencies, opened the way for flexible plant design and easy expansion, enabled precise machine control and highly focused power applications, did away with the overhead clutter, noise, and health risks, freed ceilings for installation of better illumination and ventilation, and resulted in higher labor and capital productivity (see section 12.2).

Electric motors also revolutionized passenger rail transport. The latest version of the world's first high-speed train (700 series), Japan's *Tōkaidō shinkansen*, which has been in operation since October 1964, draws electricity from catenary wires (copper or copper-clad steel) with spans of 50 m as 25 kV AC at 60 Hz, and it has 64 275-kW AC motors (four in each car) for the total of 13.2 MW. This makes frequent accelerations and decelerations, needed for relatively short interstation runs, easier, and the motors also function as dynamic brakes once they become generators driven by the wheels and exert drag on the train. Pneumatic brakes are used for speeds below 30 km/h and as a backup. The French TGV, unlike the *shinkansen*, has two locomotives (power cars) in every train set, each weighing 68 t and capable of 4.4 MW (TGVweb 2000). Electric supply and subsequent conversions are similar to the Japanese practice. A pantograph picks up AC at 25 kV and 50 Hz, a rectifier converts it to 1500 V DC, and traction inverters convert DC to variable frequency AC, which is fed to synchronous motors that are also used for dynamic braking at high speeds. Both of these trains are capable of speeds of 300 km/h.

But the modern world's great reliance on electricity has a major drawback: unacceptably high energy loss during thermal generation, particularly in coal-fired plants (see section 8.5). Only the best individual stations have efficiencies of 40%–42%. Co-generation—the use of a single primary heat source to produce simultaneously electricity and heat, saving 10%–30% of fuel in comparison with a separate generation of the two final energies—is an old technique, but its rapid commercial diffusion dates only to the 1970s (Hu 1983; E. L. Clark 1986; Boyce 2002; Petchers 2003). The technique has made substantial inroads only in Europe and Japan.

9.5 Energy Conservation: Gains and Rebounds

I agree with Rose (1986) that the term *rational and effective energy use* is preferable to *conservation* (the first law of thermodynamics dictates that energy is always conserved), but I subsume all approaches to reducing energy use under that imprecise but widely established word, used in so many post-1973 publications (Ford et al. 1975; Lovins 1976; Socolow 1977; Gibbons and Chandler 1981; Hu 1983; Rose 1986; Casten 1998; Goldemberg 2000; WEC 2006). Most of the attention has been given to improving device efficiency (for instance, compact fluorescent lights instead of light bulbs), but on an individual level such gains can be much reduced or canceled by inefficient use (leaving the light on) or lack of system benefits (light may be too far away). In terms of broad strategies, a trio of basic approaches exists: doing without, maximizing conversion efficiencies, and reducing the use of energy-intensive products through better design and extensive recycling of materials.

Doing without is an underappreciated option in affluent societies habituated to the idea of growth, but its impact is clear when comparing per capita energy use in the Western world of the early 1960s with that of the early 2000s. (I chose this span, 1965–2005, because older readers of this book will be able to call on their personal experiences in judging the gains.). Was life with 15% (the U.S. case), 40% (in Canada), or even 50% less energy (in France) so unfulfilling, so unsafe, or so unbearable? But voluntary frugality has been in short supply throughout the rich Western world, and in the absence of acute social crises democracies are averse to adopting restrictive and proscriptive measures.

Only in a minority of cases does any further improvement of conversion efficiencies run into physical limits. Thermodynamic minima have been approached in various energy-intensive chemical syntheses (see chapter 10); the best large electric motors are near-perfect converters of electricity to rotary power ($e_1 > 97\%$); and many boilers and furnaces are, in e_1 terms, more than 90% efficient. But beyond such instances there is an entire universe of wasteful conversions with many opportunities for efficiency improvements ranging from relatively modest to surprisingly large. For instance, in superinsulated houses, halving the total energy needs is not exceptional. And modest improvements, multiplied by the 10^6–10^8 units operating in modern mass consumption societies, would translate into huge savings. Further savings can come from structural changes. In comparison with a three-bedroom, single-story house, an equal-sized

two-story building has an energy efficiency gain of 15%, a two-story duplex 30%, a two-story triplex 35%, and a low-rise apartment building 40% (Burchell and Listokin 1982).

Improvements in automotive efficiency are certainly the most important instance of individually small but collectively massive gains because substantial energy savings can come from a combination of gradual adjustments and widespread applications of existing techniques. To begin with, cars are often designed to be unnecessarily powerful. Unless the driving requires unusually rapid acceleration, travel on uncommonly steep roads, or heavy towing, there is no good reason a passenger car used for urban driving should rate over 40 kW. In 2005 even a Suzuki Swift was overrated (76 kW), and a Honda Civic (87 kW) was more than twice as powerful as necessary for driving from one traffic signal to another. Weight reduction obviously lowers power requirements, but despite front-wheel drive, transversely mounted engines, and lighter-than-steel materials, even compact European cars got heavier, gaining more than 300 kg/vehicle between 1970 and 2000 (WBCSD 2004).

Reductions of aerodynamic drag still have far to go before they encounter a physical limit. Lean-burn low-friction engines, continuously variable transmissions, and greater diffusion of diesels, especially adiabatic low-friction kinds, are other components of a strategy that could see national car fleets averaging below 5 L/100 km. Only such improvements could stem the incredible waste of automotive fuels. In 2005, U.S. gasoline consumption (~13 EJ) was equal to about 60% of total primary energy use in Japan (~22 EJ). In global terms, U.S. gasoline use in 2005 accounted for nearly 4% of the total world fossil fuel consumption. In personal terms, a week's worth of gasoline for a two-car suburban U.S. family is equivalent to the total annual primary energy consumption of a fairly well-off Indian villager.

Airlines are a minor fuel consumer compared to cars and trucks (U.S. jet fuel use is equivalent to about one-sixth of gasoline use). But fuel's large share of airlines' operation costs has stimulated high-efficiency advances. Turbofan engines (see section 8.4) have realized the greatest efficiency gains, and further opportunities exist to continue this commendable trend. Modernization of lighting provides another example of energy savings achievable with existing techniques. Outdoor lighting already relies on high-performance halide and sodium lamps. But substituting fluorescent lights (available in compact form) for indoor incandescent bulbs represents a large potential for reducing electricity use. Electricity savings can also be realized by more efficient motors, especially the AC-polyphase induction machines rated between 750 W and 100 kW that dominate industrial applications (pumping, compressors, fans, blowers, machine tools). Large gains for small motors (<1 kW) and small improvements for larger machines would translate into significant aggregate gains when multiplied by millions of operating units.

Co-generation (see section 9.4) addresses a part of the qualitative mismatch between sources and final uses that is regrettably common in high-energy-use society. The most often cited example of this mismatch is the use of fossil-fuel-generated electricity for resistance heating (using inexpensive hydroelectricity, as do Norway, Quebec, and Manitoba, is a different matter). In general, rich societies need 20%–40% of their useful energy as low-temperature heat (well below 100°C), and supplying this demand by burning fossil fuels at temperatures exceeding 1000°C leads to very low e_2. This efficiency would soar with wider application of solar conversions. Repairing this qualitative energy mismatch has been one of the principal objectives of the soft energy path (Lovins 1976), but a simplistic maximization of thermodynamic efficiency should not be the sole guiding reason for restructuring the energy supply (see chapter 12).

Conservation springs from what Socolow (1977) labeled an "inverted emphasis" in a society facing the energy dilemma. Instead of a traditional concentration on enlarging the energy supply, the inverted approach embraces deliveries of particular energy needs and thus inevitably discovers huge rationalization opportunities. At the same time, one must be aware of two practical limits to conservation: the incremental nature of the gains and the effect of time horizons. Energy conservation does not offer the equivalent of spectacular oil field discoveries or dramatic single-item shortcuts. Even significant conservation efforts like mandatary efficiency standards for cars, air conditioners, or refrigerators cannot reduce national energy use by large margins over short periods of time. Moreover, successful energy conservation requires long-term commitment and a prodigious number of informed individual decisions; both can be a problem in most affluent societies.

The time factor is important in every investment decision to raise energy conversion efficiency or reduce waste. Long amortization spans justify higher energy inputs and generate greater life cycle savings. But such desirable long-term commitments are compromised by the high mobility of U.S. society and the preference, especially in North America, for lowest first cost. Other affluent but more settled societies with rising coss of living and higher energy prices have been able to promote energy conservation more readily, especially if they have traditionally high savings rates.

Energy conservation demands substantial socioeconomic adjustments in return for its long-term benefits. But its rewards go beyond energy savings. As it reduces environmental impacts, it offers the satisfaction of securing additional needed energy by restraint and careful use rather than by merely seeking an ever-greater supply.

But do conservation efforts reduce energy consumption in the long run? Both economic theory and consumer behavior lead us to expect a rebound effect as the lower costs of energy services brought about by increased conversion efficiencies stimulate consumption. The magnitude of the direct rebound effect (for instance, using a greater number of efficient lights and leaving them on overnight) has been disputed, but it typically ranges between 10% and 30% for such common energy services as space heating and cooling, lighting, and car transport (Herring 2004). Other price-elasticity effects that tend to increase the use of more efficiently delivered energies include income-related effects (spending the savings on other goods and services), product and factor substitution (more energy used in lieu of labor, time), and most important, transformational effects resulting from long-term technical and socioeconomic changes.

In the short run and at a microeconomic level, energy conservation efforts are undoubtedly financially and emotionally rewarding. Similarly, companies with energy-intensive products can realize impressive savings that can eventually extend to entire industries. But these savings do not translate into any lasting reductions in the overall use of energy in economies at large, and actually contribute to the growth of global energy use. The idea that secular advances in energy efficiency lead to declines in aggregate energy consumption was eloquently discredited for the first time by W. Stanley Jevons, the first economist to address the potential of higher efficiency for "completely neutralising the evils of scarce and costly fuel" (Jevons 1865, 137).

Jevons concluded, "It is wholly a confusion of ideas to suppose that the economical use of fuels is equivalent to a diminished consumption. The very contrary is the truth. As a rule, new modes of economy will lead to an increase of consumption according to a principle recognized in many parallel instances" (Jevons 1865, 140). His prime example was the diffusion of Watt's low-pressure steam engine and later high-pressure engines whose efficiencies were eventually more than 17 times higher than that of Savery's atmospheric machine but whose diffusion was accompanied by a huge increase in coal consumption. His conclusions have been shared and elaborated by virtually all economists who have studied the macroeconomic impacts of increased energy efficiency. Herring (2004; 2006) provides excellent surveys of these debates.

The most resolute counterarguments claim that the future elimination of large existing conversion inefficiencies and the shift toward increasingly service-based, and less material-intensive, economies can lead to stunning reductions of energy use at the national level (Lovins 1988; Hawken, Lovins, and Lovins 1999). Lovins (1988) also argued that at the consumer level the rebound effect, whereby savings accruing from more efficient energy use lead to lower prices and hence to increased consumption either in the same category (direct rebound) or for other goods and services (indirect rebound), is minimal, a position rejected by Khazzoom (1989). But others argue that improvements in efficiency are only a small part of the reason that total energy consumption may have gone up and that the overall growth of energy use is more related to increasing population, household formation, and rising incomes (Schipper and Grubb 2000).

Undoubtedly, these are all factors to consider, but there is no doubt that *relative* savings, whether measured as mineral commodities per unit of GDP or as energy and materials per specific finished products and delivered services, have often been accompanied by rising *absolute* consumption. The best examples of these trends come from lighting. By the year 2000 the efficiency of British lighting was 1,000 times that in 1800, but per capita use was 6,500 times greater and total lighting consumption was 25,000 higher (Fouquet and Pearson 2006). The efficacy of British street lighting improved about 20-fold between the 1920s and the 1990s. But more roads (overall length up by less than 50% increase) and a huge rise in average light intensity (in lumens per kilometer of road it rose more than 400 times) entirely negated these advances, and electricity consumption per kilometer of British roads increased 25-fold (Herring 2001).

Or, as another example, the typical power density of new houses is now considerably lower than right after WW II, but the houses have grown larger. The average size of new U.S. houses increased by more than 50% during the last quarter of the twentieth century, to just over 200 m² (USBC 2006). Moreover, these houses may have superefficient air conditioners, used in summer to maintain indoor temperatures that the inhabitants would consider too cold in winter. Similar contradictions have accompanied the supposed trend toward a less materials-based economy. A rapid increase in paper consumption coincided with the diffusion of the electronic "paperless office," and the short lifespans of computers and peripherals have created serious waste disposal problems because these machines contain toxic components.

National histories of energy consumption confirm that efficiency gains have not brought any long-run decline in overall energy use. The average energy intensity of the U.S. economy fell by 34% between 1980 and 2000 as the country's population increased by about 22%. But the average per capita GDP rose by more than 55%, and TPES in 2000 was about 26% higher. During the same period China experienced perhaps the greatest efficiency gain in global history as the average energy intensity of its economy fell by 70% while its population grew by less than 30%. But its high rate of GDP growth resulted in an 80% rise in TPES (NBS 2000). And during the 1990s, Japan's TPES rose by nearly 20% even though the country's GDP expanded by less than 15% and its population gained less than 4%.

Historical evidence is thus replete with examples demonstrating that on the national level substantial efficiency gains in conversion or materials use stimulated increases in fuel and electricity consumption or materials use that were far higher than the savings achieved by these innovations. Relative gains do not translate into absolute savings. As Rudin (2004) noted, efficiency disconnects the problem from the solution (e.g., lm/W from kWh used). It emphasizes output, which may be optimized but not necessarily minimized. If we are to see any actual reductions in overall energy use, we need to go beyond increased efficiency of energy conversions (see chapter 12).

10

ENERGY COSTS

Valuations and Changes

A ramp has been built into probability
the universe cannot re-ascend.
For our small span,
the sun has fuel, the moon lifts the lulling sea,
the highway shudders with stolen hydrocarbons.
How measure these inequalities
so massive and luminous
in which one's self is secreted
like a jewel mislaid in mountains of garbage?
John Updike, From "Ode to Entropy" (1985)

Industries with relatively high shares of energy costs in total operating expenditures have always closely monitored their fuel and electricity consumption, but during the era of low energy prices other enterprises, interested primarily in minimizing labor and material outlays, rarely knew how much energy it took to make a product. This neglect was promptly remedied following OPEC's first round of oil price increases with the emergence of a new

discipline, initially labeled energy analysis (IFIAS 1974; Chapman, Leach, and Slesser 1974; Verbraeck 1976; J. Thomas 1979). Its goal is to quantify direct and indirect uses of fuels and electricity in the discovery, extraction, transportation, and conversion of energies, in the extraction and processing of raw materials, in the production of food and industrial goods, and in the provision of services.

The latter category involves high shares of human labor whose physical quantification is clearly problematic. Why should an individual's labor input—food energy intake or total energy support burden, usually derived from average annual per capita commercial energy consumption—be charged toward the cost of a particular task? If she did not process insurance claims, would she not eat? Or unemployed, would she not claim a share of the nation's energy flow? Once the decision is made to account for the energy cost of labor, which approach is more rational: the minimalist choice of counting just the thermodynamic equivalent of the invested muscular exertion or the maximalist option of finding the total existential energy requirements? And how can the mental (inventive or managerial) component be reduced to a common energy denominator?

These challenges have no satisfactory solutions. Consequently, I focus first on general approaches, achievements, and limitations of energy accounting and then on the energy costs of energy (surely one of the most decisive determinants of a civilization's achievements and prospects) and on the energy costs (energy intensity or embodied energy) of basic material inputs. Food is, of course, the most indispensable human energy input, and the fuel and electricity subsidies used in modern farming are reviewed in the closing sections of this chapter. Most of the pioneering studies of energy costs and subsequent energy analyses for a large variety of industrial products are gathered in volumes by Boustead and Hancock (1979) and Brown, Hamel, and Hedman (1996). Energy analyses also are an important component of life cycle assessments (LCA) of products, structures, or services (Frankl and Rubik 2000; EEA 1998).

There are three basic approaches to the quantification of energy costs. Input-output analysis is a variant of standard econometric analysis based on input-output tables. A square sectoral matrix of national economic activity in a given year is used to extract the values of direct and indirect energy inputs, and these in turn are converted into physical energy equivalents by using prevailing fuel and electricity prices. A limitation of this approach is that sectoral aggregates may reach only heterogeneous categories (elevators, drugs, steel) and not particular products. In contrast, process energy analysis first identifies the sequence of physical operations required to produce a particular item, then accounts for all significant material and energy inputs into the process, and finally assigns energy equivalents to direct energy inputs and energy costs of materials. The first two parts of the exercise are of great heuristic value and indispensable for any successful improvements in managing the analyzed process. The third, hybrid approach combines the input-output and the process analysis approaches (Treloar 1997).

The choice of system boundaries determines the outcome. Limiting the analysis to direct energy inputs used in the final process stage may suffice in cases of simple processing. In other cases, contributing costs diminish rapidly with every successive stage, and limiting the analysis to direct energy inputs and to the best values of energy costs of all major material inputs may be satisfactory. In order to calculate the energy cost of making a polyethylene grocery bag, we do not need to know the energy cost of building a blast furnace that will operate for half a century and whose two-week's worth of pig iron output will be eventually made into steel to build an offshore platform used to extract natural gas. But in some cases, truncation errors can be large; two detailed reanalyses illustrate their extent for steel and an apartment building. Lenzen and Dey (2000) showed that an input-output analysis of the Australian steel industry yielded an average twice as large (40.1 GJ/t) as did a process analysis (19 GJ/t). And Lenzen and Treloar (2002) used input-output analysis to demonstrate that the embodied energy in a four-story Swedish apartment building should be about twice as large as the value derived by a two-stage process analysis by Börjesson and Gustavsson (2000), with the greatest differences for structural iron (~17 GJ/t vs. ~6 GJ/t) and plywood (~9 GJ/t vs. 3 GJ/t). When analyzing the energy cost in the automobile industry, it is imperative to go well beyond the cost of assembly. In order to account for energies embodied in rolled steel, aluminum engine block, or a plastic bumper, the analysis has to move to the next level, embracing the inputs needed to produces those components. Further extension includes the energies needed to produce and distribute fuels and electricity used at different stages of the process, and it could also approximate the energy costs of equipment used in manufacturing or assembly as well as in the production of key components.

Boundary dilemmas that go beyond energy analysis are well illustrated by trying to decide between a cotton and a polyester shirt. Process energy analysis shows cotton costing three times as much energy as polyester (Van Winkle et al. 1978). But this advantage is lost in the wearing; the LCA (including washing, drying, and ironing) puts the cost of the polyester shirt about 35% lower. But cotton cultivation also yields cottonseed oil, and credits for its energy make a cotton shirt only marginally more costly than the polyester one. The small remaining difference may easily be outweighed by the superior wearing qualities of cotton and by its renewability. Yet we may turn once again and note that erosion in a Texas cotton field removes annually about three times as much topsoil as is compatible with sustainable farming; that in Egypt cotton displaces food crops and leaves the country dependent on food imports; that the irrigation of the cotton crop in arid areas contributes heavily to severe soil salinization; and that residues from high doses of herbicides pollute the local environment.

Detailed production sequences and itemized inputs are available for many products, but other cases call for ingenuity and perseverance. Most analyses encounter missing information and require many unavoidable approximations, especially regarding the average lifetime of machines. Discrepancies in values for apparently identical products are due to different analytical boundaries, to studies of different techniques producing the same commodity, to comparisons of identical processes in plants of different ages and maintenance practices, and to the inclusion of marketable by-products. Net energy analysis, the study of energy required to produce energy, has its own challenges. Herendeen (1998) thought that it is conceptually fatal, particularly when several types of energy have to be considered. Its outcome is usually expressed as the quotient of marketed energy (output) and the energy needed for its production (extraction, processing, or transportation input), and the measure has become known as energy return on investment (Hall, Cleveland, and Kaufmann 1986).

EROI (more correctly it should be EROIE, energy return on invested energy), or simply net energy of primary energy supplies, must be substantially greater than 1, but in some cases it has been unclear whether the net rate is above, near, or below the breakeven point. Perhaps the best-known case of this uncertainty is the net energy return of fuel alcohol (ethanol), fermented from corn (see section 10.1). Studies of embodied energy identified energy losses and opportunities for improved conversion efficiencies and reduced energy inputs, and demonstrated energy's critical role in modern economies, which is often obscured by distorted pricing. But why stop with energy? Why not find the embodied content of other critical materials needed to produce goods and services, above all, water and land? More important, studies of embodied energy have not (contrary to some overenthusiastic opinions) displaced standard economic analyses because neither individuals nor corporations base their decisions on the minimization or optimization of energy costs.

10.1 Energy Cost of Energy: Net Gains

Given the variety of fuels and their diverse extraction, processing, and transportation modes as well as the range of options for hydro, nuclear, wind, or solar powered electricity generation, it is impossible to offer any generally valid energy costs of individual energies. Moreover, EROI completely ignores the differences in the quality of delivered energy (flexibility, convenience, portability, cleanliness at the point of use) or considers them only in terms of retail prices. And the quotient does not inform about the pace of exhausting finite resources in contrast to harnessing renewable flows. As a result, net energy analysis is an ill-defined endeavor whose outcomes are far from clear-cut and whose relative popularity during the late 1970s and the early 1980s (Hannon 1981; Hall, Cleveland, and Kaufmann 1986) rapidly ebbed after the mid-1980s. This is not to say that the approach is not heuristically valuable and that it should not be used as a part of broader, more revealing appraisals of the comparative merits of various energy resources and uses.

In a detailed account of capital equipment needed for an underground mine with annual capacity of 1.4 Mt and longwall extraction, and assuming liberally 100 GJ of embodied energy per tonne of machinery, Duda and Hemingway (1976) implied an investment of about 260 TJ/Mt of mine capacity and an electricity requirement of about 45 TJ/Mt. Even if the total capital investment were doubled, the energy cost during 30 years of operation would be about 1.9 PJ compared to at least 660 PJ of coal (assuming 22 GJ/t), implying a net energy ratio of 0.997 and EROI of 333. Sidney, Hemingway, and Berkshire (1976) reported about 2250 t of equipment needed per 1 Mt of annual capacity in a 4.4 Mt/a surface mine, and annual operating needs per 1 Mt of 51 TJ of electricity and 9 TJ of fuel. After doubling the capital

cost over 30 years, these rates translate to lifetime energy costs of 2.3 PJ compared to 600 PJ of extracted fuel, even for a lower-quality coal at 20 GJ/t, and imply EROI of 250.

Similarly, other U.S. and British analyses (Chapman, Leach, and Slesser 1974; Hayes 1976; Boustead and Hancock 1979) showed coal-mining expenditures ranging from less than 100 kJ/kg in surface mining of thick high-quality seams to about 4 MJ/kg for underground extraction of thin seams, values translating to delivered net energy as high as 0.9975 (EROI 400), typically about 0.97 (EROI 33), and no lower than about 0.83 (EROI <6). Cleveland (2005) put the typical EROI of U.S. coal extraction at 100 in 1950 and at 80 in 2000. Data on sectoral energy use prorate to about 140 kJ/kg of coal, implying a net energy ratio of at least 0.993 (for coal with just 21 MJ/kg) and EROI of no less than 140. Secondary coal-based fuels will obviously cost more, with delivered energy fractions ranging between 0.70 and 0.88 (EROI 3.3–8.3) for coke and between 0.65 and 0.81 (EROI 2.2–5.3) for manufactured gas.

Energy invested during the 1930s and 1940s in the discovery of the largest Middle Eastern oil fields was extraordinarily low, on the order of 1 MJ/t, or 0.0025% of hydrocarbons in place, and the subsequent production cost from these huge reservoirs (~0.5–5 GJ/t) yielded wellhead EROI as high as 10^3–10^4. Chapman and Hemming's (1976) calculations for two North Sea oil fields, Auk and Forties, showed net energy requirements, respectively, of about 840 MJ/t and 230 MJ/t, 2–3 OM above the lowest Middle Eastern needs. However, even in these demanding cases, the overall energy costs of oil production were repaid in oil in less than three months. Similarly, my calculations show that the record-breaking Ursa platform, built in more-than-1100-m-deep waters off New Orleans and since 1999 producing 1.25 PJ of oil and gas per day (SEPCo 2006), repaid the energy cost of its construction in less than a week of crude oil and natural gas extraction.

The net energy ratio of crude oil extraction can surpass 0.97 or even 0.995 (EROI 33–200) in rich fields, but averages in old hydrocarbon provinces are lower. Cleveland (2005) estimated that EROI for U.S. oil discovery and extraction in 1930 averaged at least 100 and calculated that average EROI of the country's hydrocarbon production (measured in thermal equivalents) rose from 17 to 25 between 1954 and 1970, declined to about 12 by 1983, rose to 20 by 1992, and then declined slightly afterwards (fig. 10.1). A similar pattern, but much lower rates of return (just above 10 during the 1990s), results when taking into account the energy quality of production inputs (electricity, refined fuels) and outputs (crude oil, unprocessed gas). My calculations based on Syncrude

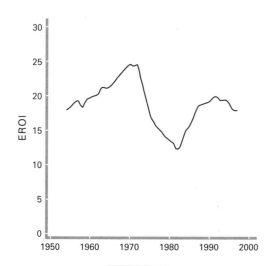

10.1 Thermal equivalent EROI for U.S. hydrocarbon production, 1954–1997. From Cleveland (2005).

Canada (2006) data indicate that mining of Alberta oil sands, extraction of bitumen, and its upgrading to light sweet crude oil has EROI around 6. Steam-assisted gravity drainage is the dominant method of oil sand extraction. It needs about 1 J of natural gas for steam generation for every 5 J of heavy bitumen, and the overall EROI of this recovery, including the cost of extraction and delivery of the gas, the cost of upgrading, and an average 15% bitumen loss, is no higher than 3.

Energy investment in oil transportation is relatively small compared to the energies that these systems deliver over their lifetimes. Energy embodied in steel, the dominant material in pipeline construction, amounts to less than 3.5 TJ/km for a 60-cm diameter pipeline, and construction adds about 1.5 TJ/km, for a total cost of about 5 TJ/km, or to less than 0.1% of energy in the oil that the pipeline will carry during its 30–40 years of service. High-capacity pipelines need on the order of 200 kJ/t-km to operate, and the size-dependent requirements for crude oil tankers range from 150 kJ (for ships <50,000 dwt) to less than 50 kJ/t-km for supertankers. Moving crude oil 1300 km through the TransAlaska pipeline thus costs only 0.5% of energy in the pumped crude (Alyeska 2003), and shipping it 3800 km by tankers from Valdez to Long Beach, California, doubles the cost to 1% of the energy contained in the oil. Similarly, a 300,000-dwt supertanker moving Saudi oil to the United States needs an equivalent of about 1% of the fuel it carries in order to travel more than 15,000 km. Energies needed for the construction of gathering pipelines, oil terminals, and tankers prorate to 1–3.5 GJ/t.

Oil refining is the most energy-intensive segment of the entire sequence of liquid fuel production because even the simplest thermal distillation claims on the order of 4% of energy content in the processed crude oil. U.S.

ENERGY COSTS

data for 2001 show a total annual throughput of about 725 Mt (30.5 EJ) of crude oil and primary energy consumption of roughly 3.5 EJ, or 4.6 GJ/t of crude oil, an equivalent of about 11% of energy in the input (Worrell and Galitsky 2003). Process rates for gasoline (most of it produced by catalytic cracking) is just over 6 GJ/t (Brown, Hamel, and Hedman 1996), implying EROI of about 7.3. Similarly, Cleveland (2005) put the EROI range for U.S. gasoline at 6–10. Energy use in refining, amounting to 4%–11% of crude oil input, or 1.7–4.6 GJ/t, lowers the fraction of initial energy delivered in refined products to as little as 0.8 and as high as 0.93; values of 0.85–0.88 (EROI 6.7–8.3) may be typical for most refined products.

U.S. data show that natural gas exploration claims energy equal to about 0.6% of the discovered fuel. This is followed by a sequence of field and transportation losses that appreciably reduce the fuel's marketable share, and whose totals are available in considerable detail in annual statistics (EIA 2005d). In 2005 flaring amounted to about 4% of global natural gas production, with extremes ranging from an equivalent of nearly 90% of annual output in Nigeria to less than 0.5% in the United States (Mouton 2005). U.S. extraction losses equal or surpass all flaring and venting (about 3.5% of total production in 2003). Lease fuel, used in well and field operations, claims about 3%, and in old gas fields repressurization is the main energy debit before the gas gets processed and distributed; in 2003 it claimed nearly 15% of dry gas extraction.

Specific energy embodied in natural gas pipelines (seamless pipes for smaller diameters and welded sheet steel for pipelines with diameters of 50 cm and larger) varies with their function (trunk lines, gathering and distribution lines) and diameter, but it always prorates to a very small share of the energy transported by the pipeline over its lifetime of at least 30 years. U.S. natural gas transmission pipelines embody on average materials worth about 1.1 TJ/km, construction 1.3 TJ/km, and engineering and maintenance 1.2 TJ/km (Meier 2002). These energies represent less than 0.1% of energy in the fuel transmitted through a pipeline during its lifetime. Centrifugal compressors that pressurize the transported gas to 1.4–10.3 MPa in stations spaced at 60–160-km intervals have been traditionally powered by natural gas, but electric motors have been making some inroads. Shares of natural gas consumed by pipeline distribution depend on the total length of a network; in Canada and the United States they are above 3%.

Leakage from U.S. natural gas pipelines averages 1.5% ± 0.5% of the transported fuel (Lelieveld et al. 2005). In contrast, Reshetnikov, Paramonova, and Shashkov (2000) found that during the early 1990s the long and aging gas pipelines in the states of the former USSR were very leaky, losing annually 47–67 Gm^3, or 6%–9% of the transported total (a loss larger than the annual natural gas consumption in Saudi Arabia and equal to annual imports to Italy). But this conclusion is not supported by Lelieveld et al. (2005), who found that the leakage from Russian pipelines, at 1%–2.5% of the transported fuel, is comparable to the U.S. losses. With all of these production and transportation uses and losses, shares of natural gas delivered to consumers may thus amount to more than 90% (EROI >12) or less than 70% (EROI <3.3) of the initially extracted fuel, with typical shares between 0.80 and 0.90 (EROI 5–10).

LNG projects require considerable investments in the liquefaction plant, a fleet of carriers and regasification and storage facilities. The total for a 5-Mt LNG/year plant was $3–4 billion in the early 2000s (Total 2005).

In energy terms, these investments translate to as much as 20 PJ of embodied energy for the entire system. About 110 MW of electricity are needed for compression and 60 MW for cooling, adding up to annual consumption of about 5 PJ of electricity, or nearly 15 PJ of primary fuel. Even so, given the very high energy density of the shipped product (25 GJ/m^3 vs. ~36 MJ/m^3 for gaseous CH_4), the lifetime cost of roughly 320 PJ over 20 years is less than 6% of the cumulative throughput of 5.5 EJ.

Published values of EROI for electricity generation range over 2 OM (fig. 10.2). Calculations of EROI for thermal generation include three major components: investment in the plant and high-voltage transmission lines, usually an equivalent of less than 5% of the plant's generation when prorated over at least 30 years of its lifespan; the plant's internal electricity use, as low as 2%, as high as 8%, with efficient fly-ash removal and flue gas desulfurization (FGD) facilities; and losses during long-distance transmission, typically 7%–9% of the generated electricity. These energy costs add up to 10%–20% of the plant's generation, resulting in EROI of 5–8 for mine mouth coal-fired stations, and 7–9 for natural gas–fired stations located near a hydrocarbon field. Long-distance gas transportation, often in excess of 1000 km, can halve EROI even where the gas is used in a combined cycle. However, Penner, Kurish, and Hannon (1980) put EROI of U.S. coal-fired generation using surface mined coal and no FGD as high as 43 for Eastern coal and 24 for Western fuel.

Alternatively, EROI of thermal electricity generation will always be considerably less than 1 if its calculation also includes all fuel inputs during a station's lifetime. This is an inevitable consequence of inherently limited

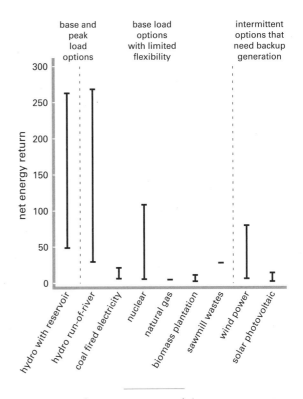

10.2   EROI for major categories of electricity generation. Based on Gagnon, Bélanger, and Uchiyama (2002).

efficiencies of thermal electricity generation. In modern stations they range mostly between 35% and 42%, can be as low as 25% for some old plants, and surpass 50% only in the best combined-cycle natural gas plants. In the last case, EROI (quotient of electricity output and all direct and indirect energy inputs) may be as high as 0.43 (Meier 2002); for most modern coal-fired stations, it will be 0.30–0.35, and for the least efficient plants as low as 0.20. But these EROIs should not be compared to the net energies of primary fuel extractions. We are willing to incur considerable losses in order to produce energy whose qualities surpass those of the primary fuels used in its generation.

Minimum returns of fission-based generation can be estimated by assuming that the highly energy-intensive production of enriched uranium fuel in gaseous diffusion plants claims 4%–5% of the station's gross annual output during 25 years of operation, and that construction costs may be equivalent to 6% of produced electricity (Chapman, Leach, and Slesser 1974). With 9%–10% taken by the plant's consumption and by distribution losses, nuclear stations would have EROI as low as 5 (excluding the energy content of the charged fuel). Meier and Kulcinski (2002) use an average of about 15. EROI for CANDU reactors using natural uranium (no need for energy-intensive enrichment) was put at 16; several other published estimates of the energy payback ratio for nuclear generation range between 20 and 60 and go (unrealistically) as high as 100 (Gagnon, Bélanger, and Uchiyama 2002; WNA 2006a). But because there is no way to include the ultimate energy cost of decommissioning and long-term disposal (10^4 years) of radioactive wastes, any arguments about the net energy of fission generation rest on grossly incomplete assumptions.

A different accounting challenge is presented by hydroelectric generation, where the only simple cases are presented by projects dedicated solely to electricity generation. In all other cases, calculations of energy costs are complicated by the multipurpose function of dams and reservoirs. Is it appropriate to charge the construction cost (including often high costs of population relocation) only against electricity generation when a reservoir is also used for flood control, irrigation, drinking water, or recreation? Further complications arise from different longevities of hydro projects (heavy silting may fill some reservoirs in less than 50 years, while reservoirs in wooded areas last for centuries), the necessity for expensive (often DC) high-voltage transmission links to distant load centers, and the fact that there is no accepted procedure to account for the renewability of hydrogeneration. Published ratios range widely for stations with reservoirs (EROI 50–260) and for run-of-river projects (EROI <50–>260); those for specific Hydro-Québec projects assessed over a period of 100 years are, respectively, 205 and 267 (Gagnon, Bélanger, and Uchiyama 2002).

Among the techniques of renewable electricity generation, the published EROI estimates (3–80) of wind turbines have ranged too much to offer any representative values. Combustion of sawmill wastes is definitely quite rewarding (EROI >25). The lowest values are, expectedly, associated with electricity generated by burning wood from intensively cultivated tree plantations (EROI <5 not counting conversion losses, and ~1.5 when including the energy content of wood) and from biomass conversions to liquid fuel. There have been many studies of the energy costs of corn-based ethanol fermentation with results ranging from net energy loss (Pimentel 1991, 2003; Keeney and DeLuca 1992) to substantial energy gains. Results of the final outcome depend on the inclusion of energy credits for by-products. Shapouri, Duffield, and Wang (2004) calculated a barely positive EROI of 1.06, but energy credits for by-products (including distillers grain and corn gluten meal) raise it to 1.67. A similar analysis by Kim and Dale (2002) found a final EROI of 1.56.

In contrast, Pimentel's studies (accounting also for energy cost of field machinery and irrigation and assuming lower by-product credits) kept confirming energy loss for the entire enterprise. The latest one (Pimentel 2003) concluded that corn production and fermentation claim nearly 30% more energy than is contained in ethanol, or EROI of 0.77. Pimentel's studies contained some exaggerated estimates (notably the energy cost of fertilizer N), but his main arguments remain intact even if energy return were unambiguously positive. Because of the attendant environmental degradation (soil erosion, N losses, mining of groundwater), U.S. corn production is not renewable, and the large-scale output needed to reduce U.S. dependence on crude oil imports would claim very large shares of U.S. farmland. Similar options based on corn or other grain crops are untenable in any land-scarce populous nation.

The controversy about corn ethanol's net gain is a perfect illustration of the inherent limitations of energy analysis. No such doubts attach to calculating the energy costs of fuel ethanol production from Brazilian sugarcane. High yields of this tropical grass (commonly in excess of 60 t/ha or more than 7 t/ha of sugar) and no need either for nitrogen fertilizer (thanks to the plant's endophytic N-fixing bacteria) or for external fuel to energize the fermentation (thanks to the combustion of bagasse, fibrous residue that is available after expressing the juice from cane stalks) combine to make the cultivation of sugarcane and ethanol production highly rewarding. A detailed account by Macedo, Leal, and da Silva (2004) shows that typical practices in the state of São Paulo have EROI 8.3, and the best operations can have EROI as high as 10.2.

LCAs for roof-top photovaltaic systems have shown EROI ranging from barely positive for some early systems to values as high as 10 for modern setups in sunny subtropical locations. Blakers and Weber (2000) calculated the overall embodied energy cost of 3.8 GJ/m^2 for a roof-top PV panel that will produce 5.5 GJ of electricity over its 10-year lifespan, implying EROI of 1.45. In contrast, Meier (2002) put the total life cycle cost of an 8-kW building-integrated PV system (1.08 GJ/m^2) at about 205 GJ and its total electrical output during 30 years at 1.165 TJ, implying EROI of 5.7. Clearly, assumptions about the system's longevity make the key difference.

10.2 Basic Materials: From Concrete to Fertilizers

Process analysis can produce fairly reliable and revealing values for energy costs of basic raw materials, metals, and synthetic compounds. Moreover, in some instances, there is an extensive historical record that allows for some fascinating secular comparisons. Given the multitude of materials for which we have either detailed process analyses or basic calculations of energy embodied in their production, I decided to focus here on energy costs of three products whose ubiquity and importance define modern civilization in material terms: concrete, a leading building material in terms of overall mass; steel, the dominant metal; and ammonia, the compound that is synthesized in greater abundance than any other industrial chemical.

Reinforced concrete is the dominant building material of modern civilization, and it is used everywhere in buildings, bridges, highways, runways, and dams. Concrete is an old Roman invention, but modern cement became available only in 1824, when Joseph Aspdin began firing limestone and clay at temperatures high enough to vitrify the alumina and silica materials and to produce a glassy clinker (Shaeffer 1992). Its grinding produced a stronger Portland cement named after limestone, whose appearance it resembled when set. Concrete, a mixture of cement, gravel, and water, is made by hydration, a reaction between cement and water that produces tight bonds and a material that is very strong in compression but has hardly any tensile strength and hence has to be reinforced for most construction uses. Such a composite material (its diffusion began after 1880) has monolithic qualities, and it can be fashioned into almost any shape.

Global cement production reached 1.86 Gt in 2003, with China accounting for 40% of the total and India and the United States being distant second and third (USGS 2006). Cement production starts with raw material preparation (crushing, drying, and raw grinding), clinker is made by firing in large kilns (the most energy-intensive part of the process), and final ball milling yields the finished material. The net theoretical minimum needed to produce clinker is about 1.75 GJ/t, and the lowest reported industrial values cluster at about 2.9 GJ/t; depending on the process and the principal fuel used (coal or natural gas), most of the published values are 3.2–7 GJ/t (DOE 1997; Sheinbaum and Ozawa 1997; Tresouthick and Mishulovich 1991). After mixing with water and aggregate (sand or gravel that costs little, ~100 MJ/t), concrete embodies 1–2 GJ/t and reinforced concrete that includes 100 kg/m^3 of steel bars embodies 2–3 GJ/t.

Energy intensity of three traditional construction materials is fairly similar. Fired clay bricks need 4–8 GJ/t (sun-dried bricks, common in the antiquity, are now used only by peasants in the poorest countries); gypsum needs as little as 3 GJ/t; and U.S. drywall (gypsum between paper boards) needs up to 6 GJ/t. The energy cost of stone quarrying is commonly less than 1 GJ/t, but transportation costs, particularly for large pieces, can multiply that value severalfold. The energy cost of lumber reflects a difference in harvested stands (climax vs. immature, native vs. plantation, hardwood vs.

softwood), harvesting techniques, and the extent of processing and drying. Costs are as low as 0.57 GJ/t and as high as 41.2 GJ/t, but most rates fall between 0.6 GJ/t and 9 GJ/t; those between 2 GJ/t and 7 GJ/t (including kiln drying) are perhaps the most representative (Glover, White, and Langrish 2002; CWC 2004; Buchanan and Honey 1994). Chipping, gluing, and compressing raises the cost of particle board to 8 GJ/t and of plywood to at least 10 GJ/t.

Iron remains by far the most important metal extracted from the crust. Worldwide iron ore production surpassed the 500-Mt mark in 1960 and reached 1.2 Gt by 2003. Smelting of pig iron topped 700 Mt in 2004, and nearly all this metal is processed into steel. With the addition of 400 Mt of steel scrap, the global production of crude steel surpassed 1 Gt in 2004 (IISI 2005). Steel provides a large part of the physical infrastructure of modern civilization as well as a high-quality, relatively inexpensive and durable material for manufacturing an enormous variety of products, including all the essential machinery for energy industries. Rapid diffusion of coke-based iron ore smelting after 1750 was accompanied by an impressive decline in specific coke consumption; a rich retrospective literature (Bell 1884; King 1948; Hyde 1977; Gold et al. 1984) enables us to follow the subsequent progress.

Technical advances lowered typical energy costs of pig iron from nearly 300 GJ/t during the late eighteenth century to less than 100 GJ/t during the 1840s and about 50 GJ/t by 1900. By 1950 the best rates were about 30 GJ/t, and during the late 1990s the net specific energy consumption of modern blast furnaces was between 12.5 GJ/t and 15 GJ/t (de Beer, Worrell, and Blok 1998), or roughly twice the theoretically lowest amount of energy (6.6 GJ/t) needed to produce iron

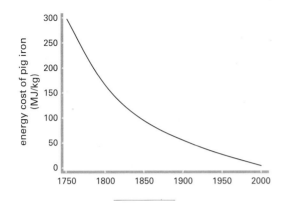

10.3   Energy cost of pig iron production, 1750–2000. Based on Smil (1994) and data in de Beer, Worrell, and Blok (1998).

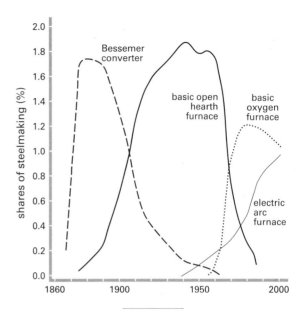

10.4   Rise and fall of successive steelmaking techniques in the United States: Bessemer, open hearth furnace, basic oxygen furnace, and electric arc furnace. From Smil (2005a).

from hematite (fig. 10.3). These efficiency gains came mainly from reduced coke consumption, increased blast temperatures, and larger furnaces. For the blast furnaces this long-lasting trend reached a plateau during the 1970s. The Kyushu Oita Works furnace (5070 m³), completed in 1976, remains the record holder. Its highest daily production of just over 12,000 t and average energy consumption of 14.5 GJ/t of hot metal translates into a power rating of 2 GW and power density of 11.9 MW/m² of hearth area (about 172 m²). The Western hemisphere's largest furnace, Inland Steel's No. 7 in Indiana Harbor (3483 m³), has energy consumption of 14 GJ/t of pig iron and power density of 11.3 MW/m² of hearth area (McManus 1981).

Modern steelmaking began with Henry Bessemer's converter, patented in 1855. It was soon supplanted by Siemens-Martin open-hearth furnaces (OHFs), patented during the late 1860s and rapidly diffused after 1879 once Thomas-Gilchrist basic converting could remove phosphorus from the metal. OHFs dominated global

steelmaking until the 1960s, when they were replaced in all but a few countries by basic oxygen furnaces (BOFs) and electric arc furnaces (EAFs). By the year 2000 a mere 4% of the world's steel came from OHFs, mostly in the Ukraine and Russia (IISI 2005). The BOF's experimental foundations were laid down by Robert Durrer before WW II, and the technique, commonly known as the Linz-Donawitz process, was first commercialized in Austria in 1953. Japan was the first large producer to embrace BOFs, but in the United States their progress was initially slow (fig. 10.4). By the year 2000 almost 60% of the world's steel came from BOFs. The entire process is slightly exothermic, with an overall gain of about 200 MJ/t of crude steel and rates of 50–60 m³ O₂/t of crude steel.

The electric arc furnace traces back to William Siemens, one of the greatest innovators of the nineteenth century, but the high cost and limited availability of electricity limited its adoption for many decades. The shift from BOFs to EAFs severed the link between steelmaking, blast furnaces, and coking and made it possible to set up smaller mills whose location did not have to take into account the supplies of coal, ore, and limestone. These minimills, with annual capacities of less than 50,000 t to as much as 600,000 t, combine EAFs with continuous casting and rolling (Szekely 1987; Hall 1997). By the year 2000 one-third of the world's steel came from EAFs. The United States, with 47% of total output, had the highest share among major producers; China's share was 16%, Japan's 29%, and Russia's less than 15% (IISI 2005). In 1965 the best furnaces needed about 630 kWh/t of crude steel; 25 years later the best rate, with the help of oxygen blowing, was down to 350 kWh/t (de Beer, Worrell, and Blok 1998), and IISI (2005) reported that during the 1990s the global mean fell from 450 to about 390 kWh/t.

The subsequent processing of steel has also undergone a fundamental change. The traditional process involved first the production of steel ingots, oblong pieces weighing 50–100 t that had to be reheated before further processing into standard semifinished products: slabs (5–25 cm thick), billets (square profiles used mainly to produce bars), and blooms (rectangular profiles wider than 20 cm used to roll beams), which were then converted by hot or cold rolling into finished plates and coils, structural pieces (bars, beams, rods), rails, and wire. This inefficient sequence often consumed as much energy as steelmaking itself, and it was eventually replaced by continuous casting of steel, the process pioneered by Siegfried Junghans, promoted by Irving Rossi, and first adopted by Japanese steelmakers (Morita and Toshihiko 2003; Luiten 2001; Tanner 1998; Fruehan 1998).

By the year 2000 more than 85% of the world's steel was cast continuously, with the shares (96%–97%) basically identical in Europe, the United States, and Japan. In China, now the world's largest steel producer, continuous casting accounted for nearly 90%; among the major steelmakers, only Russia (50%) and the Ukraine (20%) still lagged behind. The advantages of continuous casting are many: much faster production (<1 h vs. 1–2 days), higher yields of metal (up to 99% vs. less than 90% of steel to slab), energy savings of 50%–75%, and labor savings of the same magnitude. Earlier methods involving decarburization of iron in OHFs and the subsequent shaping of steel doubled the energy cost of traditionally finished products compared to modern ones; in 1950 even the best-integrated steel mills typically consumed more than 40 GJ/t.

According to Leckie, Millar, and Medley (1982), typical gains from process innovations in steelmaking and casting were (per tonne of liquid metal) more than 3 GJ from replacing OHF (4 GJ) with BOF (600 MJ) and nearly 1 GJ from substituting continuous casting (300 MJ) for rolling semifinished products from ingots (1.2 GJ). (The 600 MJ figure for BOF represents the cost of electricity to make oxygen; the smelting process itself is exothermic.) Innovations combined to bring energy costs of modern steelmaking down to 25 GJ/t by 1975 and just below 20 GJ/t during the late 1990s, with about 65% of that total used by blast furnaces. The U.S. average fell to about 19 GJ/t by the mid-1990s, and Japan's consumption averaged 19.4 GJ/t of crude steel in 2000 (AISI 2002; JISF 2003). In contrast, the most efficient EAF-based producers needed about 7 GJ/t in the late 1990s.

Continuing declines in the energy intensity of steel-making meant that in 2000 the production of roughly 850 Mt of the world's leading alloy needed about 20 EJ, or just over 6% of the world's TPES (Smil 2003). Had the energy intensity remained at the 1900 level, ferrous metallurgy would now be using almost 20% of the world's TPES. But despite these impressive efficiency gains, ferrous metallurgy remains the world's largest energy-consuming industry, claiming about 15% of all industrial energy use. In the United States, where steel-making declined from a peak of 137 Mt in 1973 to 90 Mt in 2001, this share was only about 8% of all energy used in manufacturing and 2.5% of all primary commercial energy used annually during the early 2000s (USDOE 2000).

Aluminum is the second most important structural metal of modern civilization. In 1900 the annual global output of aluminum from the primary electrolysis of alumina was 8000 t; by 1913, 65,000 t; by the end of WW II, nearly 700,000 t; and by 2000, more than 20 Mt. Global aluminum production was about 50% higher than the worldwide smelting of copper (IAI 2003). The metal is produced 99.8% pure or alloyed to be used in construction, transportation (jetliners would be impossible without it), and countless industrial products from cooking pots to computers. The Hall-Héroult process of electrolytic aluminum production was invented and rapidly commercialized during the 1880s. During the twentieth century the average cell size in Hall-Héroult plants doubled every 18 years, while energy use decreased by nearly half between 1888 and 1914 and then dropped by nearly as much by 1980 (fig. 10.5) (Smil 2005a).

The first commercial aluminum cells consumed nearly 40 kWh/kg of the metal, and continuing technical

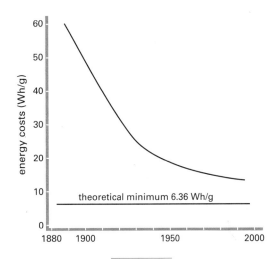

10.5 Declining specific energy cost of aluminum production, 188–199. From Smil (2005a).

advances lowered this rate to about 15 kWh/kg by 2000. The theoretical minimum requirement for carbide anode electrolysis is about 6 kWh/kg (Choate and Green 2003; IAI 2003). Adding the inevitable electricity losses in generation and transmission, the energy costs of other fuels, bauxite mining, production of alumina and carbon electrodes, and casting of the smelted metal raises the total rate to about 200 GJ/t (U.S. average in 2000 was ~196 MJ/t) and well above that for less efficient producers. Aluminum production thus remains highly energy-intensive, 1 OM above that of steel from blast furnace iron. Among structural metals, only titanium requires more energy (up to 900 GJ/t) to be produced from its ores. In 2000 about 40% of U.S. aluminum was recovered for secondary processing, whose theoretical energy requirement (0.39 kWh/kg for smelting) is less than 6.5% that of primary production and whose actual cost is about 50 GJ/t (Stodolsky et al. 1995).

ENERGY COSTS

No other compounds are now synthesized in such quantities as sulfuric acid and ammonia. H_2SO_4 has been the most important industrial chemical since the beginning of modern syntheses during the nineteenth century. In contrast, ammonia synthesis from its elements became possible only with the invention of a catalytic high-pressure process by Fritz Haber in 1909 and its rapid commercialization over the next four years by BASF, the German chemical company, under the leadership of Carl Bosch (Smil 2001). Global production was limited until 1950, nearly doubled during the 1950s, further quadrupled by 1975, and after a brief period of stagnation in the late 1980s, reached about 120 Mt by 2000. During that year, H_2SO_4 synthesis amounted to 157 Mt, but because of ammonia's lower molecular weight (17 vs. 98), ammonia is the world's leading chemical in terms of synthesized moles (nearly five times as much as H_2SO_4).

The energy requirements of chemical syntheses comprise the heat equivalents of fuels and electricity used in synthesis and the energy embodied in the feedstocks. Ammonia's lower heating value is 18.6 GJ/t, and the stoichiometric energy requirement for ammonia synthesis is 20.9 GJ/t. After taking credit for the purge gas (drawn in order to control the concentration of inert gases in the recycled flow), net feedstock inputs of modern ammonia plants are very close to this value. The reduction of energy needed for NH_3 synthesis has been impressive (Helsel 1987; Kongshaug 1998; Smil 2001). Coke-based Haber-Bosch synthesis in the first commercial plant in Oppau, Germany, needed initially more than 100 GJ/t NH_3, and typical pre–WW II plants required around 85 GJ/t NH_3 (fig. 10.6).

During the 1950s natural gas–based synthesis with low-pressure reforming (0.5–1 MPa) and with reciprocating compressors needed 50–55 GJ/t NH_3. In the

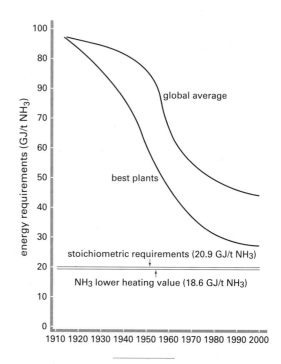

10.6   Declining specific energy cost of Haber-Bosch ammonia synthesis, 1913–2000. From Smil (2001).

early 1960s total energy consumption in plants working with reformer pressure of 1 MPa and synthesis loop pressure of 35 MPa would have added up to 46–50 GJ/t NH_3. A decade later the best large single-train plants, with high-pressure reforming (>3 MPa) and with centrifugal compressors producing converter pressure 15 MPa, reduced the energy need to 35 GJ/t NH_3. Reciprocating compressors powered by electric motors needed 520–700 kWh/t NH_3, and steam turbine–driven centrifugal compressors in much larger post-1963 plants consumed as little as 20–35 kWh/t NH_3.

By the early 1980s further gradual efficiency improvements had brought the average energy needs of ammonia plants down to just above 30 GJ/t NH_3. Another efficiency gain came when M. W. Kellogg Co. and Krupp Uhde began offering new designs requiring less than 28.9 GJ/t NH_3 (fig. 10.6). Kellogg's latest KRES design saves another 1 GJ/t NH_3, and Topsøe's heat exchange reformer cuts the need for reformer fuel by 70% and total natural gas consumption by 15%. The best designs now need about 30% less energy than in the early 1970s (M. W. Kellogg Co. 1998; Haldor Topsøe 1999). By the year 2000 the best operating plants needed only about 27 GJ/t NH_3, and rates of 25–26 GJ/t appeared to be technically feasible. The lower rate would be less than 20% above the stoichiometric requirements. Typical new plants need 30 GJ/t NH_3 with natural gas–based steam reforming, about 20% more when using partial oxidation and heavy fuel oil, and up to about 48 GJ/t NH_3 for coal-based installations (Rafiqul et al. 2005).

The average worldwide energy efficiency of ammonia synthesis, lowered by energy-intensive reforming of heavier hydrocarbons and coal, has improved as well. Before 1955 the global mean for all ammonia plants (natural gas, oil, and coal-based) was at least 80 GJ/t NH_3; by 1965 it had declined to just over 60 GJ/t; and by 1980 it was about 50 GJ/t. The best estimate for the year 2000 would be about 40 GJ/t NH_3, or roughly 48 GJ/t N. Ammonia is a gas under normal pressure and that is why urea, a fertilizer solid with the highest share of nitrogen (45%), emerged as the dominant source of N, particularly in all rice-growing countries. Ammonium nitrate (35% N) is the second most popular choice. Conversion of ammonia to urea needs 9–10 GJ/t N, which means that the total cost of this solid fertilizer (including prilling and bagging for distribution) is typically 55–58 GJ/t N.

Nitrogenous fertilizers remain energy-intensive, even with huge improvements in the synthesis of ammonia, particularly when compared with the sources of other two macronutrients. Potassium is diffused overwhelmingly as potash (KCl), whose production (nearly 25 Mt/year by 2005) involves either conventional shaft mining or solution extraction. The former process dominates worldwide, costing as little as 4–5 GJ/t; the latter needs between 15–20 GJ/t; the North American average is about 7 GJ/t. Mining of phosphate rock (about 140 Mt in 2005) costs no more than mining potash (4–5 GJ/t), but its subsequent treatment with acids multiplies the energy cost. Single superphosphate (8–9% P) is produced using H_2SO_4, nitrophosphate using nitric acid, triple superphosphate (averaging about 20% P) using phosphoric acid, and ammonium phosphates (up to 23% P) by reacting phosphoric acid and NH_3. For single superphosphate, costs are 18–20 GJ/t P, for concentrated superphosphate at least 25 GJ/t P, and for diammonium phosphate 28–33 GJ/t P.

An approximate global summation of energy costs of N, P, and K inorganic fertilizers (using averages of 55 GJ/t N, 20 GJ/t P, and 10 GJ/t K) yields about 5 EJ in the year 2000: for comparison, Kongshaug (1998) estimated 4.4 EJ for global use in 1996. The total was dominated (about 90%) by the energy cost of synthetic nitrogenous fertilizers and hence by the consumption of natural gas, and it amounted to less than 1.5% of global TPES in 2000. Few energy uses produce such a critical payoff as this feedstock and fuel, particularly in the synthesis of nitrogenous fertilizers. Higher yields resulting from their application now produce an adequate food supply for close to 40% of the world's population (Smil 2001).

ENERGY COSTS

10.3 Structures and Products: From Buildings to Computers

Process analysis of the energy embodied in buildings and manufactured products is considerably more difficult than the quantification of energy used in producing basic materials. Input-output analysis is of a limited use because many of the examined sectors produce an enormous variety of items. Process analysis must consider not only the components used in the final assembly but also the cost of long-distance transportation, which has become common in the global economy. For example, passenger cars are now assembled from components made in more than a dozen countries, and the raw materials used to make auto parts may have come from a different set of more than a dozen countries.

These realities, as well as different assumptions regarding the recycling and waste rates, analytical boundaries, and typical conversion efficiencies, result in often substantial differences in published energy costs. And the results of specific process analyses may not be applicable to other items even within the same category. This is well illustrated by focusing on three ubiquitous possessions in modern societies: houses, passenger cars, and computers. The energy cost of houses eludes easy generalizations because of very different shares of principal construction materials and the quality of interiors. Highly automated car assembly may have uniform energy costs no matter where it takes place, but at the century's turn there were more than 700 car models on the market, ranging in weight from barely 0.5 t to nearly 5 t. And even when the range is restricted to personal and office computers, these machines range from small all-purpose laptops to powerful large desktops dedicated to computer-assisted design.

The mass of modern structures is dominated by one of three main structural materials—wood, steel, or concrete—but houses include a large assortment of other components whose embodied energies range from less than 10 GJ/t (tiles) to more than 200 GJ/t (machined aluminum alloys). As a result, the energy cost of residential space in rich countries varies from 3 GJ/m^2 to 9 GJ/m^2, with floors and roofs usually the largest items (Baird 1984; Buchanan and Honey 1994; CWC 2004). These values translate to a wide range of grand totals for single-family houses, from as little as 200–300 GJ for small wooden houses and 500–700 GJ for small steel-based houses to more than 2 TJ for large houses using a mixture of materials. About 500 GJ of energy is embodied in an average three-bedroom, wood-framed North American bungalow.

Several studies have compared the embodied energy costs of identically sized family houses using the three principal structural materials. Predictably, concrete houses have the largest mass, and depending on the design, either they or the steel houses require the most energy (Glover, White, and Langrish 2002). A study of 220-m^2 family houses in the Toronto area found final embodied energies of 1.12 TJ for wood-based, 1.42 TJ for steel-based, and 1.76 TJ for concrete-based designs (CWC 2004), implying energy intensities of approximately 5, 6.5, and 8 GJ/m^2, respectively. Similarly, a process analysis of a multistory Swedish building showed the concrete-frame structure embodying 60% more energy than an identically sized wood-framed one (Börjesson and Gustavsson 2000).

Nationwide U.S. data show the expected progression for other structures: warehouses are relatively cheap (5–7 GJ/m^2), high-rise apartments take 8–9 GJ/m^2, stores, restaurants, hotels, motels, and industrial buildings need $10-13$ GJ/m^2, and hospitals and office buildings need $18-20$ GJ/m^2, mainly because of more metals in structures, elevators, and finishing (EIA 1998). A detailed analysis of two Hong Kong high-rise designs showed energy costs of $6.5-7$ GJ/m^2, with steel accounting for about 70% of the total (Chen, Burnett, and Chau 2001). As a result, a five-story building with 75 apartments embodies as little as 25 TJ, a luxury 30-story building with 1000 m^2/floor needs 300 TJ. As expected, life cycle assessments show lower shares of embodied energy for buildings in colder climates. For a large 200-m^2 Canadian house that uses about 25 W/m^2 for heating and lighting (see section 9.3) and embodies about 1.5 TJ, the ratio of construction to operation energy will be only 0.16 over a 50-year period. For detached and semidetached houses in temperate climates, the ratios are $0.3-0.4$ and rise to more than 0.75 for superinsulated structures (Mithraratne and Vale 2004; CWC 2004).

After a house, the second most valuable possession of modern households is a car. Steel components have traditionally made up the largest proportion of a car's mass and hence of its energy cost. Berry and Fels (1972) put the average late-1960s American car production cost at about 134 GJ, or roughly 85 GJ/t. Several opposing trends have been at work since that time. Continuing substitutions of steel and iron by aluminum, plastics, ceramics, and composites have introduced increasing amounts of lighter but more energy-intensive materials. Cast or machined aluminum and its alloys (see section 10.2) are at least four times, and up to eight times, more energy-intensive than most steel components, and the plastics used in cars typically require about 100 GJ/t, or two or three times as much energy as rolled steel.

The declining energy costs of steel and other basic materials have lowered the specific energies of major inputs, and robotic assembly has further lowered production costs, but these gains have been largely negated by the increased average mass and power of passenger cars. U.S. data show a steady post–WW II increase in the average mass of passenger cars to nearly 1.8 t by the mid-1970s, followed by a decline to 1.35 t by 1986, and then by steady growth to 1.49 t by 2004 (NHTSA 2005). This trend was reinforced by rapid market penetration of vans, light trucks, and SUVs (mostly in excess of 2 t/vehicle), and it has prevented any notable declines in the specific energy cost of car making.

During the 1990s total energy needed to produce car body parts from steel sheet was about 65 GJ/t for primary metal and 52 GJ/t for recycled steel (Stodolsky et al. 1995). An LCA of a 1990 Ford Taurus (1.4 t) put the vehicle's embodied energy at 120 GJ, or 86 GJ/t, a rate virtually identical with the typical value for the early 1970s (MacLean and Lave 1998). At the turn of the twentieth century, it still cost $100-125$ GJ to produce most North American passenger cars. A medium-sized (1.3 t) five-passenger car requiring 110 GJ to produce would consume (at nearly 8 L/100 km) about 50 GJ of fuel and oil annually, and hence its production cost would be only about 16% of its ten-year lifetime energy cost of about 680 GJ, including about 80 GJ for spares, repairs, garaging, and road maintenance. Less fuel-efficient cars and longer service can lower the share of embodied energy to as low as 10% of the vehicle's lifetime energy cost.

Even small European cars have become heavier as the average mass of a typical vehicle increased from about 0.8 t in 1970 to 1.2 t by 2003. In North America lighter but more energy-intensive materials have been used to

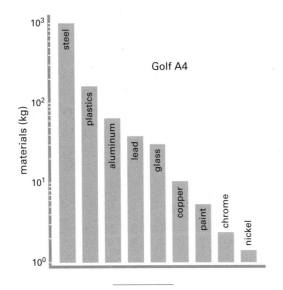

10.7 What goes into a Volkswagen Golf A4: the most important materials, according to a detailed energy cost accounting prepared by Schweimer and Levin (2000).

substitute steel parts. In 1996, Volkswagen began an LCA of its Golf automobile. A detailed account for the year 2000, based on 40 production input categories (fig. 10.7) shows specific energy costs very similar to the earlier U.S. data (Schweimer and Levin 2000). Production of an Otto cycle version of the Golf A4 (mass 1.059 t, fuel consumption 6.55 L/100 km) needed 85.6 GJ, or 80 GJ/t. Production of a slightly heavier but much more efficient diesel version (mass 1.181 t, fuel consumption 4.95 L/100 km) required 88.4 GJ, or 75 GJ/t. The LCA, assuming ten years of useful life or 150,000 km, showed overall costs of 445 GJ and 407 GJ, respectively, with production accounting for 19% and nearly 22%. With an annual output of about 40 million units, the global energy cost of passenger car production in 2000 was about 4–4.5 EJ, less than 1.5% of

the world's TPES and roughly the same as for production of all inorganic fertilizers.

The most innovative industrial sector of the last quarter of the twentieth century was the production of semiconductors. That sector has been following Moore's law, doubling the number of transistors per microchip about every 12 months until 1972 and every 18 months since that time (Intel 2003; Moore 1965). In 1972, Intel's 8008 chip had 2,500 transistors; a decade later, a single memory chip had more than 100,000 components; by 1989 the total surpassed 1 million; and by 2000, the Pentium 4 processor had 42 million transistors. Given this phenomenal growth, it is surprising that the first fairly complete energy analysis of microchip production became available only in 2002 (Williams, Ayres, and Heller 2002).

The analysis focused on the 32 MB DRAM chip and calculated that 1.6 kg of fossil fuels (all of it to generate electricity) and 72 g of chemicals (as well as 32 kg of water and 700 g N_2) were needed to produce a single 2-g device. The analysts traced energy costs for principal stages of the production chain for silicon wafers (silicon from quartz and carbon; trichlorosilane; polysilicon; single crystal ingot; wafers) and then for the subsequent fabrication and assembly of a microchip. The energy needed for the silicon wafer chain and final assembly amounted to about 41 MJ/chip, and 9.4 kg of raw Si was needed to produce 1 kg of final wafer. During its typical four-year lifetime the chip uses about 1.4 kWh of electricity, or about 15 MJ, assuming an average of 10.7 MJ/kWh. This means that unlike cars or houses, a microchip embodies nearly 75% of its lifetime energy costs, substantially higher than its operating cost. This highlights the need for increasing the longevity of computers. Using the average of 41 MJ/chip implies embodied energy of about 20 GJ/kg, 100 times more than aluminum from bauxite, 200 times more than a car, and 1,000 times more than crude steel from ore.

Extending this analysis to personal computers, E. D. Williams (2004) calculated that total energy embodied in a typical machine produced in 2000 (Pentium III processor, 30 GB hard drive, 42.5-cm monitor) amounted to 6.4 GJ, or roughly 270 MJ/kg. Given the machine's average three-year useful life, during which it will use some 420 kWh of electricity (or 1.5 GJ of primary energy), its lifetime cost is dominated by the embodied energy. Moreover, the high energy intensity of production and the rapid turnover translate into an unusually high annual life cycle energy burden of 2.6 GJ, about 30% more than for a refrigerator. But it would be misleading to use this comparison between the mass of a microchip and the mass and energy of inputs needed to produce it as the basis for judging the relative energetic and environmental merits of microchips and computers. Extension of useful life is obviously desirable, but overall energy and material impacts of microelectronics cannot be judged without also considering the savings it brings or the cost of its alternatives.

The production of paper, the carrier of civilization's memories and messages, follows a well-established route from logging to pulping to formation of the product in Fourdrinier machines, where the pulp is laid on a continuous wire mesh at the wet end of the machine, most of water is expelled in the felt press section, and the process is finished by passing paper over a series of heated cylinders. Most of the pulp is now produced by chemical (kraft) process, but semichemical and mechanical processes are also used (Ruth and Harrington 1998). A sectoral analysis for U.S. paper and paperboard production resulted in an average embodied cost of 35 GJ/t

(Brown, Hamel and Hedman 1996). The cheapest kinds (unbleached packaging paper made with mechanical pulp) need less than 20 GJ/t, but good-quality writing, typing, copying, and book paper (made with chemical, mostly sulfate, pulp) costs well over 30 GJ/t, or about as much as good-quality finished steel.

But the highest-quality paper (the coated stock most commonly encountered in art books) actually needs less energy (5%–12%, or as little as 32 GJ/t) than the production of uncoated papers (offset, standard book) of similar weight and brightness (Hein and Lower 1978). This difference is due primarily to the coated stock's lower fiber content; to produce and apply the coating materials is cheaper than to make fiber from wood. Paper from recycled stock is less energy-intensive, but the difference is reduced if the product needs de-inking (>2 GJ/t) and bleaching (~5 GJ/t). Typical printing costs are about 2 MJ/m^2. Contrary to expectations, the electronic age has boosted the demand for paper, and in the richest countries its per capita use has surpassed 200 kg or 300 kg per capita.

10.4 Crops and Animal Foods: Subsidized Diets

Extraction of coal and cheap steel introduced new labor-saving machines but could not displace animate power in fieldwork and irrigation or remove the shortages of nitrogen, the key limiting nutrient. This began to change only at the beginning of the twentieth century with the introduction of tractors and the synthesis of ammonia. Later came the first pesticides, herbicides, and high-yielding cultivars, and since the 1960s this agricultural modernization has also reached the poor world. Systematic study of the energy cost of modern food production started only in the early 1970s (Heichel 1976; Pimentel et al. 1973; G. Leach 1975), but the literature increased so rapidly that soon there was no shortage of comprehensive overviews or detailed case studies (Pimentel 1980; Fluck and Baird 1980; OECD 1982; Smil, Nachman and Long 1983; Stanhill 1984; Stout 1990). A series of six books entitled *Energy in World Agriculture*, published between 1986 and 1992, covered the entire field exhaustively (Singh 1986; Helsel 1987; McFate 1989; B. F. Parker 1991; Peart and Brook 1992; Fluck 1992), but subsequently there was a major decline in publishing on these topics (FAO 2000).

Only the internal combustion engine, a powerful but light prime mover, made it possible to mechanize field farming. The first U.S. enterprise specializing in tractor manufacture was set up in 1905. Subsequent fundamental innovations included power takeoff in 1919, mounted-type implements in 1924, power lift in 1930, low-pressure rubber tires in 1932, and hydraulic lift in 1935. Diesel tractors were introduced in 1931, LPG as a fuel a decade later (Dieffenbach and Gray 1960). In the United States the capacity of gasoline tractors surpassed that of horses by the late 1920s, and by 1950 there was nearly 1 OM difference in the total capability of these prime movers. In Europe the rapid mechanization of fieldwork began only after WW II. The process is still underway throughout the poor world, which in 2000 had less than 30% of the world's 27 million tractors, whereas the United States alone had nearly 20% of the worldwide total (FAO 2006) and even a larger share of rated tractor power.

Compared to early heavy machines (up to 450–500 g/W), modern tractors are much lighter and larger (70–80 g/W for sizes over 100 kW). In North America average maximum belt power rose from 11 kW (15 hp) in 1920 to 20 kW in 1945, 33 kW in 1970, 50 kW in 1985, and more than 70 kW by the late 1990s. The largest machines working in the huge fields of the U.S. Great Plains and the Canadian Prairies are now all diesel-powered and rate about 300 kW, with diesel tractors between 120 kW and 220 kW being fairly common (fig. 10.8). In Europe, and particularly in Asia, the sizes remain much smaller. The average annual fuel consumption of diesel-fueled tractors (in L/h) can be easily approximated from the annual Nebraska Tractor Test Data by multiplying maximum power (in kW) by 0.3. A large 175-kW (235-hp) tractor will consume roughly 55 L/h, and its annual consumption (working 500 hours) adds up to about 900 GJ. Consumption for specific operations is a function of soil, resistance, and speed.

Functional draft (soil and crop resistance) has been measured for scores of tasks with different soils. Moldboard plowing is generally the most demanding task, followed by rotary tilling. Specific needs for these tasks may be several times higher in silty clays than in sandy loams. Rolling resistance is highest with heavy implements in soft or loose soils. Typical speeds with tillage, fertilization, and seeding implements are 1.5–2.5 m/s, in grain harvesting 1–1.5 m/s. Field efficiencies are commonly up to 90% in tilling and cultivating, 65–70% in fertilizing and grain harvesting. Small plots and heavy yields will lower these levels and demand much time off for adjustments or repairs. Actual fuel consumption is thus highly variable. With diesel-fueled machines the typical ranges (in MJ/ha) are moldboard plowing 600–1200, disking 200–4900, planting 80–160, ammonia application 150–300, cultivating 100–200, and grain harvesting 250–500. For gasoline-powered machines the rates are about one-third higher.

No-till farming (row crops planted into narrow slits in undisturbed sod) and conservation tillage, which leaves at least 30% of plant residues on the soil surface and

10.8 Largest tractors working on the U.S. Great Plains and the Canadian Prairies rate about 300 kW (~400 hp). Photograph courtesy Buehler Corporation, Winnipeg.

combines tillage and planting runs in order to reduce the number of field passes and the degree of soil disturbance, has brought major declines in specific fuel needs (Phillips and Phillips 1984; Baker, Saxton, and Ritchie 1996). Additional advantages are reduced soil erosion, improved water retention, and greater flexibility of land use. Compared to conventional practices, disk-and-plant tillage needs 66% less fuel and slot-planting 75% less fuel; a further source of energy savings is the higher efficiency of nitrogen fertilization (Wittmuss, Olson, and Lane 1977). Disadvantages include the need for more herbicides, increased opportunities for pest damage, and lower soil temperatures.

Accounting for the energy cost of field machinery is much more complicated than determining its fuel con-sumption. The usual approach is to take the average weights of machines and implements used for a particular cropping cycle, estimate their typical energy costs, and then prorate this mass per hectare over a period of expected service (10–20 years). With the energy cost of tractors and major implements at 70–120 GJ/t and most conventionally grown staple crops requiring 10–30 kg/ha, these indirect energy subsidies amount annually to 0.7–4 GJ/ha. Markups of 5–15% cover the energy costs of maintenance and repairs. In 2005 there were about 27 million tractors in use worldwide. With average power of nearly 40 kW and 500 hours of work per year, their fuel consumption was about 5 EJ and the annual energy cost of building and maintaining the machines and their implements was nearly 1.5 EJ (assuming 100

MJ/kg, 5 t for a tractor and its implements, 15% markup for maintenance, and average 12 years of service).

Traditional agriculture provided the needed water by simple open-ditch irrigation fed by gravity flows or by a variety of human- or animal-powered devices (see section 6.3). Modernizing agricultures retain the inefficient ridge-and-furrow arrangements and supply them with simple mechanical pumps. Energy-intensive solid-set, big-gun, and center-pivot sprinklers—patented by Frank Zybach in 1952 and distributing water through a series of impact sprinklers by a row of mobile towers—now dominate in the United States, and trickle irrigation of high-value crops is even more efficient (Keller 2000). The total volume of water to be delivered depends on net irrigation needs (evapotranspiration plus leaching minus soil water stores plus precipitation) and on application efficiencies. These may be as high as 95%; good field practices should average 65%–75%, and furrow irrigation may be only 30%–40% efficient.

Consequently, an old diesel-driven unit (50% pump, 25% motor efficiency) in a Chinese wheat field (35% application efficiency) may consume 4.5 times more fuel than a well-run diesel-powered sprinkler in Nebraska (75% pump, 33% motor, 80% irrigation efficiency) while delivering the same effective volume of water from the same depth. With well depths of 30 m and water requirements of 30 cm/ha, electricity-powered irrigation may need about 1.5 GJ/ha, or about 4.3 GJ/ha of primary energy for thermal electricity. A natural gas–fueled engine performing the same task will use 6.3 GJ/ha, but with a 60-m-deep well the same engine would need 12.6 GJ/ha. These needs may surpass the total of all other energy subsidies. The global dependence on irrigation has trebled since the end of WW II, when about 75 million ha of cropland were watered. A generation later the total was 140 million ha, and by 2000 the figure topped 275 million ha, with three-fifths in Asia and nearly one-fifth in China alone (FAO 2006). With half the land mechanically irrigated at an average cost of 2 GJ/ha, global operating costs at the turn of the twentieth century were about 275 PJ, of which 50 PJ may go each year into installing and replacing the necessary pumps, motors, and pipes.

Chemical fertilizers represent the largest indirect energy subsidy in nonirrigated farming. No other innovation has contributed so much to increased yields as the three macronutrients N, P, and K, available in commercially produced inorganic compounds (Smil 2000b). Chilean nitrates (mostly $NaNO_3$), discovered in 1809, and rapidly depleted deposits of Pacific guano were the only source of inorganic N used to supplement legumes and organic recycling. Liebig's suggestion of treating bones with diluted H_2SO_4 increased the availability of P bound in hydroxyapatite, but the process was limited by the availability of animal skeletons. European potash deposits were abundant but undeveloped. By the second decade of the twentieth century all these limits had been removed.

The treatment of phosphate rocks by diluted H_2SO_4, pioneered by Sir John Bennett Lawes and producing ordinary superphosphate, started to diffuse by the 1870s, and rich phosphate deposits were discovered in Florida in 1888 and Morocco in 1913. Potash mining was expanded both in Europe and North America, and Haber-Bosch synthesis of ammonia removed the most important nutritional limit. WW I, the global economic slowdown of the 1930s, and WW II postponed the onset of large-scale fertilization until the early 1950s. Then the rapid diffusion of hybrid corn (begun in the United

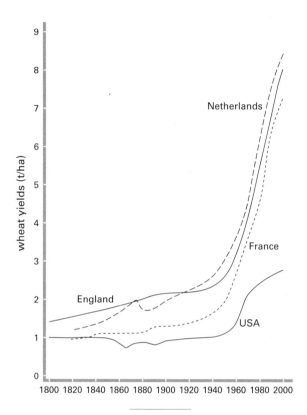

10.9 Long-term trends in staple grain production: English, Dutch, French, and U.S. wheat yields, 1820–2000. Based on Smil (1994) and FAO (2006).

States in the 1930s) and of high-yielding wheat and rice cultivars (developed at CIMMYT and IRRI during the 1950s and 1960s) could take full advantage of intensive fertilization, and the use of all macronutrients grew exponentially, as did average grain yields (fig. 10.9).

By the year 2000 average global N, P, and K applications were, about 53 kg/ha, 9 kg/ha, and 12 kg/ha, respectively, but these means concealed a wide range of values ranging from no or minimal applications in most

countries of sub-Saharan Africa to the double-cropped rice fields in Jiangsu or Hunan, where rates surpassed 500 kg N, 50 kg P, and 100 kg K/ha (NBS 2000). As already noted (see section 10.2), global production of inorganic fertilizers claimed about 5 EJ in 2000, or less than 1.5% of global TPES. And even if all energy for the Haber-Bosch synthesis came from natural gas, that claim would have equaled less than 5% of the fuel's annual global consumption. Clearly, there is little reason to worry either about current needs or future supplies of energy for nitrogen fixation. Energy equivalent to that demand could be gained by relatively small efficiency improvements of natural gas combustion in industrial boilers or household furnaces.

Post–WW II expansion of fertilization has been accompanied by increasing use of herbicides to control weeds, and insecticides and fungicides to raise the yields of new high-yielding varieties (Smil 2000b). Both weed and insect controls started in the mid-1940s, the first with preparation of 2,4-dichlorophenoxyacetic acid (2,4-D), a growth-regulating substance that selectively kills many broad-leaved plants without serious injury to crops, the other with the commercial introduction of dichloro-diphenyl-trichloroethane (DDT) (fig. 10.10). Since then more than 50,000 pesticide products have been registered to fit thousands of specific applications, but the bulk goes to only a handful of crops. Pesticide applications are highly effective and economically rewarding. Although the compounds are derived by energy-intensive processes from petrochemical feedstocks, their low application rates translate to only minor subsidies in absolute terms.

Synthesis of common active ingredients typically requires 100–200 MJ/kg, and total energy costs, including formulating, packaging, and marketing, are mostly

10.10 Molecular structures of the world's first effective herbicide, 2,4-D, and the first widely used insecticide, DDT. From Smil (2006).

200–300 MJ/kg; means of 220, 270, and 275 MJ/kg were given as representative of U.S. applications (Helsel 1987; T. A. Unger 1996). The usual recommended applications are 1–2 kg/ha, so even the use of the most energy-intensive compounds will translate to subsidies no higher than 1 GJ/ha, and the typical rates in those fields where the crop is treated by both herbicides and insecticides will be 0.5–1 GJ/ha. The annual global energy investment in pesticide production was about 500 PJ during the 1990s.

For many grain farmers, harvesting does not end energy expenditures. Long-term storage and long-distance transportation require that grains have their water content reduced below the levels allowing germination, insect propagation, and fungal growth. Modern mechanical harvesting, often done at relatively high moisture in order to minimize grain losses, requires artificial drying. This is especially the case with U.S. corn; harvested at moistures up to 28%, most of this large crop is dried. LPG and electricity are the principal energizers. Typical rates are 600–750 kJ/kg of grain dried to storage moisture of 14%. For corn this would represent costs of 3–6 GJ/ha, a small part of overall energy subsidies but a critical finishing input.

Published energy costs of individual crops are not readily comparable because of the nonuniform choice of analytical boundaries and sometimes substantial differences in input equivalents. Many analyses stop at the farm gate, others add just the leading indirect subsidies (fertilizers and pesticides), and yet others make systematic efforts to account for energies used to make and maintain field machinery. In general, energy analyses of crops will have wider error margins than those of mass-produced industrial goods. Field extremes of modern farming are marked by extensive North American wheat cropping (6–10 GJ/ha) and by establishing a vineyard for table grapes, an effort costing some 200 GJ/ha for several years before the first harvest.

Typical annual rates are 8–15 GJ/ha for dryland cereals, 20–25 GJ/ha for rain-fed, and more than 40 GJ/ha for irrigated Corn Belt corn and California rice. Nitrogen-fixing soybeans need no more than 8–15 GJ/ha, but potatoes, vegetables, and tree crops, with heavy fertilization and irrigation, need 50–100 GJ/ha, and orange groves require about 120 GJ/ha. Even these rates are dwarfed by hydroponic cultivation in greenhouses, where energy consumption is determined above all by the need for seasonal heating. In warm climates operation costs are dominated by irrigation, fertilization, and cultivation needs. Turkish rates (with some supplemental heating) range from 2.5 GJ/t of tomatoes to 4.3 GJ/t of peppers (Canakci and Akinci 2005), but in heated Dutch greenhouses the same crops may consume as much as 40 GJ/t (Dutilh and Kramer 2000) and the heating rate may be several TJ/ha. These subsidies translate to about 3–4 GJ/t of Manitoba spring wheat, 5 GJ/t of Iowa corn, and up to 7 GJ/t of rice. Although vegetables and fruits need much higher energy subsidies per hectare, their high yields translate into GJ/t rates that are similar to those for cereals.

The high metabolic requirements of heterotrophs (see chapter 4) mean that specific energy subsidies in animal husbandry have to be considerably higher than those in plant agriculture. Feedlot-fed beef, a product of large, relatively long-lived domesticated mammals consuming carbohydrate-protein mixtures, is the most heavily subsidized animal food, whereas the meat of birds and pigs, animals that are brought to slaughter weight in just a few months, is less demanding to produce. Weight gains are faster when feed consists of concentrates (cereals, leguminous grains) rather than roughages largely devoid of proteins and lipids (grasses, straw), but the cultivation of concentrates needs higher energy inputs. Eggs need less energy than meat, and milk is the least energy-intensive animal foodstuff; its low energy cost is rivaled or bested only by aquacultured herbivorous fish.

Beef production, marginal in Asia and traditionally a by-product of dairy industries in Europe, is the largest livestock undertaking in the Americas. The North American system is a combination of intensive feeding based on grain and soybeans, the continent's two main crops, and extensive Western grazing. Even beef ranching—with energy subsidies for pasture improvement, fencing, pickup trucks, fuel, and production of hay—can be fairly energy-intensive, with the published extremes of 18–130 MJ/kg of gain by weaner calves. Feedlot management

costs 3–11 MJ/kg of gain and is dwarfed by the cost of feed, which depends on the length of feeding (60–150 days) and the share of grains in the ration (no less than 30%–40% of total feed energy). The rest comes from roughages, mostly hay and silage corn.

Net energy needs during feedlot finishing of animals that weigh 400–500 kg at slaughter are 50–65 MJ/kg of daily gain, corresponding to 100–125 MJ/kg of metabolizable energy. Depending on the roughage/concentrate ratio, these gains require at least 8–10 kg dry matter equivalents of 110–140 MJ/kg of gross feed energy inputs (roughages at 8.5 MJ/kg, concentrates at 13.3 MJ/kg of metabolizable energy). When produced by extensive cropping, 1 kg of feed may need 3 MJ, and the feed would cost as little as 30 MJ/kg of beef. When coming from irrigated fields, 1 kg of feed may need 10 MJ, and each kilogram of gain is then subsidized by about 100 MJ (Smil 2000b). Case studies taking into account the needs of breeding animals have come up with total energy subsidies of 60–150 MJ/kg of feedlot beef; typical U.S. Midwest values are 80–110 MJ/kg of dressed meat.

Pork is now produced in rich countries by intensive feeding in total confinement, taking less than six months from weaning to slaughter (Pond 1991; Taverner and Dunkin 1996). Soybean meal supplies protein, grain corn supplies carbohydrates. Year-round farrowing smoothes the output, and breeding for a lower share of trimmable fat has resulted in more meaty carcasses, quite distinct from traditionally fatty Chinese pigs, which are still largely fed wastes and take more than twice as long to reach slaughter weight. Animals slaughtered after six months of feeding need 75–90 MJ (5–6 kg) of concentrates to gain 1 kg of live weight, and these gains require 15–60 MJ of energy subsidies. All other energies (for feed preparation, farrowing, feeding, and cleaning operations, equipment, and buildings) add about 10 MJ/kg, for a total of 25–70 MJ/kg. In terms of dressed meat, the U.S. mean is 40–45 MJ/kg, but Dovring (1984) put the cost of Midwestern dressed pork at 68 MJ/kg.

Broiler chickens are the most efficient converters of feed into lean meat; their efficiency of feed conversion has improved with breeding. Feeding rates for U.S. hogs have shown no declining trend, and feeding rates needed to produce milk and eggs have remained virtually identical since 1910. But by the year 2000 about 2 kg of grain were needed to add 1 kg of broiler live weight, only about 60% of the 1935 rate of 5.3 kg (Smil 2000b). Broilers thus need generally less than 10 MJ/kg of energy subsidies for their feed. Energy used for incubation, heating, ventilation, and lighting ranges from about 6 MJ/kg of live weight in buildings with good insulation to triple that rate in inferior structures.

Other costs, including feed preparation, buildings, equipment, and a 2.5–5% markup for flock mortality, add up to 10 MJ/kg. Published studies quote total costs per kilogram of live weight at 30 MJ for U.S. northern states, 27 MJ for U.S. southern states, and 33 MJ in England (Ostrander 1980); typical rates of the 1990s were 10%–20% lower. Egg production starts with rearing pullets to the point-of-lay (110–140 MJ/bird), and a laying hen requires 38–42 kg of feed per year, or 2.5–3.5 kg of feed per kilogram of eggs (annual output is 200–250 eggs). Feed costs are thus at least 230 MJ/a, and the overall rates are 450–500 MJ/a. Milk is the least energy-intensive animal food; only about 1 kg of concentrate feed is needed to produce 1 kg. Mixtures of grazing and grain feeding translate into very different energy subsidies. Other inputs, dominated by milking and water

heating, are usually no more than one-third of the total subsidy of 5–7 MJ/kg of milk.

No single measure can serve as a clear yardstick for comparing energy efficiencies of animal food production (fig. 10.11). In rich societies no animal food is eaten primarily for its overall energy content. In fact, fat, the most energy-dense component of these foods, is frequently avoided or discarded as people buy skimmed milk and low-fat cheeses and throw away trimmable fat before cooking meats. High-quality protein is the most desirable nutrient in animal foods, and young lean birds are its best source compared to pigs or cattle. But ruminants can digest cellulosic phytomass and convert otherwise unusable low-quality crop by-products and forages into high-quality protein.

Gross energy in feed is converted to dressed meat with efficiencies of about 5% in cattle and roughly 10%–15% in pork and poultry, and these respective shares drop to less than 2%, 5%, and up to 7% for cooked edible meat. Feed energy is converted to eggs with 15%–20% efficiency and to milk with 20%–25% efficiency. Production of 1 g of protein requires about 2 MJ of feed in beef, at least 500 kJ in pork, 300 kJ in poultry, eggs, and milk (Smil 2000b). These differences are not perfectly mirrored by energy subsidies going into animal protein production because the relatively high energy use in broiler and battery egg operations narrows the gap. While 1 g of beef protein needs at least 600 kJ of energy subsidies, pork protein can be grown with 400–500 kJ/g, chicken and egg protein with 300–350 kJ/g, and milk protein with 200 kJ/g. It is just a coincidence that the gross feed energy required to produce 1 g of poultry protein is identical with the energy subsidies invested in the process.

Much of ocean fish protein costs as much as beef or pork protein. Modern fishing is totally dependent on liq-

	milk	eggs	chicken	pork	beef
feed conversion (kg of feed/kg of live weight)	0.7	3.8	2.3	5.9	12.7
feed conversion (kg of feed/kg of edible weight)	0.7	4.2	4.2	10.7	31.7
protein content (% of edible weight)	3.5	13	20	14	15
protein conversion efficiency (%)	40	30	25	13	5

10.11   Efficiencies of animal food production. Based on Smil (2000b).

uid diesel fuels for powering ships and their refrigerators, and indirect energy costs for building and outfitting the vessels prorate to less than 10% of the total expense. Energy analyses of U.S. bottom trawling found rates of 35–40 GJ/t of catch, very similar to the British skippers' rule of thumb in the 1980s that it takes at least 1 t of fuel (42 GJ) to get 1 t of fish. Tyedmers (2004) reviewed direct (diesel fuel) energy costs of 29 Atlantic and Pacific fisheries and found a wide range of values, with minima of 100–140 L (3.6–5 GJ)/t for herring and mackerel, 15–30 GJ/t for British Columbia salmon, maxima of 3–3.4 kL (107–121 GJ)/t for shrimp trawling and long-line tuna fishing, and modal rates around 500 L (18 GJ)/t.

Assuming that 40% of the catch is edible and that it averages 17% protein, this approximation translates into an energy subsidy of about 265 kJ/g of protein. For comparison, Rawitscher and Mayer (1977) calculated the rates for a dozen U.S. species, ranging from 15 kJ/g of protein for sardines to 1.8 MJ/g for fresh shrimp, and Rochereau (1980) found northeastern U.S. fishery averaging just about 75 kJ/g of protein (40 kJ inshore, 174 kJ/g offshore catch). The earlier low rates were due to the small size of fishing vessels and the natural richness of the area. This had changed drastically with the collapse of the region's rich cod stocks during the 1990s. Tyedmers (2004) notes often rapid increases in energy subsidies since the 1960s, and only some reduction fisheries (catching such species as anchovetta, capelin, and herring for processing into feed meals and oils) require less than 2 GJ/t.

Large differences in the energy costs of marine protein (a spread of 2 OM) are in sharp contrast to the relative uniformity of energy subsidies in livestock production and clearly invalidate any simplistic perception of seafood as a particularly attractive energy bargain. Fishing is, after all, just intensive hunting, and it requires less energy to capture the more abundant but also less sought-after species (octopuses, krill). But even this generalization breaks down in the regions of chronic overfishing (Mediterranean) or after sudden environmental changes (El Niño's effect on Peruvian anchovetta). The same is true of aquaculture. The best efficiencies come with polycultures based on herbivorous species, particularly East Asian carp. As little as 150–200 kJ/g may be needed to produce 1 g of carp protein; aquacultured salmon needs only 100 kJ/g, but most of its feed must be of animal origin, including a significant share of fish oil.

10.5 Modern Food System: Gains, Costs, Efficiencies

The overall magnitude of agricultural energy subsidies is insignificant compared to the input of solar energy. While an intensively cultivated cornfield in Iowa may receive annually 30 GJ/ha in direct and indirect energy subsidies, solar energy that reaches it during the 150 days between planting and harvesting amounts to 30 TJ/ha, a 1,000-fold difference (Smil, Nachman, and Long 1983). The enormous solar flux is free. The costly flows of fossil fuels and electricity are limited, but without them productivity would be a fraction of the subsidized harvest even with maximum in put of animal labor. This productivity gain has not only sustained historically unprecedented population growth but also improved the quality of nutrition. Between 1900, when energy subsidies were limited to low-level fertilization and rudimentary mechanization in industrializing nations, and 2000, the world's cultivated area grew by 80%–100% (Golde-

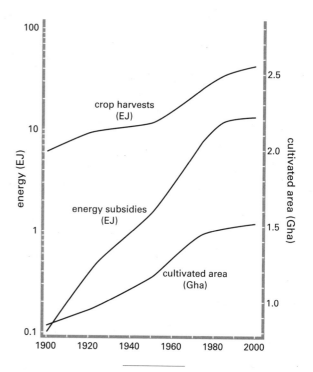

10.12 The world's cultivated area, aggregate crop harvests, and energy subsidies in modern agriculture, 1900–2000. From Smil (2000b).

wijk 2001), but the energy harvested in edible crops expanded sixfold. This greater average productivity was made possible by an 85-fold increase in energy subsidies per harvested hectare (fig. 10.12).

In 1900 1 ha of cultivated land could feed just 1.5 people, and the edible harvest (before storage and distribution losses) prorated to less than 10 MJ/day per capita. This rate offered only a slim food supply margin above the average daily requirement of about 9 MJ/day per capita. Moreover, prevailing diets were composed of just a few staples, and the relatively low crop yields greatly limited the extent of animal feeding and hence the average availability of animal foodstuffs. Modern farming increased the number of people supportable by cultivating 1 ha to about four, but with the overwhelmingly meatless diets of the early 1900s the average hectare of farmland could support at least six people, a density traditionally achieved only by intensive farming in Egypt or China.

The best modern performances are much higher. China's post-1978 farming reforms, supported by what are some of the world's highest fertilization rates, fed about 8.5 people per hectare, and in the most populous provinces the rate is in excess of 15 people per hectare. The average Chinese diet still has less animal protein than the Western one, but by the year 2000 its total energy content became the same as the mean in affluent Japan (Smil 2004a). National means of food availability at retail level are published annually by FAO in the form of food balance sheets (FAO 2006). Domestic production of crops and animal foods is adjusted for net imports; changes in stocks; harvests used for seed and industrial conversions, fed to livestock, and wasted during storage; processing; and transportation. Specific national details are available for less than half of all member nations, and many balances are calculated by FAO in Rome on the basis of numerous assumptions. Daily per capita availabilities are in excess of 15 MJ/day in the United States and many European countries and below 8.5 MJ/day in parts of sub-Saharan Africa. Remarkably, by the year 2000, Chinese and Japanese supplies became virtually identical in overall energy terms (nearly 12.5 MJ/day).

Only costly (and hence increasingly rare) household food consumption surveys can accurately establish actual daily food intakes. More common but less reliable estimates based on income and expenditure surveys or on 24-h dietary recall indicate large differences between food supply and intakes in all affluent countries (Smil 2000b). The most representative U.S. surveys indicate average daily adult male intakes of 10.3 MJ/day, adult female intakes of 6.9 MJ/day, and a mean for all individuals of all ages of 8.4 MJ/day (USDA 2006). There are three reasons why this mean is below the average daily requirement of about 9 MJ/day: the requirement mean is too high given today's largely sedentary lifestyles; widespread dieting among American females depresses the overall intake mean; and (most likely), the actual intakes (based on recall) are substantially underreported. This implies a nearly 50% gap between average consumption and average availability (15.7 MJ/day in 2003), and similarly large food losses are true for many European countries. FAO uses the food balance sheets to estimate the numbers of undernourished people. The total of 852 million in 2001–2002, with 815 million in poor countries, 220 million in India (FAO 2005b), is due overwhelmingly to internal conflicts, distribution inequalities, and poor agricultural management, not to physical shortages of food or to inability to produce sufficient amounts.

Increased productivities have thus been a universal mark of subsidized farming, but these higher yields are not due to higher net CO_2 assimilation. A rise in harvest indices from 25%–30% in traditional cultivars to as much as 50%–55% in improved varieties, that is, a redistribution of the fixed C through breeding supported by energy subsidies, has been responsible for the gains (Gifford et al. 1984; Smil 2000b). Further gains are possible, but comparisons of record yields with national averages are not a realistic indicator. In fact, staple grain yields have shown signs of leveling off, and environmental stresses will always limit the gains.

From a social point of view, the sharply decreased labor requirements of subsidized agriculture have been no less important than the higher yields. A tonne of U.S. wheat needed 137 h of labor in 1800, 56 h in 1880, and less than 1.67 h in the late 1990s. And in just two generations, between 1940 and 1980, the labor needs for the U.S. corn were reduced from 32 h/t to 2 h/t. Every affluent country had experienced a massive release of farm labor. The U.S. share went from about 64% of total labor force in 1850 to less than 2% by the year 2000, and the same process is now underway in the world's two most populous countries, China and India. But a frequently quoted increase in numbers of people supported by one U.S. farm worker—from just 4.7 in 1850 to 8 in 1900, 16 in 1950, and 95 in the year 2000—is both an underestimate and an overestimate. It is an underestimate because large shares of U.S. planted land have been devoted to export crops, and an overestimate because a proper account would have to include workers employed in the production of industrial inputs as well as in the provision of support services (education, research, extension). Such a tally could nearly double U.S. "agricultural" employment, but it would still represent only about 4% of the country's labor force.

Comparisons of national accounts of energy subsidies in agriculture are complicated by nonuniform analytical boundaries, choice of energy equivalents for major inputs, selection of farmland totals, and imports of animal feed that may originate in a number of countries on different continents and that dominate the feedstuff consumption in some countries. These realities make accurate energy accounts unlikely. Good arguments can be made for dividing the inputs by the total area involved in agriculture or by only the land under annual or permanent crops. A case-by-case approach is perhaps best. Where grazing contributes little food, the average energy subsidy can be calculated only for the cultivated area, but

elsewhere any managed pastures should be included in the total.

The different results for low-subsidy, extensive field farming and grazing in Australia or New Zealand (2–3 GJ/ha) and high-input, intensive agricultures in the Netherlands or Israel (70–80 GJ/ha) are not surprising. The most noteworthy finding is that China, the world's most populous nation, is subsidizing its farming with intensities surpassing those of the most productive North American and European agricultures. This finding is somewhat counterintuitive; one would expect the use of much more labor or organic fertilizer. But China's high fossil fuel and electricity subsidies are understandable in view of high cropping ratios (1.5 crops/ha annually), extensive irrigation (50% of all farmland), and intensive fertilization. In the late 1960s North America applied about 30 kg N/ha, and China about 5 kg N/ha. In 2005 the respective means were about 60 kg N/ha and 165 kg N/ha. Fertilizer nitrogen provides about 60% of the nutrient in China's cropping, and since over 80% of the country's protein is derived from crops, roughly half of all nitrogen in China's food comes from inorganic fertilizers.

China's population thus depends for survival on external energy subsidies, whereas the U.S. population—with less meaty diets and a much larger agricultural labor force—could be fed from the abundant farmland without any synthetic fertilizers. Detailed accounts of the global nitrogen cycle of the late 1990s indicate that about 75% of all nitrogen in food proteins available for human consumption came from arable land (the rest from pastures and aquatic species). Because synthetic fertilizers provide about 50% of all nitrogen in harvested crops, at least every third person, and more likely two people out of five, gets protein thanks to Haber-Bosch synthesis. But the fertilization process plays different roles in rich and poor countries. In affluent nations it helps to produce rich diets with plenty of animal foods and surplus food for export; in low-income countries it prevents widespread malnutrition.

Relative levels of agricultural energy subsidies depend on TPES. In affluent countries, with high energy use in households and transportation, they claim 3%–5% of TPES, whereas in some large populous nations the shares may be closer to 10% (Smil 2006; FAO 2000). Several detailed accounts of energy use in U.S. agriculture (Steinhart and Steinhart 1974; USDA 1980; Stout, Butler, and Garett 1984) found very similar shares in 1970 (3%), 1974 (2.9%), and 1981 (3.4%). By the late 1990s more efficient production and use of fertilizers, a slower growth rate in the use of pesticides, and more widespread reliance on reduced tillage lowered the share of energy subsidies in U.S. crop farming to about 2% of the country's TPES (Collins 2000). As China's TPES expanded, the share of agricultural subsidies, as high as 15% during the 1980s, fell to about 10% by the year 2000 (Smil 1992; NBS 2000).

A conservative approximation of energy subsidies used by global crop farming at the beginning of the twenty-first century is 12.8 EJ, equivalent to 300 Mtoe or 8.4 GJ/ha under annual or permanent crops. About 2 EJ goes to produce and maintain agricultural machinery (tractors, combines, implements, irrigation systems), about 5 EJ to power it, the same amount to extract, synthesize, and distribute fertilizers, 500 PJ to make pesticides and herbicides, and 300 PJ to build irrigation systems and deliver water to fields. This prorates to average global power density of about 25 mW/m². The highest regional means are 1 OM higher. Such low rates reflect the intermittent tasks of crop farming spread over large areas. The energy cost of animal husbandry (excluding the energy content of feed), aquaculture, and fishing would increase this total by no more than 4 EJ. A 17-EJ share used in global food production would represent less than 5% of the world's TPES in the year 2000. For comparison, Giampietro (2002), whose global calculation also included the cost of preparing animal feeds and energy invested in the cultivation of forages, arrived at a total of 18.2 EJ (433 Mtoe) for 1997.

Unfortunately, there is no suitable measure to evaluate the efficiency of these energy subsidies, a fact that has not stopped such efforts. Black (1971) introduced the quotient of harvested food energy and energy invested in the growing process (animate or inanimate) as an efficiency ratio of farming systems. This approach has been used repeatedly to illustrate relatively high returns of traditional cropping powered by animate energies (ratios of 10–30; see chapter 6), low and declining energy gains of modern intensive crop cultivation (grain cropping 2–8, fruit growing ~1, vegetable cultivation 0.1–1), and substantial energy losses incurred by all modern animal production systems (ratios as low as 0.05 for lean red meat and no higher than 0.5 for milk).

Critics of modern agriculture see such ratios as perfect proof of the dubious nature of subsidized farming, but the ratios are inappropriate and should be avoided (Fluck 1979; Smil, Nachman, and Long 1983). To the uninformed they misleadingly suggest a direct link, a direct conversion of input energies into food outputs. Any inference that fuel energy is converted to food is wrong. The relevant energy conversion is photosynthesis, and the subsidies merely remove or moderate some factors limiting NPP and help to channel the photosynthates into target harvest tissues (that is, maximizing NEP and the harvest index; see section 3.3). If higher energy subsidies result in higher productivities, then the conversion efficiency of a cropping system is clearly increasing. This trend can be illustrated by data on corn, the leading U.S. crop, heavily subsidized. In 1945 average subsidies of nearly 6 GJ/ha helped to produce about 2.2 t of grain (conversion efficiency merely 0.3% when assuming PAR of 11 TJ/ha). By 2003 subsidies of about 18 GJ/ha aided in harvesting 9 t/ha (conversion efficiency 1.23%). The energy subsidy rate had tripled, but the efficiency of converting solar radiation into harvested grain had more than quadrupled.

A second key error often made in deploying energy ratios is a simplistic focus on energy output. Cropping aims to maximize the productivity of particular cultivars, but not the conversion efficiency of sunlight into phytomass. If the latter were the case, we would cultivate only C_4 species: silage corn in temperate regions (whole plant harvested and used) and sugarcane in the tropics (highest producer of edible phytomass with year-round growth). Crops are clearly grown not just for their gross energy content but for their unique combinations of nutrients (carbohydrates and proteins in cereal and leguminous grains, vitamins and minerals in fruits and vegetables); processing potential (gluten-rich wheat flour has excellent dough-making properties, which are entirely absent in cornmeal); palatability, storability (cereals vs. tubers); and even the presence of indigestible but beneficial roughage.

What does it matter if grapes contain barely more energy than the fuels and electricity needed for their cultivation (Heichel 1976)? Obviously, the energies embodied in viticultural production cannot be digested, and grapes are not cultivated primarily because of their energy content. A nutritionist may point out their value as a source of vitamin C, minerals, antioxidants, and dietary roughage, but most of us eat them simply for their unique taste. This nonenergetic consideration is even more prominent when we drink wine. Only terminal alcoholics may derive most of their energy from it, but for everyone else drinking wine has nothing to do with ensuring a daily supply of food energy.

More fundamentally, such reasoning would mean that a society trying to minimize food subsidy ratios would have to run on tubers, but these plants have no, or hardly any, protein or lipids, and they spoil much faster than dry grains. For this reason, every advanced civilization was based on cereals, the Incas being the only partial exception because corn supplemented the dominant potatoes. Tuber cultivation among the New Guinean shifting gardeners has a high energy ratio of 16 (see section 6.1) in contrast to the low energy ratios (2–5) of modern cereal agricultures, but few would argue that we should choose the former practice as the energetic foundation of modern society. Energy ratios can never capture these qualitative considerations, and hence they are relevant only to systems that produce energy (wood, ethanol, methanol).

The third fundamental weakness of energy ratios is their disregard for time-energy and space-energy trade-offs. Energy subsidies in farming not only produce higher yields but do so while dramatically cutting strenuous, tedious human labor and supporting higher population densities at higher nutritional levels. Is it desirable to produce staples with energy ratios 15–30 but requiring heavy exertion involving up to 80% of the labor force, including most children? Or is it better to produce staples with ratios 2–5 and requiring mostly light work and the participation of 2%–10% of the population? These work and time gains are inevitably accompanied by a rather steep decline in efficiency ratios. Are these trade-offs acceptable, or are they sign of a fatal dependence on nonrenewable energies?

The answers depend on circumstances. Countries with relatively abundant farmland as well as those with high intakes of meat could substantially cut their energy subsidies by growing more cereal and oil crops in rotation with feed and feed legumes (alfalfa, soybeans) and by reducing their carnivory, which, in any case, does not offer advantages compared to only moderately meaty diets (Smil 2002). Populous countries with limited farmland have to rely on intensive cultivation unless they are willing to return to traditional low-yield cultivars and hence to overwhelmingly vegetarian diets. But the subsidies can be managed much more efficiently. Nitrogen losses are at least 50% and commonly 60%–70% of the applied nutrient (Cassman, Dobermann, and Walters 2002), and Asian irrigation efficiencies could be doubled.

Drastic reductions of mechanical power would be much harder to accomplish. Matching the power of U.S. tractors in the year 2000 with horses would require building up an equine stock of at least 250 million head, about ten times the record number of horses in 1918. At least 300 million ha, twice the total of U.S. arable land, would be needed to feed the animals. And doing away with insecticides would lower edible harvests by 10%–50%. This critical dependence need not be alarming. In the short run there is no shortage of conservation adjustments, and as already noted, energy inputs into farming are only a small share of TPES. They are relatively small even in comparison with energy uses in other sectors of the modern food system, which extends beyond the farm gate to include processing, storage, transportation, distribution, wholesale and retail, cooking, household refrigeration, and waste management.

None of these diverse activities can be quantified by a single mean, but the rates of 50–100 MJ/kg of retail product are common in processing and packaging; 1–3 MJ/kg for storage; up to 10 MJ/kg per week for cooling or freezing; 1–4 MJ/kg for shopping; 5–7 MJ/kg of food for home cooking; and 2–4 MJ/kg for a dishwashing event (Dutilh and Linnemann 2004). Because of a high degree of spatial concentration of modern specialized production (both nationally and internationally), energy-intensive food transport now spans increasing distances between growers and consumers. In North America fruits and vegetables commonly travel 2,500–4,000 km, and the global total of international food shipments surpassed 800 Mt in 2000, four times the mass in 1960 (Halweil 2004). As a result, energies used in food processing, distribution, and wholesale and retail can be twice as large as those consumed by field farming and animal husbandry, and food preparation takes 30%–50% of all energies used in an affluent nation's food chain.

Food processing is a major consumer of fuels and electricity (Singh 1986; Biesot and Moll 1995; Ramírez 2005). Wheat flour can be produced with as little as 1.3 MJ/kg; white rice may need no more than 1.5 MJ/kg; and pasta making takes 4–5 MJ/kg. Fermentations producing beverages with low alcoholic content consume a relatively small amount of energy: beer less than 1 MJ/L, wine about 2 MJ/L. Dutch meat processing averages less than 1.5 MJ/kg for beef, about 2 MJ/kg for pork, 3 MJ/kg for chicken, and 5.5 MJ/kg for processed meats. Sugar refining needs up to 35 MJ/kg, cheese making as little as 3.5–5 MJ/kg and as much 20 MJ/kg. Soft drinks need 20–25 MJ/kg, the distillation of liquors at least 20–25 MJ/L. Breakfast cereals need in excess of 12 MJ/kg, tomato juice about 5 MJ/kg, frozen citrus juice about 20 MJ/kg, fruit and vegetable canning about 5 MJ/kg, and oil pressing 10–15 MJ/kg.

The energy cost of bread was first traced by Johnson and Hoover (1977), who analyzed the sponge-dough process. The cost of 7.3 MJ/kg of white bread at retail was divided 2:1 between production and delivery to stores. Baking claimed only about 17% of all energy, most of it for water evaporation. Beech's (1980) audit of three English bakeries yielded 6.43–7.14 MJ/kg with direct energy use being 70%–80% of the total, packaging about 12%. Beech (1980) found striking economies of scale for home baking (three breads in oven costing about 50% less per kilogram than a single bought loaf) and a cost only slightly higher (7.8 MJ/kg) than in commercial operations. Adding the energy cost of wheat-growing (4–7 MJ/kg) and grain-milling (at least 2 MJ/kg) raises the overall cost to 14–17 MJ/kg, a comparatively modest subsidy to produce 10.6 MJ/kg of food energy with at least 8% of incomplete (lysine-deficient) protein. Assorted pastry products can embody much more energy, fruit pies as much as 35 MJ/kg (Biesot and Moll 1995).

Food packaging is fairly energy-intensive, especially with the use of plastics and aluminum and the inclusion of waste disposal costs. Modern retailing is energized largely by electricity; about 60% of its energy use is associated with refrigeration; lighting and space heating/cooling take about 15% each. Total energy needs vary mostly between 400 W/m^2 and 450 W/m^2 of selling area, and a bakery and deli section can increase the rate by up to 50%. Cooking is a rather inefficient heat transfer even in modern kitchens; its annual needs range from 3 GJ to 5 GJ per household. Throughout the poor world cooking with traditional stoves wastes most of the energy in wood or crop residues, but more efficient kerosene stoves are risky to use.

Microwavable foods and higher shares of meals eaten away from home have increased the demand for refrigeration. Home refrigerators were marketed first by Kelvinator Company in 1914, and freezers were introduced in 1940. The early models rated about 200 W/m^3, the first frost-free refrigerators needed about 350 W/m^3, and models with large freezer sections consume up to 450 W/m^3. Freezers require 18–23 MJ/kg of load. The popularity of fast food has increased demand for refrigeration, which now consumes 5%–10% of electricity in rich nations. Fortunately, better refrigerator designs offer substantial electricity savings. The energy cost of cut flowers put on a dining table can be 1 OM higher than that of common foodstuffs. A detailed Dutch study put the average energy cost at about 9 MJ/flower, ranging from more than 50 MJ/stem for a flamingo flower to 3 MJ/stem for a sword lily, with tulips at 4 MJ/stem, carnations at about 5 MJ/stem, and roses at almost 10 MJ/stem (Vringer and Blok 2000). When adjusted for lifetime optimal conditions, common lilies and carnations (400–500 kJ/stem per day) are the least demanding.

11

Environmental Consequences

Metabolism of Fossil-Fueled Civilization

Shall I not have intelligence with the earth? Am I not partly leaves and vegetable mold myself?
Henry David Thoreau, *Walden* (1854)

The metabolic parallels between the functioning of fossil-fueled civilization and heterotrophic life are clear. They both have extensive footprints: in the first case, land claimed by extraction, transportation, and conversion of fuels and generation and transmission of electricity, in the other case often large and fiercely defended home ranges. They both oxidize carbon, high-energy societies via the combustion of fuels, heterotrophs via aerobic respiration. They both require water, and both heat their surroundings. They depend on steady inflows of raw materials, structural components (metals, wood, concrete) in the first case, essential macro- and micronutrients in the second. And both play major roles in the planet's key biogeochemical cycles, fossil-fueled civilization as a growing source of worrisome interference in their functioning, heterotrophs (particularly microbes and fungi) as the principal controllers and processors of key nutrient flows in ecosystems.

These parallels also hide many substantial differences. Most important, all higher heterotrophs are incessant carbon oxidizers because their brains die in a matter of minutes without oxygen, whereas the oxidation of fossil fuels proceeds in spells that range from minutes for short car drives to more than a dozen years in modern coke-fueled blast furnaces. Even some very large heterotrophs leave only fleeting footprints on the landscape they inhabit, but many infrastructures of fossil-fueled societies change the land cover radically, some even irreparably. Heat rejection rates of fossil fuel and electricity conversions vastly surpass anything that even densely packed organisms can ever produce. Specific material inputs needed to maintain high-energy societies are far above the rates needed to support heterotrophic life. And the extent of human interference in biogeochemical cycles of C, N, and S has brought large-scale regional and even global environmental changes.

These are new phenomena. While the environmental consequences of the preindustrial quest for energy were far from negligible, they never reached continentwide or global scale. Extensive deforestation was the most obvious environmental degradation, and pollution effects were limited to poor indoor air quality (the result of low-efficiency combustion in open fireplaces or primitive stoves), seasonally high levels of atmospheric SO_2 and soot in cities, and discharges of urban wastes into streams. A century of fossil-fueled industrialization, urbanization, and subsidized farming changed both the extent and the rates of environmental intervention. By the 1960s, when environmental concerns emerged as a major preoccupation of industrial civilization, there was no doubt that energy industries and energy uses were the leading causes of environmental degradation and pollution, and hence they began receiving a great deal of research and policy-making attention.

In this chapter I assess the environmental consequences of modern energetics by concentrating on five principal categories of impacts: on land, heat rejection, water, air, and grand biospheric cycles. I examine the relation between energy and land by using the fundamental metric of power densities, first to look at the rising capacity to feed larger populations per unit of cultivated land and then to survey numerous space constraints of modern energy conversion with a particular stress on the differences between fossil-fueled and renewable energy systems. The second section is devoted to heat rejection on scales ranging from microprocessors to urban heat islands and beyond. The following two sections take a closer look at the impacts of modern energy metabolism on water resources and atmospheric pollution. Finally, I

examine the most widespread and most critical environmental consequences of modern energy harnessing and use, the large-scale impacts on the Earth's vital biogeochemical cycles of carbon, nitrogen, and sulfur whose functioning sustains the biosphere and whose alteration can have a multitude of undesirable effects.

There are many other environmental impacts whose aggregate effects elude reliable global quantification because it is impossible to generalize across so many different categories. These include material inputs needed to construct and maintain extensive infrastructures of modern energy industries. A Saudi well may need just 1 g of steel (in drilling and production equipment and in pipes) for every GJ of extracted oil, and a giant Gulf of Mexico production platform (see section 8.3) may add 10 g/GJ of produced oil. Similarly, an open-design oil-fired power plant in an arid region may need no more than 100 g of concrete/GJ, and a nuclear power station with massive foundations and reactor containment structure may require up to 450 g/GJ. Large hydro stations embody between 500 g/GJ (for relatively light-weight arches) and 20 kg/GJ of reinforced concrete (for broad-based gravity dams).

11.1 Power Densities: Energy and Land

Perceptions, valuations, and utilization of space by preindustrial civilizations were fundamentally different from the attitudes and uses of modern fossil-fueled and highly electrified societies. An excellent illustration of this is the treatment of land as a factor of production. Classical economics, born at the beginning of industrial intensification, considered land a critical natural resource. In contrast, in modern economic thought, land has been largely ignored as the production came to be seen as a synergy of labor and capital (Slesser 1978). Rising energy inputs, diffusion of distance-shrinking prime movers, and continuing intensification of agriculture are the principal reasons of this changed emphasis. But only colonization of other celestial bodies can remove the constraints that the Earth's finite area imposes on any civilization.

The intensifying use of land is highlighted by comparing typical population densities during different stages of human evolution. Even the most accomplished foragers in the most favorable environments never surpassed 1 person/km^2 of exploited area, and most hunting and gathering societies had just a few people for every 10 km^2. Pastoralism could sustain 1–2 people/km^2 of grassland or scrubland. Shifting farming pushed the rate 1 OM higher, to tens of people per square kilometer. And traditional farming brought yet another tenfold increase: several hundred people could be supported from a square kilometer of arable fields and land under permanent crops (see fig. 6.1). By the year 2005, despite the nearly universal intensification of farming, agricultural lands occupied roughly 12% of the Earth's ice-free surface (1.54 Gha under annual species and permanent crops) (FAO 2006) and supplied nearly 90% of food energy, the rest coming from grazing, forest foods, fisheries, and aquaculture.

Cultivated land thus fed on average just over 4 people/ha; China's rate was 8.5 people/ha. This means that the global average of anthropomass (live weight of humans, assuming an average of 45 kg/person for the entire population) was almost 200 kg/ha of cultivated land, and China's mean was nearly 400 kg/ha. In China's most intensively cultivated provinces the mass of humanity approached 500 kg/ha of arable land. In contrast, the average densities of the two large African primates, chimpanzees and gorillas, are mostly less than 1 kg/ha of their now so limited and disappearing habi-

tats (Bernstein and Smith 1979; Prins and Reitsma 1989; Harcourt 1996). These comparisons demonstrate the relentless ascent of the most adaptable as well as the most destructive of all heterotrophic species, but they also make it clear that this trend, predicated on rising energy subsidies, cannot continue even if the requisite amounts of energy were available.

In order to replicate the growth of yields during the twentieth century, the global annual crop harvest would have to be boosted nearly sevenfold, an achievement that would imply average yields near or above the photosynthetic maxima even with the greatest possible expansion of the cultivated land with intensive multicropping. Only as yet unavailable genetic manipulation of photosynthesis could remove these energetic limits. The intensification of energy subsidies in farming has been behind the global rise of urbanization and its attendant high residential densities. The high energy densities of fossil fuels enabled centralized mass manufacturing, but the shift of rural labor to cities could get under way only as field machinery and fertilizers began displacing animate power. By 1800 only 3% of the global population of 1.2 billion was urban; by 1900 the share was nearly 15% of 1.7 billion people; by 1950 it was roughly 30%; and by 2005 it was just above 50% of 6.5 billion. This profound transformation—from an overwhelmingly rural, decentralized, parochial, low-energy society to a predominantly urban, centralized, globalized, high-energy culture—has run its course in rich nations but is still accelerating in modernizing countries.

Urban population densities can, of course, be orders of magnitude higher than the anthropomass per unit of cultivated land because cities increase their energy and material footprints to large multiples of their areas. Similarities exist only at the extremes. Residential densities

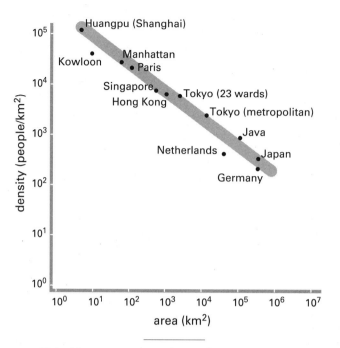

11.1 Maximum population density of countries and cities as the function of area.

in some sprawling urban districts can be lower than the farmland densities in the most intensively cultivated agricultural areas. For example, in the year 2000, West Los Angeles averaged only about 1300 people/km² compared to nearly 2000 peasants/km² of arable land in Sichuan. But human packing can reach astonishing levels (UNO 2006; Demographia 2005). Citywide means are about 13,000 people/km² for Tokyo, 16,000 for Seoul, 20,000 for Paris, and 40,000 for Manila.

Kwun Tong, Hong Kong's most densely populated district, has 50,000 people/km², and Mongkok, the heart of Hong Kong's Old Kowloon, had almost 90,000 before the area's urban renewal, about as dense as Manhattan's peak working population in 1969. In 2000 the borough's residential population was about 25,000 people/km², and its daytime density of about 55,000 people/km² is as high as that of Tokyo's four central wards. A plot of these rates indicates that every order of magnitude decrease in the area is associated with roughly tripled maximum population density (fig. 11.1). Smaller spaces can be temporarily much more crowded. The Wall Street area during working hours packs close to 250,000 people/km², and this rate is surpassed in East Asian public swimming pools or on Mediterranean beaches, where the densities exceed 500,000 people/km². Maximum permanent residential densities on the order of 50,000 people/km² translate to more than 2 kg/m² of anthropomass.

CHAPTER 11

This is a density unmatched by any other vertebrate, and it is 3 OM larger than that of large herbivorous ungulates in Africa's richest grassland ecosystems. Even more remarkably, this density surpasses even that of the combined total of all microbial (archeal, bacterial, fungal, and protozoan) biomass that is normally the dominant category of heterotrophs in natural ecosystems. Large energy subsidies for food production and transportation make the top consumer the most abundant heterotroph in the densest urban ecosystems (in terms of total biomass, not individual organisms). Densities of about 10,000 people/km^2, common in megacities, are in biomass terms 1 OM higher than the densities of invertebrate biomass in many farm soils, but they do not have any preordained effects. Crowding has no simple relation with any mechanism or response that has been measured (Cohen 1980; Freedman 1980). Unlike in animal experiments, human exposure to high density urban living rarely produces extreme social pathologies (Lepore 1994): aberrant behavior, aggression, and criminality are not at unsurpassed levels in Hong Kong or Kolkata.

Human territoriality has been elaborated to such a high degree that it is hardly recognized as such (Malmberg 1980). Obviously, the territoriality of modern urban populations has no existential energetic foundation (as all food comes from outside), but energetic reasons alone are also insufficient to explain the preindustrial quest for a defensible territory (Lopreato 1984). Even among foraging humans, food needs (in arid areas, including secure water supplies) were only one of many determinants of territoriality. In turn, needs for individual space, delimited private ground, and spatial rules for social interaction are clearly shared with many vertebrates. Privacy, identity, dignity, anonymity, diversity, status, security, and anxiety—all these powerful factors

enter the territorial imperative. Thanks to human behavioral plasticity, high residential densities do not necessarily produce an inordinate amount of social decay, and hence there is no way to argue that globally we are nearing the physical limits of urban packing. Their large-scale limits will be most likely due to intolerable metabolic consequences of extraordinarily concentrated energy conversion densities.

A revealing way to illustrate the space demands of modern energy production and use is to focus first on the disparities between the power densities of conversions that harness renewable energies and those that rely on fossil fuels. Many atmospheric, hydrospheric, and lithospheric energy conversions proceed at very high power densities: up to 10^3 W/m^2 for such spatially restricted phenomena as tornadoes, thunderstorms, and earthquakes, and about 10^2 W/m^2 for extensive latent heat releases in hurricanes and monsoons (see chapter 2). But most of these impressive natural energy flows remain completely unusable (lightning, volcanic eruptions, tsunamis, avalanches, landslides), and even the best commercial techniques can capture only small fractions of usable fluxes (see section 9.2).

In no case do the average power production densities of renewable conversions surpass 100 W/m^2. Flat plate solar heat collectors in sunny locations come close; the rates are mostly 5–15 W/m^2 for geothermal electricity, upper-course hydrogeneration (high heads, small reservoirs), and wind-powered generation; about 1 W/m^2 for most lower-course hydrogeneration (large reservoirs); and less than 1 W/m^2 for usable phytomass (see chapters 3, 6, and 10 for details on plant productivity). In contrast, extraction of fossil fuels produces coals, crude oils, and natural gases with power densities of 1–10 kW/m^2 (fig. 11.2). Modern civilization is thus energized largely

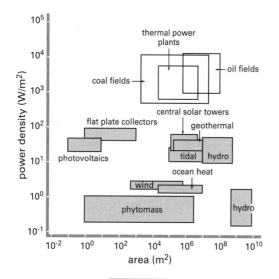

11.2 Power densities of fossil fuel extraction compared to power densities of renewable energy conversions.

its possible expansion produce a power density of only about 600 W/m². Most coal-fired power plants operate with densities of 1–3 kW/m², but proper assessments should include the entire cycle of fuel, transportation, generation, and transmission.

Results for coal-fired stations depend on the quality of the fuel, the method of its extraction, the mode and distance of its transportation, and the voltage and length of transmission lines. The adjustment may be relatively minor for mine mouth stations burning high-quality bituminous coal extracted in open mines from thick near-surface seams (followed by prompt land recultivation) and transmitting it via existing HV lines; overall power densities may be close to 2 kW_e/m². In contrast, inclusion of the land needed to extract subbituminous coal from deeper and thinner seams may halve the overall power density. And if the plant is supplied by unit trains, then the rights-of-way of a dedicated railroad may claim much more land than the plant and coal extraction, as would a new long transmission line.

This means that some coal-fired stations generate electricity with overall power densities of 100–500 W_e/m² and that for some of them the aggregate land claim would be comparable to that of solar-powered plants located in sunny regions. But the comparison shifts in favor of any thermal station once the intermittence and stochasticity of renewable flows is taken into account (hydroelectricity, with its often large storages and more predictable generation rates, is the only exception). These comparisons highlight two obvious problems inherent in any comprehensive accounts of land use for energy: qualitative differences among the areas occupied by energy infrastructures, and the duration of land claims. A cleared strip of land on a mountain slope underneath a transmission line represents an intervention incomparably

by fuels extracted with power densities 1–4 OM higher than can be delivered by renewable flows. But most of the deployed or promising renewable conversions produce electricity rather than fuel, and their comparisons with large thermal plants, particularly with coal-fired stations, show significantly lower disparities in overall power densities. Power densities of fossil-fueled electricity generation vary widely depending on the areas under consideration. Fiddler's Ferry, one of Britain's largest stations (2 GW_{ei}) provides an example of this progression. When the quotient is only the station's 70-ha core (boiler and generation buildings, cooling towers, switchyard), the density prorates to about 2.8 kW/m². Adding on-site coal storage reduces the rate to about 2.1 kW/m². Including about 165 ha destined for decades of fly ash deposits drops the density to 1.2 kW/m². And using the entire property initially reserved for the station and less intrusive than a massive coal pile or a huge sludge pond of a suburban power plant. A relatively narrow corridor of land reserved for pipeline and transmission rights-of-way sustains limited and temporary damage during construction, and afterwards all the land, except for the space occupied by towers or pumping stations, can either return to its natural state or be used for grazing, crop farming, or silviculture.

Still, even the transmission lines restrict land uses and take away farmland. For farmers, the losses of cultivable land are more than twice as large for transmission towers sitting in field headlands than for those on fence rows, and up to four times as large in row crop cultivation than in hayfields. In contrast, the land occupied by extraction facilities, buildings, and storage may never return to its original use, and in the case of nuclear waste disposal sites its occupancy may be indefinite. Land claimed by surface coal mining illustrates the range of responses. It can be completely and rapidly reclaimed as pasture, cropland, forest, or a water reservoir, or left in a derelict state for decades. Coal's land claims are further complicated by the fact most unit trains share the railway rights-of-way with other freight.

Nuclear power plants can be more compact than coal-fired stations. Pressurized water reactors fission the enriched uranium with densities of up to 300 MW/m^2 of the core's footprint, and the fenced plant sites, including cooling towers and on-site storage of radioactive wastes, rate up to 4 kW/m^2. The need for substantial low-population zones around the plants clearly limits the type of land uses in their immediate vicinity while not physically claiming any of the affected land. Moreover, it remains uncertain what areas will be ultimately claimed by long-term depositories of radioactive wastes. Although they will be located in uninhabited and difficult-

to-access regions, the unending disputes regarding the U.S. depository at Yucca Mountain in Nevada show that even in such cases there may be protracted land use conflicts (Flynn 1995).

All of these uncertainties, and even more so the considerable ranges of actual power densities for every major category of harnessing and converting commercial energies, mean that any proffered global total of the land claimed by the energy conversion and distribution can only be a very rough quantitative estimate hiding many qualitative differences. My calculations (based on liberal assumptions) result in a maximum of 300,000 km^2 during the early 2000s, an Italy-sized area, or roughly 0.25% of the Earth's ice-free land. But less than 10% of it (an area smaller than Belgium) is taken up by extraction and processing of fuels and by thermal electricity generation, and more than half is the land occupied by water reservoirs (fig. 11.3). Many of these have multiple uses (irrigation, industrial and urban supply, recreation), and a partial attribution of these large areas to electricity generation would reduce the overall claim to less than 200,000 km^2. Leaving the reservoirs and transmission rights-of-way aside, the global fossil-fueled system with throughput of about 11.5 TW and overall land claims of no more than 75,000 km^2 would average about 150 W/m^2.

A much more accurate, though still far from precise, assay can be made for the U.S. fossil-fueled energy system of the early 2000s, leaving aside not only nonfossil electricity generation but also the rights-of-way of railroads that are shared by coal trains. Despite its inevitable deficiencies this calculation is quite revealing. Extraction of all fossil fuels claims less than 500 km^2, their processing (including the refining of imported crude oil) needs about four times as much land, and rights-of-way for oil

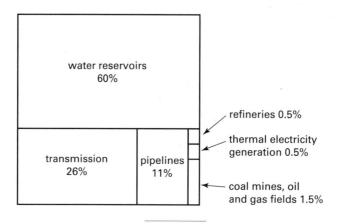

11.3 Approximate areas occupied by the global infra-structures of fuel production and transportation and by the land devoted to electricity generation and transmission.

and gas pipelines added up to less than 12,000 km². The entire primary fuel system thus occupied about 14,000 km², and with an annual throughput of roughly 2.7 TW, it operated with a power density of roughly 200 W/m². Fossil-fueled electricity generation occupied roughly as much land as fossil fuel extraction (less than 500 km²), but the land reserved for rights-of-way of HV transmission was of the same magnitude as the pipeline claims: 250,000 km of HV lines of 230+ kV and an average clearance of 35 m add up to almost 9000 km². In addition, there are also some outstanding land debts, most notably nearly 4000 km² of unreclaimed coal mine land.

The grand total of land claimed by the U.S. fossil-fueled energy system thus amounted to some 25,000 km² during the early 2000s. Net annual additions to this total are harder to estimate. Many new transmission links and pipelines use existing corridors, and most thermal power plants and refineries are expanded within their fenced areas; damage caused by exploratory drilling is usually temporary and recovery is often rapid; land losses caused by surface mining have been increasingly balanced by mandatory reclamation. As a result, net additions during the early 2000s have almost certainly been less than 1000 km² a year. Aggregate land claims of about 25,000 km² were equal to merely 0.25% of the country's area (an area the size of Vermont).

Power densities of final energy uses reach their peaks of over 10 MW/m² in large power plant boilers and in the largest blast furnaces, and a reliable, continuous fuel supply is paramount for the operation of these giant fuel converters that provide key economic inputs. Typical consumption power densities are 20–100 W/m² for houses and for low-density or low-energy-intensity manufacturing, offices, and institutional buildings. Supermarkets, multistoried factories, and office buildings use 200–400 W/m². Urban areas represent a mixture of relatively high power density use in high-rise buildings, factories, and along busy transportation corridors; medium to low power density use in suburban residential areas; and extremely low power density use in parks and other green zones. Typical urban power densities thus range from less than 10 W/m^2 in uncrowded, low-income cities to around 100 W/m^2 in affluent, crowded megacities.

Elvidge (2004) conducted a study of U.S. impervious surface area (ISA), the first of its kind. (He used a 1-km grid for the coterminous states and information from night-time light radiance and LANDSAT-derived land cover values). The study found that the total U.S. ISA, including all highways, city streets, parking lots, buildings, and other solid structures, amounted to nearly 113,000 km^2, or slightly less than area of Ohio. Permanent (roofed or paved) energy infrastructures are, of course, a part of this total, but they accounted for a very small fraction of it (less than 5%). More than 90% of the country's energy use takes place within the ISA; only trains, airplanes, ships, agricultural machinery, and off-road vehicles operate outside of it. This means that in the early 2000s, for every 5 m^2 of ISA, about 1 m^2 of land was taken up elsewhere by extraction, conversion, transportation, and transmission facilities of the U.S. fossil-fueled energy system. Taking into account much lower per capita energy use in other high-income countries (except Canada and Australia), one could thus generalize that affluent high-energy nations in temperate zones need land equivalent to 10%–20% of their ISAs for the infrastructures of their fossil-fueled energy systems.

11.2 Energy Conversions and Heat Rejection

Heat is the only inevitable product of any metabolism, but there are fundamental differences between heat that is radiated by living organisms and heat rejected by modern energy conversions. Heat rejection by autotrophs and by ectothermic heterotrophs is merely a matter of slightly delayed reradiation of absorbed solar flux. There is no additional thermal burden for the biosphere, no climatic consequences. Similarly, combustion of phytomass is only a slightly delayed (and often highly concentrated) return of solar energy that was converted into new chemical bonds months (for crop residues) to tens of years (for woody tissues) before its harvest and use. Thermoregulation is a taxing matter for many heterotrophs (see section 4.2), but heat released by endothermic bodies is only a delayed return of solar radiation resulting from the metabolism of digested food.

Heat generated by combustion of fossil fuels is in a different class of delays because up to 10^8 years may have elapsed before the conversion of solar radiation into chemical bonds and their severance and reconstitution during combustion. As shown in figure 11.4, the power densities of these anthropogenic heat releases range over 10 OM (from 10^{-3} W/m^2 for heat flux prorated over national territories to 10^7 W/m^2 of gas turbine nozzles), with power ratings of individual phenomena spanning 8 OM and their areas extending over 16 OM. Some of these power densities are high enough to disrupt local heat balances, and even regional effects may be discernible. Worldwide combustion of fossil fuels remains a negligible fraction of the Earth's thermal balance, and it causes a flux exactly 2 OM smaller than do anthropogenic greenhouse gases.

On the planetary level, primary commercial energy consumption in the year 2000 (380 EJ) prorated to a mere 0.025 W/m^2 of the Earth's surface. This heat is overwhelmingly dissipated over the continents, so a more meaningful terrestrial average was 0.081 W/m^2,

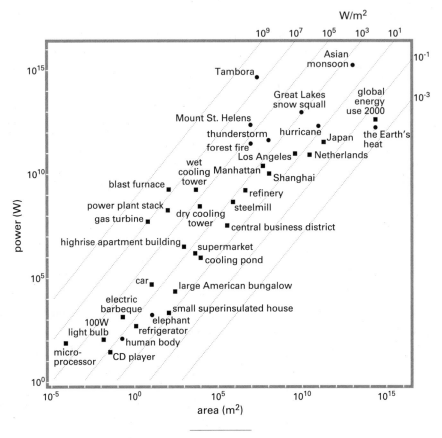

11.4 Power densities of anthropogenic and natural
heat releases.

equal to 0.04% of the global mean insolation absorbed by the continents (180 W/m²), a negligible increment whose doubling or tripling would make no discernible global difference. In contrast, by the year 2000 the cumulative effect of global emissions of greenhouse gases had already burdened the atmosphere with an additional 2.4 W/m². Speculations about future anthropogenic heat rejection depend on the dominant modes of primary energy supply. Inexpensive photovaltaic electricity would

be (thermally speaking) a mostly invariable arrangement redistributing a small share of insolation; in contrast, an affluent civilization energized mostly by nuclear fission could become a much more prodigious producer of waste heat.

The average anthropogenic heat rejection rates for such affluent, densely populated countries as Japan (1.8 W/m²) or the Netherlands (2.1 W/m²) are 2 OM above the global mean, and the waste heat fluxes keep ascending from these low rates of theoretical interest to high rates of practical engineering and environmental concern. Assuming that during the early 2000s about 90% of U.S. energy (~2.7 TW) was dissipated over impervious surfaces, whose area was roughly 113,000 km^2 (see section 11.1), the average power density would have been roughly 25 W/m^2. Clearly, densities calculated specifically for urban areas, highways, and downtowns must be of the same order of magnitude, and those of the busiest expressways and central business districts (CBDs) densely built up with skyscrapers will go 1 OM higher (see section 9.3).

Residential areas in North America rate mostly at 15–25 W/m^2; less dense CBDs may have densities of about 50 W/m^2; cores of high-rise CBDs in cold climates may easily surpass 500 W/m^2 in winter; the power densities of the busiest multilane highways may be as high as 300 W/m^2 in traffic jams; and individual skyscrapers will dissipate more than 1 kW/m^2 of their foundations. Crude oil refineries dissipate 5%–10% of energy in the processed oil as heat; with throughput densities of 1–7 kW/m^2, this translates to heat losses of 100–700 W/m^2. Heat rejection by steel mills is commonly in the same range, and high-rise buildings convert up to 3 kW/m^2 of electricity into heat (fig. 11.4). All these instances are heat dissipations by default, but there are also many carefully engineered heat rejection devices whose thermal power densities reach very high levels. Giant cooling towers and tall stacks are the most prominent objects in this category.

Cooling towers dissipate typically half of all energy consumed by thermal power plants, most of it (>70%) as latent heat. This means that large concrete-shell natural draft units designed to handle 20–40 kW$_{ei}$/m^2 (see section 8.5) will be actually rejecting (with electricity generation efficiency between 33%–40%) as much as 60 kW$_t$/m^2 of their foundations, and mechanical draft towers will dispose of 100–125 kW$_t$/m^2. For common station sizes (500 MW$_{ei}$–2 GW$_{ei}$), this means that one to four cooling towers serving a power plant will have an aggregate flux of 700 MW$_t$–3 GW$_t$. The second largest heat loss in large power plants is in the form of hot combustion gases that have not transferred their energy to water-filled tubes inside a boiler. This flux amounts typically to about 10% of energy content of the burned fuel. Part of that heat is recovered by a stack economizer (to preheat feedwater for the boiler) and by an air heater (to preheat combustion air), but until the 1950s most plants vented the rest through increasingly tall stacks (>200 m–300 m).

Because their top inside diameters are just 3–7 m (compared to 30–40 m for large natural draft towers), such stacks reject heat with densities of up to 3–5 MW$_t$/m^2 of their mouths and about 1 MW$_t$/m^2 of their foundations. But these discharges have changed with the increasingly frequent use of flue gas desulfurization (FGD) (see section 8.5). Flue gases arrive at an FGD unit at 120°C–150°C, are cooled to saturation temperature before getting stripped of SO$_2$ by reaction with alkaline compounds, and leave the stack at only 45°C–50°C. As a result, only 1%–2% of the consumed fuel can leave through a stack, or about 100 kW$_t$/m^2 of chimney foundations. Even the most efficient household natural gas furnaces now have heat loss power densities higher than FGD-equipped thermal power plants. These furnaces need no chimney as they exhaust warm CO$_2$-laden air that contains only 4%–6% of the input energy through a short plastic pipe protruding from a wall. Assuming negligible heat loss along the short pipe, the flux is about 150 W/cm^2 (1.5 MW/m^2).

ENVIRONMENTAL CONSEQUENCES

Internal combustion engines reject heat with higher power densities because they produce much hotter combustion gases. The firing temperatures of the most efficient (~40%) gas turbines are in excess of 1300°C, and the gases leave the machine usually at temperatures well above 500°C. In stationary industrial turbines this heat can be used to generate steam for steam turbines, achieving combined cycle efficiency of up to 60% and reducing the heat flux through a stack to less than 20 W/m². No such combination is possible for gas turbines in flight, and hence large turbofan jet engines like the GE90 (cruising thrust 70 kN, specific cruising fuel consumption 15.6 mg/Ns) directly reject about 25 MW/m² of hot nozzle area (1 m²) at the design speed of 0.85 M (GE Aviation 2006).

Even the most efficient automotive Otto cycle engine loses about 30% of its initial fuel input through the radiator and about 40% in exhaust gases. If all this gas-borne heat were rejected through a tailpipe, even a compact vehicle like the Honda Civic (engine rated at 85 kW, actual driving fuel consumption ~50 kW at 80–100 km) would have thermal flux close to 1 kW/cm² (10 MW/m²) of its exhaust pipe opening. In reality, most of that heat is lost by radiation before it reaches the tailpipe. Exhaust gases leave the engine at about 800°C, at a rate of 50 g/s and speed of 60 m/s, but exit the tailpipe at only about 65°C, indicating rapid heat loss along the exhaust piping. Because a large part of both exhaust and radiator heat is absorbed by other car surfaces before it is finally rejected into the atmosphere, and because the directly rejected heat is commingled with heat generated by driving, it makes sense to divide a vehicle's power by its footprint (typically 7–9 m²). At full capacity this flux is about 11 kW$_t$/m² for a Honda Civic and twice as much for a

Mercedes 600. In terms of heat rejection, cars are thus equivalent to miniature movable power plant cooling towers.

But cars are not the most ubiquitous high power density heat emitters; that primacy now belongs to microprocessors. The first microprocessor, the Intel 4004, released in 1971, had 2,300 transistors on a 135-mm² die and dissipated about 2.5 W/cm². In 1978 the Intel 8086 had 29,000 transistors and dissipated 7.6 W/cm², a rate equal to that of a kitchen hot plate (1.2 kW in a 160-cm² circle). By 2001 ultralarge-scale integration of Intel's Xeon Irwindale placed 50 million transistors on a die of 130 mm² and consumed 115–130 W; its heat rejection rate reached 100 W/cm². The Itanium and Apple's G4 have rates of about 110 W/cm², or 1.1 MW/m² (fig. 11.5) (Azar 2000; Viswanath et al. 2000; Joshi 2001). These power densities are of the same magnitude as heat rising from large power plant stacks and

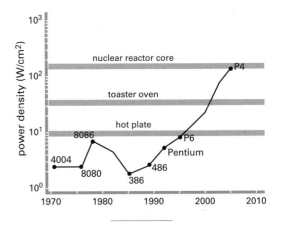

11.5 Heat rejection rates of Intel microprocessors, 1971–2005. Plotted from data in Azar (2000), Viswanath et al. (2000), and Joshi (2001).

higher than the convective heating experienced by the space shuttle's nose during reentry into the Earth's atmosphere (Laub and Venkatapathy 2003).

Heat dissipation in microprocessors takes place in very tightly confined areas where even relatively modest temperature increases can slow down and eventually shut down the electronic transfer process (temperature should be kept below 45°C for optimal functioning). Future microchips will integrate billions of transistors and may draw more than three times as much power as today's top designs, and their heat rejection rates of 10^6–10^7 W/m^2 would be equal to 10%–30% of the flux through the Sun's photosphere (64 MW/m^2). Low-power designs will help, but it is clear that standard fan-cooled heat sinks (with heat conducted into metal fins cooled by relatively bulky and noisy fans) have reached the limit of their practical performance and that new cooling designs are needed to cope with higher heat fluxes.

Cooling the microelectronic circuitry is an extraordinary engineering challenge, but the total power involved per processor is too small to have any environmental effects. Similarly, other concentrated heat rejection processes, ranging from hot plates and car exhausts to tall power plant stacks, are limited to relatively small areas. As the power density of anthropogenic energy conversions increases, the spatial extent of the more intensive heat rejection fluxes declines at a considerably faster rate. As a result, heat rejection phenomena that constitute a significant share of solar inputs (10^1 W/m^2) are limited to areas no larger than 10^8 m^2, to large cities, extensive industrial regions, and busy transportation corridors. Those heat rejection rates (10^4–10^5 W/m^2) that greatly surpass average insolation rates are restricted to areas smaller than 10^4 m^2 (100 × 100 m), and the highest anthropogenic heat fluxes (10^6–10^7 W/m^2) are limited to areas no larger than 10^2 m^2 (tall stacks) and as small as 10^{-4} m^2 (microprocessors).

Excessive heat degrades the performance of microcircuits, but the flux is negligible against the overall thermal background. Even many extensive high-density heat releases have no serious environmental impacts. Cooling towers and tall stacks generate considerable clouds and some local fogging and icing, but only infrequently do they create precipitation anomalies. But persistent heat rejection by large cities is a major factor in creating clearly discernible urban heat islands (UHI). The other reasons for this phenomenon are higher thermal capacity and lower albedo of built surfaces, a smaller sky view that hinders radiative cooling, and higher Bowen ratios due to more of the sensible and less of the latent heat flux (Taha 2004). UHIs are on average about 2°C warmer, and their cores may temporarily have temperatures up to 8°C higher, than the surrounding countryside (Taha 2004). UHIs are most readily identifiable at night when their effect on air temperature may be 1 OM greater than during the day.

Daytime heating of paved and built-up areas generate much higher surface temperatures than the heating of vegetated countryside (this difference is much smaller at night), but it will also engender much stronger urban convective flows, that is, more vigorous mixing within the atmospheric boundary layer, which minimize daytime temperature differences between city cores and the surrounding countryside (Camilloni and Barros 1997). Peterson (2003) did not find any statistically significant impact of urbanization on annual temperatures for the contiguous United States. His subsequent analysis showed that leaving 30% of the highest population stations out of the data set resulted in no statistically significant UHI impact, and that even the entire set could explain only a tiny fraction (0.048°C) of overall warming per century (Peterson and Owen 2005). Similarly, Parker (2004) concluded that UHI warming has not biased the best estimates of global warming. UHI thus has negligible large-scale climatic effects, but it has undeniable local impacts.

Well-documented effects include increased cloudiness (up to 8% more), precipitation (up to 14% more), and thunderstorms (up to 15% more) near and particularly downwind of UHI-affected areas as well as decreased relative humidity (2%–8% lower), wind speed (20%–30% lower annual mean), and insolation on horizontal surfaces (15%–20% lower due to structural shading). UHI boosts emissions of both biogenic hydrocarbons (tree isoprene) and fossil hydrocarbons stored in tanks, accelerates temperature-dependent photochemical smog reactions, and aggravates summer heat waves. In turn, excessive heat raises the frequency of premature deaths (due to heat strokes and dehydration) and reduces night-time cooling, thus making nights less tolerable for everybody without air conditioning.

The most readily quantifiable effect of urban heat islands is the increased used of electricity for additional air-conditioning (and decreased winter heating). Regression of temperature and power load data for Los Angeles showed that on warm afternoons electricity demand rises nearly 3% for every 1°C rise in the daily maximum, and the probability of smog increases much more steeply, by 5% for every 0.25°C rise (Heat Island Group 2000a). UHI increases the number of cooling degree-days by 15%–35% in large U.S. cities and by as much as 90% in Los Angeles (Taha 2004). A welcome effect has been the extension of the growing season as far as 10 km beyond the city's edges. Satellite-measured surface reflectance of seasonal changes in plant growth around eastern

North American cities with areas larger than 10 km² found that growing seasons were 15 days longer and that every 1°C of additional urban warmth during the early spring brought the first blooms three days earlier (Zhang et al. 2004).

Tree planting is the most efficient and esthetically pleasing way of reducing UHI intensity. Urban parks are 0.5°C–2.5°C cooler than their surroundings. Lighter pavements and cool roofs are also effective. Traditional roofing surfaces have albedos ranging from 5% (asphalt roofs of commercial buildings, dark house shingles) to 20% (green shingles), and more reflective light shingles with albedos of 35%–40% can reduce the temperature differences between roof and surrounding air by about 10°C (Heat Island Group 2000b). Konopacki et al. (1997) calculated that the universal adoption of light-colored roofing materials would reduce U.S. air-conditioning demand by about 10 TWh.

11.3 Energy and Water

The presence of large volumes of liquid water is one of the key material preconditions of the Earth's biosphere, and its peculiar properties have inestimable consequences for the planet's energetics (M. W. Denny 1993; see section 2.3). If water behaved like other fluids of a similar molecular weight, it would boil at −91°C and freeze at −100°C, and its liquid range would be far too much below the optima (30°C–40°C) for enzymatic energy conversions in plants and animals. Alcohols nearly match water's extraordinary departure from the general rule of decreasing boiling and freezing points with decreasing molecular weight, but they have a much lower heat capacity. Water's large specific heat (at 4.186 J/g·°C 67% higher than that of ethanol) has played two critical evolutionary roles: first, helping to maintain a relatively constant ocean temperature and hence damping the extent of global climatic fluctuations; second, greatly contributing to heat regulation of all living organisms and especially to the maintenance of homeothermy in higher animals.

Yet, despite such strong molecular coherence, water has retained high mobility. Its low viscosity, merely 1 mPa · s, means that living organisms need not spend much additional energy in pumping their body fluids and that waterborne transportation has the lowest energy cost (measured as J/kg · m). And water's high dielectric constant (78.5 at 25°C) makes for easy transport of nutrients in animate bodies (as ions remain separate) and releases more of the metabolized energy for growth and reproduction. Biospheric energetics would be very different if biota contained much less of this unusual compound. But there is a price to pay: because primary production is predicated on a highly uneven trade-off of CO_2 and H_2O (section 3.1), water needs for photosynthesis are very high. Even the most efficient C_4 species needs 20–30 t/GJ, and many C_3 plants require over 100 t of water to produce 1 GJ of phytomass. Since edible phytomass is usually no more than half of the synthesized total, the rates for plant foods are mostly 60–200 t/GJ.

As an order-of-magnitude estimate, an average of 150 t/GJ for the annual global harvest of some 40 EJ of food and feed results in worldwide agricultural consumption of at least 6000 km^3 of water. The best available estimates indicate that about two-fifths of this need are supplied by irrigation. Most of this use belongs to the consumptive category because water is lost through evapotranspiration and carried to markets with harvested and increasingly exported crops. In contrast, the largest category of water use by energy sector is in a non-consumptive process, for condensing steam in thermal electricity-generating plants. Traditional once-through cooling, whereby surface water takes a single pass through a condenser and is returned to streams, lakes, or bays, has minimum consumptive (evaporative) losses but clearly needs access to continuously large volumes of water and this will affect its quality.

About 29 t/GJ of electricity are needed with a 10°C rise, 58 t/GJ with the maximum 5°C rise. Thermal power plants in the United States relied almost solely on once-through cooling until the late 1960s. By that time about 5% of the country's total annual freshwater runoff passed through power plant cooling systems, and the practice began to cause considerable concern about the effects of thermal pollution on aquatic life. The increased temperature of discharged water may be beneficial to some aquatic species, but its deleterious effects include lower levels of dissolved O_2 and changed species composition due to reduction or demise of heat-intolerant organisms; many microorganisms (phyto- and zooplankton), eggs, larvae, and small fish are also sucked into intake water pipes or crushed against intake filters.

Amendments to the U.S. Federal Pollution Act of 1972 began a rapid shift to closed-loop cooling systems, and large wet cooling towers, common in Europe for several generations, offered an effective solution (fig. 11.6). In some arid parts of the world (beginning in South Africa) utilities began installing dry cooling towers that lower water temperature in large heat exchangers without evaporation. Cooling towers have higher consumptive water use compared to once-through arrangements. This is due to the necessity to replenish water lost to evaporation, drift (droplets entrained in the moving air), and bleeding (releasing some water in order to limit the dissolved solids). But this makeup water adds

11.6　Large European concrete wet cooling tower.

up to only about 0.03% of the throughput. The difference between the once-through and closed-loop systems is well illustrated by the data gathered every five years in USGS water use surveys.

The 2004 survey shows that in the year 2000 plants with once-through cooling (accounting for about 40% of installed thermal generation capacity) needed nearly 180 L/kWh, whereas plants with closed-loop (or air-cooled) systems withdrew only about 12 L/kWh (USGS 2004). Depending on the power plant efficiency, tower design, and environmental conditions, water losses in wet cooling towers are only 1.5–2.6 L/kWh of generated electricity. FBC plants and CC generation can lower the rate below 1 L/kWh, and in dry cooling towers water consumption is less than 0.1 L/kWh. In aggregate, U.S. thermal electricity generation has been the nation's largest user of water since 1965; it withdrew

about 270 Gm^3 in the year 2000 (91% of this for once-through cooling). This was about 48% of all withdrawals and 52% of fresh surface water use. Irrigation was the second largest user, drawing some 30% of fresh surface water. But because consumptive uses account for at least 40%–50% in irrigation (and for much more with drip irrigation) and for less than 3% of water drawn by thermoelectric plants, the actual claim of electricity generation was less than 10 Gm^3, an equivalent of about 20% of the consumptive use drawn by public waterworks.

There is insufficient information to calculate an accurate global weighted average of once-through and closed-loop cooling needs. The U.S. mean is about 80 L/kWh, but the country has an exceptionally large share of once-through cooling, and the global mean is most likely less than 40 L/kWh. Thus worldwide water withdrawals for thermoelectric plants were no higher than 500 km^3 in the year 2000, and even if consumptive use were to amount to 5% of that total, it would be no more than 25 km^3. Hydroelectric generation itself is another massive nonconsumptive use of water, but reservoirs needed for its smooth functioning lose water by evaporation, and in tropical regions, where large parts of the reservoir surfaces are covered by aquatic plants, the losses are even higher by enhanced evapotranspiration. These losses average 17 L/kWh for U.S. hydroelectricity generation and as much as 26 L/kWh in California (Gleick 1993). A liberal global mean of 20 L/kWh would translate to an annual loss of less than 60 km^3.

Most water claims in coal mining fall are 2–5 L/GJ for surface extraction and 20 L/GJ for deep mining. Oil extraction needs 5–10 L/GJ. Enhanced oil recovery in Alberta oil fields has required less than 30 L/GJ (Peachey 2005), but elsewhere the rates have been several times higher (100–200 L/GJ) and can be as much as 500–

600 L/GJ for steam injection. Similarly, natural gas extraction can consume up to 300 L/GJ. Nearly all this water can be recycled, but a higher water to oil ratio (in Alberta now >10 compared to <1 in 1970) requires more energy for pumping. Coal cleaning needs 20–50 L/GJ, oil refining at least twice as much and up to 200 L/GJ. Even the just cited upper rates would result in global withdrawals well below 100 km^3, and consumptive uses would be less than 10 km^3. The grand total of water withdrawn in 2005 by all energy-producing industries would then be less than 600 km^3, and the consumptive use would be at most about 100 km^3, more likely 60–70 km^3. In contrast, global water withdrawal amounted to about 4000 km^3 in 2005, with agriculture claiming some 70% (WRI 2005).

Energy industries thus claimed less than 15% of all water withdrawals; their share of consumptive uses cannot be pinpointed, but it was definitely less than 5% of all water consumed by agriculture. Lack of reliable information precludes calculating many global totals, notably how much energy is needed to treat the world's drinking water, pump it, and transport it by pipelines. The energy cost of drinking water varies widely: some clean supplies flow to cities by gravity and need little besides ordinary chlorination; other flows require extensive treatment and long conveyance. Somewhat smaller but still large ranges apply to wastewater treatment. Figures available for California illustrate these spreads, with the energy cost of water treatment ranging from essentially zero to 4 kWh/m^3 and wastewater treatment costing 0.3–1.3 kWh/m^3 (CEC 2005).

National and regional disparities are similarly large. The total energy cost of water supply and conveyance is more than three times higher in southern California (about 3.4 kWh/m^3) than in the northern part of the state. But even then these costs are less than the energy requirements for desalination. Desalination, most common in the Middle East (it supplies 70% of Saudi water), is highly energy-intensive. The theoretical minimum needed to desalinate normal seawater (3.45% salts) at 25°C is 0.86 kWh/m^3, but actual processes consume 1 OM more. During the 1990s multistage flash distillation (the leading process in volume terms) needed 12–24 kWh/m^3 and reverse osmosis required 5–7 kWh/m^3, but costs below 4 kWh/m^2 should be possible in the near future (*Encyclopedia of Desalination* 2006).

Energy industries are also the source of various forms of water pollution, ranging from chronic acid mine drainage from coal mines (due to oxidation of exposed FeS_2) to recurrent catastrophic spills of crude oil from tankers. Most of these releases have only localized impacts, and small spills are subject to evaporation, emulsification, sinking, auto-oxidation, and above all, microbial oxidation. These natural controls are temporarily overwhelmed only by very large oil spills following tanker accidents or offshore well blowups. The worst accidents of the last quarter of the twentieth century were the *Atlantic Express* (in 1979, off Tobago, 287,000 t) and *ABT Summer* (1991, 1000 km off Angola, 260,000 t).

Because they took place far offshore, they received much less attention than the third and fourth record spills: *Castillo de Bellver* (1981, off Saldanha Bay, South Africa, 253,000 t) and *Amoco Cadiz* (1978, Brittany beaches, 223,000 t of light crude) (ITOPF 2005). The highly publicized *Exxon Valdez* spill in Prince William Sound, Alaska, in 1989 was 37,000 t (not even among the largest 30 since the 1960s), but it killed as many as 270,000 water birds (Piper 1993). In contrast, the Mexican IXTOC 1 well in Bahia de Campeche spilled as much as 1.4 Mt in 1979–1980. Besides polluting beaches, oil spills contaminate zooplankton and benthic invertebrates (fish to a lesser extent) persist in anoxic sediments, and reduce the abundance and diversity of benthic communities.

11.4   Energy and the Atmosphere

Human actions can have no global or lasting effect on dinitrogen (N_2), the atmosphere's dominant gas. Any outflows from this immense reservoir, whether oxides produced by combustion or NH_3 synthesized by the Haber-Bosch process, will be balanced by denitrification. Nor can fossil fuel combustion reduce atmospheric O_2 to alarmingly low levels. As already noted (see section 8.3), complete combustion of 1 t C requires 2.67 t of O_2, burning 1 t CH_4 needs 4 t of O_2, and the average for 1 t of refined liquid fuels is about 3.5 t O_2. Oxidation of fuel sulfur (and sometimes a small amount of fuel N) and production of NO_x consume negligible amounts of O_2 compared to the generation of CO_2. After the subtraction of unburned fossil fuels that are used as lubricants, paving materials, and feedstocks, the global combustion of fossil fuels consumed about 25 Gt of O_2 in the year 2000, and during the twentieth century cumulative demand was at least 900 Gt of O_2.

Oxygen's atmospheric mass amounts to 1.2 Pt, and hence 25 Gt of O_2 is just 0.002%, and 900 Gt of O_2 just 0.075%, of the element's atmospheric presence. Much larger fluctuations of O_2 took place in the distant past. Modeling of the carbon and sulfur cycles indicates that oxygen's atmospheric concentrations may have varied between 15% and 30% over the past 550 Ma (Berner 1999). Elevated O_2 levels, peaking during the late Carboniferous period, could have been responsible for gigantism in insects (dragonflies with 70-cm wing spans), but there have been no appreciable changes of at-

mospheric O_2 during the course of human evolution. Oxygen is replenished by photosynthesis (mining of phosphates and their use as fertilizer is a minor source) and consumed by oxidation of organic and inorganic compounds, but its levels remain steady and globally uniform. Even complete combustion of all reserves of fossil fuels would reduce atmospheric O_2 by less than 0.3%, and the (theoretically possible) recovery of all liberally estimated fossil fuel resources would lower it by no more than 2%. In any case, levels of atmospheric oxygen have shown no significant shift from 20.95% by volume since the beginning of the twentieth century.

What has changed quite significantly are the atmospheric concentrations of particulate matter (PM; solid or liquid aerosols with diameter <500 μm) and trace gases. PM is emitted as fly ash and black carbon (soot) from coal and refined oil combustion, and the burning of biomass fuels is yet another, regionally large, source of aerosols (mainly organic carbon in smoke). Gaseous emissions from fossil fuels are dominated by SO_x (emission factors of 0.5–2 kg/GJ in coal combustion, 0.2–1 kg/GJ in oil burning), NO_x (both from fuel and from atmospheric N made available for oxidation by high-temperature breakdown of N_2; emission factors 0.1–1 kg/GJ) and volatile organic compounds (VOCs). SO_x, NO_x, and VOCs also participate in complex photochemical reactions whose products include both aerosols and highly reactive gases (O_3 and peroxyacetyl nitrate).

Particulate matter is by far the most abundant global air pollutant, and combustion of coal is a leading source of anthropogenic particulates. Their irritating effects were noted for the first time in densely inhabited medieval London (Brimblecombe 1987), and during the nineteenth century these emissions reached objectionable levels in just about every major city and industrial region of the Western world. Incombustibles in coal produce flue gas–borne fly ash (powdery, mostly spherical, specific gravity 2.1–3.0, emission rates 2.5–5.0 kg/GJ, dominant in dry-bottom boilers); bottom ash (about half of the incombustibles in wet-bottom boilers); and boiler slag (dominant in cyclone furnaces). In 2000 the U.S. annual production of these solid wastes reached, respectively, about 57 Mt, 14.4 Mt, and 2.4 Mt (Kalyoncu 2000). Worldwide production of fly ash was about 600 Mt in 2000 (Malhotra 1999), and Europe and Asia generated about 9 Mt of FGD waste.

SiO_2, Al_2O_3, Fe_2O_3, and oxides of alkaline elements (CaO, MgO, K_2O) are the principal constituents of fly ash. Trace amounts of Pb, Zn, Cr, Mn, and Ni are also present. Some coals have relatively high levels of Hg. U.S. coal-fired stations are the country's largest source of Hg emissions (about 40%), whose main environmental consequence is the accumulation of highly toxic methylmercury in fish and shellfish; 55% of fish samples collected between 1999 and 2001 had an Hg level exceeding EPA's safe limit for women (USPIRG 2004). Most coals also contain U (about 1 ppm) and Th (about 3 ppm), and their decay isotopes (Ra, Po, Bi, Pb). As a result, populations living near coal-fired plants receive higher radiation doses than those in the vicinity of nuclear stations (McBride et al. 1977). A typical U.S. 1-GW coal-fired plant releases annually about 5 t U and more than 10 t Th. Widespread use of fabric bag filters and electrostatic precipitators has eliminated visible PM in affluent countries. Precipitators capture in excess of 99.5% of PM, and their use reduced global fly ash emissions to about 30 Mt/a by 2000. Captured ash is used in concrete mixes and structural fills, as feed for cement clinker, and in pavings.

Until the late 1960s there were no commercial controls for SO_2 emissions from large-scale coal combustion. The United States was the first country to adopt FGD on a large scale. By 2005 about 102 GW_{ei} were so equipped, which is about 32% of the country's coal-fired generating capacity (EIA 2006). Germany was second, with FGD at about 50 GW_{ei} (Rubin et al. 2003). FGD processes remove flue gas sulfur (with efficiencies up to 90%) by reactions of SO_2 with basic compounds (CaO, MgO, $CaCO_3$) that produce a mixture of $CaSO_4$ (or $MgSO_4$), $CaSO_3$, fly ash, and unreacted lime or limestone. The conversion of all sulfite to sulfate and the reduction of impurities can produce gypsum suitable for wallboard manufacture. Nearly 15% of U.S. FGD waste was reused in this way in the early 2000s, but some 80% was simply landfilled, converting an air pollution problem into a land and water degradation challenge (Kalyoncu 2000).

The first significant efforts to control NO_x emissions from large stationary sources began only during the 1980s. Germany has been the leader, with controls on some 30 GW_{ei}, followed by Japan, but the global aggregate for the dominant technique, selective catalytic reduction (SCR), was just 80 GW_{ei} by the year 2000 (Rubin et al. 2003). SCR is a very expensive technique that removes 60%–90% NO_x by using Ti, V, and W oxides as catalyzers to reduce the gases by NH_3. Other control options reduce NO_x emissions less expensively but less effectively by various combustion modifications (low-excess-air firing, staged combustion, flue gas recirculation, use of low-NO_x burners). In contrast, reductions of vehicular NO_x have been much more widespread and quite effective. Three-way catalytic converters, mandatory on all passenger cars, have reduced

ENVIRONMENTAL CONSEQUENCES

NO_x emissions by 95% compared to precontrol (U.S., pre-1971) levels. They also reduced CO emissions by 96% and hydrocarbons by 99% (MVMA 2000). Atmospheric oxidation converts the emitted SO_x and NO_x rather rapidly into sulfates and nitrates whose small size (0.1–1 μ) is responsible for most of the light scattering.

There are no accurate estimates of global anthropogenic aerosol fluxes, but leaving aside dust generated by field cultivation and enhanced soil erosion, it appears that sulfates are the largest source ($\sim$150 Mt SO_4/a) followed by nitrates ($\sim$100 Mt NO_3/a), organic carbon in smoke from biomass combustion ($\sim$50 Mt/a), and fly ash from coal-fired boilers (30 Mt/a). Annual global emissions of black carbon were put at only 14 Mt, split between fossil fuel and biomass combustion (Ramanathan et al. 2001). Natural sources of aerosols, also difficult to quantify, mainly because of the total flux of desert dust, are undoubtedly much larger than anthropogenic emissions, but because they are composed of larger and hence fewer particles, the optical depths and hence global atmospheric effects of the two kinds of particulates are very similar.

Anthropogenic aerosols can have both negative (cooling) and positive (warming) temperature forcing. Those with diameters above 50 nm (sulfates in particular) are excellent condensation nuclei, and their presence increases cloud drop density (and hence cloud albedo) and produces surface cooling. In contrast, black carbon absorbs the incoming radiation and contributes to global warming, and its effect may be disproportionate to its emitted mass because it alters regional atmospheric stability, the water cycle, and climate. The leading emitters are China, mostly from inefficient coal combustion, and India, where biomass and coal combustion contribute roughly equally (Venkataraman et al. 2005). Large-scale effects of anthropogenic aerosols were noted first in northern polar latitudes, where they are the primary constituents of the Arctic haze, a thick (up to 3 km) brownish or orange pall covering a circumpolar area as large as North America (Nriagu, Coker, and Barrie 1991).

More dramatically, anthropogenic emissions, made up mostly of sulfates, organics, nitrates, black carbon, and fly ash, have been creating seasonal (January to April) anthropogenic haze above the equatorial Indian Ocean and reducing insolation by as much as 30 W/m^2 (Ramanathan et al. 2001). Global satellite measurements indicate that by 2005 direct radiative forcing by aerosols amounted to about −1.9 W/m^2, significantly stronger than standard model estimates, and that 47% $\pm$ 9% of the aerosol optical thickness over land was due to anthropogenic particles (Bellouin et al. 2005). This cooling effect is unevenly distributed, with pronounced peaks in eastern North America, Europe, and Southeast and East Asia, the regions with the highest sulfate levels. This effect also means that if aerosol emissions continue to decline (in large part because of reduced emissions from fossil and biomass fuel combustion) future atmospheric warming could be greater than currently anticipated.

Yet the decline of aerosol emissions is desirable because they have long been implicated in aggravating chronic respiratory diseases and, more recently, heart attacks. Kaiser (2005) found that aerosols with diameter less than 2.5 μm increased the risk of heart attacks and caused as many as 60,000 premature deaths per year in the United States alone. Another health risk arises from photochemical smog, whose formation requires emissions of nitrogen oxides (NO_x: NO and NO_2) generated during high-temperature combustion, which breaks the N_2 bond. NO_x originates mostly in densely populated regions of the Northern Hemisphere, which have the highest concentrations of fossil-fueled electricity-generating plants, private vehicles, and trucks. The global annual total of NO_x was about 25 Mt N in 2005.

Part of this flux contributes, after oxidation to nitrates, to the global aerosol load, but in many urban areas these gases enter into complex photochemical reactions with volatile organic compounds, released from incomplete combustion of fuels and from the processing, distribution, marketing, and combustion of petroleum products, and with CO from vehicle emissions, which culminate in the formation of photochemical smog (Colbeck and MacKenzie 1994; Mage et al. 1996; Heinsohn and Kabel 1999). The brownish haze of smoggy regions limits visibility (in extremes to just 10^1 m) and damages materials. Its health impacts range from eye irritation to increased frequency of asthmatic attacks and a higher incidence of chronic emphysema. These effects are due mostly to the formation of highly reactive O_3 and peroxyacetyl nitrate. Photochemical smog has become a recurrent seasonal presence in large cities as far as $50° N$ and a semipermanent nuisance in many subtropical and tropical megacities, including Ciudad Mexico, Bangkok, Hong Kong, and Taipei.

High smog levels have affected increasingly larger areas surrounding many major cities in North America, Western Europe, and East Asia. By the 1990s these extensive metro-agro-plexes (combining cities, industries, and intensive cropping) accounted for about 75% of global fossil fuel use and for a third of worldwide cereal production, yet their O_3 levels were above the threshold where cumulative exposures during the growing season lower crop yields (50–70 ppb). High levels of surface O_3 are most worrisome in China. The country's size makes it imperative that it remain largely self-sufficient in food production, but high O_3 concentrations may

significantly reduce yields of spring wheat, soybean, and corn (Aunan, Berntsen, and Seip 2000). Finally, it is important to note that in many traditional societies it is indoor air pollution—emissions from low-efficiency combustion of biomass fuels in unventilated or poorly ventilated rooms—that poses much higher health risks than does the outdoor contamination of air (Smith 1993).

11.5 Interference in Grand Biospheric Cycles

In spite of its large demands for oxygen, fossil-fueled civilization has a negligible effect on the planetary cycling of the element indispensable for all heterotrophic life (see section 11.4). The water needs of energy industries (see section 11.3) diminish, delay, or accelerate local or regional flows by altering surface and underground storage, runoff, and evaporation, but these changes have a minuscule impact on the ocean-dominated global water volume. Most of the nutrients essential for photosynthesis and heterotrophic metabolism do not really cycle. Insoluble minerals stay put unless moved by water or wind erosion, and soluble elements have just one-way flows, piggybacking on a segment of the water cycle as they move from continents to oceans, with temporary interruptions en route. Thus the focus must be on those elements that are introduced into the environment by fossil-fueled civilization in large quantities and that are doubly mobile, that is, both water-soluble and airborne. Only three elements are in this class: carbon, nitrogen, and sulfur. All are needed to sustain life, and each one has a unique role in the biota.

Carbon provides the basic matrix of life, accounting for nearly half of the dry living mass; without nitrogen there can be no amino acids (essential building blocks of proteins), nucleic acids, enzymes, and chlorophyll; and sulfur is an indispensable building ingredient, the fortifier responsible for the three-dimensional structure of proteins. The three elements are locked in the lithosphere and hydrosphere in carbonates, nitrates, and sulfates, and, to list just the principal members of airborne segments of their respective cycles, in CO, CO_2, CH_4, N_2O, NO, NO_2, NH_3, NO_3, SO_2, H_2S, and SO_4. During the preindustrial era human interference in the three cycles was limited to burning of biomass and conversion of natural ecosystems to cultivated lands (both essentially a locally accelerated release of plant C, N, and S), and some concentrated dumping and recycling of organic wastes.

Fossil-fueled civilization brought radically different interventions. The combustion of fossil fuels reintroduced long-dormant stores of C and S into the atmosphere and generated increasing amounts of nitrogen oxides. In addition, agricultural intensification rested on the expanding use of inorganic nitrogen fertilizers. As a result, anthropogenic fluxes of the three elements now form large shares of their total biospheric flows, especially in industrialized or intensively farmed areas (Smil 2000a). By far the most worrisome interference in the global cycle is rising atmospheric concentrations of CO_2 from combustion of fossil fuels and land use changes (fig. 11.7). These concerns arise from the critical role the gas has played in determining the biosphere's temperature.

Water vapor, the most important greenhouse gas, could not have maintained relatively stable temperatures because its changing atmospheric concentrations amplify rather than counteract departures from surface temperatures; evaporation declines with cooling and rises with warming. In addition, changes in soil moisture do little

to chemical weathering. Only long-term feedback between CO_2, surface temperature, and the weathering of silicate minerals explain the limited variability of mean tropospheric temperature. Lower temperatures and decreased rates of silicate weathering result in gradual accumulation of CO_2 and in subsequent warming (Berner 1999). A reliable record of atmospheric CO_2 is available for the past 420,000 years, thanks to the analyses of air bubbles from ice cores retrieved in Antarctica and Greenland. Preindustrial CO_2 levels were never below 180 ppm and never above 300 ppm (fig. 11.8) (Raynaud et al. 1993; Petit et al. 1999). And between the beginnings of the first civilizations 5000–6000 years ago and the onset of the fossil-fueled era, these levels fluctuated narrowly between 250 ppm and 290 ppm.

The post-1850 rise of fossil fuel combustion (including relatively small contributions by cement production and natural gas flaring) brought global carbon emissions (1 t C = 3.66 t CO_2) from less than 0.5 Gt in 1900 to 1.5 Gt in 1950, and to over 6.5 Gt C by the year 2000, with about 35% originating from coal and 60% from hydrocarbons (Marland, Boden, and Andres 2005). Many studies have also evaluated life cycle emissions of CO_2, or more precisely, CO_2 equivalents of other greenhouse gases (Lenzen 1999; Meier 2002; Gagnon, Bélanger, and Uchiyama 2002; IHA 2003). The lowest values (rounded to avoid unwarranted impressions of accuracy and expressed in CO_2-equivalent t/GW_eh) are for wind ($\sim$10), nuclear fission and large-scale hydrogeneration ($\sim$15, with the latter up to 50). Biomass-generated electricity rates just over 100, combined cycle gas turbines at nearly 500, diesel generators at almost 800, and conventional fossil-fueled plants around 1,000. Analyses of solar-thermal and photovaltaic generation are rela-

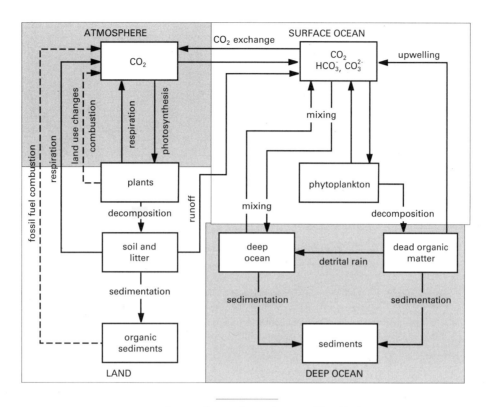

11.7 Biospheric carbon cycle: combustion of fossil fuels and land use changes (above all, tropical deforestation) are the largest human interferences.

tively uncertain, with rates ranging between 13 and more than 700 CO_2-equivalent t/GW$_e$h.

The cumulative total of emissions from fossil fuel combustion was about 280 Gt C between 1850 and 2000 (Marland, Boden, and Andres 2005). To this must be added emissions from conversions of natural ecosystems (mainly tropical deforestation) to other uses. This flux was dominant until the early 1950s, and its cumulative pre-2000 total most likely amounted to 150–200 Gt C

(Warneck 2000). During the first years of the twenty-first century the two fluxes totaled, respectively, nearly 7 Gt and perhaps as much as 2 Gt C, but less than half of this annual input stays in the atmosphere (the rest is absorbed by the ocean or enhances photosynthesis). Annual increases of atmospheric CO_2 have averaged about 1.2 ppm (1 Gt C = 0.47 ppm) since continuous background measurements began in 1958, far from major anthropogenic sources and from forested areas, at Mauna

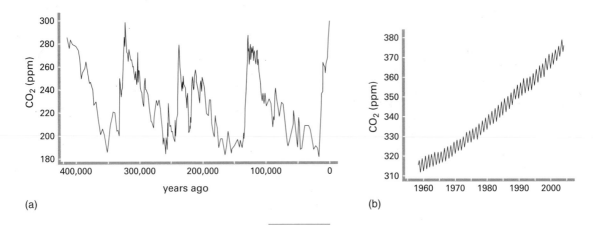

11.8 Atmospheric CO_2 concentrations, (a) during the past 420,000 years, derived from ice core bubbles (Petit et al. 1999), and (b) since 1958 as monthly averages (tracing the annual biospheric breath) of continuous measurements at Mauna Loa. Plotted from data in Keeling (1998) and Carbon Dioxide Information Analysis Center.

Loa and at the South Pole (Keeling 1998). In 1958 the CO_2 level at Mauna Loa averaged 320 ppm; by 2000 it surpassed 370 ppm (fig. 11.8).

To this must be added the thermal burden of other anthropogenic greenhouse gases (GHG) whose combined forcing is now roughly equal to that of CO_2 (Hansen et al. 2000). While the atmospheric concentrations of these gases are much lower than those of CO_2, the specific absorption rates of the outgoing infrared radiation (that is, their global warming potential, GWP) greatly surpass that of CO_2. The natural gas industry (field losses during extraction, flaring, pipeline leaks) and coal mining are major sources of methane, the second most important anthropogenic GHG (GWP = 270), which is also produced by anaerobic fermentation in rice fields and landfills and by enteric fermentation of ruminant live-

stock. As a result, CH_4 levels have roughly doubled since 1850, to about 1.7 ppm by the year 2000. Emissions of N_2O, the third most important anthropogenic GHG (GWP = 60), originate mainly from denitrification and fossil fuel combustion.

Compared to 1880, the combined forcing of all GHGs reached nearly 3 W/m^2 by 2000 and 3.05 W/m^2 by 2003 (fig. 11.9) (Hansen et al. 2000; 2005). Black carbon added nearly 0.5 W/m^2, but the net effect of anthropogenic aerosols was to cool the Earth at -1.39 W/m^2. Adding the effects of land use and snow albedo produced overall forcing of 1.8 (± 0.85) W/m^2 relative to 1880 and climate sensitivity of 0.6°C–0.7°C/W · m^2. This is actually the observed 1880–2003 warming in full response to nearly 1 W/m^2 of past forcing, while about 0.85 W/m^2 is yet to have its effect; additional warming

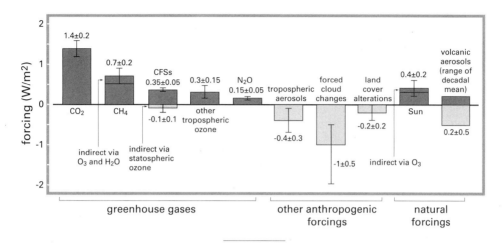

11.9   Climate forcing by greenhouse gases and other atmospheric constituents (W/m^2), 1880–2000. From Hansen, Sato, et al. (2000).

of about 0.6°C is pending unless it is counteracted by complex feedbacks. The eventual doubling of the preindustrial CO_2, which was about 280 ppm in 1850, would most likely raise average tropospheric temperatures by 2°C–4.5°C and cause relatively rapid global climate change, including unevenly distributed higher seasonal temperatures (faster warming in polar regions), accelerated water cycle, changed precipitation patterns, and rising ocean level (Alley et al. 2007). As the surface ocean absorbs additional CO_2, its pH falls; this may decline by as much as 0.4 units by 2100 (Orr et al. 2006). The potentially destabilizing environmental, health, and socioeconomic consequences of these changes have become a major topic of interdisciplinary research and public policy concern (Houghton et al. 2001; Epstein and Mills 2005).

In contrast to omnipresent carbon, nitrogen is a relatively rare element in the biosphere. It is absent in cellulose and lignin, the two dominant molecules of terrestrial life, and only plant seeds and lean animal tissues have high protein content (overall 10%–25%, soybeans to 40%). Although nitrogen is an indispensable ingredient of all enzymes, its total reservoir in living matter is less than 2% of carbon's huge stores, and whereas C cycling includes large flows entirely or largely unconnected to the biota, every major link in nitrogen's biospheric cycle is mediated by bacteria (Smil 1985; 2000b). Human interference in the nitrogen cycle is largely due to food production: recycling of organic wastes, planting of leguminous crops (whose symbiotic bacteria can fix N), and above all, applications of ammonia-based N fertilizers that eliminated the nitrogen limit on crop production and allowed for the expansion of human population (Smil 2001). As already noted (section 10.2), the Haber-Bosch synthesis of NH_3 and subsequent production of solid and liquid N fertilizers consumes annually about 5 EJ, or roughly 1.5% of the world's TPES.

ENVIRONMENTAL CONSEQUENCES

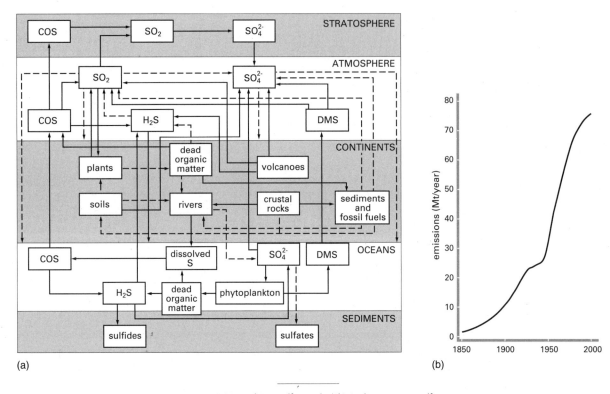

(a)

(b)

11.10   (a) Biospheric sulfur cycle. (b) Anthropogenic sulfur
emissions, 1850–2000. From Smil (2002).

Applications of N compounds amounted to about 85 Mt N in 2005, and it is unlikely that organic recycling and legume crops added more than 35–40 Mt N/a (Smil 1999b). The third most important human interference in the global nitrogen cycle is the high-temperature combustion of fossil fuels: it released more than 30 Mt N, mostly as nitrogen oxides (NO_x), in 2005. In aggregate, anthropogenic mobilization of some 150 Mt N/a is roughly equal to the annual rate of natural biofixation by symbiotic and free-living bacteria (Galloway and Cowling 2002). An excess of reactive N acidifies precipi-

tation, lowers soil pH, causes eutrophication of fresh and coastal waters, and raises concerns about the long-term effects on biodiversity and productivity of grasslands and forests (Bergström and Jansson 2006; Phoenix et al. 2006). As already mentioned, NO_x are key precursors of photochemical smog, and N_2O, released during denitrification, is an important greenhouse gas.

Sulfur is an even rarer component of living molecules than is nitrogen. Only 2 of the 20 amino acids that make up proteins (methionine and cysteine) have the element as a part of their molecules, but every protein needs disulfide bridges in order to make long three-dimensional polypeptide chains whose complex folds allow proteins to be engaged in countless biochemical reactions. Sea spray is the largest (140–180 Mt S) natural input of the element into the atmosphere, but 90% of this mass is promptly redeposited in the ocean (fig. 11.10). Some volcanic eruptions are very large sources of SO_2, others have S-poor veils. The long-term average is about 20 Mt S/a, and dust, mainly desert gypsum, may contribute as much as that. Biogenic sulfur flows may be as low as 15 Mt S/a and as high as 40 Mt S/a; they are produced on land by both sulfur-oxidizing and sulfate-reducing bacteria present in waters, muds, and hot springs.

Fossil fuel combustion generates more than 90% of all anthropogenic sulfur (fig. 11.10). Its global emissions rose from 5 Mt in 1900 to about 80 Mt in 2000, matching or perhaps even surpassing the natural volcanic, dust, and biogenic flux (Lefohn, Husar, and Husar 1999). The remainder is emitted largely by the smelters of color metals (mainly Cu, Zn, Pb). Emitted SO_2 is rapidly oxidized to sulfates whose deposition is the leading source of acid deposition (dry or as rain, snow, and fog). Before 1950 emissions from households and from low industrial chimneys caused only local acidification. Tall stacks of post-1950 coal-fired plants emitted hot flue gases that could rise into the mid-troposphere and be carried considerable distances downwind. During their time aloft (up to three or four days, average <40 h) the acidifying gases can travel 10^2–10^3 km before being deposited on distant ecosystems. Because of the short atmospheric residence time of S compounds, the element does not have a true global cycle as do carbon and nitrogen.

Deposited sulfates and nitrates acidify aquatic ecosystems and reduce or eliminate sensitive species of fish, amphibians, gastropods, crustaceans, and invertebrates. Chronic acidification dissolves aluminum hydroxide, and Al^{3+} irritates fish gills and destroys their protective mucus. Acidification also mobilizes abnormally high levels of all heavy metals. However, the role of acid deposition in reduced productivity and die-back of some forests is not as clear (Tomlinson 1990; Godbold and Hüttermann 1994). Acid deposition also accelerates corrosion of metals and deterioration of mortar, limestone, and marble (the Parthenon and the Taj Mahal are two notable examples). But prevailing levels of acid deposition in North America and Europe have not caused any measurable reductions of crop yields. Europe and eastern North America have successfully reduced SO_2 emissions, but the Asian flux has been rising. The cooling effect of airborne sulfates has been noted; it is most pronounced over eastern North America, Europe, and East Asia, the regions with the highest sulfate levels.

The 16-fold increase in global commercial energy consumption during the twentieth century has been the most important cause of human interference in the biospheric cycles of doubly mobile elements. By 2000 the combustion of fossil fuels generated at least 75% of all CO_2 emissions, 90% of the anthropogenic mobilization of sulfur, and about 20% of the anthropogenic releases of reactive nitrogen. Further substantial growth of global TPES means that we must consider an unprecedented possibility: that future limits on human energy use may arise, not from resource shortage but from the necessity to keep these cycles compatible with the long-term habitability of the biosphere.

12

ENERGETIC CORRELATES

Complexities of High-Energy Civilization

Man, nevertheless, being human, needs some external prosperity.... We must not however suppose, that, because one cannot be happy without some external goods, a great variety of such goods is necessary for happiness. For neither self-sufficiency nor moral action demands excess of such things.
Aristotle (384–322 B.C.E.), *Nicomachean Ethics*

Economists are traditionally concerned with capital and labor, and more recently with technical innovation. But some scientists and environmentalists now insist that energy use should be the primary standard of value by which all things and actions are judged. This attitude leads to an austere ethics of an energy-invariant society that functions only on the basis of renewable solar income. Before commenting on these radical reformist views, I look at basic macroeconomic realities linking energy use and national economic performance. Energy use is a fundamental means to attain a variety of desired ends. Quality of life should head the list of such desiderata, and

that is why I examine the fascinating evidence linking various aspects of well-being with energy conversions.

Regrettably, substantial shares of energy use have been repeatedly used for wars. The twentieth century intensified this pattern, and the aggregate death toll mounted with the deployment of such highly destructive weapons as machine guns, tanks, heavy artillery, and bombers. Nuclear bombs and intercontinental ballistic missiles provided the ultimate mass killing techniques. Their production and deployment has been predicated on energy-intensive industrial processes and modes of delivery. (advanced metallurgy, complex chemical syntheses; land- and sea-based missiles, fighter planes, battle tanks). After a brief look at energy and war, the chapter closes with some musings on energy futures.

12.1   Energy and the Economy

Frederick Soddy's (1933, 56) conviction that "the flow of energy should be the primary concern of economics" has had no appeal for mainstream economists. For

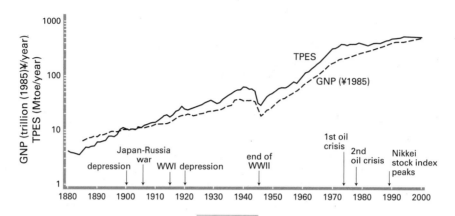

12.1 An example of a close relation between the growth of
economic product and TPES, Japan, 1880–2000. Plotted from
data published annually by the Institute of Energy Economics,
Tokyo.

decades, energy economics existed only as a segment of resource economics, itself a field outside the mainstream of the discipline. Matters did not change much even after 1973. Energy as an economic factor got more attention in microeconomic terms, but general treatments remained relatively rare (Georgescu-Roegen 1976; Slesser 1978; Gordon 1981; Bohi 1989; Banks 2000; Stevens 2000; Buenstorf 2004), and only ecological economics takes energy as a central concern (Hall, Cleveland, and Kaufmann 1986; Ruth 1993; Ayres, Ayres, and Warr 2003; D. I. Stern 2004). The link between TPES and economic growth in particular has attracted a great deal of analytical attention.

When expressed in constant monies, and with national data converted using purchasing power parities (PPPs), the gross world economic product has been marching in lockstep with the global commercial TPES. Between 1900 and 2000 energy use rose about 17-fold (from 22 EJ to 380 EJ), and the economic product (in constant

1990 dollars) increased 16-fold (from about $2 trillion to $32 trillion), indicating a highly stable elasticity near 1.0 (Maddison 1995; World Bank 2001). The closeness of the link is also revealed by very high correlations between national per capita GDP and TPES. For any given year the correlation between the two variables for all countries surpasses 0.95, as close a link as one may find in the unruly realm of economic affairs. Similarly high correlations can be found for the secular link between the two variables for a single country (fig. 12.1).

A closer look reveals some nontrivial problems. On the energy side is the exclusion of biomass fuels from all TPES data (an omission that greatly underrates actual energy use in many poor countries), inaccuracies in converting fuels to a common energy denominator, debatable conversions of primary electricity (see section 1.2), and complete neglect of energy quality. Substantial differences in market prices of various energies recognize these energy quality, but the common denominator in

TPES ignores it. On the economics side are not only questionable currency conversions (official exchange rates undervalue GDPs of low-income economies; PPPs overvalue them) but the fundamental question of what GDP measures and hence attempts to consider more revealing alternatives (OECD 2006). Nothing has changed since Rose (1974, 359) noted, "So far, increasingly large amounts of energy have been used to turn resources into junk, from which activity we derive ephemeral benefit and pleasure; the track record is not too good." And the balance looks even worse when the extensive environmental destruction caused by this consumption is included in the overall appraisal of this record.

Quality considerations are best illustrated by the fact that no fuel can confer economic benefits like efficiency, productivity, and flexibility better than electricity (NRC 1986; Smil 2005a). Its high final conversion efficiencies, precise control, focused applications, and fractional uses offer an incomparable combination of advantages. Modern mass production demanding flexibility, precision, and expansion is unthinkable without electricity, as are modern health services, household comforts, and entertainment. Only a small (but steadily rising) fraction of electricity is used to design, manage, route, regulate, and improve the information that now suffuses modern civilization, and the formulation of the equivalence between energy and information opened the way for rigorous quantitative studies of this critical energy flux (Shannon and Weaver 1949). But applying this approach to economies at large is neither easy nor necessarily useful because the effort to minimize the energy to information ratio (J/bit) keeps colliding with inherent, often culturally driven inefficiencies of human behavior.

A closer look at energy-GDP links also reveals that a given level of economic well-being does not require a fixed level of TPES. Different energy intensities (EIs) of national economies (annual TPES/GDP) show considerable scatter within both low- and high-income groups. British EI shows a steady secular decline following the rapid rise caused by the adoption of steam engines and railways between 1830 and 1850 (Humphrey and Stanislaw 1979). Canadian and U.S. EIs followed the British trend with a lag of 60–70 years (fig. 12.2). Between 1955 and 1973 the U.S. EI was flat (fluctuating just $\pm 2\%$) while the real GDP grew 2.5-fold, but then the ratio's decline resumed, and by 2000 it was below the 1950 level. In contrast, Japanese EI rose until 1970, as did the Chinese rate, which since that time has fallen faster than at any previous time (fig. 12.2).

In the year 2000 the EIs of the world's most important economies spanned a considerable range of values. In G7 countries they ranged from less than 7 MJ/1000 $ for Italy and Japan to more than 13 MJ/1000 $ for the United States and over 18 MJ/1000 $ for Canada. In contrast, EIs were in excess of 30 MJ/1000 $ for India and China. National EIs confirm and reinforce some of the widely held snapshot notions: efficient Japan, relatively wasteful United States, China with a long modernization road ahead. Advanced extraction, processing, and manufacturing techniques reduce national EIs, and countries with relatively low intensities should enjoy some important economic, social, and environmental advantages.

But these distillations of complex realities are simplistic and misleading if interpreted in a naive, ahistorical fashion. Explanations of national EI differences are multifactorial, ranging from climate to recreational habits, but most of the gap can be accounted for by the makeup of primary energy consumption and by the structure and efficiency of final conversions. Higher

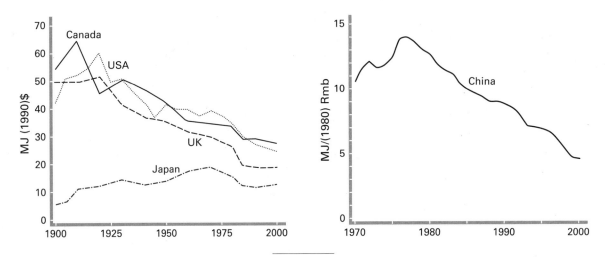

12.2 Long-term trends of energy intensities in Canada, the United States, Britain, Japan, and China. From Smil (2003; 2004a).

shares of energy-intensive industries (e.g., Canada's large metallurgical sector) will boost national EI even if those operations are highly efficient (Smil 2003). Kaufmann (1992) found that most of the post-1950 decline of EIs in affluent economies was not due to technical change but was associated with shifts in the kind of energies used and the type of goods and services produced and consumed.

In contrast, secular declines, or at least remarkable constancy, of energy prices illustrate the combined power of technical innovation, economies of scale, and competitive markets. U.S. electricity has become a particularly great bargain: its average price fell by nearly 98% between 1900 and 2000 (fig. 12.3). With average per capita disposable incomes about five times as large, and conversion efficiencies up to 1 OM higher, a unit of service provided by electricity in the United States was 200–600 times more affordable in 2000 than in 1900. Similarly, the affordability of lighting services in Britain was about 160 times higher in 2000 than in 1900 (Fouquet and Pearson 2006). Inflation-adjusted prices of coal and oil fluctuated a great deal but have remained remarkably constant in the long run. The average (constant dollar) price of U.S. bituminous coal in 2000 was almost exactly the same as in 1950 or in 1920, and the price of imported Middle Eastern crude oil was as cheap as the average price of U.S.-produced crude oil during the first years of the twentieth century. Only natural gas has shown a gradual increase in inflation-adjusted prices (fig. 12.3).

But it would be naive to see these or any other energy prices either as outcomes of free market competition or as values closely reflecting the real cost of energy. A long history of price-distorting practices has brought not only unnecessarily higher but also patently lower prices due to

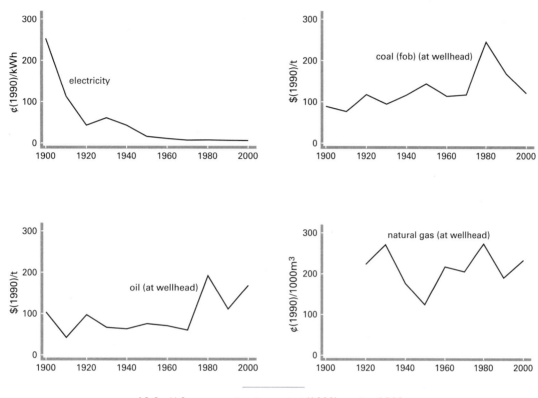

12.3   U.S. energy prices in constant (1990) monies, 1900–
2000. From Smil (2003).

subsidies and special regulations favoring one form of supply over other forms of energy (Gordon 1991; D. H. Martin 1998; NIRS 1999; Von Moltke, McKee, and Morgan 2004). The U.S. nuclear industry received more than 96% of $145 billion (in 1998 dollars) disbursed by the U.S. Congress between 1947 and 1998 (NIRS 1999). Moreover, in 1954 the Price-Anderson Act reduced private liability by guaranteeing public compensation in the event of a catastrophic accident in commercial nuclear generation (USDOE 2001). No other industry has enjoyed such sweeping state protection.

And the United States is not alone in favoring nuclear energy. Of the roughly $9.4 billion spent in 2004 on energy R&D by all IEA countries, $3.1 billion went for fission, and fusion received $700 million, nearly twice as much as photovoltaics (IEA 2006).

Special tax treatment has been enjoyed for even longer by U.S. oil companies, which can immediately write off so-called intangible drilling costs. All independent oil and gas producers are allowed to deduct 15% of their gross revenue, and this depletion deduction can greatly exceed actual costs (McIntyre 2001). More recently generous tax benefits have been given to the developers of some renewable energies (wind, ethanol). At the same time, power producers have been heavily regulated in many countries, and Gordon (1994) singled out the general underpricing of electricity caused by this regulation as a key market failure. Until March 1971, U.S. oil prices were controlled through production quotas by a very effective cartel, the Texas Railroad Commission, and since 1973, OPEC—although its power has never been such as to entirely overrule the market (Mabro 1992)—used production quotas in order to repeatedly raise the price (Adelman 1997). One constant has remained throughout these shifts: the world price of crude oil bears virtually no relation to the actual cost of extracting the fuel.

Excellent arguments can be made for even lower prices of energy than those that have prevailed since WW I. At the same time, most existing prices do not fully reflect the real costs of fossil fuels and electricity, and the inclusion of numerous environmental, health and safety, and other externalities would push them higher (Hubbard 1991; Smil 2003). After the U.S. coal industry internalized many of its externalities (through regulations of dust and methane levels, mine ventilation, and suppression of dust), its annual fatality rate fell impressively, and it is now less than 1% of China's dismal record (MSHA 2000; Fridley et al. 2001). U.S. mining companies must also contribute to the disability and compensation funds put in place to ease the suffering of miners afflicted with black-lung disease (Derickson 1998). Electrostatic precipitators, FGD, and NO_x removal (and the first steps toward carbon taxes in some EU countries) brought coal-fired generation much closer to the real cost of thermal electricity. But the true cost of fuels and electricity remains a matter of contention.

Hohmeyer (1989) and Hohmeyer and Ottinger (1991) calculated the external costs of German electricity production to be of the same order of magnitude as the internalized costs, but Friedrich and Voss (1993) judged the methodologies used for those calculations unsuitable and the estimates derived from them too high. A European study concluded that the cost of producing electricity from coal or oil would double and that natural gas–generated electricity would cost 30% more if environmental and health impacts (excluding those attributable to global warming) were taken into account (ExternE 2001). But because the specific impact was dominated by effects on human health (hence by total populations affected), 1 t of particulates was calculated to produce damage worth up to €57,000 in Paris but only €1,300 in Finland.

An ORNL study found that damages from the U.S. coal fuel cycle amount to merely 0.1 cent/kWh (Lee et al. 1995), a negligible fraction of the average cost of just over 6 cents/kWh and a total many magnitudes apart from Cullen's (1993) estimate of a tenfold increase in actual fuel costs. Similarly, generalized damage values adopted by some public utility commissions in the United States during the 1990s not only differed a great deal (e.g., more than a 30-fold range for damages due to NO_x) but they were typically much higher than damages found by studies that modeled actual dispersion and the impact of pollutants for specific sites (Martin 1995; Lee, Krupnik, and Burtraw 1995).

Health impacts generally dominate the cost of externalities but are most difficult to monetize because of lifelong impacts of pollutants, dubious designs of epidemiological studies, and problems of separating the effects of individual pollutants (e.g., sulfates from general particulate matter) when people are exposed to complex, varying mixtures of harmful substances. Cost-benefit studies of air pollution control remain a matter of scientific controversy (Lipfert and Morris 1991; Phalen 2002). Long-term impacts on vegetation (acid rain, nitrogen enrichment) are also difficult to quantify, and an even greater challenge is to monetize eventual multifaceted consequences of relatively rapid global warming.

An even more contentious matter is accounting for military expenditures attributable to securing energy supplies. Problems of attribution (what shares of outlays should be charged to continuing military presence or to military actions?) and of spatial and temporal boundaries deployed in such accounting have no unequivocal solutions. Should the cost of Desert Shield/Storm in 1990–1991 be charged entirely against the price of Persian Gulf oil? If not, what share? On narrow accounting grounds the war was actually profitable for the United States. Total pledges of $48.3 billion of foreign help were almost $800 million above the Office of Management and Budget's estimate of U.S. funding requirements (GAO 1991). In contrast, there is a huge cost of the aftermath of the 2003 Iraq war and an even more intractable problem of deciding what part of it should be attributed to the real post-2003 price of oil.

One conclusion remains clear: rising disposable incomes have made all forms of energy relatively inexpensive in all affluent economies and particularly in the United States. U.S. expenditures for food bought for home consumption fell from about 25% after WW II to less than 10% of total outlays by 2003, and household and transportation energy purchases accounted for nearly identical and astonishingly low shares of about 2.5% (USBC 2006). How could there be any strongly elastic market response, any public clamor for real change, any determined policies to reduce overall energy consump-

tion when annual household expenditures for electricity are lower than for purchase of video and audio products, or when the spending on gasoline is less than half of the total spent on eating out?

12.2 Energy and Value

Although the mainstream economics has remained curiously detached from the challenges just described, ecological economics has done more than just being concerned with energy as a key driver of economic production. Since the 1970s a number of attempts have been made to elevate energy to the dominant standard of value and to explain economic growth within ecological and thermodynamic frameworks. The most radical approach to valuing human affairs in energy terms is to put thermodynamic considerations in command of economics. Nicholas Georgescu-Roegen (1971; 1976) was the most passionate proponent of this radical shift. He called the second law of thermodynamics "the most economic of all physical laws," and argued that civilization's foremost goal should be to minimize entropic degradation. This led him to single out accessible material low entropy as the most critical bioeconomic element.

Inevitably, this position leads to a rejection of steady-state civilization, a concept promoted by other eco-economists (Daly 1973) in response to concerns about continued rapid growth (Meadows et al. 1972). The only thermodynamically acceptable size of global population is one supportable by purely solar agriculture. This implies sizable reduction of the current population total because such a large share of humanity exists due to high fossil energy subsidies (see chapter 10). An entropically guided civilization would require declining population willing to live by a new ethics. What are the chances of constraining (or largely eliminating) the human addiction to exosomatic comfort? Georgescu-Roegen's (1975, 379) answer was not hopeful: "Perhaps, the destiny of man is to have a short, but fiery, exciting and extravagant life rather than a long, uneventful and vegetative existence. Let other species—the amoebas, for example—which have no spiritual ambitions inherit the earth still bathed in plenty of sunshine."

Such musings appear to be incontestable. Unless our species leaves this planet, the only possible strategy to maximize the duration of our terrestrial tenure is to minimize entropic drift, a strategy that may require a gradually declining population in order to channel solar radiation into increasingly energy-intensive procurement of materials. And yet the exaltation of entropy misses some key points. On the most fundamental level, Brooks and Wiley (1986) argue, life's evolutionary entropic behavior is not determined by energy flow because this cannot explain the existence of organisms, their variability or structure. Life's evolution depends on mutations, and there is no link between them and energy flow analogous to the role energy plays in organizing nonliving systems. Life's intrinsic properties determine how energy flows, not the other way around. Epigenetic information channels energy into maintenance, growth, differentiation, and reproduction; these irreversible transformations dissipate both matter and energy.

Evolution is thus inevitably entropic, but the availability of energy is not its guiding force. Layzer (1988) extended the caveat to material flow as well. Its free flow, much like energy flow, is essential, but neither one drives the evolutionary process because the ability of organisms to mobilize free energy and to organize matter stems from the reproductive instability of genetic material. In practical terms, entropic concerns extend far beyond any rational planning horizon of half a century. During

that time we are much more likely to be constrained by environmental changes than by scarcities of low-entropy energies and materials. And in longer run there is an enormous potential for material substitutability (Goeller and Weinberg 1976). We cannot exclude the eventual possibility of virtually unlimited energy from extraterrestrial capture of solar radiation or from advanced nuclear techniques. In any case, our inability to forecast and to comprehend complex wholes relegates any scenarios of distant futures to the category of speculation.

It turns out that even rational attempts to guide societies by energy-based valuations rather than by admittedly inadequate or misleading monetary appraisals are rather impractical. The impulse behind such attempts is understandable. The inadequacies of standard monetary accounts are indisputable: omissions of subsistence food production, barter, and black market activities; distortions of prevailing terms of trade and exchange rates; uncertain purchasing power parities. In addition, monetization becomes challenging when valuing nonrenewable resources, degradation or destruction of public goods, and the true worth of indispensable biospheric services. These fundamentals are either entirely beyond the realm of standard pricing, or their valuations have failed to capture their finite nature and low entropy. In contrast, energy-based valuation has the appeal of rigor and universality for those who approach economics via the natural sciences or engineering.

The line of these advocates extends from Ostwald (1909); to the Technical Alliance, later Technocracy, formed by Howard Scott in 1918 (Technocracy 1937); to Soddy (1926), Cottrell (1955), and Odum (1971; 1975); and to some proponents of energy analysis during the 1970s (see chapter 10). The link with ecological thinking has been a strong part of this movement. By setting energy as the standard of value, society would realign itself with natural systems, where energy's importance could be never in doubt. That is why Technocracy (1937) advocated the radical step of replacing money by energy certificates within a system based on a centralized continuous registry of energy conversions that would record "the desires of every citizen in his choice of consumable goods and available services" and allow individuals "the widest latitude of choice" in consuming their share of the annually allotted net energy total. This hybrid of egalitarian utopianism and mind-boggling superbureaucracy had no chance of implementation.

Calls for energy-based valuation came back in the wake of the energy "crisis" of the 1970s. Ritchie-Calder (1979) argued that energy accounting rather than monetary accounting might straighten out the energy problem and set the real value of materials and commodities. Odum (1975, 99) concluded, "Money ... does not flow along the paths to the energy sources or the energy interactions of the environment. Thus money cannot be used to evaluate either energy sources or environmental impacts. Energy analysis is required." There was even a new call for national and personal energy budgets: "The annual budget would represent a portion—dictated by our value of the future—of the proven energy reserves.... The flow of the currency would be regulated by the amount of energy budgeted for a given period. If less energy existed at the end of the period, then currency flow would have to be reduced proportionately during the next period" (Hannon 1973, 153). Some of these ideas were actually enacted. The Nonnuclear Energy Research and Development Act of December 31, 1974, required net energy analysis for any conversion technique entering commercial application (U.S. Congress 1974).

A more fundamental approach includes exergy-based valuations (Sciubba 2004; Sciubba and Ulgiati 2005; Tonon et al. 2006). Because exergy is a property of all energies and materials, it can provide an all-encompassing biophysical framework by quantifying all inputs as well as resulting wastes. Ayres, Ayres, and Warr (2003) identified the declining price of useful work (product of exergy and conversion efficiency) as the growth engine of the U.S. economy during the twentieth century (fig. 12.4). Solow (1957) credited technical change with some 88% of the increase in the productivity of the U.S. economy, and Denison (1985) linked most of this share to advances in knowledge and the rest to

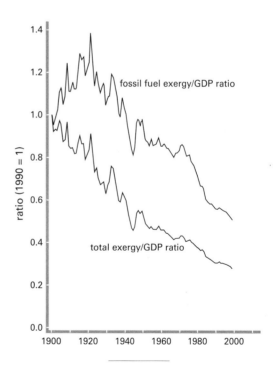

12.4 Declining ratios of exergy inputs to the U.S. economy, 1900–2000. From Ayres, Ayres, and Warr (2003).

ENERGETIC CORRELATES

improved resource allocation and economies of scale. Ayres, Ayres, and Warr (2003) hypothesized that most of Solow's residual may be accounted for by thermodynamic efficiency improvements in the production of primary work (and, where measurable, of secondary work).

Howard T. Odum found all theoretical or highly aggregate approaches to energy-based valuation fundamentally mistaken because they exclude solar radiation as the initial universal input and do not convert different energy flows to equivalent energy costs. In order to carry out consistent and complete net energy analyses he proposed the use of energy quality (concentration) factors or transformation ratios for a variety of biospheric and anthropogenic inputs (Odum 1973; 1983). During the early 1980s the ratios acquired both a new name, solar transformities, and new units, solar emergy (embodied energy), or solar emjoules (sej), per unit of energy, sej/J (Odum 1988). Odum pursued these valuations for the rest of his life (Odum 1996; 2000).

The reasoning behind exergy or emergy valuations is intuitively appealing. Energy is the only unsubstitutable *and* unrecyclable input into every human activity, and as such it is the ultimate limiting factor of development. Fossil-fueled civilization relies on energies whose supplies are finite. Consequently, the energy theory of value, best applied through net energy analysis, appears to be the most fundamental and most revealing approach to assessing our endeavors as well as our prospects. Critical examination shows many weaknesses with this reasoning and many problems with the execution of the idea. On the most general level, one must agree with Rose (1986) that all single-item theories of value suffer from selective inattention to the complexity of civilizations and to the interconnectedness of things, and hence no single-variable valuation can be satisfactory.

As for energy's professed uniqueness, there were other such claims on behalf of land among eighteenth-century physiocrats, and on behalf of labor among Marxists. Is energy different? Costanza (1980), while admitting that energy valuation could have parallels in other theories of value, maintained that no one would seriously suggest that labor can create sunlight. But neither would anyone seriously suggest that the Earth's biosphere could function without relying on the geotectonic recycling of crustal elements—for both its fundamental biochemical functions (P in ATP, S in proteins, Co and Mo in nitrogenase) and structures (Si in plant stems, Ca in animal skeletons)—whose prime mover has nothing to do with sunlight (see section 2.5).

And surely time cannot be treated as a derivative of energy. Indian swamis may be oblivious to its existence, but in modern Western civilization it is an obviously scarce entity, and its valuation and management very often take clear precedence over the levels and efficiencies of energy use (Spreng 1978). Even the gurus of eco-economics travel by airplane; they, too, value time and do not base all their choices on energy valuations. Such trade-offs are ubiquitous, and many of them have come to define the very fabric of modern society, although we may no longer think of them that way: a refrigerator is primarily a time-saving device, as is a lightbulb.

Treating all nonenergy entities as energy transforms, and pricing everything according to embodied energy content, is forcing multifaceted reality into one-dimensional confines. This approach is clearly inappropriate in a world where geophysical, biophysical, technical, social, and moral concerns are intertwined. Management of a civilization is far from being merely a matter of energy conversions. Nor is any valuation just a matter of supply. In the real world there are always many relative scarcities, and only demand can accomplish their effective valuation. Market-driven innovation reduces embodied energy, and the resulting energy savings may greatly surpass any gains brought by efforts aimed solely at minimizing energy inputs. The demand for speed and capacity in electronic computing has been accompanied by dramatic reductions of both embodied and operating energies; the deployment of highly efficient turbofans was driven by the need to accommodate more passengers on longer flights.

Net energy assessments encounter their most frustrating problems in the choice of boundaries and the treatment of mental labor (see chapter 10). Decisions about where to stop the analysis have no acceptable universal solutions. Misleading results are especially likely if the goal of the exercise is the energy-based valuation of all cascading consequences. Energy flows may have a fractal structure, and hence there may be no finite net energy cost. No less vexing is the necessity of converting to a common denominator. Excluding sunlight from an all-encompassing net energy analysis is indefensible; sunlight should count, but to count it coherently is daunting. In Odum's accounting, using more emergy does more real work and leads to a higher standard of life. But does the fact that items with longer turnover times necessarily have higher transformities imply that we should value oak trees up to 10^7 more than bacteria? The biosphere can prosper without oaks but not without bacteria. Why, then, should oaks be valued so much more than bacteria?

And how can we express geotectonic processes (whose priming has nothing to do with solar radiation) in sej? Similarly, how should we estimate the solar energy embodied in fossil fuels? Should we include just the solar radiation that energized the synthesis of original phytomass, or the radiation that energized erosion and sedi-mentation that buried and transformed the phytomass? What portions of those flows are attributable to fuel formation, and how can the inputs of tectonic (nonsolar) energies on fuel formation be included? Incredibly, Odum favored a single transformity for each of the three main fossil fuels, a gross distortion of reality because some of them required 10^3 years to form and others 10^8 years to form. Single sej values are thus inconsistent with the technique's key premise of quality varying with embodied energy. In any case, in real world quality is determined not by the age of the original phytomass but by the presence of sulfur, ash, and moisture in the fuel (see sections 8.1 and 8.2), and the market t value is strongly affected by the mode of extraction and the cost of transport, variables that have nothing to do with ancient solar flux.

The main criticism of exergy is that it does not capture such qualitative attributes as energy density, cleanliness, ease of conversion, and relative environmental impacts. For materials, exergy does not capture such qualitative factors as tensile strength, heat and corrosion resistance, ductility, and conductivity (Cleveland, Kaufmann, and Stern 2000). In practical terms, actual calculation of exergies in a national economy would be extremely challenging. In sum, the intent of energy-based valuations may be laudable, but their execution must remain unsatisfactory. Energy-based valuations have brought more heat than light to our understanding of economic and social values (Mirowski 1988). They have been subject to a number of fundamental methodological problems typical of the social sciences (Reaven 1984).

True understanding calls for a multidimensional approach. We should definitely pay attention to embodied and net energy, but we must also realize that even such a fundamental entity as energy (be it under the exergy or the emergy label) cannot be an adequate surrogate for valuing space, time, qualitative attributes of materials, biodiversity, mental labor, ideas, social order, cultural riches, and morality. At the same time, it is clear that money-based valuation cannot encompass most of these qualities either, because it relegates them to the realm of externalities, a grievous mistake when routinely applied to the environment.

12.3 Energy and the Quality of Life

Modern societies act as if economic growth and high per capita energy use were the goals of an all-consuming quest rather than a means of securing a high quality of life, a concept that subsumes not only the satisfaction of basic physical needs but also the development of human intellect. Because quality of life is a multidimensional concept embracing narrow personal well-being (health, nutrition), wider environmental and social setting (natural and human-caused risks), and the vast intellectual aspect of human development (basic education, individual freedoms), it cannot have a single revealing indicator, but a surprisingly small number of variables are its sensitive markers. Infant mortality (IM; deaths per 1,000 live births) and life expectancy are perhaps the two best (unambiguous) indicators of the physical quality of life. Infant mortality is an excellent surrogate measure, a highly sensitive reflection of the complex effects of nutrition, health care, and environmental exposures on the most vulnerable group in any population, and life expectancy subsumes long-term effects of these critical variables.

At the beginning of the twenty-first century the lowest IM rates were in the most affluent parts of the modern world—in Japan (4), in Western Europe, North America, and Oceania (5–7)—and the highest rates (>100,

even >150) were in African (mostly sub-Saharan) countries as well as in Afghanistan and Cambodia (UNDP 2005). Leaving the low Sri Lankan rate aside, acceptable IM rates (<30) corresponded to annual per capita energy use of at least 30–40 GJ. But fairly low IM rates (<20) prevailed only in countries consuming at least 60 GJ per capita, and the lowest rates (<10) were not found in any country using less than about 110 GJ per capita (fig. 12.5). Increased energy use beyond this level is not associated with any further declines of IM.

In every society, female life expectancy (LE) at birth is on average 3–5 years longer than the male rate. The lowest female LE (<45 years) is in Africa's poorest countries, the highest (>80 years) in Japan, Canada, and nearly half of EU nations. Again, as in the case of IM, the correlation with average per capita energy use explains less than half of the variance. Leaving Sri Lanka aside, high female LE (>70 years) requires at least 45–50 GJ per capita; the 75-year threshold is surpassed at about 60 GJ; and averages above 80 years go with no less than about 110 GJ per capita (fig. 12.5).

Average per capita availability of food energy is not a particularly useful indicator. Food rationing can provide adequate nutrition even in a poor nation, whereas excessive food supply results in enormous food waste and a higher incidence of obesity (Smil 2000b). National means of food energy availability should thus be seen only as indicators of relative abundance and variety of food. Minima of adequate supply and variety are above 12 MJ/day, the rates corresponding to average per capita consumption of 40–50 GJ (fig. 12.5). There is no benefit in raising average daily food supply above 13 MJ per capita.

Education and literacy data are not easy to interpret. Enrollment ratios tell little about actual literacy or

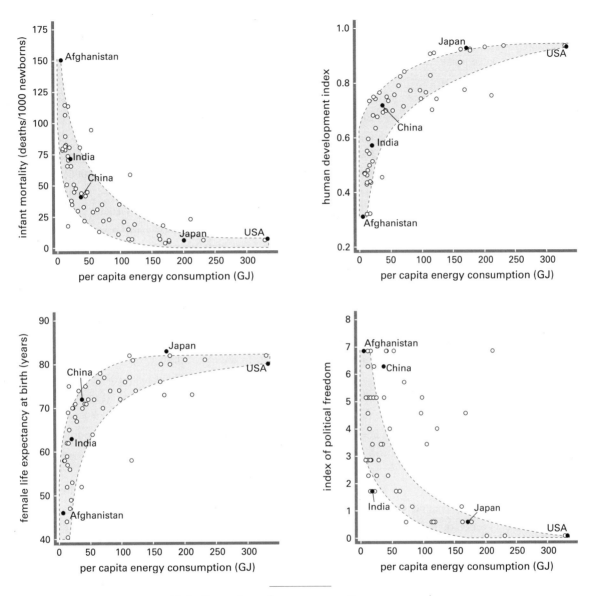

12.5 Comparisons of average per capita energy use and infant mortality, life expectancy at birth, Human Development Index, and Freedom Index show clear inflection points at relatively modest consumption levels, followed by rapid saturations. From Smil (2003).

ENERGETIC CORRELATES

numeracy, and high university attendance reflects low standards rather than high intellectual achievements. High enrollment in primary and secondary schools (>80% of eligible cohorts) have been attained in some countries with energy use as low as 40–50 GJ per capita, and relatively high university enrollments (20%–25% of all young adults receiving postsecondary education) have been associated with energy consumption of at least 70 GJ per capita (UNDP 2005). The United Nations Development Programme uses life expectancy at birth, adult literacy, combined educational enrollment, and per capita GDP to construct the Human Development Index (HDI). There is little difference at the top: in 2005 Norway's HDI was 0.939, followed by Australia, Canada, and Sweden (all with 0.936), but the value for Italy, in twentieth place, was still 0.909. The lowest values (<0.35) are shared by sub-Saharan countries. Data plots show a nonlinear trend with a high HDI (>0.8) reached with as little as 65 GJ per capita, and with minimal or no gains above 110 GJ (fig. 12.5).

The weakest of important links between energy consumption and the quality of life are the political arrangements that guarantee personal freedoms. Fundamental personal freedoms and institutions of participatory democracy were introduced and codified by our ancestors generations before the emergence of modern high-energy civilization, when average per capita primary energy use was a mere fraction of the late twentieth-century levels. The only key exception was women's suffrage. The U.S. federal law guaranteeing women's right to vote was passed only in 1920, and an analogous British act made it through Parliament only in 1928 (Hannam, Holden, and Auchterlonie 2000). Events in the twentieth century showed that suppression or cultivation of freedoms was not dictated by energy use. Such free-doms thrived in the energy-rich United States and in energy-poor India alike, and they were repressed in the energy-rich Stalinist USSR as they still are in energy-scarce North Korea.

Consequently, the ranks of free countries (Freedom House ratings between 1 and 2.5) contain not only all high-energy Western democracies but also such mid- to low-level energy users as South Africa, Thailand, Philippines, and India (Freedom House 2005). Countries with the lowest freedom rating (6.5–7) include not only energy-poor Afghanistan and Vietnam but also oil-rich Libya and Saudi Arabia. Basic personal freedoms are thus compatible with societies using as little as 20 GJ per capita (Ghana, India). Chile and Argentina are populous nations with high freedom rankings (1–1.5) using low amounts of commercial fuels and electricity (<75 GJ) (fig. 12.5).

These realities translate into some fascinating conclusions. In the early 2000s a society concerned about equity and willing to channel its resources into securing adequate diets, good health care, and basic schooling could guarantee a decent quality of life (high life expectancy, varied nutrition, educational opportunities) with annual TPES of as little as 40–50 GJ per capita. A better performance (IM <20, female LE >75, HDI >0.8) requires at least 60–65 GJ per capita. The best rates (IM <10, female LE >80, HDI >0.9) need no less than 110 GJ per capita. All these variables relate to average per capita energy use in a distinctly nonlinear manner, with clear inflections evident at 50–70 GJ per capita, diminishing returns afterwards, and basically no additional gains accompanying consumption above 110 GJ per capita. Political freedoms have little to do with any increases in energy use above existential minima. And because of steadily increasing conversion efficiencies the same quality of life could be secured within a generation (20–25 years) with consumption rates 10%–25% lower than in 2005.

Finally, I note the absence of correlation between average energy use and economic performance on one hand and feelings of personal and economic security, optimism about the future, and general satisfaction with life on the other. During the late 1990s average per capita energy use in Germany (175 GJ) was only half the U.S. rate (340 GJ), and in Thailand (40 GJ) one-eighth the U.S. rate; the U.S. PPP-adjusted GDP was 34% above the German mean and 5.2 times the Thai average (UNDP 2005). But 74% of Germans and Thais were satisfied with their personal lives, compared to 72% of Americans (Moore and Newport 1995). Such findings are not surprising because personal assessments involve strong individual emotions and perceptions that may be largely unrelated to objective measures. Many studies have shown little connection between subjective appraisals of quality of life and personal satisfaction on one hand and objective socioeconomic indicators on the other (Nader and Beckerman 1978; Diener, Suh, and Oishi 1997; Layard 2005; Bruni and Porta 2005).

Labor and leisure implications arising from higher rates of energy use have been profound, proceeding in two grand waves. Energy subsidies in farming reduced labor requirements and released rural residents to cities. By the year 2000 agriculture employed less than 5% of the Western labor force, urban populations were dominant in Latin America, and urbanization was accelerating throughout Asia. Early urbanization was based on manufacturing that still required a large share of heavy exertion and drudgery and extensive child labor; in 1900 at least 1.75 million U.S. children still labored in factories. Only

the second major labor shift, from industries to services, resulted in a much lightened labor burden. Because services generally have lower energy intensity than industries, economic growth proceeds with reduced environmental impact.

Electricity has played the critical role in these transformations. Its unrivaled role is unforgettably illustrated in Robert Caro's (1983) biography of Lyndon Johnson. As Caro pointed out, it was not the shortage of energy that made life in Texas Hill County so hard (households had plenty of wood and kerosene) but the absence of electricity. In an almost physically painful account Caro describes the drudgery and danger of ironing with heavy wedges of metal heated on wood stoves, pumping and carrying of water, feed grinding, and sawing. This situation changed only when transmission lines reached the county as electricity brought a near-miraculous liberation. Millions of peasants in poor countries are still waiting for such liberation. Women benefited most from household electrification, but time allocation studies show that reduced heavy labor was not accompanied by significantly decreased total labor time or more leisure hours (Minge-Klewana 1980). Because of the forgone labor of children; women work more in modern households than they did in traditional families.

The automobile must also be singled out when looking at energy-related time-saving and leisure activities, because of its contradictory contribution and its inordinately high social valuation. In Boulding's (1974) memorable analogy, a car turns a driver into a knight riding a mechanical steed, making it hard to go back to being a peasant. This rapidly acquired dependence on cars led to extensive spatial and social restructuring of modern society. At the same time, it presented a formidable obstacle

ENERGETIC CORRELATES

to rationalization of energy use and to the introduction of alternatives, and it has become increasingly counterproductive. Illich (1974) calculated that after taking into account the time needed to earn money for the purchase of the car, fuel, maintenance, and insurance, the average speed of U.S. car travel amounted to less than 8 km/h.

I recalculated the rate for the typical urban U.S. situation of the early 2000s and found the speed no higher than 5 km/h. In some cities nearly permanent traffic congestion has reduced driving speeds to levels not much superior to those achieved before 1900 with horse-drawn omnibuses and electric streetcars. Moreover, the enormous energy losses (rolling efficiencies at best 5% of initial crude oil input) mean that even cars with highly efficient catalytic converters are major contributors to local and regional environmental degradation. Driving also tops the list of all energy-related accidental death causes. In 2003 the global total was 1.3 million fatalities and about 40 million injuries, whose social and economic impact is aggravated by the youth of the victims (Evans 2002).

The realm of energy-related mishaps and exposures is immense, ranging from the frequent but mostly nonfatal presence of *Legionella* bacteria in air-conditioning ducts to such rare but potentially catastrophic events as fires in refineries, failures of large dams, and releases of radiation from nuclear power plants. In addition, there are many grave consequences of terrorist attacks whose assessments exemplify the challenge of appraising potential risks purely on the basis of complex assumptions. For example, increasing reliance on LNG shipments (fig. 12.6) makes it impossible to ignore the risk of catastrophic fires, above all, the uncontainable pool fires of the gas on water (Havens 2003). A 5-min burn of a single 25,000-m^3 LNG tank (tankers usually carry five) would release energy equivalent to about 10 Hiroshima bombs (Fay 1980), but experimental spills have involved LNG volumes 2 OM smaller ($<$50 m^3).

Concerns about a terrorist attack on a nuclear facilities increased after 9/11 (Chapin et al. 2002), as did worries about exploding small nuclear devices in cities. An explosion of a small nuclear device would have the greatest potential impact if a bomb were placed in a float at a major football game (Willrich and Taylor 1974); this would produce a nearly 100% casualty rate among as many as 100,000 spectators, most of them men of productive age. The Three Mile Island mishap in 1979 and the Chernobyl accident in 1986 caused profound reappraisals of the probabilities of risks posed by commercial nuclear plants (Hohenemser 1988). During the late 1980s estimates of the chances of core meltdown in a U.S. nuclear power plant during a 20-year period differed by a factor of 200 (Hively 1988), but the subsequent operating record has shown no deterioration.

Assessments of lifelong, even intergenerational, health consequences, are extremely difficult. As a result, some conclusions put the risks of nuclear generation much lower than the total health costs of coal-generated electricity, whereas others see the nuclear industry as an intolerably risky enterprise. The long-term effects of air pollution generated by large-scale coal combustion are similarly unclear. Attempts to quantify the number of premature deaths caused by emissions from a 1-GW coal-fired power plant produced totals between 0.07 and 400,000 (Ricci and Molton 1986). Such a range is clearly useless for rational decision making, and these uncertainties must be kept in mind when reviewing numerous calculations of energy-related mortality and morbidity risks (Travis and Etnier 1982; IAEA 1984; Fremlin 1987; Sharma 1990).

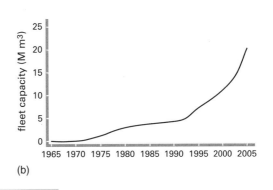

fleet capacity (M m³)

1965 1970 1975 1980 1985 1990 1995 2000 2005

(a)　　　　　　　　　　　　　　(b)

12.6　(a) Modern LNG tanker (photograph by Alstom Transport). (b) Aggregate growth of global LNG fleet (based on a graph in EIA 2004).

A rationally appraised technique should not be judged unacceptable if its risks are several orders of magnitude below those of natural disasters, yet the reverse is true because of the fundamental divide in people's tolerance of voluntary and involuntary exposures (Starr 1969; Starr, Rudman, and Whipple 1976). People are willing to assume voluntary risks about 3 OM higher than risks from exposures perceived as involuntary, such as a nearby siting of a nuclear power plant or an LNG terminal. Natural mortality (the risk of dying is about 10^{-6}/ person $\cdot$ hour of exposure) acts as an important subconscious yardstick in determining the acceptability of everyday risks. Several voluntary activities (driving, commercial flight) carry risks nearly as large as just living, but involuntary air pollution or radiation exposures, whose risks are no higher than those of common natural hazards (earthquakes, hurricanes, tornadoes), that is, 3–5 OM lower than general disease mortality, are often seen as totally unacceptable.

12.4　Energy and War

Wars demand an extraordinary mobilization of energy resources, and modern wars, made possible by energy-intensive weapons, represent the most concentrated and the most devastating release of destructive power. In addition to casualties they bring major disruption of energy supplies in regions or countries that are affected by combat or subjected to prolonged bombing. Given these realities, it is inexplicable that wars have received so little attention as energy phenomena. At the same time, there is a fairly common perception—greatly reinforced by the U.S. intervention in Iraq in 2003—that energy is often the main reason why nations go to war. I address all these issues.

Weapons are the prime movers of war. They are designed to inflict damage through a sudden release of kinetic energy (all handheld weapons, projectiles, explosives) or heat, or a combination both. Nuclear weapons kill almost instantly by combined blast and thermal radiation, and also cause delayed deaths and sickness due to exposure to ionizing radiation. All prehistoric, classical, and early medieval warfare was powered only by human and animal muscles. The invention of gunpowder—clear directions for its preparation were published in China in 1040, and the proportions for its mixing eventually settled at 75% saltpeter (KNO_3), 15% charcoal, and 10% S—led to a rapid diffusion of initially clumsy front- and breach-loading rifles and to much more powerful field and ship guns (Smil 1994). While ordinary combustion must draw oxygen from the surrounding air, the ignited KNO_3 provides it internally, and gunpowder undergoes a rapid expansion to about 3,000 times its volume in gas. The first true guns were cast in China before the end of the thirteenth century, and Europe was just a few decades behind.

Gunpowder raised the destructiveness of weapons and radically changed the conduct of both land and maritime battles. When confined and directed in rifle barrels, gunpowder imparts to bullets kinetic energy 1 OM higher than that of a heavy arrow shot from a crossbow gun (1 kJ vs. 100 J); the kinetic energy of iron balls fired from

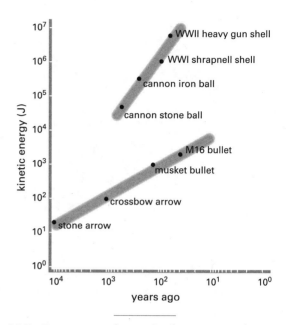

12.7 Kinetic energy of projectiles, from stone-tipped arrows to heavy gun shells. Plotted from data in Smil (2004c).

cannons was 3 OM higher. Increasingly accurate gunfire eliminated the defensive value of moats and walls, and the impact of guns was even greater in maritime engagements (fig. 12.7). Gunned ships equipped with two other Chinese innovations, compass and rudder, as well as with better sails, projected empire-building European power (Cipolla 1966; McNeill 1989). The dominance of these ships ended only with the introduction of naval steam engines during the nineteenth century.

The next weapons era began with the formulation of high explosives prepared by the nitration of such organic compounds as cellulose, glycerine, phenol, and toluene. Ascanio Sobrero prepared nitroglycerin in 1846, but its practical use began only after Alfred Nobel mixed it with an inert porous substance (diatomaceous earth) to create dynamite and after he introduced a practical detonator (Smil 2005a). Dynamite's velocity of detonation is as much as 6.6 km/s, compared to about 400 m/s for gunpowder (Urbanski 1967). Slower-acting and preferably smokeless propellants needed for weapons were introduced before 1900: gelatinized and extruded nitrocellulose by Paul Vieille in 1884; Nobel's ballistite in 1887; and the most powerful of all prenuclear explosives, cyclotrimethylenetrinitramine (cyclonite, commonly known as RDX, Royal Demolition eXplosive, with a detonation velocity of 8.7 km/s), made by Hans Henning in 1899.

Other new highly destructive weapons whose use contributed to the unprecedented casualties of WW I included machine guns, tanks, submarines, the first military planes, including light bombers, and poisonous gases. The two decades between the world wars brought rapid development of battle tanks, fighter planes, and long-range bombers and aircraft carriers; these were the decisive weapons of WW II. The closing months of WW II saw the deployment of two new prime movers—the first gas turbines in flight and the first rocket engines, in the German ballistic missile V-2—and of an entirely new class of weapons, the first fission bomb, tested in July 1945. The increase in maximum destructive power between 1914 and 1945 was astonishing; 100-kg shells fired by large WW I artillery guns had explosive energy of about 400 MJ, and the two nuclear bombs dropped on Japan in early August 1945 released, respectively, 52.5 TJ and 92.4 TJ (CCMD 1981).

Jet propulsion enabled the fastest fighter aircraft to surpass the speed of sound in 1947 and eventually to reach maximum velocities in excess of Mach 3. The postwar arms race between the United States and the USSR began with the assembly of more powerful fission bombs to be carried by strategic bombers. The first fusion bombs were tested in 1952 and 1953, and by the early 1960s the two antagonists were engaged in a spiraling accumulation of intercontinental ballistic missiles. These were not weapons of war. Their real purpose was to deter use by the other side. But in order to achieve this objective, did the superpowers have to amass more than 20,000 nuclear warheads, some of them 3 OM more powerful than the 1945 bombs (Smil 2006)? When expressed in common units of TNT equivalents, the Hiroshima and Nagasaki bombs rated, respectively, 12.5 kt and 22 kt. The most powerful thermonuclear bomb tested by the USSR over the Novaya Zemlya on October 30, 1961, rated 58 Mt, equal to 4,600 Hiroshima bombs. By 1990 the total power of U.S. and Soviet nuclear warheads surpassed 10 Gt.

Modern wars are waged with weaponry whose construction requires some of the most energy-intensive materials and whose deployment relies on continuous flows of gasoline, kerosene, and electricity in order to energize the machines that carry them, and to equip and provision the troops who operate them. The production of special steels in heavy armored equipment typically needs 40–50 MJ/kg, and the use of depleted uranium for armor-piercing shells and enhanced armor protection is much more energy-intensive. Aluminum, titanium, and composite fibers, the principal construction materials of modern aircraft, embody, respectively, 170–250 MJ/kg, as much as 450 MJ/kg, and typically 100–150 MJ/kg.

The most powerful modern war machines are designed for maximum performance, not for minimum energy consumption. For example, the U.S. 60-t M1/A1 Abrams main battle tank, powered by a 1.1-MW AGT-1500 Honeywell gas turbine, needs (depending on mission, terrain, and weather) 400–800 L/100 km. For comparison, a large Mercedes S600 consumes about 15 L/100 km, and a Honda Civic needs 8 L/100 km. And jet fuel requirements of such highly maneuverable supersonic combat aircraft as F-16 (Lockheed Falcon) and F-18 (McDonnell Douglas Hornet) are so high that no extended mission is possible without in-flight refueling from large tanker planes (KC-10, KC-135).

There are no detailed, reasoned studies of the energy cost of modern wars. This is not surprising because it is difficult to calculate even their financial costs. Available aggregates (Smil 2004c) show total U.S. expenditures on major twentieth-century conflicts, in constant 2000 dollars, at about $250 billion for WW I, $2.75 trillion for WW II, and $450 billion (1964–1972) for the Vietnam War. Multiplying the totals by adjusted averages of respective energy intensities of U.S. GDP during those periods sets the minimum energy costs of these conflicts. With conservative multiples of 1.5 for WW I, 2 for WW II, and 3 for the Vietnam War to capture the higher energy intensity of war-related industrial production and transportation, the most likely magnitudes of the energy burden would have been about 15% of total U.S. energy consumption for WW I, about 40% for WW II, but definitely less than 4% for the Vietnam War.

Given the enormous U.S. energy consumption total, the direct peacetime use of fuel and electricity by the U.S. military is a tiny fraction of the whole. In 2005 the Department of Defense claimed less than 1% of the country's TPES (EIA 2006a). But in absolute terms, the direct U.S. military energy use of about 25 Mtoe is roughly equal to the annual TPES of Switzerland or Austria, or more remarkably, higher than the commercial energy consumption in nearly two-thirds of the world's countries. My conservative estimate is that at least 5% of all U.S. and Soviet commercial energy consumed

between 1950 and 1990 was claimed by developing and deploying weapons and their means of delivery, and the burden continues with expensive safeguarding and cleanup of contaminated production sites. Even when limited to targeting strategic facilities, the direct effects of blast, fire, and ionizing radiation would have killed at least 27 million and up to 59 million people (Hippel et al. 1988).

The sudden demise of the USSR did not usher in an era of extended peace but, ironically, one of even more acute security threats. In a complete reversal of dominant concerns, the most common and generally most feared weapons in the new war of terror are both inexpensive and easily available. A few kilograms of high explosives spiked with metal bits and fastened to the bodies of suicide bombers can cause dozens of deaths and gruesome injuries, and create mass psychosis among the attacked population. Simple car bombs are much more devastating. These devices, some weighing 10^2 kg and able to destroy massive buildings and kill hundreds of people, are made from a mixture of two readily available materials, ammonium nitrate (a common solid fertilizer that can be purchased or stolen at thousands of locations around the world) and fuel oil (even more widely available).

The most shocking weapons were fashioned on September 11, 2001, by 19 Islamic hijackers simply by commandeering rapidly moving massive objects, Boeings 757 and 767, and steering two of them into the World Trade Center towers and one into the Pentagon. The WTC towers were designed to absorb an impact of a slow-flying Boeing 707 lost in the fog and searching for a landing at JFK or Newark airport. The gross weight and fuel capacity of such an airplane are just slightly smaller (respectively, 15% and 5%) than the specifications for the

Boeing 767-200, and the structures performed as intended, even though the impact velocity of the hijacked planes was more than three times higher (262 m/s vs. 80 m/s) than that of a slowly flying plane close to landing.

As a result, the kinetic energy at impact was about 11 times greater than envisaged in the original design, or roughly 4.3 GJ vs. 390 MJ. But because each tower had a mass of more than 2,500 times that of the impacting aircraft, the enormous concentrated kinetic energy of the planes acted much like a bullet hitting a massive tree. It penetrated instead of pushing; it was absorbed by bending, tearing, and distortion of structural steel and concrete; and the perimeter tube design redistributed lost loads to nearby columns. Consequently, it was the more gradual flux of the ignited fuel rather than an instantaneous massive kinetic insult that weakened the columns of structural steel. (Each 767 airplane carried more than 50 t of kerosene, whose heat content was more than 2 TJ.)

Unfortunately, the eventuality of such a fire had not been considered in the original WTC design. Moreover, no fireproofing systems were available at that time to control such fires. Once the jet fuel spilled into the building, it ignited an even larger mass of combustible materials (mainly paper and plastics) inside the structures, and the fires burned with diffuse flames at low power densities of less than 10 W/cm^2. The fuel-rich, open-air fire could not reach the 1,500°C needed to actually melt the metal. This left enough time for most people to leave the buildings before the thermally weakened structural steel, thermal gradients on outside columns, and nonuniform heating of long floor joists precipitated the staggered floor collapse that soon reached free-fall speed (hitting bottom at 200 km/h) as the towers fell in only about 10 seconds (Eagar and Musso 2001).

But massive mobilization of energies is no guarantee of the outcome. Rapid deployment of U.S. economic might, energized by a 46% increase in the total use of fuels and primary electricity between 1939 and 1944, was clearly instrumental in winning WW II. In contrast, the Vietnam War showed that in order to win it is not enough to have advanced weapons and to use an enormous amount of explosives (nearly three times as much as all bombs dropped by the U.S. Air Force on Germany and Japan in WW II). And the attacks of September 11, 2001, illustrate the perils of the aptly named asymmetrical threats as enormous damage is inflicted by expending relatively little energy. Nineteen Islamic terrorists, at the cost of their lives and an investment of perhaps less than $100,000, caused about 3,000 virtually instantaneous deaths as well as direct and indirect economic dislocations that led to costly and open-ended deployments of military and covert power.

Finally, a few notes on energy resources as a justification for war. Japan's attack on the United States in December 1941 is often cited as a classic case of a country's going to war to preserve its access to energy resources (R. Stern 2006). In July 1940, President Roosevelt terminated export licenses for aviation gasoline, and the attack on Pearl Harbor was said to clear the way for Japan's control of Sumatran and Burmese oil fields. Declining oil supplies undeniably figured in the decision to attack the United States, but it is indefensible to depict Japan's aggression as solely an energy-driven quest. The attack at Pearl Harbor was preceded by nearly a decade of expansive Japanese militarism, (beginning with the conquest of Manchuria in 1933 and escalated by the attack on China in 1937). Jansen (2000) writes about the peculiarly self-inflicted nature of the entire confrontation with the United States. Further, no convincing arguments can be made for energy supply concerns as a justification for Hitler's serial aggressions against Czechoslovakia (1938, 1939), Poland (1939), Western Europe (1939, 1940), and the USSR (1941) or for his genocidal war against the Jews.

The same is true about the genesis of the Korean War, the war in Vietnam (waged by the French until 1954, by the United States thereafter), the Soviet occupation of Afghanistan (1979–1989), the U.S. war against the Taliban, nearly all cross-border conflicts (Sino-Indian, Indo-Pakistani, Eritrean-Ethiopian), and numerous post-1950 civil wars. And while it could be argued that Nigeria's war with secessionist Biafra (1967–1970) and Sudan's endless civil war had clear oil supply components, both were undeniably precipitated by ethnic differences, and the second began decades before any oil was discovered in central Sudan.

On the other hand, there have been various indirect foreign interventions in Middle Eastern countries (arms sales, military training, covert actions) that aimed at stabilizing or subverting governments in the oil-rich region. Their clearest manifestations during the Cold War were the sales or transfers of Soviet arms to Egypt, Syria, Libya, and Iraq, and U.S. arms sales to Iran (before 1979), Saudi Arabia, and the Gulf states. During the 1980s these actions also included Western support of Iraq during its war with Iran (1980–1988). These interventions culminated in two wars where energy resources were widely seen as the real cause of the conflicts. By invading Kuwait in August 1990, Iraq not only doubled crude oil reserves under its control, raising them to about 20% of the global total, but it also directly threatened the nearby Saudi oil fields and hence the very survival of the monarchy that controls 25% of the world's oil reserves.

Yet even in this seemingly clear-cut case there were other compelling reasons to roll back the Iraqi expansion (Lesser 1991). Hussein's quest for nuclear and other nonconventional weapons with which the country could dominate and destabilize the entire region, implications of this shift for the security of U.S. allies, and risks of another Iraqi-Iranian or Arab-Israeli war (recall the Iraqi missile attacks on Israel design to provoke such a conflict) mattered a great deal. And if the control of oil was the sole, or at least the primary, objective of the 1991 Gulf War, why was the victorious army ordered to stop its uncheckable progress as the routed Iraqi divisions were fleeing north, and why it did not occupy at least Iraq's southern oil fields? By 2005 nobody could deny that if the U.S. objective were gaining access to Iraqi oil, it would have been much cheaper (and casualty-free) to give Saddam Hussein interest-free loans to boost Iraq's production capacity rather than to occupy the country in 2003.

12.5 Energy and the Future

I recognize the value of exploratory forecasts to stimulate critical thinking and as tools for formulating and criticizing new ideas. But most energy forecasts fall into a much less exalted category of plain failures (Smil 2003). The reasons for these failures are several. On the most general level is a longstanding Western intellectual tradition that favors a catastrophic view of the future. Fear of running out of resources in general, and of energy in particular, is a relatively recent addition to this genre. Its great nineteenth-century classic foresaw the exhaustion of Britain's coal resources, leading to inevitable decline of the country's economic and political supremacy (Jevons 1865).

The British empire did dissolve during the twentieth century, but exhaustion of coal had nothing to do with it; coal remains in the ground but is hardly produced.

In the early 1920s, U.S. geologists predicted an early end to the oil era, and Alderson (1920) believed that retorting of oil from shale and the establishment of a large-scale oil shale industry was the decade's key energy challenge as "the finger of fate points towards it." But fate manifested itself with the discovery of a supergiant East Texas oil field in 1930, and oil shale dreams were revived (and abandoned) only during the 1970s. Putnam (1954) predicted the global depletion of recoverable fossil fuel reserves between 2000 and 2025, and 25 years later the CIA warned that "the world does not have years in which to make a smooth transition to alternative energy resources" (CIA 1979, iii). Did they really believe that such a shift could be done in a few months? Two decades after this prediction, oil's global R/P ratio stood at a near-record level, but this did not prevent end-of-millennium fears of the immediate peaking of oil extraction (see section 8.2).

Another powerful factor molding our perceptions is the vision of the future as a replica of the recent past. This influence was particularly strong after 1950, when sustained global economic growth and tight energy/GDP ratios provided a seemingly perfect foundation for trend forecasts. Mesmerized by the regularity of such indicators (looking further back would have shaken this faith), forecasters ended up with exponentially rising curves and enormous future consumption totals. U.S. forecasts prepared between 1960 and 1970 fared particularly badly. Their mean value for the year 2000 was just over 170 EJ (maxima >200 EJ), whereas the actual total was less than 100 EJ, a mean overestimate of nearly 80% (fig. 12.8) (Smil 2003).

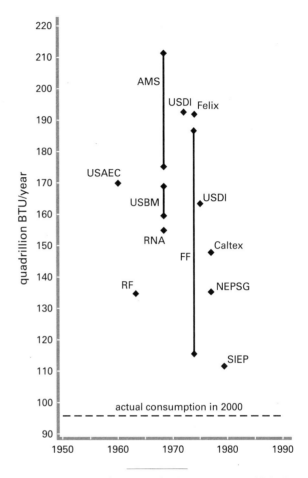

12.8 Forecasts of U.S. TPES for the year 2000, published between 1960 and 1980, averaged nearly 80% above the actual demand. Forecasts, identified by acronyms of the issuing organizations, are plotted as originally published. From Smil (2003).

Reasons for large numbers of wrong forecasts can be found in the herdlike behavior of forecasters smitten by prevailing moods (Leydon 1987; Smil 2003). In the mid-1970s forecasts of global oil demand for the year 2000 suggested levels between 185–220 EJ; in the mid-1980s they dropped to just 118–133 EJ, lower than the actual 1980 consumption (actual total in 2000 was 152 EJ). The human propensity to conceive of the future as an extension of the immediate past indicates a striking incapacity to learn from previous shifts and discontinuities. Forecasters may keep ignoring incontrovertible evidence of past trend shifts, but they share infatuation with new energy sources and conversions.

Basalla (1982) illustrated this propensity by pointing out the recurrent myth of new energy sources as ultimate solutions. Coal's mystique was transferred first to hydroelectricity (white coal), whose large-scale development appealed as much to Lenin ("Communism is the Soviet power plus electrification!") as it did to Roosevelt. Nuclear delusions began during the 1950s with promises of electricity too cheap to meter, and Seaborg (1972) foresaw not only electricity generation dominated by fission (and increasingly by fast breeder reactors) by 2000 but also commercial fusion and nuclear-propelled spaceships ferrying people to Mars. Nuclear explosives were to be used in mining and blasting new harbors and canals (Kirsch 2005). The U.S. Atomic Energy Commission's 1974 forecast had 260 GW installed in the United States by 1985, and 1.2 TW in 2000. The actual 2000 total was 81.5 GW, and there were no clear prospects for fusion.

The same adjectives used to extol nuclear generation—inexhaustible, cheap, nonpolluting—reappeared in glowing descriptions of renewable energetics published during since the 1970s as the advocates of small-scale, decentralized energy production promised a new,

morally superior millennium devoid of nuclear and fossil fuel sins. The virtues of hydrogen-powered society have been extolled for decades (Hoffmann 1981; Rifkin 2002; NRC 2004), but costs and complications of producing this secondary energy carrier and setting up the requisite distribution infrastructures keep pushing the dates of a hydrogen economy farther into the future. And then there are utterly unrealistic ideas, ranging from Goela's (1979) wind harnessing with huge kites to the tapping of meltwaters from the Eastern Greenland glaciers (Partl 1977). On even grander scales is electricity generation by letting the Mediterranean Sea waters to fall into the 18,000-km^2 evaporative lake formed in Libya's Qattara depression, extraterrestrial capture of solar radiation by geostationary satellites (Glaser 1968), and Moon-based photovoltaic plants beaming microwave energy to the Earth (Criswell 2000).

In the real world we have to accept the limits of our understanding. Thus far, fossil-fueled civilization has enjoyed relatively easy access to resources recoverable with high net energy returns. Indeed, the ease of discovery, cheapness of extraction, and size of reserves of many supergiant oil and gas fields provided extraordinary energetic boons. Clearly, a civilization coping with falling net energy returns will have to make many adjustments. But this reality in itself may not be intolerably restrictive. What we do with the available energy will matter more than its net costs. And even a perfect prevision of particular energy techniques would be of limited help. The challenge is much harder: to envisage what the whole society will be like. Given our powers, this should mean what we aspire it to look like and how much we are willing to adapt (dare I say sacrifice?) in order to bring it about. There are many possible futures within the confines of resource availability and thermodynamic imperatives. Civilization's course is not preordained but remains open to our choices.

Not everybody would agree with this conclusion. We may be embarked on an irreversible course of destruction, the only uncertainty being about the eventual causes of our demise: environmental degradation, nuclear war, or a virulent pandemic. Gentler, but no less comforting, scenarios of human demise are possible. Wesley (1974) put forth a reasoned case for an early elimination of people by machines. He argued that the point of no return came around 1830, when the steam engine became more efficient than muscles. Since then humankind has become critically dependent on machines, whose evolution (including vigorous reproduction and selection for higher fitness, increased speciation, improved ecological efficiency, and rapidly growing mass) has destroyed a great deal of the Earth's life. My calculations show that by 2000 the global car fleet alone was 1 OM heavier than the anthropomass ($\sim$1.5 Gt vs. $\sim$100 Mt) and that its carbon needs were much higher and are rising much faster (humans harvest $\sim$1.3 Gt C as food; almost 2.5 Gt C goes into car making and fuels).

Eventually machine mass could surpass total planetary biomass. Machines could become autonomous, in which case Wesley (1974) would allow no more than one more century before the complete disappearance of *Homo sapiens*. More recently Moravec (1999) and Kurzweil (1990) suggested a much faster takeover by superintelligent computers, claiming that perhaps as soon as 2040 thinking machines cut loose, develop hyperintelligence, and bring about our demise. These may be incredibly prescient visions or just self-destructive wishful thinking. In any case, such multidecadal forecasts are always questionable, and truly long-term perspectives (10^2–10^4 years, the latter equal to the time it took us to move from Neo-

lithic foraging to fossil-fueled civilization) remain utterly elusive.

Hubbert (1962) popularized the notion of fossil-fueled civilization as a brief interlude between the solar past and the solar (nuclear?) future. Many believe that rapid climate change may imperil civilization long before any real shortages of relatively expensive fossil fuels, perhaps even before 2100. Four or five generations is a short span even in historic terms, but what could be accomplished during that brief period is best appreciated by trying to look at the end of the twentieth century from the vantage point of 1900. Crises and discontinuities provide the challenge and drive adaptive change, and the inevitable transition from fossil-fueled societies to a new global system will not be marked by applying ancient precepts to unprecedented tasks.

Energy decisions involve often irreconcilable considerations of thermodynamic efficiency, personal comfort, resource depletion, economic well-being, environmental degradation, national security, social stability, and democratic values. This reality is incompatible with any strategic optima; it merely admits a choice of practical alternatives. Optimal allocation of finite resources would not be possible unless we knew the demand over the entire future. Hotelling (1931) knew this, but thousands of modelers apparently do not. And often there is an intuitive sense of an optimum where none exist. During pre-industrial millennia, the horse provided the fastest individual land transport, but there is no best way to sit on a horse (Thomson 1987). True believers sell optimal photovoltaic, fusion, or hydrogen futures, but the safest bet is that there is no single best solution of the global energy challenge.

All we can do with some semblance of rationality is to map our tasks for the next few generations. The

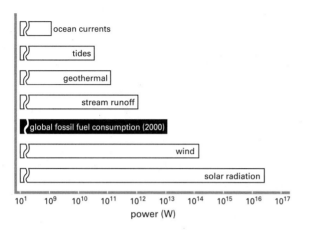

12.9 Comparison of global renewable energy flows and the world's annual TPES.

imperatives of ultimate resource availability, infrastructural inertia, and contrasts between conversion and consumption power densities limit our choices. Solar radiation is the only truly unlimited renewable resource, 3 OM larger than global TPES at the beginning of the twenty-first century. Other fluxes could supply only a fraction of existing demand even if they were exploited to the limit of technical capacities (fig. 12.9). At the same time, enormous infrastructural investment in our cities, industries, transportation, and energy networks creates a highly inertial socioeconomic arrangement. In the long run we may deurbanize; smaller, stabilized populations with access to advanced communication techniques may make large cities obsolete. But during the twenty-first century we will still have to cater to scores of megacities and conurbations, which are growing rapidly in most poor countries.

The further extension of these inherently energy-intensive systems makes large-scale high-power-density conversions producing electricity and liquid fuels indispensable for decades to come. Any rapid substitutions by low-power-density renewable flows are illusory without dismantling existing urban societies. Escaping the imperatives of scale built into the world's energy system by more than a century of fossil fuel combustion and electricity generation will not be easy. In fossil-fueled civilization we have been shifting downward, producing fuels and thermal electricity with power densities 1–3 OM higher than the common power densities of energy use in buildings, factories, and cities.

As a result, transportation and transmission rights-of-way vastly surpass the land claims of fixed extraction and conversion infrastructures. In a solar civilization inheriting today's urban and industrial systems, we would at best harness energies with the same power densities with which they are used, and more often we would have to concentrate diffuse flows, bridging power density gaps of 2–3 OM. This would not only increase fixed land requirements but also require more extensive transmission rights-of-way. The inflexible location of electricity generation based on renewable flows is another key concern. But a solar society would face by far the greatest land requirements if it were to supplant all crude oil–derived liquid fuels used in modern transportation by phytomass-derived ethanol.

During the first years of the twenty-first century global consumption of gasoline and diesel fuel in land and marine transport, and kerosene in flying, was about 75 EJ. Even if the most productive solar alternative (Brazilian ethanol from sugarcane at 0.45 W/m^2) could be replicated throughout the tropics, the aggregate land requirements for producing transportation ethanol would reach about 550 Mha, slightly more than one-third of the world's cultivated land or nearly all agricultural land in the tropics. Thus global transportation fuel demand cannot be filled by even the most productive alcohol production. In the United States corn-derived ethanol provided an equivalent of less than 1.5% of gasoline on an energy basis in 2005, and its output should double by 2012 (Farrell et al. 2006). Ethanol's power density of 0.22 W/m^2 means that about 390 Mha (slightly more than twice the country's entire cultivated area) would be needed to satisfy the U.S. demand for liquid transportation fuel. The power densities of a fully solar operation (fueling the machinery with ethanol, distilling with heat derived by the combustion of crop residues) would drop, even with the highest claimed EROI, to about 0.07 W/m^2.

The United States would then require 1.2 Gha, more than six times its entire arable area and about 75% of the world's cultivated land, planted to corn destined for fermentation. The prospect does not change radically by using crop residues to produce cellulosic ethanol because only a part of these residues should be removed from fields in order to maintain key ecosystemic services of recycling organic mater and nitrogen, retaining moisture, and preventing soil erosion (Smil 1999a). Moreover, even large efficiency improvements in alcohol fermentation or car performance will not make up for the inherently low power densities of cropping. The U.S. transportation sector, three times more efficient than it was in 2000, would still claim some 75% of the country's farmland if it ran solely on ethanol produced at rates prevailing in 2005.

Nor would it be easy to supplant the most important metallurgic use of fossil fuels in iron and steel production. A return to charcoal would be the only practical choice. Using the best Brazilian smelting practices (0.725 t of charcoal/t of pig iron) and yields of 10 t/a for tropical eucalyptus (Ferreira 2000) would require (for nearly 600 Mt of pig iron smelted annually in the early 2000s) about 250 Mha of tree plantations. Half of Brazil's total forested area in 2000 would have to be devoted to growing wood for the world's metallurgical charcoal, a most unlikely proposition. And it would be even more difficult to solarize the production of nitrogenous fertilizer. Haber-Bosch synthesis uses mostly natural gas as a source of hydrogen and as a fuel (oil and coal are more cumbersome choices), and no large-scale nonfossil alternatives to this technique are commercially available (Smil 2001).

Solarization will be challenging even in the case of the best match of conversion and final utilization densities. A nearly perfect power density overlap between small-scale solar conversions and household energy needs means that mature, reliable flat plate and photovoltaic techniques could cover significant portions of domestic energy needs in warmer, sunny regions without any fundamental changes of current land use. And on a sunny January day, enough radiation could be captured even at $50°\text{N}$ to heat a superinsulated single-story house by rooftop collectors. But heating all houses through the entire winter in temperate latitudes would be very difficult without interseasonal energy storage and major modifications of both neighborhoods and energy needs. Photovoltaic cells, even on ideally oriented (SW exposure) roofs, would not suffice to cover peak daily needs or heating loads during cold but overcast days. Two- and three-story houses would be especially disadvantaged, as would many old houses with insufficient insulation. For such dwellings any retrofits of expensive energy storage would be prohibitive. In reality, unsuitably oriented roofs are common, as is partial or total shading of houses by surrounding trees and buildings.

Clearly, the most efficient solar housing must be planned and built *de novo*. On the other hand, many large-roofed, single-story buildings with low energy needs (warehouses, many offices, factories) could use their roofs to generate surplus heat or electricity for neighboring structures. But it may be impractical to solarize an energy-intensive factory in a decentralized manner, and impossible to do so for high-rises or downtowns: they would have to rely on transmitted electricity supply from central photovoltaic stations. These realities call for gradual but fundamental adjustments. A combination of very low population growth (even a decline), enormous opportunities for higher conversion efficiencies, and continuing shifts toward lower energy intensities make it possible to imagine not only stationary but even declining TPES among the world's richest 1 billion people. Moderation of extraordinarily meaty Western diets would greatly reduce farming energy subsidies, and concerns about climate change and the safety of nuclear generation should favor rational uses of energy.

In contrast, the still growing populations in Asia, Latin America, and Africa need higher TPES. Their traditional reliance on biomass energies has been one of the principal causes of deforestation, desertification, and erosion. Some of their cooking needs can be met by cultivation of suitable fuel wood species and by diffusion of improved stoves, but the use of crop residues for fuel should be generally discouraged, and the production of biomass-derived ethanol makes sense only in some tropical land-rich countries (even there, such conversion could reduce the food harvests). Some countries still have considerable untapped hydro potential, and wind and geothermal electricity can make important local and regional contributions, but both the resource magnitudes and power densities dictate that solar radiation must eventually be the largest component of renewable energetics.

It will take long time before renewable conversions supply the energies needed for farming subsidies or for most industries and transportation. And even small- and medium-scale uses of new renewables will be complicated by their seasonality, intermittence, and uneven spatial distribution. These realities belie any simplistic notions about the near-term reach of large-scale renewable conversions. Lovins (1976) anticipated that in the United States renewable conversions would supply about 32 EJ by the year 2000 (30% of the total), but after the subtraction of large-scale hydrogeneration, new renewables contributed just 3.2 EJ, only one-tenth of Lovins's forecast (fig. 12.10). Sørensen (1980) put the share at

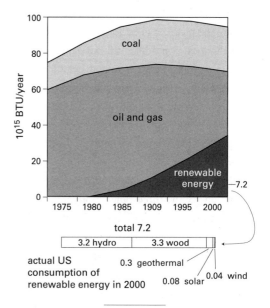

12.10 Amory Lovins's 1976 forecast for the year 2000 of the share of renewable energies in the U.S. TPES. The actual achievement was 1 OM lower.

CHAPTER 12

49% by the year 2005, with bio-gas and wind each supplying 5%, and photovoltaics 11%, of the total. Actual shares were 0% for bio-gas, 0.04% for wind, and 0.08% for photovoltaics, so the forecast was off by at least 2 OM.

At the same time, we should resist any return to visions of massively centralized energy schemes such as those promoted during the late 1970s (ESPG 1981). Extreme views of energy futures lead to unrealistic global expectations. By the year 2030 the soft path was to reduce the world's 1980 energy use by half, and the hard option was to lead to as much as a fourfold rise in total primary energy conversions. Despite their huge differences both visions shared the bias of a preferred technical fix, a misplaced and mistaken faith in a particular set of techniques as *the* solution to complex energy challenges (Smil 1987). We must shun grand designs. We need workable, reliable, economical, and environmentally acceptable approaches. We cannot precipitately abandon our primary energy sources without profoundly reshaping our way of life. But we also need to begin, with vigor and determination, the inevitable transition to the post–fossil fuel world.

Inevitably, there is apprehension—and there is hope. On the debit side are the concerns about the availability of fossil fuels extractible with high EROI, environmental consequences of energy conversions, and bridging the enormous gap between rich and poor economies. On the credit side is our better understanding of the biosphere, our technical ingenuity, and our social adaptability. The key ingredients of successful long-term strategies are clear. Above all, we should not encourage questionable demands, equate them with needs, and fill the gap by providing more energy. And we should not believe that the genuinely higher need for energy services must result in a higher supply of energy. Most of our ingenuity should be devoted to the reduction of final uses rather than to the expansion of primary supply.

It would take monumental intellectual arrogance to maintain that we can decide a priori what will work best. As Sophocles knew, "One must learn by doing the thing, for though you think you know it, you have no certainty until you try." The need to tolerate uncertainty, that essential openness of the human future, is both uncomfortable and promising. John von Neumann (1955, 152) summarized the task perfectly:

The one solid fact is that the difficulties are due to an evolution that, while useful and constructive, is also dangerous. Can we produce the required adjustments with the necessary speed? The most hopeful answer is that the human species has been subjected to similar tests before and seems to have a congenital ability to come through, after varying amounts of trouble. To ask in advance for a complete recipe would be unreasonable. We can specify only the human qualities required: patience, flexibility, intelligence.

13

GRAND PATTERNS

Energetic and Other Essentials

Constantly consider how all things such as they now are, in time past also were, and consider that they will be the same again. And place before thy eyes entire dramas and stages of the same form, whatever thou hast learned from thy experience or from older history.
Marcus Aurelius (121–180), *Meditations*

Everything in the observable universe can be seen, analyzed, and explained in energy terms. The evolution of life on Earth has been energized by radiation streaming from the closest star and also, in a minor but fundamental way, by the planet's internal heat. The Earth's biosphere is an intricate, interactive assembly of energy stores and flows. Life maintains itself negentropically in constant disequilibrium with its surroundings through incessant imports and conversions of external energies. The development of civilization has been a quest for higher energy throughputs transformed into larger anthropomass and greater complexity. And yet this exis-

tential prime mover is an abstraction. Energy is an intellectual construct, a concept evolved by a small group of nineteenth-century scientists in order to analyze and explain a variety of natural phenomena ranging from the hue of arterial blood to the efficiency of mechanical engines. They elucidated energy's permanence as well as its changing quality and diminishing availability inherent in all of its conversions.

Since then science has delved into myriads of energetic phenomena (reaching all the way into the ephemeral world of subatomic particles to the structure of galaxies), and applied research and engineering have introduced an astonishing variety of practical energy conversions. This has resulted in an overwhelming wealth of particularistic understanding, but broadly based interdisciplinary syntheses aimed at fairly comprehensive presentations of general energetics, or at least of its principal constituent parts, have been rare. This is not surprising, given the predilection of modern science for ever more specialized inquiries and compartmentalization of knowledge. This final, integrative, chapter does the very opposite; it aggregates the most important findings of general energetics.

13.1 Energy in the Biosphere

The Sun floods the Earth with a surfeit of energy. Of 174 PW intercepted by the planet, about 122 PW (240 W/m^2) is absorbed by the biosphere. The fate of this absorbed radiation is strongly influenced by a highly unlikely atmospheric composition that makes it possible to have huge amounts of liquid water. In turn, water's unique energetic attributes—high specific heat, high heat capacity, and high heat of vaporization—are decisive in storing and redistributing absorbed solar radiation. Slightly more than one-third of the absorbed flux (45 PW) drives the planetary water cycle, and only a small share of it (3.5 PW) is needed to keep the atmosphere in motion (fig. 13.1). Power densities of solar flows range from 10^3 W/m^2 for peak (noontime) direct radiation to 10^0 W/m^2 in ocean waves, river runoff, and winds. The kinetic energies of air and water can be highly destructive because vertical power densities of cyclonic flows and floods commonly surpass 10 kW/m^2 and reach up to 1 MW/m^2. These violent kinetic events have shaped the evolution of ecosystems, yet in aggregate terms all rain-carrying cyclones release much more energy as latent heat. But the kinetic energies of precipitation and the potential energies of water have been key agents of geomorphic and ecosystemic change.

The Earth's solid crust is made up of huge rigid plates subjected to slow cycles of creation and destruction, which are energized by basal cooling of the planet and by radioactive decay of several isotopes. Lunar and solar gravity are responsible for tidal friction. In comparison with solar flows, both the aggregate terrestrial heat flux (~42 TW) and its average power density (80 mW/m^2) are minuscule. But these low-power flows have been reshaping ocean floor, breaking apart and reassembling the continents, building mountain ranges, and recycling plates into the mantle. Dramatic displays of concentrated geothermal releases, earthquakes and volcanic eruptions, have also had a great impact on the biosphere's evolution.

There is a surfeit of free energy in the biosphere because only a fraction of available solar radiation is needed to energize the water cycle and atmospheric motion, and tectonic processes require only a very small part of planetary heat. This is also true as far as photosynthesis is concerned. Photosynthesis is not, with obvious exceptions of high-latitude winters and plants in dark forest undergrowth, generally limited by the availability of solar energy but rather by a relatively low concentration of atmospheric CO_2, by temperature, and by water and nutrient supply. Reaction losses and autotrophic respiration limit photosynthetic efficiency to no more than about 5%, even under optimum conditions for the most efficient plants. Species with suppressed photorespiration (using the C_4 photosynthetic pathway) have an efficiency advantage. C_4 plants also have substantially higher water use efficiencies. Unfortunately, most of the major crops are C_3 species. The CAM pathway is an excellent adaptation to extreme temperature and water stresses, but its productivity is necessarily very low.

Assessments of phytomass stores (dry matter energy densities 17–20 kJ/g) can be only approximate. The global aggregate is about 1 Tt, or 20 ZJ, with more than 80% of the total in woody tissues. Continuing destruction of forests is the fastest way to reduce the Earth's phytomass and, in the case of species-rich tropical ecosystems, to lower dramatically the diversity of life. Small

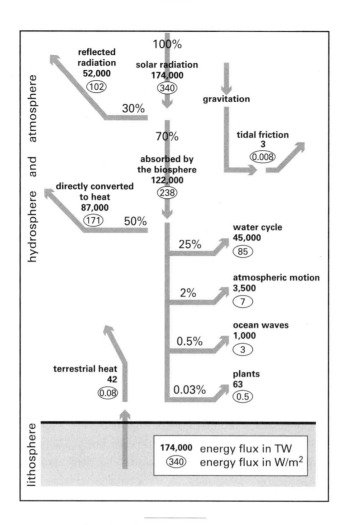

13.1   Global fluxes and reservoirs of energies.

sizes and rapid turnovers mean that oceanic phytomass equals less than 1/1000 of continental stores. Best estimates of global terrestrial NPP are 50–60 Gt C, with forests contributing nearly half, grasslands almost one-third. Marine NPP (45–50 Gt C) is almost as large as the continental total. NPP efficiencies are highest in wetlands (>2%) and rarely surpass 1.5% in forests. The terrestrial mean is about 0.3%; the oceanic average is 1 OM lower because of the nutrient-poor euphotic zone. Net ecosystem productivity is highest in grasslands and croplands, lowest in forests. The energetic imperatives of forest growth favor evergreens in arid and boreal environments (where the cost of annual replacement of leaves would be prohibitive), limit the number of surviving mature trees per unit area (self-thinning process), and determine the efficiency of nutrient use.

Heterotrophs found it profitable to feed on the abundant polymers synthesized by autotrophs or to eat other heterotrophs. Of their two principal metabolic choices, aerobic respiration has a decisive energetic advantage (1 OM higher energy gain per unit of matter) over anaerobic fermentation; all higher organisms use it. Complex plant carbohydrates are the most abundant source of feed energy for heterotrophs; proteins (critical building blocks of animal tissues) are used as energy sources only when the supply of carbohydrates and lipids is limited. As with autotrophs, ATP is the energy carrier of the intricate heterotrophic metabolism, whose regularities on the organismic level are best expressed by many allometric equations (exponential relations related to body mass). Most heterotrophic BMRs scale as body mass raised to powers of 0.66 or 0.75, and disputes continue regarding the value and universality of the exponents and the reasons for their regularity.

The requirements of reproduction, growth, and locomotion raise the total energy need substantially above BMR, but thermoregulation is generally the highest constant endothermic energy burden. Ectothermic thermoregulation is overwhelmingly behavioral (choice of suitable environments), although some insects can become temporarily endothermic, and tuna and sharks have specialized endothermic muscles. Ectotherms cannot be active in thermally extreme environments, nor can they be as competitive in optimal temperatures as endotherms (fig. 13.2). Carrying a uniform thermal environment confers enormous survival and competitive advantages on endotherms, and they have radiated to virtually every terrestrial niche. But this evolutionary strategy has a high energetic cost. In order to be above the ambient temperature most of the time (to facilitate cooling through cutaneous evaporation and sweating), the body temperatures of endotherms are relatively high 36°C–42°C), close to the thermal decay threshold of proteins. Maintenance of these temperatures requires much higher feeding rates than in ectotherms and limits the share of energy that can be diverted to reproduction and growth.

Endothermy also limits the size of the smallest animals because the rising specific metabolic rates with falling mass require higher frequency of feeding in smaller creatures. Larger animals with sufficient fat stores go through extreme cold with only minimum lowering of body temperatures or remain active thanks to their superior insulation, but they face greater difficulties in coping with extreme heat. Endothermic feed assimilation efficiencies are generally higher than in ectotherms (70%–90%), but endotherms pay an energetic price in terms of much lower growth and ecological growth efficiencies. These rates are invariably much below 5% compared with ecto-

energetic characteristics	ectotherms	endotherms
body mass (g)	10^0-10^5	10^1-10^7
basal metabolic rates (mW/g)	0.25-0.30	1-10
metabolic scopes	2-10	6-30
growth efficiencies (%)	20-40	1-4
maximum predator/prey ratios (%)	10-80	1-3
daily movement distances (km)	0.1-1	0.5-12
social behavoir	weak	developed

(a)

energetic characteristics	large terrestrial mammals	*Homo sapiens*
adult body weights (kg)	10-2700	40-100
gestation periods (days)	60-450	266-294
basal metabolic rates (mW/g)	0.8-4.6	1-2.7
metabolic scopes	10-32	15-25
maximum perspiration rates (g/m²-h)	100-250	300-500
encephalization quotients	0.3-2.4	7.3-7.7
longevity (years)	1-50	50-80

(b)

13.2 (a) Major energetic characteristics of ectothermic and endothermic vertebrates. (b) Major energetic characteristics of large terrestrial mammals and *Homo sapiens*.

thermic rates of over 30% and over 10%, respectively. Most ectotherms also pour much of their production into copious offspring. This strategy limits parental survival and precludes repeated reproduction, but it makes all such species both very efficient colonizers and persistent pests. In contrast, endotherms reproduce more slowly but repeatedly.

Endotherms are unsurpassed masters of both rapid and long-distance locomotion. Their metabolic scopes are mostly around 10 for mammals (>30 for some species) and up to 15 for birds. The maximum aerobic power of endotherms is commonly 1 OM higher than in ectotherms. Larger animals have lower specific locomotion costs, but both massive mammals and tiny birds undertake long seasonal migrations powered by lipid stores. In terms of minimal transportation cost, swimming is the most efficient way of locomotion, running the most demanding. Flying falls in between, but energetic considerations limit the size of flyers to less than 15 kg. Most of the time spent in motion is in search of food, but foraging, even in the absence of predators, is not simply a matter of minimized energy expenditures. The objective is selection of specific nutrients rather than maximization of net energy returns.

Herbivores are not generally energy-limited; their numbers are kept in check by predators whose existence is clearly food-limited. In the most favorable environments herbivores consume up to 60% of NPP, but typical grazing rates are below 10% and less than 1% for vertebrates. Energy transfer efficiency between plants and primary consumers is 1%–15%, whereas carnivores consume 5%–25% of available prey. Inevitably, large losses during energy transfers between successive trophic levels limit the biomasses of heterotrophs to small fractions of standing phytomass. An inverse relation between body size and density means that there may be a few MJ/m² of decomposer biomass (bacteria and fungi) but not even 1 kJ/m² of birds and insectivorous and carnivorous mammals. Large carnivores must be rare; even in the richest ecosystems their mass does not surpass 1%–2% of all herbivorous zoomass. The global total of heterotrophic biomass is highly uncertain, mainly because of the dominant but invisible or hidden bacterial and fungal mass. Wild vertebrates add up to just over 0.1% of continental zoomass, 1 OM less than domesticated animals and also considerably less than planetary anthropomass.

Biospheric energy flows cannot be realistically understood without the appreciation of numerous limiting factors. Low conversion efficiencies in photosynthesis (commonly 1 OM below the genetic potential) are largely due to nutrient deficiencies and limited water availability. Hence PAR alone is a very poor predictor of the spatial distribution of low-productivity ecosystems. Behavior is frequently a major limiting factor for heterotrophs, especially because increasing densities promote territorial strife and aggression reduces the fecundity of many fish, bird, and mammalian species. Physical features of the environment (above all, availability of cover and nesting or denning sites) also limit the spacing, density, and dispersion of vertebrate species even in the presence of abundant food. The diversity of heterotrophs also demonstrates that there are many ways to ensure evolutionary competitiveness.

Although endotherms have an undisputed adaptive edge, ectotherms are both very abundant and highly diversified. And whereas larger bodies entail both thermoregulatory and locomotive advantages, energy harvested daily per unit area is independent of the unit mass of feeding heterotrophs, and hence no herbivorous species can become more successful only because of its bigger size. Energetic considerations alone are also insufficient to explain the spatial behavior of heterotrophs, and the degree of their explanatory power clearly decreases with the advancing complexity of behavior. The evolution of heterotrophs provides clear evidence of deviation-amplifying changes. There was a span of some 2.5 Ga between the emergence of the first prokaryotic cells and more complex eukaryota, but Metazoa were present only 300 Ma later. Similarly, in human evolution there is a huge disparity between the duration of stone-tool cultures (~2 Ma) and the time between first agricultures and industrial civilization (<10 ka).

The rapid ascent of *Homo sapiens*, one of whose consequences has been the decline of wild vertebrate zoomass, is an impressive but worrisome testimony to the success of the most versatile and the most adaptive of all heterotrophs (fig. 13.2). Many vertebrate species surpass us in particular functions, but among terrestrial mammals, humans have no equal as generalists (only muscle-powered flight is beyond our ability). Decoupling of the metabolic cost of running from speed, high metabolic scopes (up to 25 for exceptional individuals), and high perspiration capacity mean that humans are unsurpassed endurance runners. Our unrivaled encephalization enabled us to construct and use a myriad of exosomatic aids that have elevated our existence far above the plane of mammalian heterotrophy and that have conquered all of the planet's terrestrial environments. But human energetics is still far from perfectly understood. Even intake recommendations for some essential nutrients required for human growth, tissue maintenance, and activity are uncertain.

Lipids have the highest energy density, 38 kJ/g, and carbohydrates and proteins provide 17 kJ/g. Food requirements are a complex function of age, gender, body size, activity, climate, and individual BMR. Adult BMRs range from 60 W to 90 W and show considerable variation among equally massive individuals. Kidneys, heart, liver, and brain account for almost two-thirds of BMR, but hard mental work requires a negligible energy markup. The early peak of specific metabolic rates, reaching 2.7 W/kg during the first six months of life, is followed by a steady decline to about 1 W/kg by the age of 70 years. Minimum survival needs are about 1.25

BMR, and normal activities raise the multiple to 1.6–2.1. The energy cost of growth is 15–35 kJ/g of lean tissue (about 20 kJ/g in young children). Pregnancy costs about 27 kJ/g of new tissue. Lactation is energized by fat deposited during pregnancy but, as with pregnancy, its cost among some Asian and African populations may be minimal. Superior metabolic efficiency in nutritionally limited circumstances is the only plausible explanation.

Energy expenditure for walking is a U-shaped function of speed, with minima at 5–6 km/h, but the cost of running remains nearly identical for a wide range of speeds. This impressive feat is the result of human bipedalism and efficient heat dissipation. Core temperature is maintained by dilation of peripheral vessels and shifting of blood to feet and hands and, above all, by copious active sweating. Human sweating rates surpass those of other efficient perspirers (camels and horses) and can sustain prolonged exertions of up to 600 W without any rise in core temperature. Acclimatized individuals can remove over 1.3 kW of body heat by sweating. In contrast, thermoregulation in cold environment relies on increased BMR and pronounced vasoconstriction in extremities. The limits of human physical performance are set by hydrolysis of high-energy compounds.

Anaerobic processes support very high but necessarily brief exertions, rating up to 8–12 kW for trained individuals. All long-term efforts are energized overwhelmingly by aerobic recharge, and the limits of human power are thus largely a function of maximum oxygen intakes. Most people can support rates of 600–900 W, and best performers can go up to 2 kW, or about 25 times the basal metabolic rate. Among mammals this metabolic scope is surpassed only by canids. Efficiency of converting digested food to work is 10%–13% in anaerobic processes and 15%–20% in aerobic exertions. This means that during an 8-h spell the maximum sustained power output of 300–350 W translates into 1.5–2 MJ of useful work at 50–70 W.

Humans have spent more than 99% of their existence as foragers or shifting crop cultivators. The earliest foragers were most likely only opportunistic scavengers, and most gatherers and hunters experienced repeated food shortages. Only those groups that exploited highly productive environments spent a fraction of the time needed by workers of industrial society in order to secure an adequate diet. Prehistoric foragers favored gathering seeds and nuts (energy density of up to 25 kJ/g) and often prolific, though less nutritious, roots and tubers. This was energetically a highly rewarding strategy (net returns 5–15, up to 40). But in most ecosystems seasonal fluctuations of phytomass availability and a variety of natural hazards made the foraging a deanding experience.

The great diversity of habitats and subsistence patterns precludes sweeping generalizations, but several energetic imperatives are obvious. All foraging societies were omnivorous, but except for maritime groups, plant foods were dominant. Much of the plant gathering fits the optimal foraging pattern, but other considerations (the need to secure water, vitamins, and minerals as well as the presence of large carnivores and competing foragers) complicated the process. Gathering in forests, where most of the phytomass is in inedible tissues, required more frequent residential moves. In contrast, the seasonal surfeit of grains, nuts, and roots in grasslands and shrublands allowed for fewer, but longer, camp relocations.

Hunting was largely limited to herbivores, and its net energy returns were often very low, especially in tropical forests with their small folivorous fauna. Larger grasslands herbivores were also preferred because of their higher fat content (lean wild meat has only 6 kJ/g).

GRAND PATTERNS

Group hunting improved the chances of success and hence the rates of energy return. The population densities of foragers ranged over 2 OM, from 1–2/100 km² in forests to about 100/km² in maritime groups, which were among the first societies to adopt sedentary living. The evolution of sedentism, with its food storage, increase of private property, and more complex social structures, was a multifocal and protracted process; agricultures coexisted for millennia with gathering and hunting. This gradual shift from simple, opportunistic, omnivorous heterotrophs to increasingly more adept exploiters of photosynthesis through cropping did not come because of the energetic advantages of farming but because intensifying crop cultivation is the only way to sustain higher population densities. That shift also began civilization's continuing quest for higher energy inputs.

13.2 Energy and Civilization

With agriculture humans ceased to be simple heterotrophs and became increasingly sophisticated manipulators of solar flows and builders of complex societies. Although energy returns in early farming were typically no better (and often somewhat lower) than gains in foraging, agriculture could support larger populations from smaller areas of land. Higher population densities promoted further intensification of farming, that is, higher energy subsidies with attendant technical and organizational advances. The resulting growth of permanent settlements, diversification of manufactures, accumulation of knowledge and private possessions, and evolution of hierarchical structures were key ingredients in the development of civilizations.

The most extensive forms of managed food production—nomadic pastoralism and shifting agriculture —have persisted for millennia alongside intensifying farming, clear testimony to the appeal of minimized energy inputs. For pastoralists, the production of milk, blood, and (secondarily) meat required very low labor inputs, often delegated to children, and supported highly variable but always low population densities. With stocking rates of 0.05–0.15 animal units/ha, grasslands can carry 0.8–2.7 people/km². Shifting agriculture follows a sequence of clearing natural or secondary growth, burning the undergrowth, planting a variety (12–50) of food, feed, and medicinal species, weeding (often also protecting) the plots or gardens, and staggered harvesting. It has typical energy inputs of 1–2 GJ/ha and high net energy returns of 15–30, and it supports 10–30 people/km², 1 OM more than pastoralism.

Continuing population increases shortened the fallow periods of shifting agriculture and eventually led to intensive field cropping. Plowing, requiring animal draft, was its energetic hallmark. Working animals, from donkeys to large horses, delivered drafts of 30–80 kg at speeds of 0.6–1 m/s for effective sustained power of 100–800 W. Horses were the most powerful choice only when properly harnessed and adequately fed. Only the universal adoption of the collar harness and feeding of grains opened the way for heavy horse labor. A good horse could easily do the work of six to eight men but its feed of 110–120 MJ/day meant that relatively large shares of farmland (as much as 25% in the United States) had to be devoted to pastures and feed crops for the animals. In agricultures relying on oxen and water buffalo the shares were considerably lower (5%–7% in India and China). Improved designs and diversification of animal-drawn implements came in the eighteenth century. Steel plows, seeders, reapers, mechanical harvesters, and com-

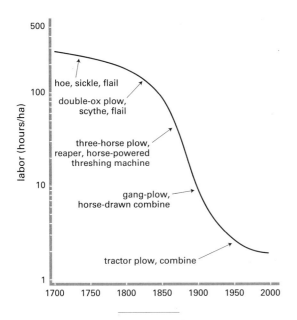

13.3   Declining agricultural labor, illustrated with average inputs into U.S. wheat production.

bines revolutionized traditional farming and sharply reduced the inputs of human labor (fig. 13.3).

Irrigation was the first component of intensified field farming. Its antique origins are attested to by a variety of simple water-lifting machines (*shādūf*, Archimedean screw, paddle wheel, rope and bucket lift, *sāqīya*, *noria*) that were tediously powered by humans or animals. Lifts were only 0.5–2.5 m, efficiencies no better than 20%–30%, capacities 3–15 m^3/h, and energy costs 100–250 kJ/m^3 for human-powered lifts and 4.5–6.5 MJ/m^3 for animal water raising. But energy returns in additional crop yields were high, 10–20-fold. Fertilization relied solely on the recycling of organic matter. Nitrogen was nearly always the limiting nutrient, but all common organic fertilizers (crop residues, manures, and human wastes) have just 0.5%–1.5% of the element. Consequently, effective fertilization required laborious applications of huge amounts of organic matter (annual maxima up to 40 t/ha), but energy gained in additional yields was 10–20 times the energy cost of fertilization.

Crop rotation was the third key ingredient of farming intensification. Leguminous species, fixing their own nitrogen and enriching the soil for subsequent crops, were rotated with staple grains by every traditional agriculture. Soybeans, beans, peas, and lentils were indispensable sources of protein. The least intensive traditional practices could support just 1–2 people/ha of arable land, and typical subsistence agricultures stagnated for centuries at 2–3 people/ha. In contrast, the most intensive agricultures of China and Western Europe, using irrigation, recycling, crop rotation, and multicropping, could support more than 7 people/ha, or with modest vegetarian diets, 12 people/ha. Intensive agricultures had yields up to nearly 50 GJ/ha and overall energy returns of 15–20. Despite enormous cultural differences, traditional peasant societies shared a strong preference for a subsistence compromise by which mimimum levels of material welfare and nutritional safety were acquired with the least expenditure of energy inputs. By minimizing labor inputs, peasants left themselves vulnerable to recurrent famines.

Traditional societies relied on human and animal muscles and on the combustion of biomass fuels. Human labor dominated all ancient societies, from dynastic Egypt to Rome. Classical Greece was no exception. The civilization of philosophers who articulated the concept of individual freedom was energized by plentiful and cheap slave labor. Animals (*tetrapoda*) were more expensive to keep and less flexible to use than *andrapoda* (literally "man-footed," Homer's term in the *Iliad*), mostly war captives used as unskilled labor. The dismal labor productivity and static techniques of ancient Greece were undoubtedly rooted in this reality. Small-scale manufacturing, construction, and land transport (people carrying back-loads, pushing wheelbarrows, or pulling carts, animals pulling wagons) were dominated by inputs of 50–100 W, the sustainable work rate of adults; concentrated animate labor (moving heavy objects, rowing large ships) could deliver briefly 10–30 kW. Considerable ingenuity went into the design of mechanical devices (lever, inclined plane, pulley, treadwheel), but maximum sustained useful power inputs were raised substantially only with the widespread diffusion of waterwheels and windmills (fig. 13.4).

The spreading use of water and wind power, rather than any extraordinary advances in animate power, set medieval societies above their ancient ancestors. The kinetic energies of water and wind harnessed by increasingly more efficient machines provided unprecedented concentrations of power for scores of incipient industrial applications. The unit ratings of waterwheels remained limited for centuries (even in the eighteenth century the typical size did not exceed 4 kW), but multiple installations delivered 10–20 kW at a single site during Roman times and close to 100 kW after the year 1700. Major technical improvements during the nineteenth century led to units of up to 400 kW. Grain milling was the most frequent early use, followed later by many processing and manufacturing tasks. Medieval post windmills delivered just a few kilowatts of useful power, but tower and smock mills of the eighteenth and nineteenth centuries supplied 6–14 kW, much less than contemporary hydraulic installations. In many windy regions windmills played a crucial role in incipient industrialization.

Fossil fuels were known (and sporadically used) by some preindustrial civilizations, but their thermal energy needs came principally from combustion of phytomass. Absolutely dry wood has 17–21 MJ/kg (softwoods have the highest energy density), and crop residues (mainly cereal straws) 17–18 MJ/kg. When burned (air-dried), these fuels usually have about 15 MJ/kg, but most of their heat content was wasted in open fires and simple fireplaces, which deliver no more than 5%–10% of useful energy. In warm climates the annual consumption of phytomass fuels was less than 10 GJ per capita, in the mid-latitudes (and in manufacturing uses) up to 50 GJ per capita, and in the most advanced wooden-age society, the United States in the mid-nineteenth century, close to 100 GJ per capita. Since the late Middle Ages production of high-energy-density charcoal (~30 MJ/

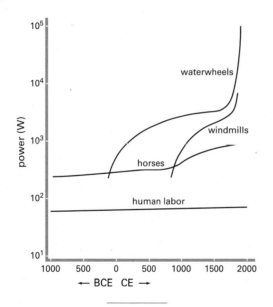

13.4   Maximum unit power of preindustrial prime movers.

kg) for iron smelting has accounted for an increasing share of consumed phytomass.

Traditional charcoaling was very wasteful (5–6 kg of wood/kg of charcoal), as was primitive iron smelting (up to 600 MJ/kg of metal). These demands led to extensive deforestation throughout Europe and Asia. By the end of the eighteenth century the iron smelting rate was below 250 MJ/kg, and the best charcoal-fueled blast furnaces of the twentieth century needed less than 30 MJ/kg. The introduction of coke was one of the two major reasons for rapid expansion of coal mining; widespread adoption of the steam engine was the other. This was a profound change. Earlier societies had been energized by virtually instantaneous solar flows, whereas coal mining laid the foundation for a civilization supported by accumulated stores of fossil fuels. Withdrawals of this fuel capital have energized the exponential increase in global population through unprecedented improvements of agricultural productivity, and have led to large increases in material affluence among the richest one-fifth of humanity as well as to the globalization of human affairs.

In 1800 global phytomass combustion surpassed the burning of coal tenfold; by 1900 the gross energy content of biomass energies was equal to that of fossil fuels; and by 2000 fossil fuels contributed 1 OM more than biomass (fig. 13.5). Because of the superior conversion efficiencies of fossil fuel combustion (on the average at least four times that of preindustrial burning of phytomass), coals, oils, and gases have come to contribute about 40 times more useful energy than plant matter. This dominance has had far-reaching effects on the structure and functioning of human societies. Ours is a fossil-fueled civilization, and its dependence on coals and hydrocarbons cannot be shed without profoundly reshaping the entire society.

Resources that molded our civilization are mineraloids of organic origin (though perhaps some are abiogenic) containing variable shares of moisture, trace elements (notably S and N), and incombustible ash. Energy density is just 8 MJ/kg for the poorest lignites, and it goes up to 36 MJ/kg for the best anthracites. Standard bituminous coal has 29 MJ/kg, most steam coals 20–25 MJ/kg. Complex hydrocarbons give more homogeneity to crude oils; their energy density is 42–44 MJ/kg, and that of natural gases 30–45 MJ/m^3 (3 OM less than that of liquid fuels). Total resources of these fuels are uncertain, but in 2005 proved recoverable stores of all conventional fossil fuels added up to some 30 ZJ, and ultimate recovery estimates were more than twice as large.

Global coal reserves recoverable with current techniques amount to some 160 years of 2005 production. In 2005 the global R/P ratios for crude oil and natural gas were about 40 and nearly 70 years respectively, but considerable uncertainties regarding ultimately recoverable hydrocarbon reserves make it impossible to pinpoint

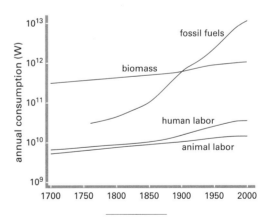

13.5   Global consumption of biomass and fossil fuels, 1700–2000. From Smil (2003).

the time of worldwide peak extraction and the eventual duration of the fossil fuel era. But even the most liberal assumptions regarding ultimately recoverable resources do not change a basic fact: the overall duration of the fossil fuel era cannot last for many centuries. Spatial distribution of fossil fuels is highly unequal, with three nations (United States, Russia, and China) claiming more than two-thirds of all coal and the Persian Gulf–Zagros Basin storing more than half of all recoverable hydrocarbons.

Coal's ascent was based on an enormous investment of hard and dangerous human labor. The mechanization of underground mining spread widely only after 1945; it remains low in Chinese and Indian mines. Daily productivities of 0.5–2 t/miner were substantially lifted by a gradual shift to surface mining. Its relative safety, high coal recovery rate (>90% compared to 50% for the traditional room-and-pillar method), and high productivity (>20 t/shift) make it superior even to long-wall extraction, the best modern underground technique. The power densities of coal extraction range from 1–2 kW/m^2 in underground mines to 20 kW/m^2 in large surface operations. China, the United States, Australia, and Russia produce nearly three-quarters of global output. In 2005 only about 15% of hard coal's global extraction was exported, but crude oil is the world's most valuable traded commodity; 40% of its output was exported.

The slow rise of crude oil production during the late nineteenth century, when it was used mostly for illumination and lubricants, was followed by worldwide growth driven above all by the diffusion of internal combustion engines and the expansion of petrochemical industries. Improvements in drilling (cable-tool drills displaced by rotaries, horizontal wells), transportation (extensive construction of larger and longer pipelines, increased tanker

sizes peaking at just over 500,000 dwt), and processing (larger, more efficient refineries with catalytic cracking) made oil the world's most important source of primary energy. Its production densities are 10–20 kW/m^2, and about 10% of it goes for nonenergy uses (lubricants, petrochemical feedstocks). The Middle Eastern oil fields dominate global extraction.

Outside the United States natural gas became an important fuel only after WW II, with the advent of long-distance, large-diameter, seamless pipelines. Trunk gas lines now connect Canada with the United States, and Europe with Western Siberia, the North Sea, and North Africa. Since the 1960s, LNG exports have been steadily expanding, with Indonesia, Malaysia, Qatar, Trinidad, and Tobago being the top exporters and Japan, South Korea, the United States, and Spain the leading buyers. Natural gas is also an important feedstock (above all, it supplies H for the Haber-Bosch synthesis of ammonia from its elements), and its cleanliness makes it the preferred choice for household and urban uses. Russia and the United States are by far the largest producers.

Since the 1890s increasing amounts of fossil fuels (more than 30% by the year 2000) have been used indirectly as electricity. Edison's brilliant creation of a whole new energy generation, distribution, and conversion system started a still continuing expansion of the most convenient, cleanest (at the point of use), and productively most rewarding source of energy. Growth rates of electricity production have consistently outperformed increases in fossil fuel extraction. Most of the installed global capacity remains in fossil-fueled, especially coal-fired, power plants, but almost 25% of total output comes from hydrogeneration and nearly 20% from nuclear fission. Efficiencies of thermal generation rose from 5% to slightly over 40% in the best plants, long-distance transmission links (often DC) move GW-sized quanta of electricity over distances exceeding 1000 km, and electric lighting and motors are as indispensable in industries as they are in households, services, and agriculture.

The extraction of fossil fuels provides high net energy returns, and in the case of thermal electricity generation we accept substantial exergy loss in return for a high-quality, clean, flexible, precisely controllable energy flow. The highest EROI for Middle Eastern crude oils is about 10^4; typical rates for old oil provinces are 10–20. EROI for refined fuel at the consumer level is less than 10, as are the rates for delivered natural gas. Coal extraction has a wide range of net energy returns, from more than 200 to less than 50, and virtually all renewable energy conversions have EROI less than 10. Electricity generated from fossil fuels contains only 35%–40% of the energy in the charged fuel, but EROI for nuclear electricity is definitely higher than for photovoltaic or wind-based generation.

Rising demand for fossil energies has been accompanied as well as driven by improved performance of new prime movers: increase in unit sizes, decline in power intensities, and higher conversion efficiencies. These advances brought high concentrations of production, processing, and conversion capacities, unprecedented personal mobility, the growth of global trade, and a revolution in agricultural productivity. In turn, these changes led to rapid urbanization, increasing affluence, and continuing integration of the global economy. For millennia animate power limited unit work inputs to 10^2 W, and waterwheels and windmills raised that, locally and sporadically, to 10^3 W. The steam engine era began during the early eighteenth century and lasted for 200 years; by 1905 there could be no doubt about the supremacy of steam turbines. Nearly a century later they still remain

the most powerful continuous-load prime movers. The largest steam turbines, delivering up to 1.5 GW, are about 20,000 times more powerful than was the largest prime mover two centuries ago (fig. 13.6).

The evolution of power intensities (weight to power ratios) progressed in the opposite direction (fig. 13.6). People and draft animals need at least 500 g/W of useful power, and the earliest steam engines were no better. Eventually locomotive steam engines, the prime movers of the first transportation revolution, rated well below 100 g/W. Their performance was vastly surpassed by internal combustion engines, whose power intensities dropped by more than 2 OM in less than a century, to 1 g/W for the best automotive engines and to nearly half that value for the best aeroengines. These innovations ushered in the automotive revolution in land transport and made it possible to realize the ancient dream of powered flight. As the reciprocating aeroengines neared their performance limits during the 1940s, gas turbines took over. Their power intensities fell below 0.1 g/W, opening the way to mass air travel and freight. Gas turbines powering trains and ships also transformed important segments of land and water transport.

Improvements of conversion efficiencies raised the performance of steam engines about 40-fold between Newcomen's top model of 1712 and the best triple-expansion machines of 1900. An efficiency improvement of 1 OM accompanied the development of thermal electricity generation, from 4% during the 1880s to 40% for the best units a century later. Increased unit ratings, lower power intensities, and higher efficiencies of prime movers led to impressive growth of machines and industrial plants and to enormous concentration of extractive and processing facilities. Coal mines grew from small pits producing a few hundred kilowatts to surface giants that extract over

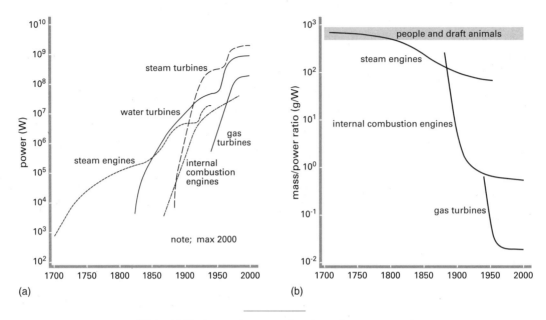

13.6 (a) Maximum unit powers of modern prime movers, and
(b) their weight/power ratios.

1 GW of coal. Thousands of oil fields extract more than 100 MW, some Middle Eastern giants over 50 GW. The largest thermal electricity generating plants surpass 5 GW, the largest hydrostations 10 GW.

The importance of personal transportation is reflected by the exponential growth of car ownership and its vigorous diffusion beyond the once heavily dominant United States. This comes at a major energy cost. Building a car takes 80–150 MJ/kg and running it typically requires 2.5–3.5 MJ/km, considerably down from 6 MJ/km a generation ago but still far from the best commercially achievable rates of just over 1 MJ/km. Declining costs and better performance of heavy-duty prime movers (diesel engines, gas turbines) were critical for the expansion of air travel and a sustained increase in global economic integration. By 2005 foreign trade accounted for about 20% of GWP, compared to less than 1% in 1945. Considering their large structural mass, machines, designed to minimize friction, have been highly energy-efficient providers of transportation (fig. 13.7).

The availability of powerful, lightweight, and efficient prime movers also revolutionized field farming through rising energy subsidies. Where heavily irrigated crops are grown, the cost of irrigation is the largest energy input (up to 20 GJ/ha). In nonirrigated fields fertilizers are by far the largest energy subsidy. The mining and relatively simple processing of potash (energy costs 4–10 MJ/kg K) and phosphates (20–30 MJ/kg P) use only a fraction of the energy costs of Haber-Bosch ammonia synthesis from atmospheric nitrogen and hydrogen (derived mostly

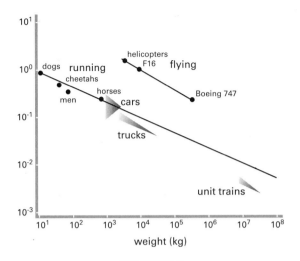

13.7 Minimum cost of transport is calculated by dividing the product of heterotrophic or machine mass (in kg), gravitational constant, and optimum speed (in m/s) into the power input (in W). Cost of flying is about 1 OM higher than that of land locomotion, and automobiles fit right along the line for large mammalian runners.

from CH_4). The process was commercialized in 1913, and its initially very high energy costs (>100 MJ/kg N) were reduced by a series of innovations to less than 33 MJ/kg N by 2000. Nitrogen fertilizers subsequently synthesized from ammonia cost 50–100 MJ/kg, and intensive fertilizer applications represent embodied energy costs of up to 40 GJ/ha.

Agricultural machinery is the third indispensable component of farming energy subsidies. The first trial of a gasoline tractor took place in 1892, and by 2000 there were nearly 30 million of these machines worldwide, averaging 30–50 kW (maxima ~300 kW), costing 70–120 MJ/kg to make and consuming 1.5–3.5 GJ/ha in a variety of field operations. Pesticides are the latest, post–WW II, major subsidy, but because they are applied in relatively small amounts, their high unit energy costs (150–250 MJ/kg) do not add up to a high energy burden. Higher crop yields are not a result of improved photosynthesis. Photosynthetic rates are fixed, but new high-yielding cultivars, especially short-stalked varieties of wheat and rice, have much higher harvest indices (shares of the photosynthate stored in the grain) than traditional varieties, 45%–55% vs. 25%–30%, respectively. The new cultivars respond vigorously to timely water, nutrient, and pesticide applications, which reduce the amount of photosynthate channeled into roots and into replacement of damaged and destroyed tissues. But the peak national harvests of staple cereals (~100 GJ/ha in terms of edible grain) are still only a fraction of record yields.

Synthetic fertilizers account for slightly more than half of all global farming energy subsidies, which prorate from just a few GJ/ha in extensive cereal cropping to about 10 GJ/ha in the United States, nearly 30 GJ/ha in China, and 80 GJ/ha in Israel. The global total of these energy subsidies (~13 EJ in 2000) has expanded nearly 90-fold since 1900, and this has raised average yields nearly fivefold, to more than 26 GJ/ha. Without these subsidies global population would have to decline even if the rich nations were to revert to largely vegetarian diets. With the subsidies the world has a surplus of agricultural production (daily average of about 18 MJ per capita, compared to the minimum need of about 9 MJ). Food shortages and malnutrition are products of unequal distribution and national mismanagement. Rising energy subsidies in farming released the rural labor force to move into the growing manufacturing sector and started a rapid transition to urban society, which is now essentially complete in the rich countries.

13.3 The Challenges Ahead

Global energy use rose nearly tenfold during the twentieth century, driven by a 16-fold increase in the extraction of fossil fuels. Modern, high-energy civilization—marked by megacities, globalized economy, unprecedented levels of affluence, intensive transportation, instant communication, a surfeit of food, and the amassment of possessions—could not have arisen without high energy densities of fossil fuels, portability of refined oil products, and superior flexibility of electricity. The global distribution of energy consumption is skewed. The highest levels of annual per capita use occur in the United States and Canada (>300 GJ) and in the European Union (120–180 GJ). The richest 10% of global population accounts for more than 40% of the world's TPES, whereas the poorest half of humanity consumes just 10% of it (see fig. 9.7). This enormous inequality has been reduced only modestly since 1950. Although there is no need for annual per capita rates to exceed 100 GJ in order to enjoy a good quality of life, there is a clear need to at least double the poor world's rate, which was about 20 GJ in 2005.

Rich nations show no inclination to reduce their energy demand, yet the poor world's demand must increase if the growing populations are to secure a decent physical quality of life and at least a modicum of intellectual advancement. Therefore, the most likely prospect is for substantially higher global energy demand. Modern civilization thus appears to follow the law of maximized energy flows, which Lotka (1922, 148) singled out as a key evolutionary trend: "In every instance considered, natural selection will so operate as to increase the total mass of the organic system, to increase the rate of circulation of matter through the system, and to increase the total energy flux through the system, so long as there is presented an underutilized residue of matter and available energy."

Ensuring the future energy supply will be more challenging than were the extractions and conversions of the twentieth century, and not because of any imminent physical shortages of dominant fossil fuels. The available resource base guarantees that the first half of the twenty-first century can be comfortably energized by high-quality fossil fuels, and the energy intensity of economies should continue to decline, even with business-as-usual practices, because of potential efficiency gains in all sectors. Improvements commonly range between 15% and 40% in modernizing countries (Goldemberg 2000). Most environmental impacts caused by extraction and conversion of fossil fuels are local or regional (surface mining, hot water discharges, visibility reduction, acid deposition) and, with investment, amenable to technical solutions. Similarly, soil erosion, the most widespread environmental cost of energy-intensive farming, is manageable by agronomic measures ranging from windbreak planting to reduced tillage.

Nor is the increased generation of waste heat a major worry. The total anthropogenic power flux of nearly 13 TW in 2005 is less than 0.01% of the solar radiation absorbed by the biosphere. This flux is too small to influence global climate directly. Potentially the most worrisome aspect of fossil fuel combustion is the human interference in the biospheric carbon cycle evinced by the accumulation of tropospheric CO_2. Its atmospheric level exceeded 380 ppm in 2005, and it is rising by 1.2–2.2 ppm/a. The warming effect is exacerbated by rising concentrations of other greenhouse gases, above all, CH_4 and N_2O. This global environmental challenge has no clear and ready technical fix. Complete combustion yields CO_2, little can be done to stop natural methanogenic fermentation, and fortunately, there is no way to prevent biosphere-wide bacterial denitrification (if there were, the biosphere would run out of nitrogen).

In the short run it should be possible to moderate the rate of greenhouse gas emissions (the aim of the Kyoto Protocol). But the effort would be truly effective only if many options were pursued vigorously and simultaneously with effective international cooperation and virtually universal commitment. This seems unlikely during the coming one or two generations, and that is why most strategies aimed at moderating carbon emissions (Pacala and Socolow 2004) will fall short of their goals. In the long run only substantially reduced rates of fossil fuel combustion (stabilization would not suffice) and limits on deforestation, fertilizer applications, and ruminant livestock can break the secular trend of steady increases in greenhouse gas emissions. There is no shortage of bold proposals aimed at making the use of fossil fuels more efficient and their environmental impacts less intrusive, but even the most likely combination of these techniques will not prevent further substantial increases in emissions.

Suggested innovations range from high-efficiency, zero-emissions coal-fired power plants to large-scale sequestration of CO_2. Expensive alterations can make coal-fired plants much cleaner, but contrary to a new-found enthusiasm for CO_2 capture and storage (IPCC 2006), any realistic assessment must see carbon sequestration as nothing but a marginal effort. The scaling challenge is immense: in 2005 the annual storage of the three experimental projects in oil and gas fields rated 1–2 Mt CO_2, and fossil fuel combustion generated more than 7 Gt CO_2. Even if the gas were stored entirely in the supercritical form (CO_2 density 0.468 g/mL at pressure of 71.4 MPa), putting away just 10% of its global flux would require annual handling of a volume equivalent to the worldwide extraction of crude oil.

Nor is there any early possibility of a hydrogen-based system. Undeniably, energy transitions have been steadily decarbonizing the global supply as average atomic H/C ratios rose from 0.1 for wood to 1 for coal, 2 for crude oil, and 4 for methane. As a result, a logistic growth process points to a methane-dominated global economy after 2030, but a hydrogen-dominated economy, requiring production of large volumes of the gas without fossil energy, could take shape only during the closing decades of the twenty-first century (Ausubel 1996). Mass production of hydrogen could eventually take place in facilities energized by advanced forms of nuclear fission or efficient photovoltaics. But hopes for an early reliance on hydrogen are just that (Mazza and Hammerschlag 2004). There is no inexpensive way to produce this high-energy density carrier and no realistic prospects for a hydrogen economy to materialize for decades (Service 2004). A methanol economy may be a better, although also very uncertain, alternative (Olah, Goeppert, and Prakash 2006). There will also be no rapid and massive adoption of fuel cell vehicles because they do not offer any significant efficiency advantage over hybrid cars in city driving (Demirdöven and Deutch 2004).

Prospects for more efficient fission designs remain highly uncertain. Public acceptance of nuclear generation and final disposal of radioactive wastes are the key obstacles to massive expansion. And it is extremely unlikely that nuclear fusion can be part of a solution before 2050, if at all. U.S. spending on fusion has averaged about $250 million a year for the past 50 years with nothing practical to show for it, and the engineering challenges of a viable plant design (heat removal, size and radiation damage to the containment vessel, maintenance of vacuum integrity) mean that the technique has virtually no chance of making any substantial contribution to global TPES in the next 50 years (Parkins 2006).

Gradual transition to a civilization running once again on solar radiation and its rapid transforms (but now converted with superior efficiency) is the most obvious solution to energy-induced global environmental change. Only secondarily, and in a much longer run, it is also the necessity dictated by the limited crustal stores of fossil fuels. Such a transition cannot be fast or easy because it will amount to an unprecedented test of worldwide socioeconomic arrangements. Moreover, this need for a costly and complicated transition to solar-based energetics may become more acute if faster-than-expected global warming forces an earlier, more aggressive reduction of fossil fuel use and an accelerated transition to a nonfossil system. The advantages of this shift are clear (renewability, invariable thermal load, minimized greenhouse gas emissions), but the challenges, limits, and adjustments of such a transition have been commonly underestimated. The shift will present a huge challenge for a still centralizing, rapidly urbanizing, high-energy-density civilization.

Every society is molded by energies it consumes and embodies. A different set of primary energizers must necessarily remold structures and mores in many profound ways. In order to sustain their high power consumption densities, fossil-fueled societies are diffusing concentrated energy flows as they produce and store nonrenewable fuels and generate thermal electricity with unprecedented power densities that are 1–3 OM higher than the typical final-use densities in buildings, factories, and cities. Space taken up by extraction and conversion of fuels is relatively small in comparison with transportation and transmission rights-of-way that are required to deliver fuels and electricity to consumers. Even so, affluent high-energy nations of temperate zone need land equal to at least 10% and up to 20% of their impervious surface area for their fossil-based energy infrastructures. Obviously, these ratios would have to change substantially with transformation to a purely solar civilization.

Insolation is the only renewable flux whose magnitude (122 PW) is almost 4 OM greater than the world's TPES of nearly 13 TW in 2005 (see fig. 12.9). No less important, direct solar radiation is the only renewable energy flux available with power densities of 10^2 W/m^2 (global mean ~170 W/m^2), which means that increasing the efficiencies of its conversion (above all, better photovoltaics) could harness it with effective densities of several 10^1 W/m^2 (best all-day rates in 2005 were ~30 W/m^2). But direct solar conversions would share two key drawbacks with other renewables: loss of location flexibility of electricity-generating plants and inherent stochasticity of energy flows. The second reality poses a particularly great challenge to any conversion system aimed at a steady and reliable supply of energy required by modern industrial, commercial, and residential infrastructures.

Terrestrial NPP of 55–60 TW is nearly five times as large as global TPES in 2005, but proposals for massive biomass energy schemes are among the most regrettable examples of wishful thinking and ignorance of ecosystemic realities. Humans already appropriate 30%–40% of all NPP as food, feed, fiber, and fuel, with wood and crop residues supplying about 10% of TPES. An additional claim of more than 10 TW would require a further 20% of the biosphere's NPP, and it would push the overall claim close to or even above 50% of all net terrestrial photosynthesis. Moreover, the highly unequal distribution of human NPP use means that phytomass appropriation ratios are more than 60% in East Asia and more than 70% in Western Europe (Imhoff et al. 2004).

Claims that simple and cost-effective biomass could provide 50% of the world's TPES by 2050 or that 1–2 Gt of crop residues can be burned every year (Breeze 2004) would put the human appropriation of phytomass close to or above 50% of terrestrial NPP. This would further reduce the phytomass available for microbes and wild heterotrophs, eliminate or weaken many ecosystemic services, and reduce the recycling of organic matter in agriculture. Moreover, the average power densities of NPP are minuscule (about 450 mW/m² of ice-free land). Even the most productive fuel crops or tree plantations have gross yields of less than 1 W/m², and subsequent conversions to electricity and liquid fuels prorate to less than 0.5 W/m².

No other renewable energy resource can provide more than 10 TW. Generous estimates of technically feasible maxima (economically acceptable rates may be much lower) are less than 10 TW for wind, less than 5 TW for ocean waves, less than 2 TW for hydroelectricity, and less than 1 TW for geothermal and tidal energy and for ocean currents (fig. 12.9). These flows can be tapped with densities generally no higher than 10^0–10^1 W/m² and as low as 10^{-1} W/m². In order to energize the existing residential, industrial, and transportation infrastructures inherited from the fossil-fueled era, a solar-based society would have to concentrate diffuse flows to bridge power density gaps of 2–3 OM (fig. 13.8).

The mismatch between the low power densities of renewable energy flows and the relatively high power densities of modern final energy uses means that a solar-based system will require a profound spatial restructuring

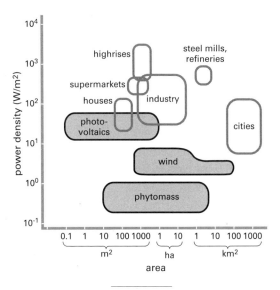

13.8 Mismatch of typical power densities of renewable energy conversions and common energy uses in modern societies.

with major environmental and socioeconomic consequences. Most notably, there would be vastly increased fixed land requirements for primary conversions, and some of these new arrangements would also necessitate more extensive transmission rights-of-way. Efficient and economical means of hydrogen production from renewably generated electricity (or by direct microbial conversion) would be an effective high-energy-density solution.

Only solar energy transformed into phytomass through heavily subsidized photosynthesis (in order to eliminate any water and nutrient shortages and to protect the harvest from pest attack) can be harvested and used predictably. Without massive storage none of the prospective renewable kinetic flows could provide the large base loads required by modern electricity-intensive societies. Yet voluminous water reservoirs have been the only practical means of storing large quanta of almost instantly deployable energy. In spite of more than a century of diligent efforts to develop other effective storages, all other options remain either inefficient or inadequate (fig. 13.9). Opportunities for building large pumped storages are limited. By 2005 there were only two large compressed air energy storage facilities (290-MW Huntorf in Germany and 110-MW McIntosh in Alabama), but several larger projects were in development (Van der Linden 2006). Capacities of flow batteries (typically up to 15 MW), flywheels (containerized 1-MW systems), and superconducting magnetic energy storage (average rating 3 MW) remain limited.

Agricultural adjustments in fully solar societies would be equally profound. Intensive field cropping is a space reduction technique made possible by rising energy subsidies in order to support increasing population densities. In 1900 the nonsolar subsidies in global agriculture added to about 100 PJ, and they helped to produce harvests of about 6 EJ. A century later the subsidies rose to at least 13 EJ, and the output reached about 35 EJ. A purely solar society would have to replace all crude oil–based liquid fuels needed for fieldwork and irrigation, and it would face an additional challenge of using renewably generated electricity to decompose water in order to get hydrogen for ammonia synthesis (the element now comes from natural gas). On the other hand, advanced (and impossible to predict) genetic modification could become a powerful factor in reducing energy inputs. And in the long run it may be even possible to engineer photosynthetic systems for direct production of hydrogen (Smith, Friedman, and Venter 2003), but for neither of these possibilities are there any time and scale specifics.

There are no decision-making shortcuts to aid in selecting and optimizing strategies leading in the right direction. The ultimate makeup of a new global energy system that may dominate in the second half of the twenty-first century will not resemble currently fashionable scenarios. Moreover, prevailing energy pricing does not send the correct signals for the needed change. There is no doubt that energy prices have ignored (or at best only partially internalized) many environmental, health, and strategic externalities, but it is not clear that economic honesty will push us inexorably toward renewable energy (Mintzer, Miller, and Serchuk 1995). Calculating the real cost of energy is a complex challenge that may yield only an unhelpful, too broad range of outcomes rather than a single convincing value.

Careful process energy analyses are a valuable management tool, but thermodynamic efficiency should not become the overriding arbiter in social decisions. Our monetary valuations are very unsatisfactory on many scores, but substituting energy valuations would merely install another misleading denominator. Calculations of net energy returns are desirable (although not necessarily decisive) in assessing the costs of energy supply, but they are irrelevant when they ignore qualitative considerations and nonenergy benefits, for example, food eaten for its protein, vitamins, and minerals or for its taste or cachet, or electricity chosen for its cleanliness, adjustable control, and versatility.

I strongly believe that the key to managing future global energy needs is to break with the current expectation of unrestrained energy use in affluent societies. Of course, Ethiopia or China needs more energy services and hence an efficiently expanded supply. But most of the world's low-entropy flux is used by nations that could derive great benefits from seriously examining their longstanding pursuit of higher energy inputs. At the beginning of the twentieth century Ostwald tied the avail-

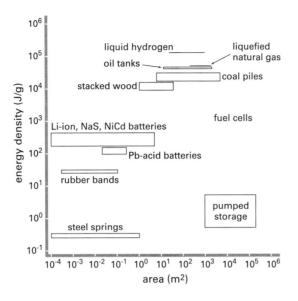

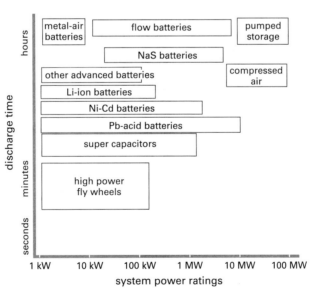

13.9   Densities and discharge times of energy storages. Fossil fuels have energy densities 2 OM higher than various batteries and about 4 OM higher than pumped storages. Discharge times range from minutes to many hours.

GRAND PATTERNS

ability of energy, substitution of labor by mechanical, prime movers, and increased efficiency of energy conversions to cultural progress. And the extension of Lotka's principle of maximized energy flows to human affairs would mean that the most competitive societies would strive for the highest possible energy fluxes.

Historical perspectives cast doubts on the validity of the maximized power stratagem in civilization. Expansions of empires may be seen as perfect examples of the striving for maximized power flows, but societies commanding prodigious energy flows—be it late imperial Rome or the early-twenty-first-century United States—are limited by their very reach and complexity. They depend on energy and material imports, are vulnerable to internal malaise, and display social drift and the loss of direction that is incompatible with the resources at their command. And even at the peak of their physical powers these high-energy societies may not be able to deal with assailants (be they Germanic tribes, Vietnamese peasants, or Islamic terrorists) whose determination more than makes up for their low-energy status.

Higher energy use does not guarantee anything except greater environmental burdens. Higher energy use does not make a country more secure. The Soviet case, with nearly doubled post–WW II per capita energy use but with a crippling share channeled into armaments, was perhaps the most striking example during the latter half of the twentieth century. Enormous energy use could not prevent economic prostration, a fundamental reappraisal of the Soviet strategic posture, and Mikhail Gorbachev's initiation of long-overdue changes. All of this was too little too late, and by 1991 the Soviet empire, at that time the world's largest energy producer, disintegrated.

Ever higher energy use is not the precondition for greater economic prosperity. Higher energy use in farming does not guarantee prosperous agriculture. Increased energy subsidies may be used with very poor efficiency in irrigation and fertilization, may support unhealthy diets leading to obesity, or may be responsible for severe environmental degradation incompatible with permanent farming (higher soil erosion, irrigation-induced salinization, pesticide residues). Higher energy use in industry does not lead automatically to modernization in poor nations. Stalinist USSR and Maoist China are examples of misallocation of energies into inefficient, militarized economies.

Higher energy use does not bring greater cultural flowering. If this self-evident fact needs illustrating, it is enough to juxtapose the Greek urban civilization of 450 B.C.E. with today's Athens, or Florence of the late fifteenth century with Los Angeles of the early twenty-first century. In both comparisons there is a difference of 1 OM in per capita use of primary energy and an immeasurably large *inverse* disparity in terms of respective cultural legacies. Higher energy use beyond the desirable annual energy consumption minima does not create a superior quality of life. Higher energy flows actually erode quality of life, first for populations that are immediately affected by extraction or conversion of energies, eventually for everyone through worrisome global environmental changes.

Higher energy use does not promote social stability. Just the reverse is true: it tends to be accompanied by greater social disintegration, demoralization, and malaise. None of the social dysfunctions—the abuse of children and women, violent crime, widespread alcohol and drug use—has ebbed in affluent societies, and many of them have only grown worse. Higher energy use does not bring necessarily higher system efficiencies. Some impressive efficiency increases of individual prime movers and fuel and electricity converters of the nineteenth and twentieth centuries brought about rapid technical advances (Smil 2005a; 2006). But as a large part of TPES goes into short-lived disposable junk and into dubious pleasures and thrills promoted by mindless advertising, the overall ecological efficiency of modern high-energy societies is hardly an improvement on the earlier state of human development.

Higher energy use also does not bring any meaningful increase in civilization's diversity. In natural ecosystems the link between useful energy throughputs and species diversity is clear. But it would be misleading to interpret an overwhelming choice of consumer goods and the expanding availability of services as signs of admirable diversity in modern high-energy societies. Rather, with rampant (and often crass) materialism, increasing numbers of functionally illiterate and innumerate people, and mass media that promote the lowest common denominator of taste, human intellectual diversity may be at an historically unrivaled low point. Finally, there is no obvious link between satisfaction with life, individual happiness, and per capita energy use.

The gains that elevate humanity, that make us more secure and more hopeful about the future, cannot be bought solely by rising energy use. National security is not primarily a matter of energy-intensive weaponry. It is unattainable without social cohesion, without purposeful striving for a more fulfilling future, and without a sound economy. Economic security comes when nations do not live beyond their means. True quality of life arises from awareness of history, from strong cultural values, and from preservation of nature's irreplaceable services rather than from profligate extraction of its goods and accumulation of ephemeral acquisitions. Social stability rests above all on the cohesiveness of family, on the sense of belonging, on shared moral values. Satisfactory performance in agriculture comes from farming without excess. Wise investment of energy in a nation's modernization requires diversification, flexibility, and avoidance of shameful disparities. And true human diversity and satisfaction with life is impossible without elevating human efforts above mindless consumerism.

At the beginning of the twenty-first century a purposeful society could guarantee a decent level of physical well-being and longevity, varied nutrition, basic educational opportunities, and respect for individual freedoms with annual TPES of 50–70 GJ per capita. Remarkably, the global mean of per capita energy consumption at the beginning of the twenty-first century, 58 GJ/a, is almost exactly in the middle of this range. Equitable sharing would thus provide the world's entire population with enough energy to lead healthy, long, and active lives enriched by more than a basic level of education and the exercise of individual liberties. We could do much better within a single generation. The global economy has been lowering its energy intensity by about 1% per year, and a continuation of this trend would mean that by 2025 the mean 2000 TPES of 58 GJ per capita would be able to energize the production of goods and services for which we now need about 75 GJ. Conversely, energy services provided by 58 GJ in 2000 required about 70 GJ per capita of initial inputs during the early 1970s, and that rate was the French mean of the early 1960s and the Japanese mean of the late 1960s.

This simple comparison demonstrates that an impressively high worldwide standard of living could be achieved with virtually unchanged global energy consumption. Billions of today's poor people would be happy to experience by 2025 the quality of life that was enjoyed by people in Lyon or Kyoto during the 1960s. It would be an immense improvement, a gain that would elevate them from barely adequate subsistence to incipient affluence. But lowering the rich world's average TPES (as well as that of a few hundreds of millions of rich urbanites in the poor world) seems to be an utterly unrealistic proposition. Leaving aside the accumulation of ephemeral junk, what is so precious about our gains through high energy use that we seem unwilling even to contemplate a return to lower, but still (by any reasonable standard) generous, levels of fuel and electricity consumption?

There is no benefit in pushing food supply above 13 MJ/day; waste and spreading obesity are the only "rewards." And the activity that has been shown to be most beneficial in preventing the foremost cause of death in Western populations is a brisk 30–60-min walk most days (Haennel and Lemire 2002), not living in a virtual electronic universe. As unrealistic as reductions of more than 50% of average per capita TPES appear to be, the comparison should be on our minds as we think about the future energy consumption, bridging the gap between rich and poor worlds, and establishing a more secure global civilization. After all, North America's levels of consumption cannot become global means. Extending the pattern of 5% of the world's population consuming 25% of global TPES would call for a quintupling of global energy use. And perpetuation of existing inequalities only aggravates endless global strife.

We must realize that the quest for maximization of energy flows that has marked the ascent of fossil-fueled civilization is not an inevitable evolutionary trend. We must hope that during the twenty-first century humanity will work out a new balance between adequate energy use to sustain a decent quality of life and the imperative of not affecting the biosphere in ways inimical to human survival. Achieving this grand compromise is not inevitable or certain. Possibilities of other futures easily come to mind, and there is no shortage of dark visions (Rees 2003; Meadows, Randers, and Meadows 2004; Kunstler 2005). Our best hope is that we will find the determination to make choices that would confirm the Linnaean designation of our species—*sapiens*.

APPENDIX

Table A.1 Basic SI Units

Quantity	Name	Symbol
Length	meter	m
Mass	kilogram	kg
Time	second	s
Electric current	ampere	A
Temperature	kelvin	K
Amount of substance	mole	mol
Luminous intensity	candela	cd

Table A.2 Energy, Power and Associated Units

Quantity	Name	Symbol
Electric potential	volt	V
Electric resistance	ohm	Ω
Energy	joule	J
Force	newton	N
Frequency	hertz	Hz
Luminous flux	lumen	lm
Power	watt	W
electric		W_e
installed		W_{ei}
peak		W_p
Pressure	pascal	Pa
Temperature	degree Celsius	°C

Table A.3 Multiples

Prefix	Symbol	Factor
deka	da	10^1
hecto	h	10^2
kilo	k	10^3
mega	M	10^6
giga	G	10^9
tera	T	10^{12}
peta	P	10^{15}
exa	E	10^{18}
zeta	Z	10^{21}
yota	Y	10^{24}

Table A.4 Submultiples

Prefix	Symbol	Factor
deci	d	10^{-1}
centi	c	10^{-2}
milli	m	10^{-3}
micro	μ	10^{-6}
nano	n	10^{-9}
pico	p	10^{-12}
femto	f	10^{-15}
atto	a	10^{-18}
zepto	z	10^{-21}
yocto	y	10^{-24}

APPENDIX

Table A.5 Common Energy Conversions

barrel (bbl)	0.159 m^3
calorie (cal)	4.187 J
horsepower (hp)	745.7 W
lumen (lm)	1.496 mW
tonne of coal equivalent (tce)	29.3 GJ
tonne of oil equivalent (toe)	41.9 GJ
tonne of TNT (t TNT)	4.184 GJ

Table A.6 Energy Content of Fuels

	Heating Value (MJ/kg)	
Fuel	Higher	Lower
Hydrogen	142.1	120.1
Methane	55.5	50.0
Propane	50.1	45.6
Gasolines	43.6–47.3	41.7–44.1
Diesel fuel	44.6–46.4	41.8–44.1
Crude oils	42.0–44.0	40.0–41.0
Natural gases (MJ/m^3)	32.5–39.5	29.3–35.7
Liquefied natural gas	54.5	49.3
Anthracite	29.0–31.5	29.0–31.0
Ethanol	30.6	27.7
Standard bituminous coal	29.3	28.0
Coke	28.5	27.5
Typical steam coals	22.0–24.0	19.0–21.0
Lignites	11.0–20.0	8.0–17.0
Wood, air dry	15.0–16.5	12.0–14.5
Crop residues, air dry	14.0–16.5	10.0–14.5
Dung, air dry	11.0–12.0	7.0–10.5

Table A.7 Energy Content of Nutrients, Foodstuffs, and Metabolic Products

	Energy Content (MJ/kg)	
Nutrients	Total	Digestible
Carbohydrates	17.0	17.0
Proteins	23.0	17.0
Lipids	39.0	38.0
Ethanol	29.6	a
Foodstuffs	**As Purchased**	**Cooked**
Cereal grains	15.2–15.4	5.0–10.0
Legume grains	14.0–14.5	4.4–4.9
Potatoes	3.2–4.8	3.2–4.9
Sugar	16.1	16.1
Plant oils	37.0	37.0
Vegetables	0.6–1.8	0.6–1.5
Fruits	1.5–4.0	1.5–4.1
Red meats	5.6–23.1	7.0–18.0
Poultry	4.9–13.6	6.9–13.3
Milk	1.5–2.9	1.5–2.9
Butter	30.0	30.0
Eggs	6.8–8.0	6.8–8.0
Fish	2.9–9.3	2.9–7.6
Alcoholic beverages	1.7–12.3	b
Metabolic Products	**Wet Mass**	**Dry Mass**
Muscle	5.9–9.5	23.8
Fat	31.0–37.0	36.0–38.0
Urine	0.1–0.2	c
Feces	1.8–3.0	7.2–12.0

a. Ethanol can be digested only in the liver at a maximum hourly rate of 0.1 g/kg of body weight.

b. Cooking may evaporate most or all of the ethanol present in these liquids.

c. Urea, the most important component of dry matter in urine, has an energy content of 10.6 MJ/kg.

Table A.8 Energy Flows and Stores: 31 orders of Magnitude

Energy Flow or Store	Energy (J)	Order of Magnitude
Solar radiation intercepted by the Earth[a,c]	5.5	24
Global coal resources[b]	2.0	23
Global plant mass[b]	2.0	22
Global net photosynthesis[a,c]	2.0	21
Global fossil fuel production[a,c]	3.0	20
Typical Caribbean hurricane[a]	3.8	19
Global lightning[a,c]	3.2	18
Largest H-bomb tested (1961)[a]	2.4	17
Global zodiacal light[a,c]	6.3	16
Latent heat of a thunderstorm[b]	5.0	15
Kinetic energy of a thunderstorm[a]	1.0	14
Hiroshima bomb (1945)[a]	8.4	13
Coal in a 100-t hopper car[b]	2.5	12
Good grain corn harvest (8 t/ha)[b]	1.2	11
Gasoline for a compact car[b,c]	4.0	10
Barrel of crude oil[b]	6.5	9
Basal metabolism of a large horse[a]	1.0	8
Daily adult food intake[a]	1.0	7
Bottle of white table wine[b]	2.6	6
Large hen egg[b]	4.0	5
Vole's daily basal metabolism[a]	5.0	4
Small chickpea[b]	5.0	3
Baseball (140 g) pitched at 40 m/s[a]	1.1	2
Tennis ball (50 g) served at 25 m/s[a]	1.5	1
Full teacup (300 g) held in hand[d]	2.6	0
Falling 2-cm hailstone[a]	2.0	−1
Striking a typewriter key[a]	2.0	−2
Fly on a kitchen table[d]	9.0	−3

Table A.8 (continued)

Energy Flow or Store	Energy (J)	Order of Magnitude
Small bird's 5-s song[a]	5.0	−4
A 2-mm raindrop falling at 6 m/s[a]	7.5	−5
The same drop on a blade of grass[d]	4.0	−6
Flea hop[a]	1.0	−7

ha = hectare.

a. Flows of mechanical, chemical, electrical, and heat energy.

b. Stores of chemical and heat energy.

c. Annual totals.

d. Potential energy.

APPENDIX

Table A.9 Power of Continuous Phenomena

Energy Flow	Power (W)	Order of Magnitude
Global intercept of solar radiation	1.7	17
Wind-generated waves on the ocean	9.0	16
Solar radiation received by China	2.0	15
Global gross primary productivity	1.0	14
Global Earth heat flow	4.2	13
U.S. primary energy consumption in 2005	3.1	12
Global earthquake activity	3.0	11
Florida current between Miami and Bimini	2.0	10
Rotating 1-GW turbogenerator	1.0	9
Midsize nuclear reactor	5.0	8
Gas pipeline compressor	2.0	7
Electricity for a 30-story high-rise building	1.5	6
Energy needs of a typical supermarket	2.0	5
Large 18th-century waterwheel	1.0	4
Japanese per capita energy use in 2005	5.5	3
Emergency exit light	2.0	2
Basal metabolism of a 70-kg person	8.0	1
Net productivity per m^2 of tropical forest	1.0	0
Metabolic rate of neonatal heart	4.0	−1
Mean global rate of erosion per m^2	5.0	−2

Table A.10 Power of Ephemeral Phenomena

Energy Flow	Order of Magnitude	Duration (s)	Power (W)
Magnitude 9 earthquake	15	30	1.6
Large volcanic eruption	14	10^4	1.0
Giant lightning	13	10^{-5}	2.0
Rainstorm's latent heat	12	1200	1.0
Thunderstorm's kinetic energy	11	1200	1.0
Large WW II bombing raid	10	3600	2.0
Average U.S. tornado	9	160	1.7
Mount St. Helen's seismic waves	8	10^4	5.0
Small avalanche with 500-m drop	7	20	1.1
Large coal unit-train shuttle	6	10^3	5.0
Intercity truck trip	5	10^4	3.0
Gasoline for a 20-km car drive	4	1200	4.0
Running 100-m dash	3	10	1.3
Machine-washing laundry	2	1500	5.0
CD player spinning Mozart's last symphony (K. 551)	1	2238	2.5
Small candle burning to the end	0	1800	3.0
Hummingbird flight	−1	300	7.0

Table A.11 Efficiencies of Common Energy Conversions

Converter	Conversion	Efficiency (%)
Large electricity generator	m ⇒ e	98–99
Large power plant boiler	c ⇒ t	90–98
Large electric motor	e ⇒ m	90–97
Best household natural gas furnace	c ⇒ t	90–97
Dry-cell battery	c ⇒ e	85–95
Human lactation	c ⇒ c	75–85
Overshot waterwheel	m ⇒ m	60–85
Small electric motor	e ⇒ m	65–80
Most efficient bacterial growth	c ⇒ c	50–65
Glycolysis maxima	c ⇒ c	50–60
Large steam turbine	t ⇒ m	40–45
Improved wood stove	c ⇒ t	25–45
Large gas turbine	c ⇒ m	35–40
Diesel engine	c ⇒ m	30–35
Mammalian postnatal growth	c ⇒ c	30–35
Best PV cell	r ⇒ e	20–30
Best large steam engine	c ⇒ m	20–25
Internal combustion engine	c ⇒ m	15–25
High-pressure sodium lamp	e ⇒ r	15–20
Mammalian muscles	c ⇒ m	15–20
Milk production	c ⇒ c	15–20
Pregnancy	c ⇒ c	10–20
Broiler production	c ⇒ c	10–15
Traditional stove	c ⇒ t	10–15
Fluorescent light	e ⇒ r	10–12
Beef production	c ⇒ c	5–10
Steam locomotive	c ⇒ m	3–6
Peak photosynthetic rate	r ⇒ c	4–5
Incandescent light bulb	e ⇒ r	2–5
Paraffin candle	c ⇒ r	1–2
Most productive ecosystem	r ⇒ c	1–2
Global photosynthetic mean	r ⇒ c	0.3

c = chemical energy; e = electrical energy; m = mechanical (kinetic) energy; r = radiant (electromagnetic) energy; and t = thermal energy.

Table A.12 Typical Energy Cost of Common Materials

Material	Energy Cost (MJ/kg)	Source
Aluminum	190–230	Bauxite
Aluminum	10–40	Recycled metal
Bricks	2–5	Fired clay
Cement	5–9	Raw materials
Ceramics	3–7	Raw materials
Concrete	1–3	Cement and aggregate
Copper	60–150	Ore
Explosives	10–70	Raw materials
Glass	15–30	Raw materials
Gravel	<0.1	Quarries, rivers
Hydrogen	192–252	Electrolysis of water
Iron	20–25	Ore
Lead	30–50	Ore
Lime	10–12	Limestone
Newsprint	8–10	Wood pulp
Oxygen	6–14	Air
Nitrogen	1.5–1.9	Air
Paints	90–100	Raw materials
Paper, packaging	10–15	Kraft process
Paper, high quality	25–35	Wood pulp
Polyethylene	75–115	Crude oil
Polyvinylchloride	75–100	Crude oil
Sand	<0.1	Excavated
Silicon	1400–4100	Single crystal from silica
Steel, ordinary	20–25	Pig iron
Steel, specialty alloy	30–60	Raw materials
Stone	<1	Quarried
Sulfuric acid	2–3	Sulfur
Timber	1–3	Standing wood
Titanium	900–1000	Ore concentrate
Water	<0.01	Streams, reservoirs

Table A.13 Global Harvests, Energy Subsidies, and Population Densities, 1900–2000

	1900	1925	1950	1975	2000
Harvested area (Gha)	1.10	1.20	1.25	1.46	1.53
Annual harvest (EJ)	6.0	9.0	12.0	25.0	40.0
Annual yield (GJ/ha)	5.5	7.5	9.6	17.1	26.1
Annual energy subsidies (EJ)	0.1	0.5	1.5	0	12.8
Subsidy density (GJ/ha)	0.1	0.4	1.2	5.5	8.4
Population (10^9)	1.7	2.0	2.5	4.0	6.1
Population density (people/ha)	1.5	1.7	2.0	2.7	4.0
Per capita harvest (MJ/day)	9.7	12.3	13.2	17.1	18.0

Compiled and calculated from statistical yearbooks published by the League of Nations and the United Nations Food and Agriculture Organization.

Gha = 10^9 hectares; ha = hectare.

Table A.14 Comparison of Natural, Personal, and Energy-Related Risks

Risk of Dying from	Deaths/person/hour of exposure	
Heavy smoking	−6	$2-3 \times 10^{-6}$
Disease	−6	$1-2 \times 10^{-6}$
Scheduled air travel	−6	$1-2 \times 10^{-6}$
Underground coal mining	−6	$1-2 \times 10^{-6}$
Driving	−6	$1-2 \times 10^{-6}$
Surface coal mining	−7	$1-2 \times 10^{-7}$
Railway travel	−8	$3-6 \times 10^{-8}$
Murder	−9	$4-9 \times 10^{-9}$
Electrical shock	−9	$1-3 \times 10^{-9}$
Hydrocarbon extraction	−9	$1-3 \times 10^{-9}$
Tornado	−10	$2-3 \times 10^{-10}$
Flood	−10	$2-5 \times 10^{-10}$
Earthquake	−10	$1-5 \times 10^{-10}$
Hurricane	−11	$5-9 \times 10^{-11}$
Falling meteorite	−13	$1-2 \times 10^{-13}$

Calculated from detailed U.S. mortality data for the 1990s and from a variety of European mortality statistics.

Table A.15 Population and Primary Energy, 1500–2005

	Population (10^9)	Primary Energy (EJ)	
		Phytomass	Fossil Fuels and Primary Electricity
1500	0.48	6	0
1700	0.63	10	0
1800	0.98	12.0	1.5
1850	1.29	13.0	2.5
1860	1.34	14.0	3.9
1870	1.38	15.0	6.2
1880	1.47	16.0	9.8
1890	1.57	18.0	15.0
1900	1.69	21.0	22.5
1910	1.84	23.0	35.1
1920	1.98	24.0	41.0
1930	2.17	25.0	49.0
1940	2.40	27.0	57.0
1950	2.52	28.0	74.0
1960	3.02	32.0	126.0
1970	3.70	36.0	211.0
1980	4.42	38.0	279.0
1990	5.28	40.0	341.0
2000	6.09	42.0	390.0
2005	6.45	45.0	443.0

Population totals from NEAA (2005); phytomass totals are estimates by author; fossil fuels and primary electricity from UNO (1956), Etemad et al. (1991), and BP (2006).

Table A.16 Global Reserves, Resources, and Fluxes of Energies

Energy Stores	Resources (ZJ)	Reserves (2005) (ZJ)
Fossil fuels	>1,000	<40
Coals	200	20
Crude oils	15	7
Oil sands and shale	>200	2
Natural gases	15	7
Clathrates	>500	–
Fissionable elements		
Uranium on land	1,000	500
Uranium in seawater	363,000	–

Energy Fluxes	Total Flux (TW)	Usable Flux (GW)	Used in 2005 (GW)
Solar radiation	174,260	–	2[a]
Water	11.7	1,700	750[b]
Wind	2,000	6,000	60[c]
Ocean waves	2	–	0
Ocean currents	0.1	<5	0
OTEC	10	<0.1	0
Tides	2.5	<10	–
Photosynthesis	110	<3	1.5
Terrestrial	60	<5,000	<2,000[d]
Oceanic	50	<1,000	0
Geothermal	42	<3,000	15[b]

a. Excludes photosynthesis; includes only the total for peak PV electricity-generating capacity (W_p).

b. Total for the installed hydroelectricity-generating capacity.

c. Total for the peak installed capacity of wind turbines (W_p).

d. Only wood, charcoal, and crop residues for energy and fuel ethanol.

Selected Abbreviations, Acronyms, and Symbols

a	year (annum), as in /a, per year. *See also* bp, Ga.		DC	direct current
α	albedo		DE	digestible energy
AC	alternating current		DEE	daily energy expenditure
ATP	adenosine triphosphate		DLW	doubly labeled water
AU	astronomical unit		DM	dry matter
AW	all-wave		DMD	daily movement distance
			DRI	dietary reference intake
B	Bubnoff unit		DVI	dust veil index
bbl	barrel		dwt	deadweight ton
B.C.E.	before common era			
BMR	basal metabolic rate		*E*	energy
BOF	basic oxygen furnace		EAF	electric arc furnace
bp	before the present time		ECT	ecological cost of transport
BWR	boiling water reactor		EI	energy intensity
			EROI	energy return on investment
CAFE	corporate automobile fuel efficiency		EU	European Union
cal	calorie			
CAM	crassulacean acid metabolism		*F*	radiant flux
CBD	central business district		FBC	fluidized bed combustion
CCGT	combined cycle gas turbine		FGD	flue gas desulfurization
C.E.	common era		FMR	field metabolic rate
COT	cost of transport		Fr	Froude number

g	gravitational constant		NPP	net primary productivity
Ga	10^9 years			
GCR	gas-cooled reactor		OHF	open-hearth furnace
GDP	gross domestic product		OM	order of magnitude
GE	gross energy		OPEC	Organization of the Petroleum Exporting Countries
Gha	10^9 hectares		OTEC	ocean thermal energy conversion
GHG	greenhouse gas			
GPP	gross primary productivity		PAL	physical activity level
GWP	global warming potential		PAR	photosynthetically active radiation
GWP	gross world product		PHWR	pressurized heavy water reactor
			ppm	part per million
h	height		PPP	purchasing power parity
H	enthalpy (heat content)		PV	photovoltaics
ha	hectare		PWR	pressurized water reactor
HDI	human development index			
hp	horsepower		r	radius
HRM	heart rate monitoring		R	respiration
HV	high voltage		ρ	density
			Re	Reynolds number
ICL	incremental cost of locomotion		RMR	resting metabolic rate
IM	infant mortality			
IR	infrared		S	entropy
ISA	impervious surface area		sej	solar emjoule, a unit of emergy
			SI	Système international d'unités (International System of Units)
LAI	leaf area index			
LCA	life cycle analysis		St	Strouhal number
LE	life expectancy		SW	shortwave
lm	lumen			
LNG	liquefied natural gas		t	tonne (metric ton)
LW	longwave		toe	tonne of oil equivalent
ly	light-year		T	temperature
			TEE	total energy expenditure
M	body mass		TNT	trinitrotoluene
M	earthquake magnitude		TPES	total primary energy supply
ME	metabolizable energy			
MET	metabolic equivalent		UV	ultraviolet
MOC	meridional overturning circulation			
			v	velocity
NDVI	normalized difference vegetation index		VEI	volcanic explosivity index
NEP	net ecosystem productivity		WW I	World War I
NME	net metabolizable energy		WW II	World War II

SELECTED ABBREVIATIONS, ACRONYMS, AND SYMBOLS

REFERENCES

Abel, W. 1962. *Geschichte der deutschen Landwirtschaft von frühen Mittelalter bis zum 19 Jahrhundert.* Stuttgart: Ulmer.

Adair, L. S., and E. Pollitt. 1982. Energy balance during pregnancy and lactation. *The Lancet* 2: 219.

Adam, J.-P. 1994. *Roman Building: Materials and Techniques.* Bloomington: Indiana University Press.

Adelman, M. 1997. My education in mineral (especially oil) economics. *Annual Review of Energy and the Environment* 22: 13–46.

Adshead, S. A. M. 1992. *Salt and Civilization.* New York: St. Martin's Press.

Agricola, G. 1556. *De Re Metallica.* Trans. H. C. Hoover and L. H. Hoover. New York: Dover Publications, 1950.

Ahern, J. E. 1980. *The Exergy Method of Energy Systems Analysis.* New York: Wiley.

Ahlbrandt, T. S., R. R. Charpentier, T. R. Klett, J. W. Schmoker, C. J. Schenk, and G. F. Ulmishek. 2006. *Global Resource Estimates from Total Petroleum Systems.* Tulsa, Okla.: American Association of Petroleum Geologists.

Aiello, L. C., and J. C. K. Wells. 2002. Energetics and the evolution of the genus *Homo. Annual Review of Anthropology* 31: 323–338.

Aiello, L. C., and P. Wheeler. 1995. The expensive-tissue hypothesis. *Current Anthropology* 36: 199–221.

Ainsworth, B. E. 2002. *The Compendium of Physical Activities Tracking Guide.* University of South Carolina. ⟨http://prevention.sph.sc.edu/tools/docs/documents_compendium.pdf⟩.

AISI (American Iron and Steel Institute). 2002. *Perspective: American Steel and Domestic Manufacturing.* ⟨http://www.steel.org/⟩.

Alderson, V. C. 1920. *The Oil Shale Industry.* New York: F. A. Stokes.

Alerstam, T. 2006. Conflicting evidence about long-distance animal navigation. *Science* 313: 791–794.

Alerstam, T., A. Hedenström, and S. Åkesson. 2003. Long-distance migration: Evolution and determinants. *Oikos* 103: 247–260.

Alexander, R. M. 1984. Elastic energy stores in running verte-brates. *American Zoologist* 24: 85–94.

———. 1999a. *Energy for Animal Life.* Oxford: Oxford University Press.

———. 1999b. One price to run, swim or fly? *Nature* 397: 651–653.

———. 2003. *Principles of Animal Locomotion.* Princeton, N.J.: Princeton University Press.

———. 2005. Models and the scaling of energy cost of locomotion. *Journal of Experimental Biology* 208: 1645–1652.

Alford, M. H. 2003. Redistribution of energy available for ocean mixing by long-range propagation of internal waves. *Nature* 423: 159–162.

Allaby, M. 2004. *Tornadoes.* New York: Facts On File.

Allan, W. 1965. *The African Husbandman.* Edinburgh: Oliver and Boyd.

Allaud, L., and M. Martin. 1976. *Schlumberger: Histoire d'une technique.* Paris: Berger-Levrault.

Alley, R., T. Berntsen, N. L. Bindoff, Z. Chen, A. Chidthaisong, P. Friedlingstein, J. Gregory et al. 2007. *Climate Change 2007: The Physical Science Basis.* Geneva: IPCC.

Alvard, M. S., and L. Kuznar. 2001. Deferred harvests: The transition from hunting to animal husbandry. *American Anthropologist* 103: 295–311.

Alyeska Pipeline Service Company. 2003. *Pipeline Facts.* ⟨http://www.alyeska-pipeline.com/pipelinefacts.html⟩.

Ambrose, S. H. 1998. Late Pleistocene human population bottlenecks: Volcanic winter and the differentiation of modern humans. *Journal of Human Evolution* 34: 623–651.

Ammann, C. M., and P. Naveau. 2003. Statistical analysis of tropical explosive volcanism occurrences over the past six centuries. *Geophysical Research Letters* 30: 14/1–14/4.

Ammon, C. J., J. Chen, H.-K. Thio, D. Robinson, S. Ni, V. Hjorleifsdottir, H. Kanamori et al. 2005. Rupture process of the 2004 Sumatra-Andaman earthquake. *Science* 308: 1133–1139.

Amthor, J. S. et al. 1998. *Terrestrial Responses to Global Change.* Oak Ridge, Tenn.: Oak Ridge National Laboratory.

Amthor, J. S., and D. D. Baldocchi. 2001. Terrestrial higher plant respiration and net primary production. In *Terrestrial Global Productivity*, ed. J. Roy, B. Saugier, and H. A. Mooney, 33–59. San Diego, Calif.: Academic Press.

An, Z., J. E. Kutzbach, W. W. Prell, and S. C. Porter. 2001. Evolution of Asian monsoons and phased uplift of the Himalaya-Tibetan plateau since Late Miocene times. *Nature* 411: 62–66.

Anderson, A. B., P. H. May, and M. Balick. 1991. *The Subsidy from Nature: Palm Forests, Peasantry, and Development on an Amazon Frontier.* New York: Columbia University Press.

Anderson, D. L. 2002. Plate tectonics as a far-from-equilibrium self-organized system. In *Plate Boundary Zones*, ed. S. Stein and J. T. Freymueller, 411–425. Washington, D.C.: American Geophysical Union.

Anderson, G. B. 1980. *One Hundred Booming Years: A History of Bucyrus-Erie Company, 1880–1980.* South Milwaukee, Wisc.: Bucyrus-Erie.

Anderson, W. J. 1902. *The Architecture of Greece and Rome: A Sketch of Its Historic Development.* London: B. T. Batsford.

Angell, J. K., and J. Korshover. 1985. Surface temperature changes following the six major volcanic episodes between 1780 and 1980. *Journal of Climate and Applied Meteorology* 24: 937–951.

Anthony, D., D. Y. Telegin, and D. Brown. 1991. The origin of horseback riding. *Scientific American* 265 (6): 94–100.

Antoine, D., J.-M. André, and A. Morel. 1996. Oceanic primary production. 2. Estimation at global scale from satellite (coastal zone color scanner) chlorophyll. *Global Biogeochemical Cycles* 10: 57–69.

REFERENCES

Archer, C. L., and M. Z. Jacobson. 2005. Evaluation of global wind power. *Journal of Geophysical Research* 110: doi:10.1029/2004JD005462.

Archer, M. D., and R. Hill, eds. 2001. *Clean Electricity from Photovoltaics*. London: Imperial College Press.

Ardrey, R. L. 1894. *American Agricultural Implements*. Chicago: R. L. Ardrey. Reissued New York: Arno Press, 1972.

Armelagos, G. J., and K. N. Harper. 2005. Genomics at the origins of agriculture. *Evolutionary Anthropology* 14: 68–77.

Armstrong, R. 1969. *The Merchantmen*. London: Ernest Benn.

Arrhenius, S. 1896. On the influence of carbonic acid in the air upon the temperature of the ground. *Philosophical Magazine* 41: 237–276.

Arsac, L. M., and E. Locatelli. 2002. Modeling the energetics of 100-m running by using speed curves of world champions. *Journal of Applied Physiology* 92: 1781–1788.

Ashton, T. S., and J. Sykes. 1929. *The Coal Industry of the Eighteenth Century*. Manchester, UK: Manchester University Press.

Assmann, E. 1970. *The Principles of Forest Yield Study*. Oxford: Pergamon Press.

Atkins, P. W. 1984. *The Second Law*. New York: Scientific American Library.

Atkins, S. E. 2000. *Historical Encyclopedia of Atomic Energy*. New York: Greenwood Press.

Atwater, W. O., and F. G. Benedict. 1899. *Experiments on Metabolism of Matter and Energy in the Human Body*. Bulletin 29. U.S. Department of Agriculture. ⟨http://www.ars.usda.gov/is/timeline/nutrition.htm⟩.

Atwater, W. O., and C. D. Woods. 1896. *The Chemical Composition of American Food Materials*. Bulletin 28. U.S. Department of Agriculture. ⟨http://www.ars.usda.gov/is/timeline/nutrition.htm⟩.

Aunan, K., T. K. Berntsen, and H. M. Seip. 2000. Surface ozone in China and its possible impact on agricultural crop yields. *Ambio* 29: 294–301.

Australian Government. 2005. *Australia's Export Coal Industry*. Canberra: Department of Industry, Tourism and Resources.

Ausubel, J. H. 1996. Can technology spare the Earth? *American Scientist* 84: 166–178.

AWEA. 2007. *Global Wind Energy Markets Continue to Boom— 2006 Another Record Year*. ⟨http://www.awea.org/newsroom/pdf/070202_GWEC_Global_Market_Annual_Statistics.pdf⟩.

Ayala, F. J. 1998. Is sex better? Parasites say "no." *Proceedings of the National Academy of Sciences* 95: 3346–3348.

Ayres, R. U., L. W. Ayres, and B. Warr. 2003. Exergy, power and work in the US economy, 1900–1998. *Energy* 28: 219–273.

Azar, K. 2000. The history of power dissipation. *Electronics Cooling* 6 (1): 1–10. ⟨http://www.electronics-cooling.com/html/2000_jan_a2.html⟩.

Baeyer, H. C. von. 1999. *Warmth Disperses and Time Passes: The History of Heat*. New York: Modern Library. Originally published as *Maxwell's Demon*. New York: Random House, 1998.

Bailey, L. H., ed. 1908. *Cyclopedia of American Agriculture*. New York: Macmillan.

Bailey, R. C., G. Head, M. Jenike, B. Owen, R. Rechtman, and E. Zechenter. 1989. Hunting and gathering in tropical rain forest: Is it possible? *American Anthropologist* 91: 59–82.

Bailey, S. M. 1982. Absolute and relative sex differences in body composition. In *Sexual Dimorphism in Homo Sapiens*, ed. R. L. Hall, 363–390. New York: Praeger.

Bailis, R. 2004. Wood in household energy use. In *Encyclopedia of Energy*. Vol. 6, 509–526.

Baird, G. 1984. *Energy Performance of Buildings*. Boca Raton, Fla.: CRC Press.

Baker, C. J., K. E. Saxton, and W. R. Ritchie. 1996. *No-Tillage Seeding: Science and Practice*. Oxford: Oxford University Press.

Baker, T. L. 1985. *A Field Guide to American Windmills*. Norman: University of Oklahoma Press.

REFERENCES

Baker, V. R., ed. 1981. *Catastrophic Flooding: The Origin of the Channeled Scabland*. Stroudsburg, Pa.: Dowden, Hutchinson and Ross.

Baldwin, G. C. 1977. *Pyramids of the New World*. New York: G. P. Putnam's Sons.

Balter, M. 1998. Why settle down? The mystery of communities. *Science* 282: 1442–1445.

Banavar, J., A. Maritan, and A. Rinaldo. 1999. Size and form in efficient transportation networks. *Nature* 399: 130–132.

Banerjee, P., F. F. Pollitz, and R. Bürgmann. 2005. The size and duration of the Sumatra-Andaman earthquake from far-field static offsets. *Science* 308: 1769–1772.

Banks, F. E. 2000. *Energy Economics: A Modern Introduction*. Boston: Kluwer.

Barczak, T. M. 1992. *The History and Future of Longwall Mining in the United States*. U.S. Bureau of Mines.

Barnard, A., ed. 2004. *Hunter-Gatherers in History, Archaeology and Anthropology*. Oxford: Berg.

Barnes, B. V., D. R. Zak, S. R. Denton, and S. H. Spurr. 1998. *Forest Ecology*. 4th ed. New York: Wiley.

Bartok, W. 1991. *Fossil Fuel Combustion: A Source Book*. New York: Wiley.

Basalla, G. 1982. Some persistent energy myths. In *Energy and Transport*, ed. G. H. Daniels and M. H. Rose, 27–38. Beverly Hills, Calif.: Sage.

———. 1988. *The Evolution of Technology*. Cambridge: Cambridge University Press.

Bascom, W. 1959. Ocean waves. *Scientific American* 201 (2): 74–84.

Bassham, J. A., and M. Calvin. 1957. *The Path of Carbon in Photosynthesis*. Englewood Cliffs, N.J.: Prentice-Hall.

Bateman, R. M., P. R. Crane, W. A. DiMichele, P. R. Kenrick, N. P. Rowe, T. Speck, and W. E. Stein. 1998. Early evolution of land plants: Phylogeny, physiology, and ecology of the primary terrestrial radiation. *Annual Review of Ecology and Systematics* 29: 263–292.

Baviere, M., ed. 1991. *Basic Concepts in Enhanced Oil Recovery Processes*. Amsterdam: Elsevier.

Bazilevich, N. I., L. Rodin, and N. N. Rozov. 1971. Geographical aspects of biological productivity. *Soviet Geography* 12: 293–317.

Bebout, G. E., D. W. Scholl, S. H. Kirby, and J. P. Platt, eds. 1996. *Subduction Top to Bottom*. Washington, D.C.: American Geophysical Union.

Beech, G. A. 1980. Energy use in bread baking. *Journal of Science of Food and Agriculture* 31: 289–298.

Beedell, S. M. 1975. *Windmills*. Newton Abbot, UK: David and Charles.

Behrenfeld, M. J., E. Boss, D. A. Siegel, and D. M. Shea. 2005. Carbon-based ocean productivity and phytoplankton physiology from space. *Global Biogeochemical Cycles* 19: doi:10.1029/2004GB002299 GB1006.

Behrenfeld, M. J., and P. G. Falkowski. 1997. Photosynthetic rates derived from satellite-based chlorophyll concentration. *Limnology and Oceanography* 42: 1–20.

Belgrano, A., A. P. Allen, B. J. Enquist, and J. F. Gillooly. 2002. Allometric scaling of maximum population density: A common rule for marine phytoplankton and terrestrial plants. *Ecology Letters* 5: 611–613.

Bell, I. Lothian. 1884. *Principles of the Manufacture of Iron and Steel*. London: George Routledge & Sons.

Bellouin, N., O. Boucher, J. Haywood, and M. S. Reddy. 2005. Global estimate of aerosol direct radiative forcing from satellite measurements. *Nature* 438: 1138–1141.

Bellwood, P. 2004. *First Farmer: The Origins of Agricultural Societies*. Oxford: Blackwell.

Benedict, F. G., and E. P. Cathcart. 1913. *Muscular Work*. Washington, D.C.: Carnegie Institution.

REFERENCES

Benefice, E., S. Chevassus-Agnes, and H. Barral. 1984. Nutritional situation and seasonal variations for pastoralist populations of the Sahel (Senegalese Ferlo). *Ecology of Food and Nutrition* 14: 229–247.

Bennett, M. K. 1935. British wheat yield per acre for seven centuries. *Economic History* 3 (10): 12–29.

Bennett, R., and J. Elton. 1898. *History of Corn Milling.* London: Simpkin Marshall.

Bercovici, D., and S. Karato. 2003. Whole-mantle convection and the transition-zone water filter. *Nature* 425: 39–44.

Bergström, A.-K., and M. Jansson. 2006. Atmospheric nitrogen has caused nutrient enhancement and eutrophication of lakes in the Northern Hemisphere. *Change Biology* 12: 635–643.

Berkowitz, N. 1985. *The Chemistry of Coal.* New York: Elsevier.

———. 1997. *Fossil Hydorcarbons: Chemistry and Technology.* San Diego, Calif.: Academic Press.

Bernal, D., J. M. Donley, R. E. Shadwick, and D. A. Syme. 2005. Mammal-like muscles power swimming in a cold-water shark. *Nature* 437: 1349–1352.

Berner, R. A. 1999. Atmospheric oxygen over Phanerozoic time. *Proceedings of the National Academy of Sciences* 96: 10955–10957.

Bernstein, I. S., and E. O. Smith, eds. 1979. *Primate Ecology and Human Origins: Ecological Influences on Social Organization.* New York: Garland STPM Press.

Berry, R. S., and M. F. Fels. 1972. *The Production and Consumption of Automobiles.* Chicago: University of Chicago Department of Chemistry.

Bertalanffy, L. von. 1932–1942. *Theoretische Biologie.* 2 vols. Bonn: Borntraeger.

———. 1968. *General System Theory.* New York: George Braziller.

Best, R. W. B. 1979. Limits to wind-power. *Energy Conversion* 19: 71–72.

Bethe, H. A. 1939. Energy production in stars. *Physical Review* 55: 434–456.

Bierman, P. R., and K. K. Nichols. 2004. Rock to sediment-slope to sea with [10]Be-rates of landscape change. *Annual Review of Earth and Planetary Sciences* 32: 215–255.

Biesot, W., and H. C. Moll, eds. 1995. *Reduction of CO_2 Emissions by Lifestyle Changes.* Groningen: Rijksuniversitet.

Biewener, A. A. 2003. *Animal Locomotion.* Oxford: Oxford University Press.

Bilham, R. 2005. A flying start, then a slow slip. *Science* 308: 1126–1127.

BIPM (Bureau International des Poids et Mesures). 2006. *Le Système international d'unités (SI).* ⟨http://www.bipm.fr/si/⟩.

Biringuccio, V. 1540. *Pirotechnia.* Trans. C. S. Smith and M. T. Gnudi. New York: Dover Publications, 1990.

Bister, M., and K. A. Emanuel. 1998. Dissipative heating and hurricane intensity. *Meteorology and Atmospheric Physics* 65: 233–240.

Black, A. E., W. A. Coward, T. J. Cole, and A. M. Prentice. 1996. Human energy expenditure in affluent societies: An analysis of 547 doubly-labelled water measurements. *European Journal of Clinical Nutrition* 50: 72–92.

Black, J. N. 1971. Energy relations in crop production—A preliminary survey. *Annals of Applied Biology* 67: 272–278.

Black, R. A., and J. Hallett. 1998. The mystery of cloud electrification. *American Scientist* 86: 526–534.

Blackman, D. R., and A. T. Hodge. 2001. *Frontinus' Legacy: Essays on Frontinus' de aquis urbis Romae.* Ann Arbor: University of Michigan Press.

Blackwell, M. 2000. Terrestrial life: Fungal from the start? *Science* 289: 1884–1885.

Blakers, A., and K. Weber. 2000. The energy intensity of photovoltaic systems. ⟨http://www.ecotopia.com/apollo2/pvepbtoz.htm⟩.

Bland, P. A., and N. A. Artemieva. 2003. Efficient disruption of small asteroids by Earth's atmosphere. *Nature* 424: 288–291.

Blem, C. R. 1980. The energetics of migration. In *Animal Migration, Orientation and Navigation*, ed. S. A. Gauthreaux, 175–224. New York: Academic Press.

Block, L. 2003. *To Harness the Wind: A Short History of the Development of Sails.* Annapolis, Md.: Naval Institute Press.

Bloomfield, L. A. 1997. *How Things Work: The Physics of Everyday Life.* New York: Wiley.

Blumberg, M. S. 2002. *Body Heat: Temperature and Life on Earth.* Cambridge, Mass.: Harvard University Press.

Blumenschine, R. J., and J. A. Cavallo. 1992. Scavenging and human evolution. *Scientific American* 267 (4): 90–95.

Boatwright, J. L., and G. L. Choy. 1986. Teleseismic estimates of the energy radiated by shallow earthquakes. *Journal of Geophysical Research* 91: 2095–2112.

Bohi, D. 1989. *Energy Price Shocks and Macroeconomic Performance.* Washington, D.C.: Resources for the Future.

Bokma, F. 2004. Evidence against universal metabolic allometry. *Functional Ecology* 18: 184–187.

Boonenburg, K. 1949. *De windmolens.* Amsterdam: A. de Lange.

———. 1952. *Windmills in Holland.* The Hague: Netherlands Government Information Service.

Börjesson, P., and L. Gustavsson. 2000. Greenhouse gas balances in building construction: Wood versus concrete from life-cycle and forest land-use perspective. *Energy Policy* 28: 575–588.

Borlaug, N. 1970. The Green Revolution, Peace, and Humanity. Speech at awarding of the 1970 Nobel Peace Prize in Oslo, Norway, December 11. ⟨http://nobelprize.org/nobel_prizes/peace/laureates/1970/borlaug-lecture.html⟩.

Borrini, G., and S. Margen. 1985. *Human Energetics.* Ottawa: International Development Research Centre.

Boserup, E. 1965. *The Conditions of Agricultural Growth: The Economics of Agrarian Change under Population Pressure.* Chicago: Aldine.

———. 1976. Environment, population, and technology in primitive societies. *Population and Development Review* 2: 21–36.

Bouchard, C., R. Malina, and L. Perusse. 1997. *Genetics of Fitness and Physical Performance.* Champaign, Ill.: Human Kinetics.

Boulding, K. E. 1974. The social system and the energy crisis. *Science* 184: 255–257.

Boustead, I., and G. F. Hancock. 1979. *Handbook of Industrial Energy Analysis.* Chichester, UK: Ellis Horwood.

Bowen, H. J. M. 1966. *Trace Elements in Biochemistry.* London: Academic Press.

Boyce, M. P. 2002. *Handbook for Cogeneration and Combined Cycle Power Plants.* New York: ASME Press.

BP (British Petroleum). 2006. *Statistical Review of World Energy.* ⟨http://www.bp.com/worldenergy⟩.

Bramble, D. M., and D. E. Lieberman. 2004. Endurance running and the evolution of *Homo. Nature* 432: 345–352.

Brandon, N., and D. Thompsett. 2005. *Fuel Cells Compendium.* Amsterdam: Elsevier.

Brandstetter, T. 2005. "The most wonderful piece of machinery the world can boast of": The water-works at Marly, 1680–1830. *History and Technology* 21: 205–220.

Brantly, J. E. 1971. *History of Oil Well Drilling.* Houston: Gulf Publishing.

Braun, G. W., and D. R. Smith. 1992. Commercial wind power: Recent experience in the United States. *Annual Review of Energy and the Environment* 17: 97–121.

Bray, F. 1984. *Science and Civilisation in China.* Vol. 6, Pt 2: *Agriculture.* Cambridge: Cambridge University Press.

Bray, W. 1977. From foraging to farming in early Mexico. In *Hunters, Gatherers and First Farmers Beyond Europe*, ed.

J. V. S. Megaw, 225–250. Leicester, UK: Leicester University Press.

Breeze, P. 2004. *The Future of Biomass Power Generation*. London: Business Insights.

Bresse, M. 1876. *Water-Wheels or Hydraulic Motors*. New York: Wiley.

Bretz, J. H. 1923. The channeled scablands of the Columbia Plateau. *Journal of Geology* 31: 617–649.

Brian, M. V., ed. 1978. *Production Ecology of Ants and Termites*. Cambridge: Cambridge University Press.

Briffa, K. R., P. D. Jones, F. H. Schweingruber, and T. Osborn. 1998. Influence of volcanic eruptions on Northern Hemisphere summer temperature over the past 600 years. *Nature* 393: 450–454.

Briggle, L. W. 1980. Origin and botany of wheat. In *Wheat*, 6–13. Basel: CIBA-Geigy.

Brimblecombe, P. 1987. *The Big Smoke*. London: Routledge.

Broda, E. 1975. *The Evolution of the Bioenergetic Processes*. Oxford: Pergamon Press.

Brody, S. 1945. *Bioenergetics and Growth*. New York: Reinhold.

Brooks, D. R., and E. O. Wiley. 1986. *Evolution as Entropy*. Chicago: University of Chicago Press.

Brooks, J., ed. 1990. *Classic Petroleum Provinces*. London: Geological Society.

Brown, G. I. 1999. *Count Rumford: The Extraordinary Life of a Scientific Genius*. Stroud, UK: Sutton Publishing.

Brown, H. L., B. B. Hamel, and B. A. Hedman. 1996. *Energy Analysis of 108 Industrial Products*. Lilburn, Ga.: Fairmont Press.

Brown, J. H., and G. B. West, eds. 2000. *Scaling in Biology*. Oxford: Oxford University Press.

Brown, P., and D. F. Tuzin. 1983. *The Ethnography of Cannibalism*. Washington, D.C.: Society of Psychological Anthropology.

Browning, K. A., and R. J. Gurney, eds. 1999. *Global Energy and Water Cycles*. Cambridge: Cambridge University Press.

Bruni, B., and P. L. Porta, eds. 2005. *Economics and Happiness: Framing the Analysis*. New York: Oxford University Press.

Buchanan, A. H., and B. G. Honey. 1994. Energy and carbon dioxide implications of building construction. *Energy and Buildings* 20: 205–217.

Buchdahl, H. A. 1966. *The Concepts of Classical Thermodynamics*. Cambridge: Cambridge University Press.

Buck, J. L. 1930. *Chinese Farm Economy*. Nanking: University of Nanking.

———. 1937. *Land Utilization in China*. Nanking: University of Nanking.

Budiansky, S. 1997. *The Nature of Horses: Exploring Equine Evolution, Intelligence and Behavior*. New York: Free Press.

Budyko, M. I., ed. 1963. *Atlas of the Heat Balance of the Earth*. Moscow: MGK.

Buenstorf, G. 2004. *The Economics of Energy and the Production Process: An Evolutionary Approach*. Cheltenham, UK: E. Elgar.

Buliet, R. W. 1975. *The Camel and the Wheel*. Cambridge, Mass.: Harvard University Press.

Burbank, D. W., A. E. Blythe, J. Putkonen, B. Pratt-Sitaula, E. Gabet, M. Oskin, A. Barros, and T. P. Ojha. 2003. Decoupling erosion and precipitation in the Himalayas. *Nature* 426: 652–655.

Burbank, D. W., J. Leland, E. Fielding, R. S. Anderson, N. Brozovic, M. R. Reid, and C. Duncan. 1996. Bedrock incision, rock uplift, and threshold hillslopes in the northwestern Himalayas. *Nature* 379: 505–510.

Burchell, R. W., and D. Listokin. 1982. *Energy and Land Use*. Piscataway, N.J.: Center for Urban Policy Research.

Burstall, A. F. 1963. *A History of Mechanical Engineering*. London: Faber and Faber.

REFERENCES

———. 1968. *Simple Working Models of Historic Machines.* Cambridge, Mass.: MIT Press.

Butler, P. J., J. A. Green, I. L. Boyd, and J. R. Speakman. 2004. Measuring metabolic rate in the field: The pros and cons of the doubly labelled water and heart rate methods. *Functional Ecology* 18: 168–183.

Butzer, K. W. 1976. *Early Hydraulic Civilization in Egypt.* Chicago: University of Chicago Press.

———. 1984. Long-term Nile flood variation and political discontinuities in Pharaonic Egypt. In *From Hunters to Farmers*, ed. J. D. Clark and S. A. Brandt, 102–112. Berkeley, Calif.: University of California Press.

Byrne, R. 1987. Climatic change and the origins of agriculture. *Gordon Childe Colloquium* 5: 21–34.

Cachel, S. 1997. Dietary shifts and the European Upper Paleolithic transition. *Current Anthropology* 38: 579–603.

Caine, N. 1976. A uniform measure of subaerial erosion. *Geological Society of America Bulletin* 87: 137–140.

Calder, W. A. 1978. The kiwi. *Scientific American* 239 (1): 132–142.

———. 1983. Ecological scaling: Mammals and birds. *Annual Review of Ecology and Systematics* 14: 213–230.

Caldwell, J. C. 1976. Toward a restatement of demographic transition theory. *Population and Development Review* 2: 321–366.

Calow, P. 1977. Conversion efficiencies in heterotrophic organisms. *Biological Reviews* 52: 385–409.

Calvin, M. 1989. Forty years of photosynthesis and related activities. *Photosynthesis Research* 211: 3–16.

Camilloni, I., and V. Barros. 1997. On the urban heat island effect dependence on temperature trends. *Climatic Change* 37: 665–681.

Campbell, C. J. 1997. *The Coming Oil Crisis.* Brentwood, UK: Petroconsultants and Multi-Science Publishing.

Campbell, C. J., and J. Laherrère. 1998. The end of cheap oil. *Scientific American* 278 (3): 78–83.

Campbell, D. L., H. Z. Martin, E. V. Murphree, and C. W. Tyson. 1948. U.S. patent 2,451,804: A method of and apparatus for contacting solids and gases. ⟨http://www.uspto.gov/⟩.

Campbell, H. R. 1907. *The Manufacture and Properties of Iron and Steel.* New York: Hill.

Campbell, M. S., and M. Overton. 1993. A new perspective on medieval and early modern agriculture: Six centuries of Norfolk farming c.1250–c.1850. *Past and Present* 141: 38–105.

Canakci, M., and I. Akinci. 2005. Energy use pattern analyses of greenhouse vegetable production. *Energy* 31: 1243–1256.

Caneva, K. L. 1993. *Robert Mayer and the Conservation of Energy.* Princeton, N.J.: Princeton University Press.

Cao, M., and F. I. Woodward. 1998. Net primary and ecosystem production and carbon stocks of terrestrial ecosystems and their response to climate change. *Global Change Biology* 4: 185–198.

CAPP (Canadian Association of Petroleum Producers). 2005. *Oil Sands.* ⟨http://www.capp.ca/⟩.

Carbon Dioxide Information Analysis Center. ⟨http://cdiac.ornl.gov/⟩.

Carbone, C., and J. L. Gittleman. 2002. A common rule for the scaling of carnivore density. *Science* 295: 2273–2276.

Carcoana, A. 1992. *Applied Enhanced Oil Recovery.* Englewood Cliffs, N.J.: Prentice-Hall.

Cardwell, D. S. L. 1971. *From Watt to Clausius: The Rise of Thermodynamics in the Early Industrial Age.* Ithaca, N.Y.: Cornell University Press.

———. 1991. *Turning Points in Western Technology: A Study of Technology, Science and History.* Canton, Mass.: Science History Publications.

Carnot, S. 1824. *Réflexions sur la puissance motrice du feu et sur les machines propres à développer cette puissance.* Paris: Bachelier.

Trans. R. H. Thurston, 1890. Also in *Reflections on the Motive Power of Fire by Sadi Carnot and Other Papers on the Second Law of Thermodynamics*, ed. E. Mendoza. New York: Dover, 1960.

Caro, R. A. 1983. *The Years of Lyndon Johnson: The Path to Power*. New York: Knopf.

Carpenter, A. M. 1988. *Coal Classification*. London: IEA Coal Research.

Carrier, D. R. 1984. The energetic paradox of human running and hominid evolution. *Current Anthropology* 25: 483–495.

Carter, G. F. 1977. A hypothesis suggesting a single origin of agriculture. In *Origins of Agriculture*, ed. C. Reed, 123–138. The Hague: Mouton.

Carter, R. A. 2000. *Buffalo Bill Cody: The Man Behind the Legend*. New York: Wiley.

Carter, W. E. 1969. *New Lands and Old Traditions: Kekchi Cultivators in the Guatemala Lowlands*. Gainesville: University of Florida Press.

Cary, S. C., T. Shank, and J. Stein. 1998. Worms bask in extreme temperatures. *Nature* 391: 545–546.

Cassman, K. G., A. Dobermann, and D. T. Walters. 2002. Agroecosystems, nitrogen-use efficiency, and nitrogen management. *Ambio* 31: 132–140.

Casten, T. R. 1998. *Turning off the Heat: Why America Must Double Energy Efficiency to Save Money and Reduce Global Warming*. Amherst, N.Y.: Prometheus Books.

CCMD (Committee for the Compilation of Materials on Damage Caused by the Atomic Bombs in Hiroshima and Nagasaki). 1981. *Hiroshima and Nagasaki*. New York: Basic Books.

CEC (California Energy Commission). 2005. *California's Water-Energy Relationship*. Sacramento: California Energy Commission.

CERES (Clouds and the Earth's Radiant Energy System). 2002. ⟨http://asd-www.larc.nasa.gov/ceres/⟩.

Chabot, B. F., and D. J. Hicks. 1982. The ecology of leaf life spans. *Annual Review of Ecology and Systematics* 13: 229–259.

Chadwick, J. 1932. Possible existence of a neutron. *Nature* 129: 312.

Chai, P., and R. Dudley. 1995. Limits to vertebrate locomotor energetics suggested by hummingbird hovering in heliox. *Nature* 377: 722–725.

Chapin, D. M., K. P. Cohen, W. K. Davis, E. E. Kintner, L. J. Koch, J. W. Landis, M. Levenson et al. 2002. Nuclear power plants and their fuel as terrorist targets. *Science* 297: 1997–1999.

Chapman, J. L., and M. R. Reiss. 1999. *Ecology*. Cambridge: Cambridge University Press.

Chapman, P. F., and D. F. Hemming. 1976. Energy requirements of some energy sources. In *The Energy Accounting of Materials, Products, Processes, and Services*, ed. A. Verbraeck, 119–140. Rotterdam: TNO (Netherlands Institute for Applied Scientific Research).

Chapman, P. F., G. Leach, and M. Slesser. 1974. The energy cost of fuels. *Energy Policy* 2: 231–243.

Chapman, R. F. 1998. *The Insects*. Cambridge: Cambridge University Press.

Charkoudian, N. 2003. Skin blood flow in adult human thermoregulation: How it works, when it does not, and why. *Mayo Clinic Proceedings* 78: 603–612.

Charlson, R. J., F. P. J. Valero, and J. H. Seinfeld. 2005. In search of balance. *Science* 308: 806–807.

Chatterson, E. K. 1977. *Sailing Ships: The Story of Their Development from the Earliest Times to the Present*. New York: Gordon Press.

Chen, J., B. E. Carlson, and A. D. Del Genio. 2002. Evidence for strengthening of the tropical general circulation in the 1990s. *Science* 295: 838–841.

Chen, X. 1981. Studies of energy intakes, expenditures and requirements in China. In *Protein-Energy Requirements of Developing Countries: Evaluation of New Data*, ed. B. Torun, V. R.

REFERENCES

Young, and W. R. Rand, 150–158. Tokyo: United Nations University.

Chen, Y. Y., J. Burnett, and C. K. Chau. 2001. Analysis of embodied energy use in the residential building of Hong Kong. *Energy* 26: 323–340.

Chevalier, R. 1976. *Roman Roads.* Berkeley: University of California Press.

Chi, C. 1936. *Key Economic Areas in Chinese History, as Revealed in the Development of Public Works for Water Control.* London: Allen and Unwin.

Childe, V. G. 1951. The Neolithic revolution. In *Man Makes Himself,* ed. V. G. Childe, 67–72. London: C. A. Watts.

Chisholm, S. W., R. J. Olson, E. R. Zettler, R. Goericke, J. B. Waterbury, and N. A. Welschmeyer. 1988. A novel free-living prochlorophyte abundant in the oceanic euphotic zone. *Nature* 334: 340–343.

Choate, W. T., and J. A. S. Green. 2003. *U.S. Energy Requirements for Aluminum Production: Historical Perspectives, Theoretical Limits and New Opportunities.* ⟨http://www.seca.net/docs/resources/US_Energy_Requirements_for_Aluminum_Production.pdf⟩.

Chorley, G. P. H. 1981. The agricultural revolution in Northern Europe, 1750–1880: Nitrogen, legumes, and crop productivity. *Economic History* 34: 71–93.

Choudhury, P. C., ed. 1734. *Hastividyarnava.* Guwahati, India: Publication Board of Assam. Reissued 1975.

Choy, G. L., and J. L. Boatwright. 1995. Global patterns of radiated seismic energy and apparent stress. *Journal of Geophysical Research* 100: 18205–18228.

Christ, K. 1984. *The Romans.* Berkeley: University of California Press.

Christian, H. J., R. J. Blakeslee, D. J. Boccippio, W. L. Boeck, D. E. Buechler, K. T. Driscoll, S. J. Goodman et al. 2003. Global frequency and distribution of lightning as observed from space by the Optical Transient Detector. *Journal of Geophysical Research* 108: 4005–4019.

Chung, R. M., D. B. Ballantyne, E. Comeau, T. L. Holzer, D. Madrzykowski, A. J. Schiff, W. C. Stone et al. 1996. *The January 17, 1995 Hyogoken-Nanbu (Kobe) Earthquake: Performance of Structures, Lifelines, and Fire Protection Systems.* SN003-003-03412-6. Washington, D.C.: Government Printing Office.

CIA (Central Intelligence Agency). 1979. *The World Oil Market in the Years Ahead.* ⟨https://www.cia.gov/cia/publications/mapspub/201.shtml⟩.

Ciais, P., M. Reichstein, N. Viovy, A. Granier, J. Ogee, V. Allard, M. Aubinet et al. 2005. Europe-wide reduction in primary productivity caused by the heat and drought in 2003. *Nature* 437: 529–533.

Cipolla, C. M. 1966. *Guns, Sails and Empires.* New York: Pantheon.

Clark, C., and M. Haswell. 1970. *The Economics of Subsistence Agriculture.* London: Macmillan.

Clark, D. A., and D. B. Clark. 1984. Spacing dynamics of a tropical rain forest tree: Evaluation of the Janzen-Connell model. *American Naturalist* 124: 769–788.

Clark, E. L. 1986. Cogeneration: Efficient energy source. *Annual Review of Energy* 11: 275–294.

Clark, G. B. 1981. Basic properties of ammonium nitrate fuel oil explosives (ANFO). *Colorado School of Mines Quarterly* 76: 1–32.

Clark, J. D., and S. A. Brandt, eds. 1984. *From Hunters to Farmers.* Berkeley: University of California Press.

Clark, P. U., N. G. Pisias, T. F. Stocker, and A. J. Weaveret. 2002. The role of thermohaline circulation in abrupt climate change. *Nature* 415: 863–869.

Clarke, A. 2004. Is there a universal temperature dependence of metabolism? *Functional Ecology* 18: 252–256.

Clarke, A., and K. P. P. Fraser. 2004. Why does metabolism scale with temperature? *Functional Ecology* 18: 243–251.

REFERENCES

Clarke, G., D. Leverington, J. Teller, and A. Dyke. 2003. Superlakes, megafloods, and abrupt climatic change. *Science* 301: 922–923.

Clausius, R. 1850. Über die bewegende Kraft der Wärme. *Annalen der Physik und Chemie* 79: 368–397+.

———. 1867. *Abhandlungen über die mechanische Wärmetheorie.* Braunschweig: F. Vieweg.

Clavering, E. 1995. The coal mills of Northeast England: The use of waterwheels for draining coal mines, 1600–1750. *Technology and Culture* 36: 211–241.

Clayton, K. M. 1997. The rate of denudation of some British lowland landscapes. *Earth Surface Processes and Landforms* 22: 721–731.

Clerk, D. 1911. Oil engine. In *Encyclopædia Britannica.* 11th ed. Vol. 20, 25–43. Cambridge: Cambridge University Press.

Cleveland, C. J. 2005. Net energy from the extraction of oil and gas in the United States. *Energy* 30: 769–782.

Cleveland, C. J., R. Costanza, C. A. S. Hall, and R. K. Kaufmann. 1984. Energy and the U.S. economy: A biophysical perspective. *Science* 225: 890–897.

Cleveland, C. J., R. K. Kaufmann, and D. I. Stern. 2000. Aggregation and the role of energy in the economy. *Ecological Economics* 32: 301–317.

CLIMAP (Climate: Long-Range Investigation, Mapping, and Prediction). 1976. The surface of the ice-age Earth. *Science* 191: 1131–1137.

Clutton-Brock, J. 1992. *Horse Power: A History of Horse and the Donkey in Human Societies.* Cambridge, Mass.: Harvard University Press.

Coats, R. R. 1962. Magma type and crustal structure in the Aleutian Arc. In *The Crust of the Pacific Basin,* ed. G. Macdonald and H. Kuno, 92–109. Washington, D.C.: American Geophysical Union.

Cockrill, W. R., ed. 1974. *The Husbandry and Health of the Domestic Buffalo.* Rome: FAO.

Cohen, M. N. 1977. *The Food Crisis in Prehistory: Overpopulation and the Origins of Agriculture.* New Haven, Conn.: Yale University Press.

———. 1980. *Biosocial Mechanisms of Population Regulation.* New Haven, Conn.: Yale University Press.

Colbeck, I., and A. R. MacKenzie. 1994. *Air Pollution by Photochemical Oxidants.* Amsterdam: Elsevier.

Coleman, D. C., and D. A. Crossley. 1996. *Fundamentals of Soil Ecology.* San Diego, Calif.: Academic Press.

Coles, J. M., and E. S. Higgs. 1969. *The Archaeology of Early Man.* London: Faber.

Collins, E. V., and A. B. Caine. 1926. Testing draft horses. *Iowa Experiment Station Bulletin* 240: 193–223.

Collins, K. 2000. Statement of Keith Collins, Chief Economist, U.S. Department of Agriculture, before the U.S. Senate Committee on Agriculture, Nutrition, and Forestry, July 20. ⟨http://agriculture.senate.gov/Hearings/Hearings_2000/July_20_2000/00720col.htm⟩.

Coltman, J. W. 1988. The transformer. *Scientific American* 258 (1): 86–95.

Condie, K. C. 1997. *Plate Tectonics and Crustal Evolution.* Oxford: Oxford University Press.

Conklin, H. C. 1957. *Hanunoo Agriculture.* Rome: FAO.

Connell, J. H., and M. D. Lowman. 1989. Low-diversity tropical rain forests: Some possible mechanisms for their existence. *American Naturalist* 134: 88–119.

Conrad, C. P., and C. Lithgow-Bertelloni. 2002. How mantle slabs drive plate tectonics. *Science* 298: 207–209.

Constant, E. W. 1981. *The Origins of Turbojet Revolution.* Baltimore, Md.: Johns Hopkins University Press.

Cook, E. 1976. *Man, Energy, Society.* San Francisco: W. H. Freeman.

Cooper, G. A. 1994. Directional drilling. *Scientific American* 270 (5): 82–87.

REFERENCES

Corkhill, M. 1975. *LNG Carriers: The Ships and Their Market.* London: Fairplay Publications.

Corum, J. 2003. *The Strouhal Number in Cruising Flight.* ⟨http://www.style.org/strouhalflight/⟩.

Costanza, R. 1980. Embodied energy and economic valuation. *Science* 210: 1219–1224.

Cottrell, F. 1955. *Energy and Society.* New York: McGraw-Hill.

Coughenour, M. B., J. E. Ellis, D. M. Swift, D. L. Coppock, K. Galvin, J. T. McCabe, and T. C. Hart. 1985. Energy extraction and use in a nomadic pastoral ecosystem. *Science* 230: 619–624.

Coulton, J. J. 1977. *Ancient Greek Architects at Work: Problems of Structure and Design.* Ithaca, N.Y.: Cornell University Press.

Coupland, R. T., ed. 1979. *Grassland Ecosystems of the World: Analysis of Grasslands and Their Uses.* Cambridge: Cambridge University Press.

Courtenay, L. T. 1997. *The Engineering of Medieval Cathedrals.* Aldershot: Ashgate.

Courtillot, V., A. Davaille, J. Besse, and J. Stock. 2003. Three different types of hotspots in the Earth's mantle. *Earth and Planetary Science Letters* 205: 295–308.

Cowan, C., and P. J. Watson, eds. 1992. *The Origins of Agriculture.* Washington, D.C.: Smithsonian Institution.

Cowan, R. 1990. Nuclear power reactors: A study in technological lock-in. *Journal of Economic History* 50: 541–567.

Cox, A. N., W. C. Livingston, and M. S. Matthews. 1991. *Solar Interior and Atmosphere.* Tucson: University of Arizona Press.

Crafts, N., and T. C. Mills. 2004. Was the nineteenth century British growth steam-powered? The climacteric revisited. *Explorations in Economic History* 41: 156–171.

Cramer, W., D. W. Kicklighter, A. Bondeau, B. Moore III, G. Churkina, B. Nemry, A. Ruimy et al. 1999. Comparing global models of terrestrial net primary productivity (NPP): Overview and key results. *Global Change Biology* 5: 1–15.

Crawley, M. J. 1983. *Herbivory: The Dynamics of Animal-Plant Interactions.* Berkeley: University of California Press.

Creel, H. G. 1965. The role of the horse in Chinese history. *American Historical Review* 70: 647–672.

Criswell, D. 2000. Lunar solar power system: Review of the technology base of an operational LSP system. *Acta Astronautica* 46: 531–540.

Cullen, R. 1993. The true cost of coal. *Atlantic Monthly* 272 (6): 38–52.

CWC (Canadian Wood Council). 2004. *Energy and the Environment in Residential Construction.* ⟨http://www.cwc.ca/pdfs/EnergyAndEnvironment.pdf⟩.

Czaya, E. 1981. *Rivers of the World.* New York: Van Nostrand Reinhold.

da Rosa, A. V. 2005. *Fundamentals of Renewable Energy Processes.* Amsterdam: Elsevier.

Dabberdt, W. F., D. H. Lenschow, T. W. Horst, P. R. Zimmerman, S. P. Oncley, and A. C. Delany. 1993. Atmosphere-surface exchange measurements. *Science* 260: 1472–1481.

Daly, H. E., ed. 1973. *Toward a Steady-State Economy.* San Francisco: W. H. Freeman.

Damuth, J. 1981. Population density and body size in mammals. *Nature* 290: 699–700.

———. 1993. Cope's rule, the island rule and the scaling of mammalian population density. *Nature* 365: 748–750.

Daumas, M., ed. 1969. *A History of Technology and Invention.* New York: Crown.

Davies, G. F. 1980. Review of oceanic and global heat flow estimates. *Reviews of Geophysics and Space Physics* 18: 718–722.

———. 1999. *Dynamic Earth: Plates, Plumes and Mantle Convection.* Cambridge: Cambridge University Press.

Davis, L. S., K. N. Johnson, P. Bettinger, and T. E. Howard. 2001. *Forest Management: To Sustain Ecological, Economic, and Social Values.* 4th ed. New York: McGraw-Hill.

Dawson, O. L. 1970. *Communist China's Agriculture*. New York: Praeger.

Dawson, T. J. 1977. Kangaroos. *Scientific American* 237 (2): 78–89.

de Beaune, S. A., and R. White. 1993. Ice age lamps. *Scientific American* 266 (3): 108–113.

de Beer, J. G., E. Worrell, and K. Blok. 1998. Future technologies for energy-efficient iron and steel making. *Annual Review of Energy and Environment* 23: 123–205.

de Duve, C. 1984. *The Living Cell*. New York: Scientific American Library.

De la Cruz-Reyna, S. 1991. Poisson-distributed patterns of explosive eruptive activity. *Bulletin of Volcanology* 54: 57–67.

de Mora, S., S. Demers, and M. Vernet, eds. 2000. *The Effects of UV Radiation in the Marine Environment*. New York: Cambridge University Press.

De Niro, M. J. 1987. Stable isotopy and archaeology. *American Scientist* 75: 182–191.

de Toma, G., O. R. White, G. A. Chapman, and S. R. Walton. 2004. Solar irradiance variability: Progress in measurement and empirical analysis. *Advances in Space Research* 34: 237–242.

De Vooys, C. G. N. 1979. Primary production in aquatic systems. In *The Global Carbon Cycle*, ed. B. Bolin, E. T. Degens, S. Kempe, and P. Ketner, 259–292. SCOPE 13. New York: Wiley.

Decker, R., and B. Decker. 1981. The eruptions of Mount St. Helens. *Scientific American* 244 (3): 68–80.

Deffeyes, K. S. 2001. *Hubbert's Peak: The Impending World Oil Shortage*. Princeton, N.J.: Princeton University Press.

DeFries, R. S., C. B. Field, I. Fung, G. J. Collatz, and L. Bounoua. 1999. Combining satellite data and biogeochemical models to estimate global effects of human-induced land cover change on carbon emissions and primary productivity. *Global Biogeochemical Cycles* 13: 803–815.

DeLany, J. P. 1997. Doubly labeled water for energy expenditure. In *Emerging Technologies for Nutrition Research*, ed. S. J. Carlson-Newberry and R. B. Costell, 281–296. Washington, D.C.: National Academies Press.

Demirdöven, N., and J. Deutch. 2004. Hybrid cars now, fuel cell cars later. *Science* 305: 974–976.

Demographia. 2005. *World Urban Atlas*. 〈http://www.demographia.com/db-worldua.pdf〉.

Denison, E. F. 1985. *Trends in American Economic Growth, 1929–1982*. Washington, D.C.: Brookings Institution.

Denning, R. S. 1985. The Three Mile Island unit's core: A post-mortem examination. *Annual Review of Energy* 10: 35–52.

Denny, M. 2004. The efficiency of overshot and undershot waterwheels. *European Journal of Physics* 25: 193–202.

Denny, M. W. 1993. *Air and Water: The Biology and Physics of Life's Media*. Princeton, N.J.: Princeton University Press.

Dent, A. 1974. *The Horse*. New York: Holt, Rinehart and Winston.

Derickson, A. 1998. *Black Lung: Anatomy of a Public Health Disaster*. Ithaca, N.Y.: Cornell University Press.

Des Marais, D. J. 2000. When did photosynthesis emerge on Earth? *Science* 289: 1703–1705.

des Noëttes, R. J. 1931. *L'Attelage et le cheval de selle a travers les ages: Contribution à l'histoire de l'esclavage*. Paris: Picard.

Devereux, S. 1999. *Drilling Technology in Nontechnical Language*. Tulsa, Okla.: PennWell Publications.

Devine, W. D. 1983. From shafts to wires: Historical perspective on electrification. *Journal of Economic History* 63: 347–372.

Dewey, K. G. 1997. Energy and protein requirements during lactation. *Annual Review of Nutrition* 17: 19–36.

DeZeeuw, J. W. 1978. Peat and the Dutch Golden Age. *AAG Bijdragen* 21: 3–31.

REFERENCES

Dick, H. J. B., J. Lian, and H. Schouten. 2003. An ultraslow-spreading class of ocean ridge. *Nature* 426: 405–412.

Dickinson, H. W. 1939. *A Short History of the Steam Engine.* Cambridge: Cambridge University Press.

Dickson, B., I. Yashayaev, J. Meincke, B. Turrell, S. Dye, and J. Holfort. 2002. Rapid freshening of the deep North Atlantic Ocean over the past four decades. *Nature* 416: 832–836.

Diderot, D., and J. L. d'Alembert. 1751–1772. *Encyclopédie ou Dictionnaire raisonné des sciences, des arts et des métiers.* 28 vols. Paris. 〈http://www.lib.uchicago.edu/efts/ARTFL/projects/encyc/〉.

Dieffenbach, E. M., and R. B. Gray. 1960. The development of the tractor. In *USDA Agricultural Yearbook 1960,* 24–45. Washington, D.C.: USDA.

Diener, E., E. Suh, and S. Oishi. 1997. Recent findings on subjective well-being. *Indian Journal of Clinical Psychology 24: 25–41.*

Dillon, W. 1992. *Gas (Methane) Hydrates: A New Frontier.* 〈http://marine.er.usgs.gov/fact-sheets/gas-hydrates/title.html〉.

Dinneen, G. U., and G. L. Cook. 1974. Oil shale and the energy crisis. *Technology Review* 76 (3): 26–33.

Donald, C. M., and J. Hamblin. 1976. The biological yield and harvest index of cereals as agronomic and plant breeding criteria. *Advances in Agronomy* 28: 361–405.

Donelan, J. M., R. Kram, and A. D. Kuo. 2002. Mechanical work for step-to-step transitions is a major determinant of the metabolic cost of human walking. *Journal of Experimental Biology* 205: 3717–3727.

Donnison, J. R., and D. F. Mikulskis. 1992. Three-body orbital stability criteria for circular orbits. *Monthly Notes of the Royal Astronomical Society* 254: 21–26.

Doorenbos, J., and A. H. Kassam. 1979. *Yield Response to Water.* Irrigation and Drainage Paper 33. Rome: FAO.

Downey, M. W. 2001. *Petroleum Provinces of the Twenty-first Century.* Washington, D.C.: American Association of Petroleum Geologists.

Drela, M., and J. S. Langford. 1985. Human-powered flight. *Scientific American* 253 (5): 144–151.

Duby, G. 1968. *Rural Economy and Country Life in the Medieval West.* Trans. C. Postan. London: Edward Arnold. Reissued Philadelphia: University of Pennsylvania Press, 1998.

Duchesne, L. C., and D. W. Larson. 1989. Cellulose and the evolution of plant life. *BioScience* 39: 238–241.

Duda, J. R., and E. L. Hemingway. 1976. *Basic Estimated Capital Investment and Operating Costs for Underground Bituminous Coal Mines Developed for Longwall Mining.* U.S. Department of the Interior.

Dukes, J. S. 2003. Burning buried sunshine: Human consumption of ancient solar energy. *Climatic Change* 61: 31–44.

Duman, J. G. 2001. Antifreeze and ice nucleator proteins in terrestrial arthropods. *Annual Review of Physiology* 63: 327–357.

Dumas, J.-B., and J.-B. Boussingault. 1842. *The Chemical and Physiological Balance of Organic Nature.* New York: Saxton and Miles.

Duncan-Jones, R. 1990. *Structure and Scale in the Roman Economy.* Cambridge: Cambridge University Press.

Düring, I. 1966. *Aristotele: Darstellung und Interpretation seines Denken.* Heidelberg: Carl Winter.

Durnin, J. V. G., and R. Passmore. 1967. *Energy, Work and Leisure.* London: Heinemann.

Dutilh, C. E., and K. J. Kramer. 2000. Energy consumption in the food chain. *Ambio* 29: 98–101.

Dutilh, C. E., and A. R. Linnemann. 2004. Food system, energy use in. In *Encyclopedia of Energy.* Vol. 2, 719–726.

DWIA (Danish Wind Industry Association). 2005. 〈http://www.windpower.org/〉.

REFERENCES

Dyson-Hudson, N. 1980. Strategies of resource exploitation among East African savanna pastoralists. In *Human Ecology in Savanna Environments*, ed. D. R. Harris, 171–184. New York: Academic Press.

Eagar, T. W., and C. Musso. 2001. Why did the World Trade Center collapse? Science, engineering, and speculation. *JOM* 53: 8–11. ⟨http://www.tms.org/pubs/journals/JOM/0112/Eagar/Eagar-0112.html⟩.

Earl, D. 1973. *Charcoal and Forest Management*. Oxford: Oxford University Press.

Eaton, R. L. 1974. *The Cheetah*. New York: Van Nostrand Reinhold.

Ecosystems Working Group. 1998. *Terrestrial Ecosystem Responses to Global Change*. Environmental Sciences Division. Publ. 4821. Oak Ridge National Laboratory. ⟨http://www.ornl.gov/~webworks/cpr/rpt/1001.pdf⟩.

Edgerton, D. 2007. *The Shock of the Old: Technology and Global History since 1900*. New York: Oxford University Press.

Edgerton, S. Y. 1961. Heat and style: Eighteenth-century house warming by stoves. *Journal of the Society of Architectural Historians* 20: 20–26.

Edmonds, R. L., ed. 1982. *Analysis of Coniferous Forest Ecosystems in the Western United States*. New York: Van Nostrand Reinhold.

Edwards, G., and D. Walker. 1983. *C₃, C₄: Mechanisms and Cellular and Environmental Regulation of Photosynthesis*. Berkeley: University of California Press.

Edwards, J. F. 2003. Building the Great Pyramid: Probable construction methods employed at Giza. *Technology and Culture* 44: 340–354.

EEA (European Environment Agency). 1998. *Life Cycle Assessment (LCA): A Guide to Approaches, Experiences and Information Sources*. ⟨http://reports.eea.europa.eu/GH-07-97-595-EN-C/en⟩.

Egbert, G. D., and R. D. Ray. 2000: Significant dissipation of tidal energy in the deep ocean inferred from satellite altimeter data. *Nature* 405: 775–778.

Ehleringer, J., and I. Forseth. 1980. Solar tracking by plants. *Science* 210: 1094–1098.

EIA (Energy Information Administration). 1998. *A Look at Commercial Buildings in 1995: Characteristics, Energy Consumption, and Energy Expenditures*. ⟨http://www.eia.doe.gov/emeu/cbecs/report_1995.html⟩.

———. 2001a. *2001 Residential Energy Consumption Survey*. ⟨http://www.eia.doe.gov/emeu/recs/contents.html⟩.

———. 2001b. *End-Use Consumption of Electricity 2001*. ⟨http://www.eia.doe.gov/emeu/recs/contents.html⟩.

———. 2004. *World LNG Shipping Capacity Expanding*. ⟨http://www.eia.doe.gov/oiaf/analysispaper/global/worldlng.html⟩.

———. 2005a. *Energy Use in Commercial Buildings*. ⟨http://www.eia.doe.gov/emeu/cbecs/⟩.

———. 2005b. *International Energy Annual*. ⟨http://www.eia.doe.gov/iea/⟩.

———. 2005c. *Major U.S. Coal Mines*. ⟨http://www.eia.doe.gov/cneaf/coal/page/acr/table9.html⟩.

———. 2005d. *Natural Gas Annual 2004*. ⟨http://tonto.eia.doe.gov/FTPROOT/natgas/013104.pdf⟩.

———. 2006a. *Annual Energy Review*. ⟨http://www.eia.doe.gov/emeu/aer/contents.html⟩.

———. 2006b. *U.S. Coal Supply and Demand: 2005 Review*. ⟨http://www.eia.doe.gov/cneaf/coal/page/special/feature.html⟩.

Eiler, J., ed. 2003. *Inside the Subduction Factory*. Washington, D.C.: American Geophysical Union.

Einstein, A. 1905. Zur Elektrodynamik bewegter Körper. *Annalen der Physik* 17: 891–921.

REFERENCES

————. 1907. Über das Relativitätsprinzip und die aus demselben gezogenen Folgerungen. In *Jahrbuch der Radioaktivität* 4: 411–462.

Elder, J. 1976. *The Bowels of the Earth*. Oxford: Oxford University Press.

Elderfield, H., and A. Schultz. 1996. Mid-ocean ridge hydrothermal fluxes and the chemical composition of the ocean. *Annual Review of Earth and Planetary Sciences* 24: 191–224.

Elliott, D. 2003. *Energy, Society, and Environment: Technology for a Sustainable Future*. London: Routledge.

Elliott, D. L., and M. N. Schwartz. 1993. *Wind Energy Potential in the United States*. ⟨http://www.ece.umr.edu/power/Energy_Course/energy/potential.html⟩.

Ellwood, M. D. F., and W. A. Foster. 2004. Doubling the estimate of invertebrate biomass in a rainforest canopy. *Nature* 429: 549–551.

Elton, C. 1927. *Animal Ecology*. New York: Macmillan.

Elvidge, C. D. 2004. U.S. constructed area approaches the size of Ohio. *EOS* 85: 233–234.

Emanuel, K. A. 1998. The power of a hurricane: An example of reckless driving on the information superhighway. *Weather* 54: 107–108.

————. 2003. Tropical cyclones. *Annual Review of Earth and Planetary Sciences* 31: 75–104.

Emanuel, W. R., H. H. Shugart, and M. Stevenson. 1985. Climatic change and the broad-scale distribution of ecosystem complexes. *Climatic Change* 7: 29–43.

Emilio, M., J. R. Kuhn, R. I. Bush, and P. Scherrer. 2000. On the constancy of the solar diameter. *Astrophysical Journal* 543: 1007–1010.

Encyclopedia of Desalination and Water Resources (DESWARE). 2006. ⟨http://desware.net/⟩.

Encyclopedia of Energy. 2004. 6 vols. Amsterdam: Elsevier.

Enquist, B. J., J. H. Brown, and G. B. West. 1998. Allometric scaling of plant energetics and population density. *Nature* 395: 163–165.

Enquist, B. J., and K. J. Niklas. 2002. Global allocation rules for patterns of biomass partitioning in seed plants. *Science* 295: 1517–1520.

Enquist, B. J., G. B. West, E. L. Charnov, and J. H. Brown. 1999. Allometric scaling of production and life-history variation in vascular plants. *Nature* 401: 907–911.

Epstein, P. R., and E. Mills, eds. 2005. *Climate Change Futures: Health, Ecological and Economic Consequences*. Cambridge, Mass.: Harvard Medical School.

Erasmus, F. C. 1975. Die Entwicklung des Steinkohlenbergbaus im Ruhrrevier in den siebziger Jahren. *Glückauf* 11: 311–318.

ERBE (Earth Radiation Budget Experiment). 2005. *The Earth Radiation Budget Experiment*. NASA. ⟨http://asd-www.larc.nasa.gov/erbe/ASDerbe.html⟩.

Erlande-Brandenburg, A. 1995. *Cathedrals and Castles: Building in the Middle Ages*. New York: H. N. Abrams.

Erman, A. 1894. *Life in Ancient Egypt*. London: Macmillan.

Erwin, D., J. Valentine, and D. Jablonski. 1997. The origin of animal body plans. *American Scientist* 85: 126–137.

ESA (Electricity Storage Association). 2001. *Large Scale Electricity Storage Technologies*. ⟨http://www.electricitystorage.org/⟩.

Esmay, M. L., and C. W. Hall, eds. 1968. *Agricultural Mechanization in Developing Countries*. Tokyo: Shin-Norinsha.

ESPG (Energy Systems Program Group). 1981. *Energy in a Finite World: A Global Systems Analysis*. Cambridge, Mass.: Ballinger.

Etemad, B., J. Luciani, P. Bairoch, and J.-C. Toutain. 1991. *World Energy Production: 1800–1985*. Geneva: Librairie DROZ.

Evangelou, P. 1984. *Livestock Development in Kenya's Maasailand*. Boulder, Colo.: Westview Press.

Evans, H. B. 1994. *Water Distribution in Ancient Rome*. Ann Arbor: University of Michigan Press.

Evans, L. 2002. Traffic crashes. *American Scientist* 90: 244–253.

Evelyn, J. 1664. *Sylva, or a Discourse of Forest Trees*. London: Royal Society.

Ewbank, T. 1870. *A Descriptive and Historical Account of Hydraulic and Other Machines for Raising Water*. New York: Scribner.

ExternE: Externalities of Energy. 2001. Luxembourg: European Commission. ⟨http://externe.jrc.es/infosys.html⟩.

Falk, J. H. 1980. The primary productivity of lawns in a temperate environment. *Journal of Applied Ecology* 17: 689–695.

Falkenstein, A. 1939. *Zehnter vorläufiger Bericht über die von der Notgemeinschaft der deutschen Wissenschaft in Uruk-Warka unternommen Ausgrabungen*. Berlin: Akademie der Wissenschaften.

Falkowski, P. G., R. T. Barber, and V. V. Smetacek. 1998. Biogeochemical controls and feedbacks on ocean primary production. *Science* 281: 200–205.

Falkowski, P. G., and J. A. Raven. 1997. *Aquatic Photosynthesis*. Oxford: Blackwell.

FAO (Food and Agriculture Organization). 1957. *Calorie Requirements*. Rome: FAO.

———. 1973. *Energy and Protein Requirements: Report of a Joint FAO/WHO Ad Hoc Expert Committee*. Rome: FAO.

———. 1980. *A Global Reconnaissance Survey of Fuelwood Supply/Requirement Situation*. Rome: FAO.

———. 1985. *Energy and Protein Requirements: Report of a Joint FAO/WHO/UNU Expert Consultation*. Rome: FAO.

———. 1999. *State of the World's Forests*. ⟨http://www.fao.org/docrep/W9950E/W9950E00.htm⟩.

———. 2000. *The Energy and Agriculture Nexus*. ⟨http://www.fao.org/docrep/003/X8054e/x8054e00.htm⟩.

———. 2003. *Food Energy: Methods of Analysis and Conversion Factors*. ⟨http://www.fao.org/docrep/006/Y5022e/Y5022e00.htm⟩.

———. 2004. *Human Energy Requirements: Report of a Joint FAO/WHO/UNU Consultation*. ⟨http://www.fao.org/docrep/007/y5686e/y5686e00.htm⟩.

———. 2005a. *Global Forest Resources Assessment*. ⟨http://www.fao.org/forestry/site/fra2005/en/⟩.

———. 2005b. *The State of Food Insecurity in the World 2005*. ⟨http://www.fao.org/docrep/008/a0200e/a0200e00.htm⟩.

———. 2006. FAOSTAT. ⟨http://app.sfao.org/⟩.

Farey, J. 1827. *A Treatise on the Steam Engine*. London: Longman, Rees.

Farrell, A. E., R. J. Plevin, B. T. Turner, A. D. Jones, M. O'Hare, and D. M. Kammen. 2006. Ethanol can contribute to energy and environmental goals. *Science* 311: 506–508.

Fay, J. A. 1980. Risks of LNG and LPG. *Annual Review of Energy* 5: 89–105.

Ferguson, E. F. 1971. The measurement of the "man-day". *Scientific American* 225 (4): 96–103.

Ferreira, O. C. 2000. The future of charcoal in metallurgy. *Energy and Economy* 21: 1–5.

Ferry, J. G., ed. 1993. *Methanogenesis: Ecology, Physiology, Biochemistry and Genetics*. New York: Chapman and Hall.

Fettweis, G. B. 1979. *World Coal Resources*. Amsterdam: Elsevier.

Feynman, R. 1963. *The Feynman Lectures on Physics*. 3 vols. Redwood City, Calif.: Addison-Wesley.

Field, C. B., M. J. Behrenfeld, J. T. Randerson, and P. Falkowski. 1998. Primary production of the biosphere: Integrating terrestrial and oceanic components. *Science* 281: 237–240.

Filippone, A. 2003. *Atmospheric Flight*. ⟨http://aerodyn.org/Atm-flight/sregime.html⟩.

REFERENCES

Fitchen, J. 1961. *The Construction of Gothic Cathedrals.* Chicago: University of Chicago Press.

Fletcher, G. L., C. L. Hew, and P. L. Davies. 2001. Antifreeze proteins of teleost fishes. *Annual Review of Physiology* 63: 359–390.

Fluck, R. C., ed. 1992. *Energy in Farm Production.* Amsterdam: Elsevier.

Fluck, R. C., and C. D. Baird. 1980. *Agricultural Energetics.* Westport, Conn.: Avi Publishing.

Flynn, J. 1995. *One Hundred Centuries of Solitude: Redirecting America's High-level Nuclear Waste Policy.* Boulder, Colo.: Westview Press.

FNB (Food and Nutrition Board). 1989. *Recommended Dietary Allowances.* Washington, D.C.: National Academies Press.

———. 2005. *Dietary Reference Intakes for Energy, Carbohydrate, Fiber, Fat, Fatty Acids, Cholesterol, Protein, and Amino Acids (Macronutrients).* Washington, D.C.: National Academies Press.

Foley, R. 2001. The evolutionary consequences of increased carnivory in hominids. In *Meat-Eating and Human Evolution,* ed. C. B. Stanford and H. T. Bunn, 305–331. Oxford: Oxford University Press.

Foley, R. A., and P. C. Lee. 1991. Ecology and energetics of encephalization in hominid evolution. *Philosophical Transactions of the Royal Society* B334: 223–232.

Folk, G. E. 1976. *Textbook of Environmental Physiology.* Philadelphia: Lea & Febiger.

Fontana, L., T. E. Meyer, S. Klein, and J. O. Holloszy. 2004. Long-term calorie restriction is highly effective in reducing the risk of atherosclerosis in humans. *Proceedings of the National Academy of Sciences* 101: 6659–6663.

Forbes, R. J. 1958. Power to 1850. In *A History of Technology.* Vol. 4, 148–167. Oxford: Clarendon Press.

———. 1964. Bitumen and petroleum in antiquity. In *Studies in Ancient Technology.* Vol. 1, 1–24. Leiden, Netherlands: E. J. Brill.

———. 1965. Irrigation and drainage. In *Studies in Ancient Technology.* Vol. 2, 1–79. Leiden, Netherlands: E. J. Brill.

———. 1966. Heat and heating. In *Studies in Ancient Technology.* Vol. 6, 1–103. Leiden, Netherlands: E. J. Brill.

Forbes, R. J., ed. 1964–1972. *Studies in Ancient Technology.* 9 vols. Leiden, Netherlands: E. J. Brill.

Ford, K. W., G. J. Rochlin, M. H. Ross, and R. H. Socolow, eds. 1975. *Efficient Use of Energy.* New York: American Institute of Physics.

Foukal, P. 1990. The variable sun. *Scientific American* 262 (2): 34–41.

Foukal, P., C. Frohlich, H. Spruit, and T. M. L. Wigley. 2006. Variations in solar luminosity and their effect on the Earth's climate. *Nature* 443: 161–166.

Foulger, G. R., and J. H. Natland. 2003. Is "hotspot" volcanism a consequence of plate tectonics? *Science* 300: 921–922.

Fouquet, R., and P. J. G. Pearson. 2006. Seven centuries of energy services: the price and use of light in the United Kingdom (1300–2000). *Energy Journal* 27: 139–177.

Fox, R. 1971. *The Caloric Theory of Gases: From Lavoisier to Regnault.* Oxford: Clarendon Press.

Fox, W., B. Brooks, and J. Tyrwhitt. 1976. *The Mill.* Toronto: McClelland and Stewart.

Fraenkel, P. 1986. *Water Lifting Devices.* Rome: FAO.

Francis, D. 1990. *The Great Chase: A History of World Whaling.* Toronto: Penguin.

Francis, W., and M. C. Peters, eds. 1980. *Fuels and Fuel Technology.* New York: Pergamon Press.

Frankl, P., and F. Rubik. 2000. *Life Cycle Assessment in Industry and Business: Adoption Patterns, Applications, and Implications.* Berlin: Springer.

REFERENCES

Frazier, K. 1979. *Violent Face of Nature.* New York: William Morrow.

Freedman, J. L. 1980. *Crowding Behavior.* San Francisco: W. H. Freeman.

Freedom House. 2005. Freedom in the World: The Annual Survey of Political Rights and Civil Liberties. ⟨http://www.freedomhouse.org/⟩.

Freese, S. 1957. *Windmills and Millwrighting.* Newton Abbot, UK: David and Charles.

Fremlin, J. H. 1987. *Power Production: What Are the Risks?* Oxford: Oxford University Press.

French, A. R. 1988. The patterns of mammalian hibernation. *American Scientist* 76: 569–575.

Fridley, D., J. Sinton, R. Lehman, J. Lewis, L. Jieming, Z. Fengqi, and L. Ji, eds. 2001. *China Energy Databook.* 5th rev. ed. Berkeley, Calif.: Lawrence Berkeley National Laboratory.

Friedrich, R., and A. Voss. 1993. External costs of electricity generation. *Energy Policy* 21: 14–122.

Friedrich, W. L. 2000. *Fire in the Sea: The Santorini Volcano: Natural History and the Legend of Atlantis.* Trans. A. R. McBirney. Cambridge: Cambridge University Press.

Frison, G. C. 2004. *Survival by Hunting: Prehistoric Human Predators and Animal Prey.* Berkeley: University of California Press.

Fröhlich, C. 1987. Variability of the solar "constant" on time scales of minutes to years. *Journal of Geophysical Research* 92: 796–800.

Fruehan, R., ed. 1998. *The Making, Shaping, and Treating of Steel.* 11th ed. *Steelmaking and Refining.* Pittsburgh, Pa.: Association for Iron and Steel Technology.

Fussell, G. E. 1972. *The Classical Tradition in West European Farming.* Rutherford, N.J.: Fairleigh Dickinson University Press.

GACGC (German Advisory Council on Global Change). 2004. *World in Transition Towards Sustainable Energy Systems.* London: Earthscan.

Gadgil, S. 2003. The Indian monsoon and its variability. *Annual Review of Earth and Planetary Sciences* 31: 429–467.

Gagnon, L., C. Bélanger, and Y. Uchiyama. 2002. Life-cycle assessment of electricity generation options: The status of research in year 2001. *Energy Policy* 30: 1267–1278.

Galaty, J. G., and P. C. Salzman, eds. 1981. *Change and Development in Nomadic and Pastoral Societies.* Leiden, Netherlands: E. J. Brill.

Galloway, J. A., D. Keene, and M. Murphy. 1996. Fuelling the city: Production and distribution of firewood and fuel in London's region, 1290–1400. *Economic History Review* 49: 447–472.

Galloway, J. N., and E. B. Cowling. 2002. Reactive nitrogen and the world: 200 years of change. *Ambio* 3: 64–71.

Ganachaud, A., and C. Wunsch. 2003. Large-scale ocean heat and freshwater transports during the World Ocean Circulation Experiment. *Journal of Climate* 16: 696–705.

GAO (General Accounting Office). 1991. *Persian Gulf: Allied Burden Sharing Efforts.* Washington, D.C.: Government Printing Office.

Garland, T. 1983. Scaling the ecological cost of transport to body mass in terrestrial mammals. *American Naturalist* 121: 571–587.

Garnero, E. J. 2000. Heterogeneity of the lowermost mantle. *Annual Review of Earth and Planetary Sciences* 28: 509–537.

Gartner, B. L., ed. 1995. *Plant Stems: Physiology and Functional Morphology.* San Diego, Calif.: Academic Press.

Gary, J. H., and G. E. Handwerk. 1984. *Petroleum Refining.* 2d ed. New York: Marcel Dekker.

Gauthier-Pilters, H., and A. I. Dagg. 1981. *The Camel.* Chicago: University of Chicago Press.

Gawell, K. M., M. Reed, and P. M. Wright. 1999. *Geothermal Energy: The Potential for Clean Power from the Earth.* Washington, D.C.: Geothermal Energy Association.

GE Aviation. 2006. *The GE90 Engine Family.* 〈http://www.geae.com/engines/commercial/ge90/〉.

Gehring, W. J., and R. Wehner. 1995. Heat shock protein synthesis and thermotolerance in *Cataglyphis,* an ant from the Sahara desert. *Proceedings of the National Academy of Sciences* 92: 2994–2998.

Geider, R. J., E. H. Delucia, P. G. Falkowski, A. C. Finzi, J. P. Grime, J. Grace, T. M. Kana et al. 2001. Primary productivity of planet earth: Biological determinants and physical constraints in terrestrial and aquatic habitats. *Global Change Biology* 7: 849–882.

Georgescu-Roegen, N. 1971. *The Entropy Law and the Economic Process.* Cambridge, Mass.: Harvard University Press.

———. 1975. Energy and economic myths. *Southern Economic Journal* 41: 347–381.

———. 1976. *Energy and Economic Myths: Institutional and Analytical Economic Essays.* New York: Pergamon Press.

GERB (Geostationary Earth Radiation Budget Experiment) Consortium. 2005. *The GERB Project.* 〈http://www.ssd.rl.ac.uk/gerb/〉.

German, J. M. 2004. Hybrid electric vehicles. In *Encyclopedia of Energy.* Vol. 3, 197–213.

Gerstenberger, M. C., S. Wiemer, L. M. Jones, and P. A. Reasenberg. 2005. Real-time forecasts of tomorrow's earthquakes in California. *Nature* 435: 328–331.

Ghadiri, H. 2004. Crater formation in soils by raindrop impact. *Earth Surface Processes and Landforms* 29: 77–89.

Giampietro, M. 2002. Fossil energy in world agriculture. In *Encyclopedia of Life Sciences.* London: Wiley.

Gibbons, J., and W. H. Chandler. 1981. *Energy: The Conservation Revolution.* New York: Plenum Press.

Gibbs, J. W. 1906. *The Scientific Papers of J. Willard Gibbs.* London: Longmans, Green.

Gifford, R. M. 1976. An overview of fuel used for crops and national agricultural systems. *Search* 7: 412–417.

Gifford, R. M., J. H. Thorne, W. D. Hitz, and R. T. Giaquinta. 1984. Crop productivity and photoassimilate partitioning. *Science* 225: 801–808.

Gille, B., ed. 1978. *Histoire des Techniques.* Paris: Gallimard.

Gillooly, J. F., J. H. Brown, G. B. West, V. M. Savage, and E. L. Charnov. 2001. Effects of size and temperature on metabolic rate. *Science* 293: 2248–2251.

Ginouvès, R. 1962. *Balaneutikè: Recherches sur le bain dans l'antiquité grecque.* Paris: de Boccard.

Glaser, P. E. 1968. Power from the sun. *Science* 162: 957–961.

Gleick, P. H., ed. 1993. *Water in Crisis: A Guide to the World's Fresh Water Resources.* New York: Oxford University Press.

Glover, J., D. O. White, and T. A. G. Langrish. 2002. Wood versus concrete and steel in house construction: A life cycle assessment. *Journal of Forestry* 100 (8): 34–41.

Glover, T. O., M. E. Hinkle, and H. L. Riley. 1970. *Unit Train Transportation of Coal: Technology and Description of Nine Representative Operations.* Circular 8444. U.S. Bureau of Mines.

Gluyas, J. G. 2003. *Petroleum Geoscience.* Oxford: Blackwell.

Godbold, D. L., and A. Hüttermann. 1994. *Effects of Acid Precipitation on Forest Processes.* New York: Wiley-Liss.

Goela, J. S. 1979. Wind power through kites. *Mechanical Engineering* 101 (6): 42–43.

Goeller, H. E., and A. M. Weinberg. 1976. The age of substitutability. *Science* 191: 683–689.

Goes, J. I., P. G. Thoppil, H. Gomes, and J. T. Fasullo. 2005. Warming of the Eurasian landmass is making the Arabian Sea more productive. *Science* 308: 545–547.

Gold, B., W. S. Peirce, G. Rosegger, and M. Perlman, eds. 1984. *Technological Progress and Industrial Leadership: The Growth of the U.S. Steel Industry, 1900–1970.* Lexington, Mass.: Lexington Books.

Gold, T. 1999. *The Deep Hot Biosphere.* New York: Copernicus.

REFERENCES

Goldemberg, J., ed. 2000. *World Energy Assessment: Energy and the Challenge of Sustainability.* New York: United Nations Development Programme.

Goldewijk, K. K. 2001. Estimating global land use change over the past 300 years: The HYDE database. *Global Biogeochemical Cycles* 15: 417–433.

Goldsmith, F. B., ed. 1998. *Tropical Rain Forest: A Wider Perspective.* London: Chapman and Hall.

Golley, F., and E. Medina, eds. 1975. *Tropical Ecological Systems.* New York: Springer.

Gonzalo, R. 2006. *Energy Efficient Architecture: Basics for Planning and Construction.* Boston: Birkhäuser.

Goodland, R. 1995. *Distinguishing Better Hydros from Worse: The Environmental Sustainability Challenge for the Hydro Industry.* Washington, D.C.: World Bank.

Gordon, A. L., and R. A. Fine. 1996. Pathways of water between the Pacific and Indian oceans in the Indonesian sea. *Nature* 379: 146–149.

Gordon, A. L., R. D. Susanto, and K. Vranes. 2003. Throughflow as a consequence of restricted surface layer flow. *Nature* 425: 824–828.

Gordon, R. L. 1981. *An Economic Analysis of World Energy Problems.* Cambridge, Mass.: MIT Press.

———. 1991. Depoliticizing energy: The lessons of Desert Storm. *Earth and Mineral Sciences* 60 (3): 55–58.

———. 1994. Energy, exhaustion, environmentalism, and etatism. *Energy Journal* 15: 1–16.

Goren-Inbar, N., N. Alperson, M. E. Kislev, O. Simchoni, Y. Melamed, A. Ben-Nun, and E. Werker. 2004. Evidence of hominid control of fire at Gesher Benot Ya'aqov, Israel. *Science* 304: 725–727.

Gottschall, J. S., and R. Kram. 2003. Energy cost and muscular activity required for propulsion during walking. *Journal of Applied Physiology* 94: 1766–1772.

Goudie, A. 1984. *The Nature of the Environment.* Oxford: Blackwell.

Goudsblom, J. 1992. *Fire and Civilization.* London: Allen Lane.

Gowdy, J. M., ed. 1998. *Limited Wants, Unlimited Means: A Reader on Hunter-Gatherer Economics and the Environment.* Washington, D.C.: Island Press.

Gräber, P., and G. Milazzo. 1997. *Bioenergetics.* Basel: Birkhäuser.

Grace, J. 2001. Has the primary productivity of the planet been underestimated? *Global Change Biology* 7: 869–870.

Greenwood, W. H. 1907. *Iron.* London: Cassell.

Greven, H. 1980. *Die Bärtierchen: Tardigrada.* Wittenberg: A. Ziemsen.

Griffin, T. M., and R. Kram. 2000. Penguin waddling is not wasteful. *Nature* 408: 929.

Grigg, D. B. 1974. *The Agricultural Systems of the World.* Cambridge: Cambridge University Press.

———. 1992. *The Transformation of Agriculture in the West.* Oxford: Blackwell.

Grimal, N. 1992. *A History of Ancient Egypt.* Oxford: Blackwell.

Grubb, M. J., and N. I. Meyer. 1993. Wind energy: Resources, systems, and regional strategies. In *Renewable Energy: Sources for Fuel and Electricity,* ed. T. B. Johansson, H. Kelly, A. K. N. Reddy, R. H. Williams, 157–212. Washington, D.C.: Island Press.

Grübler, A. 2004. Transitions in energy use. In *Encyclopedia of Energy.* Vol. 6, 163–177.

Gu, L., D. D. Baldocchi, S. C. Wofsy, J. W. Munger, J. J. Michalsky, S. P. Urbanski, and T. A. Boden. 2003. Response of a deciduous forest to the Mount Pinatubo eruption: Enhanced photosynthesis. *Science* 299: 2035–2038.

Gunston, B. 1986. *World Encyclopaedia of Aero Engines.* Wellingborough, UK: Patrick Stephens.

REFERENCES

Gurney, J. 1997. Migration or replenishment in the Gulf. *Petroleum Review* (May): 200–203.

Gutenberg, B., and C. F. Richter. 1942. Earthquake magnitude, intensity, energy and acceleration. *Bulletin of the Seismological Society of America* 32: 163–191.

Gutman, G., and A. Ignatov. 1995. Global land monitoring from AVHRR: Potential and limitations. *International Journal of Remote Sensing* 16: 2301–2309.

Gutman, G., D. Tarpley, A. Ignatov, and S. Olson. 1995. The enhanced NOAA global data set from Advanced Very High Resolution Radiometer. *Bulletin of the American Meteorological Society* 76: 1141–1156.

Hadfield, C. 1986. *World Canals: Inland Navigation Past and Present.* Newton Abbot, UK: David and Charles.

Haennel, R. G., and F. Lemire. 2002. Physical activity to prevent cardiovascular disease: How much is enough? *Canadian Family Physician* 48: 65–71.

Hahn, O., and F. Strassman. 1939. Über den Nachweis und das Verhalten der bei der Bestrahlung des Urans mittels Neutronen entstehenden Erdalkalimetalle. *Naturwissenschaften* 27: 11–15.

Hairston, N. G., F. E. Smith, and L. B. Slobodkin. 1960. Community structure, population control, and competition. *American Naturalist* 94: 421–425.

Haldor Topsøe. 1999. *Topsøe's Position as Process Licensor for Ammonia Plants.* Lyngby, Denmark: Haldor Topsøe.

Haley, J. E. 1959. *Erle P. Halliburton, Genius with Cement.* Duncan, Okla.: Halliburton Oil Well Cementing Co..

Hall, C. A. S., C. J. Cleveland, and R. Kaufmann. 1986. *Energy and Resource Quality: The Ecology of Economic Process.* New York: Wiley.

Hall, C. G. L. 1997. *Steel Phoenix: The Fall and Rise of the U.S. Steel Industry.* New York: St. Martin's Press.

Hall, D. O., G. W. Barnard, and P. A. Moss. 1982. *Biomass for Energy in the Developing Countries.* Oxford: Pergamon Press.

Hall, D. O., and K. K. Rao. 1999. *Photosynthesis.* New York: Cambridge University Press.

Hall, D. O., J. M. O. Scurlock, H. O. Bolhar-Nordenkampf, R. C. Leegood, and S. P. Long, eds. 1993. *Photosynthesis and Production in a Changing Environment.* Amsterdam: Elsevier.

Hall, S. J., and D. Raffaelli. 1991. Food-web patterns: Lessons from a species-rich web. *Journal of Animal Ecology* 60: 823–842.

Halweil, B. 2004. *Eat Here: Reclaiming Homegrown Pleasures in a Global Supermarket.* New York: W. W. Norton.

Handrich, Y., R. M. Bevan, J.-B. Charrassin, P. J. Butler, K. Ptz, A. J. Woakes, J. Lage, and Y. Le Maho. Hypothermia in foraging king penguins. 1997. *Nature* 388: 64–67.

Hanks, T. C., and H. Kanamori. 1979. A moment magnitude scale. *Journal of Geophysical Research* 84: 2348–2350.

Hanna, J. M., and D. E. Brown. 1983. Human heat tolerance: An anthropological perspective. *Annual Review of Anthropology,* 12: 259–284.

Hannam, J., K. Holden, and M. Auchterlonie, eds. 2000. *International Encyclopedia of Women's Suffrage.* Santa Barbara, Calif.: ABC-CLIO.

Hannon, B. 1973. Energy standard of value. *Annals of the American Academy of Political and Social Science* 410: 139–153.

———. 1981. The energy cost of energy. In *Energy, Economics, and the Environment: Conflicting Views of an Essential Interrelationship,* ed. H. E. Daly and A. F. Umaña, 81–107. Boulder, Colo.: Westview Press.

Hansen, J., L. Nazarenko, R. Ruedy, M. Sato, J. Willis, A. Del Genio, D. Koch et al. 2005. Earth's energy imbalance: Confirmation and implications. *Science* 308: 1431–1435.

Hansen, J., M. Sato, R. Ruedy, A. Lacis, and V. Oinas. 2000. Global warming in the twenty-first century: An alternative scenario. *Proceedings of the National Academy of Sciences* 97: 9875–9880.

REFERENCES

Hanson, H., N. E. Borlaug, and R. G. Anderson. 1982. *Wheat in the Third World*. Boulder, Colo.: Westview Press.

Harako, R. 1981. The cultural ecology of hunting behavior among Mbuti Pygmies of the Ituri Forest, Zaire. In *Omnivorous Primates*, ed. R. S. O. Harding and G. Teleki, 499–555. New York: Columbia University Press.

Harcourt, A. H. 1996. Is the gorilla a threatened species? How should we judge? *Biological Conservation* 75: 165–176.

Harding, R. S. O., and G. Teleki, eds. 1981. *Omnivorous Primates*. New York: Columbia University Press.

Hardy, J. D., J. A. Stolwijk, and A. P. Gagge. 1971. Man. In *Comparative Physiology of Thermoregulation*, ed. G. C. Whittow. Vol. 3, 327–380. New York: Academic Press.

Harris, J. R. 1974. The rise of coal technology. *Scientific American* 233 (2): 92–97.

———. 1988. *The British Iron Industry, 1700–1850*. London: Macmillan.

Harris, M. 1966. The cultural ecology of India's sacred cattle. *Current Anthropology* 7: 51–66.

Hart, C. 1985. *The Prehistory of Flight*. Berkeley: University of California Press.

Hartenstein, R. 1986. Earthworm biotechnology and global biogeochemistry. *Advances in Ecological Research* 15: 379–409.

Harverson, M. 1991. *Persian Windmills*. The Hague: International Molinological Society. ⟨http://www.timsmills.info/⟩.

Hassan, F. A. 1984. Environment and subsistence in predynastic Egypt. In *From Hunters to Farmers*, ed. J. D. Clark and S. A. Brandt, 57–64. Berkeley: University of California Press.

Hatch, M. D. 1992. C_4 photosynthesis: An unlikely process full of surprises. *Plant Cell Physiology* 4: 333–342.

Hatcher, H., M. Flinn, R. Church, B. Supple, and W. Ashworth, eds. *The History of the British Coal Industry*. 1984–1993. 5 vols. Oxford: Clarendon Press.

Hatzianastassiou, N., and I. Vardavas. 1999. The net radiation budget of the Northern Hemisphere. *Journal of Geophysical Research* 104: 27341–27359.

Haudricourt, A. G., and M. J. B. Delamarre. 1955. *L'Homme et la Charrue travers le Monde*. Paris: Gallimard.

Havens, J. 2003. Terrorism: Ready to blow? *Bulletin of the Atomic Scientists* 59 (4): 16–18.

Hawken, P., A. Lovins, and L. H. Lovins. 1999. *Natural Capitalism*. Boston: Little, Brown.

Hay, R. K. M. 1995. Harvest index: A review of its use in plant crop physiology. *Annals of Applied Biology* 126: 197–216.

Hay, W. W. 1998. Detrital sediment fluxes from continents to oceans. *Chemical Geology* 145: 287–323.

Hayden, B. 1981. Subsistence and ecological adaptations of modern hunter/gatherers. In *Omnivorous Primates*, ed. R. S. O. Harding and G. Teleki, 344–421. New York: Columbia University Press.

Hayes, E. T. 1976. Energy implications of materials processing. *Science* 191: 661–665.

Haynie, D. T. 2001. *Biological Thermodynamics*. Cambridge: Cambridge University Press.

Heat Island Group. 2000a. *Cool Roofs*. ⟨http://eetd.lbl.gov/HeatIsland/CoolRoofs/⟩.

———. 2000b. *Energy Use*. ⟨http://eetd.lbl.gov/HeatIsland/EnergyUse/⟩.

Hedervari, P. 1963. On the energy and magnitude of volcanic eruptions. *Bulletin Volcanologique* 25: 373–385.

Heichel, G. 1976. Agricultural production and energy resources. *American Scientist* 64: 64–72.

Heilprin, A. 1903. *Mont Pelée and the Tragedy of Martinique: A Study of the Great Catastrophes of 1902, with Observations and Experiences in the Field*. Philadelphia: J. B. Lippincott.

Hein, G. M., and B. L. Lower. 1978. Energy requirements for coated and uncoated papers. *TAPPI Journal* 61: 3536.

REFERENCES

Heinrich, B. 1993. *The Hot-Blooded Insects: Strategies and Mechanisms of Thermoregulation*. Cambridge, Mass.: Harvard University Press.

————. 2001. *Racing the Antelope: What Animals Can Teach Us About Running and Life*. New York: HarperCollins.

Heinsohn, R. J., and R. L. Kabel. 1999. *Sources and Control of Air Pollution*. Upper Saddle River, N.J.: Prentice-Hall.

Helffrich, G. R., and B. J. Wood. 2001. The Earth's mantle. *Nature* 412: 501–507.

Helland, J. 1980. *Five Essays on the Study of Pastoralists and the Development of Pastoralism*. Bergen: Universitetet i Bergen.

Helmholtz, H. 1847. *Über die Erhaltung der Kraft*. Berlin: G. Reimer.

Helsel, Z. R. 1987. Energy and alternatives for fertilizer and pesticide use. In *Energy in Plant Nutrition and Pest Control*, ed. Z. R. Helsel, 177–201. Amsterdam: Elsevier.

Hemmingsen, A. M. 1960. Energy metabolism as related to body size and respiratory surfaces, and its evolution. *Reports of Steno Memorial Hospital* 9: 1–110.

Henry, C. J. K., and D. G. Rees. 1991. New predictive equations for the estimation of basal metabolic rate in tropical peoples. *European Journal of Clinical Nutrition* 45: 177–185.

Henzel, D. S., B. A. Laseke, E. O. Smith, and D. O. Swenson. 1982. *Handbook for Flue Gas Desulfurization Scrubbing with Limestone*. Park Ridge, N.J.: Noyes Data Corp.

Herbert, R. A., and R. J. Sharp, eds. 1992. *Molecular Biology and Biotechnology of Extremophiles*. Glasgow: Blackie.

Herendeen, R. 1998. Embodied energy, embodied everything... now what? In *Advances in EnergyStudies: Energy Flows in Ecology and Economy (Proceedings of the International Workshop, Porto Venere, Italy)*, ed. S. Ulgiati, 13–48. Rome: MUSIS (Museo della Scienza e dell'informazione Scientifica).

Hermann, W. A. 2006. Quantifying global exergy resources. *Energy* 31: 1349–1366.

Herring, H. 2001. Why energy efficiency is not enough. In *Advances in Energy Studies*, ed. S. Ulgiati, 349–359. Padua, Italy: SGE.

————. 2004. Rebound effect in energy conservation. In *Encyclopedia of Energy*. Vol. 5, 411–423.

————. 2006. Energy efficiency: A critical view. *Energy* 31: 10–20.

Hess, H. H. 1962. History of the ocean basins. In *Petrological Studies: Buddington Memorial Volume*, 599–620. New York: Geological Society of America.

Heston, A. 1971. An approach to the sacred cow of India. *Current Anthropology* 12: 191–209.

Heun, M., R. Schafer-Pregl, D. Klawan, R. Castagna, M. Accerbi, B. Borghi, and F. Salamini. 1997. Site of einkorn wheat domestication identified by DNA fingerprinting. *Science* 278: 1312–1314.

Heusner, A. A. 1982. Energy metabolism and body size. *Respiratory Physiology* 48: 1–12.

Hibberd, J. M., and W. P. Quick. 2002. Characteristics of C_4 photosynthesis in stems and petioles of C_3 flowering plants. *Nature* 415: 451–454.

Hicks, J., and G. Allen. 1999. A Century of Change: Trends in UK Statistics since 1900. Research paper 99/111. House of Commons Library. ⟨http://www.parliament.uk/commons/lib/research/rp99/rp99-111.pdf⟩.

Hill, D. 1984. *A History of Engineering in Classical and Medieval Times*. La Salle, Ill.: Open Court Publishing.

Hindle, B., ed. 1975. *America's Wooden Age: Aspects of Its Early Technology*. Tarrytown, N.Y.: Sleepy Hollow Restorations.

Hippel, F. von, B. G. Levi, T. Postol, and W. H. Daugherty. 1988. Civilian casualties from counterforce attacks. *Scientific American* 259 (3): 36–42.

Hirano, N., E. Takahashi, J. Yamamoto, N. Abe, S. P. Ingle, I. Kaneoka, T. Hirata et al. 2006. Volcanism in response to plate flexure. *Science* 313: doi:10.1126/science:1128235.

REFERENCES

Hirn, A., and M. Laigle. 2004. Silent heralds of megathrust earthquake? *Science* 305: 1917–1918.

Hitchcock, R. K., and J. I. Ebert. 1984. Foraging and food production among Kalahari hunter/gatherers. In *From Hunters to Farmers*, ed. J. D. Clark and S. A. Brandt, 328–348. Berkeley: University of California Press.

Hively, W. 1988. Nuclear power at risk. *American Scientist* 76: 341–343.

Ho, P. 1975. *The Cradle of the East*. Hong Kong: Chinese University of Hong Kong Press.

Hochachka, P. W., and G. N. Somero. 2002. *Biochemical Adaptation: Mechanism and Process in Physiological Evolution*. Oxford: Oxford University Press.

Hocker, F. M., and C. A. Ward, eds. 2004. *The Philosophy of Shipbuilding: Conceptual Approaches to the Study of Wooden Ships*. College Station: Texas A and M University Press.

Hodge, A. T. 1985. Siphons in Roman aqueducts. *Scientific American* 252 (6): 114–119.

———. 2002. *Roman Aqueducts and Water Supply*. London: Duckworth.

Hodges, P. 1989. *How the Pyramids Were Built*. Dorset: Element Books.

Hoffmann, P. 1981. *The Forever Fuel*. Boulder, Colo.: Westview Press.

Hofmann, A. W. 2003. Just add water. *Nature* 425: 24–25.

Hohenemser, C. 1988. The accident at Chernobyl: Health and environmental consequences and the implications from risk management. *Annual Review of Energy* 13: 383–428.

Hohmeyer, O. 1989. *Social Costs of Energy Consumption*. Berlin: Springer.

Hohmeyer, O., and R. L. Ottinger, eds. 1991. *External Environmental Costs of Electric Power*. Berlin: Springer.

Hollingsworth, J. A. 1966. *History of Development of Strip Mining Machines*. Milwaukee, Wisc.: Bucyrus-Erie Company.

Holt, R. A. 1988. *The Mills of Medieval England*. Oxford: Oxford University Press.

Homen, T. 1897. Der tägliche Wärmehaushalt im Boden und die Wärmestrahlung zwischen Himmel und Erde. *Acta Societatis Scientiarum Fennicae* 23: 1–147.

Honig, J. M. 1999. *Thermodynamics*. San Diego, Calif.: Academic Press.

Hopfen, H. J. 1969. Farm Implements for Arid and Tropical Regions. Rome: FAO.

Hoppeler, H., and E. R. Weibel. 2005. Scaling functions to body size: Theories and facts. *Journal of Experimental Biology* 208: 1573–1574.

Hossli, W. 1969. Steam turbines. *Scientific American* 220 (4). Also in *Scientific Technology and Social Change*, ed. G. Rochlin. San Francisco: Freeman, 1974.

Hotelling, H. 1931. The economics of exhaustible resources. *Journal of Political Economy* 39: 137–175.

Houdry, E. 1931. U.S. patent 1,837,963: A process for the manufacture of liquid fuels. December 22. ⟨http://www.uspto.gov⟩.

Houghton, J. T. 1984. *The Global Climate*. Cambridge: Cambridge University Press.

Houghton, J. T., Y. Ding, D. J. Griggs, M. Noguer, P. S. van der Linden, X. Dai, K. Maskell, and C. A. Johnson, eds. 2001. *Climate Change 2001: The Scientific Basis*. New York: Cambridge University Press.

Houghton, R. A. 2005. Above ground forest biomass and the global carbon balance. *Global Change Biology* 11: 945–958.

Houghton, R. A., and J. L. Hackler. 2002. *Carbon Flux to the Atmosphere from Land-Use Changes*. ⟨http://cdiac.esd.ornl.gov/trends/landuse/houghton/houghton.html⟩.

Howell, B. F. 1990. *An Introduction to Seismological Research: History and Development*. Cambridge: Cambridge University Press.

REFERENCES

Hu, S. D. 1983. *Handbook of Industrial Energy Conservation.* New York: Van Nostrand Reinhold.

Hubbard, H. M. 1991. The real cost of energy. *Scientific American* 264 (4): 36–42.

Hubbert, M. K. 1962. *Energy Resources. A Report to the Committee on Natural Resources of the National Academy of Sciences-National Research Council.* Publ. 1000-D. Washington: National Academies Press.

Huber, P. W., and M. P. Mills. 2005. *The Bottomless Well: The Twilight of Fuel, the Virtue of Waste, and Why We Will Never Run Out of Energy.* New York: Basic Books.

Huey, R. B., E. R. Pianka, and T. W. Schoener, eds. 1983. *Lizard Ecology.* Cambridge, Mass.: Harvard University Press.

Hughes, T. P. 1983. *Networks of Power: Electrification in Western Society, 1880–1930.* Baltimore: Johns Hopkins University Press.

Huixian, L., G. W. Housner, X. Lili, and H. Duxin. 2002. *The Great Tangshan Earthquake of 1976.* Pasadena: California Institute of Technology.

Humayun, M., L. Qin, and M. D. Norman. 2004. Geochemical evidence for excess iron in the mantle beneath Hawaii. *Science* 306: 91–94.

Humphrey, J. W., J. P. Oleson, and A. N. Sherwood. 1998. *Greek and Roman Technology: A Sourcebook.* London: Routledge.

Humphrey, W. S., and J. Stanislaw. 1979. Economic growth and energy consumption in the UK, 1700–1975. *Energy Policy* 7 (1): 29–43.

Humphreys, W. F. 1979. Production and respiration in animal communities. *Journal of Animal Ecology* 48: 427–453.

Hunt, J. M. 1996. *Petroleum Geochemistry and Geology.* San Francisco: W. H. Freeman.

Hunter, L. C. 1975. Water power in the century of steam. In *America's Wooden Age: Aspects of Its Early Technology,* ed. B. Hindle, 160–192. Tarrytown, N.Y.: Sleepy Hollow Restorations.

Husslage, G. 1965. *Windmolens: Een overzicht van de verschillende molensoorten en hun werkwijze.* Amsterdam: Heijnis.

Hutchinson, J. R., D. Famini, R. Lair, and R. Kram. 2003. Are fast-moving elephants really running? *Nature* 422: 493–494.

Hyde, C. K. 1977. *Technological Change and the British Iron Industry, 1700–1870.* Princeton, N.J.: Princeton University Press.

Hyland, A. 1990. *Equus: The Horse in the Roman World.* New Haven, Conn.: Yale University Press.

———. 2003. *The Horse in the Ancient World.* New York: Praeger.

Hytten, F. E. 1980. Nutrition. In *Clinical Physiology in Obstetrics,* ed. F. E. Hytten and G. Chamberlain, 163–192. Oxford: Blackwell.

IAAF (International Association of Athletics Federations). 2006. *World Records and Top Lists.* ⟨http://www.iaaf.org/statistics/index.html⟩.

IAEA (International Atomic Energy Agency). 1984. *Proceedings of an International Symposium on the Risks and Benefits of Energy Systems.* ⟨http://www.iaea.org/⟩.

———. 2006. *Power Reactor Information System.* ⟨http://www.iaea.org/DataCenter/datasystems.html⟩.

IAI (International Aluminium Institute). 2003. *History of Aluminium.* London: IAI. ⟨http://www.world-aluminum.org/history/⟩.

ICAO (International Civil Aviation Organization). 2001. *Annual Report.* ⟨http://www.icao.int/icao/en/pub/rp.htm⟩.

———. 2006. Special report: Annual review of civil aviation. *ICAO Journal* 61 (5): 6–16.

ICOLD (International Commission on Large Dams). 1998. *World Register of Dams.* ⟨http://www.icold-cigb.net/⟩.

REFERENCES

IEA (International Energy Agency). 1994. *Natural Gas Technologies, Energy Security, Environment and Economic Development.* ⟨http://www.iea.org/Textbase/publications/free_new_Desc.asp?PUBS_ID=1515⟩.

———. 2001. *Key World Energy Statistics.* Paris: IEA.

———. 2005. *LNG: Making Gas Markets Global.* ⟨http://www.iea.org/textbase/work/workshopdetail.asp?WS_ID=221⟩.

———. 2006. *RDandD Budgets.* ⟨http://www.iea.org/⟩.

IFIAS (International Federation of Institutes for Advanced Study). 1974. *Proceedings of the Energy Analysis Workshop on Methodology and Conventions.* Stockholm: IFIAS.

IGA (International Geothermal Association). 2005. *Installed Generating Capacity.* ⟨http://iga.igg.cnr.it/⟩.

IHA (International Hydropower Association). 2000. *Hydropower and the World's Energy Future.* ⟨http://www.ieahydro.org/reports/Hydrofut.pdf⟩.

———. 2003. *The Role of Hydropower in Sustainable Development.* ⟨http://www.hydropower.org/⟩.

IHFL (International Heat Flow Commission). 2005. *The Global Heat Flow Database.* Fargo: University of North Dakota. ⟨http://www.heatflow.und.edu/index2.html⟩.

IISI (International Iron and Steel Institute). 2005. *World Steel in Figures.* ⟨http://www.worldsteel.org⟩.

Ikoku, C. U. 1984. *Natural Gas Production Engineering.* New York: Wiley.

Illich, I. 1974. *Energy and Equity.* London: Calder and Boyars.

IMCS (Institute of Marine and Coastal Science, Rutgers University). 2000. *Ocean Primary Productivity Study.* ⟨http://marine.rutgers.edu/opp/⟩.

Imhoff, M. L., L. Bounoua, T. Ricketts, C. Loucks, R. Harriss, and W. T. Lawrence. 2004. Global patterns in human consumption of net primary production. *Nature* 429: 870–873.

IMMDA (International Marathon Medical Directors Association). 2001. *IMMDA Advisory Statement on Guidelines for Fluid Replacement During Marathon Running.* ⟨http://www.usatf.org/groups/Coaches/library/hydration/IMMDAAdvisoryStatement.pdf⟩.

Intel. 2003. Moore's law. ⟨http://www.intel.com/technology/mooreslaw/index.htm⟩.

IPCC (Intergovernmental Panel on Climate Change). 2006. *Carbon Dioxide Capture and Storage.* ⟨http://arch.rivm.nl/env/int/ipcc/pages_media/SRCCS-final/IPCCSpecialReportonCarbondioxideCaptureandStorage.htm⟩.

Irons, W., and N. Dyson-Hudson, eds. 1972. *Perspective on Nomadism.* Leiden, Netherlands: E. J. Brill.

Irving, P. M., ed. 1991. *Acidic Deposition: State of Science and Technology.* Washington, D.C.: National Acid Precipitation Assessment Program.

Isaacs, J. D., and W. R. Schmitt. 1980. Ocean energy: Forms and prospects. *Science* 207: 265–273.

Ishii, M., P. M. Shearer, H. Houston, and J. E. Vidale. 2005. Extent, duration and speed of the 2004 Sumatra-Andaman earthquake imaged by the Hi-Net Array. *Nature* 435: 933–936.

Islas, J. 1999. The gas turbine: A new technological paradigm in electricity generation. *Technological Forecasting and Social Change* 60: 129–148.

Israel, Paul. 1998. *Edison: A Life of Invention.* New York: Wiley.

Israelsen, O. W., V. E. Hansen, and G. E. Stringham. 1980. *Irrigation Principles and Practices.* 4th rev. ed. New York: Wiley.

ITOPF (International Tanker Owners Pollution Federation). 2005. *Historical Data.* ⟨http://www.itopf.com/⟩.

Jackson, I., ed. 1998. *The Earth's Mantle: Composition, Structure, and Evolution.* Cambridge: Cambridge University Press.

James, T. G. H. 1984. *Pharaoh's People.* Chicago: University of Chicago Press.

REFERENCES

James, W. P. T., and E. C. Schofield. 1990. *Human Energy Requirements. A Manual for Planners and Nutritionists.* Oxford: Oxford Medical Publications.

Jansen, M. B. 2000. *The Making of Modern Japan.* Cambridge, Mass.: Belknap Press.

Janvier, P. 1996. *Early Vertebrates.* New York: Oxford University Press.

Jenike, M. R. 2001. Nutritional ecology: Diet, physical activity and body size. In *Hunter-Gatherers: An Interdisciplinary Perspective*, ed. C. Panter-Brick, R. H. Layton, and P. Rowley-Conwy, 205–238. Cambridge: Cambridge University Press.

Jensen, H. 1969. *Sign, Symbol and Script.* New York: G. P. Putnam's Sons.

Jetz, W., C. Carbone, J. Fulford, and J. H. Brown. 2004. The scaling of animal space use. *Science* 306: 266–268.

Jevons, W. S. 1865. *The Coal Question: An Inquiry Concerning the Progress of the Nation, and the Probable Exhaustion of Our Coal Mines.* London: Macmillan.

JISF (Japan Iron and Steel Federation). 2003. *Energy Consumption by the Steel Industry at 2,013 PJ.* ⟨http://www.jisf.or.jp/en/index.html⟩.

Johannsen, O. 1953. *Geschichte des Eisens.* Düsseldorf: Stahleisen.

Johnson, L. A., and W. J. Hoover. 1977. Energy use in baking bread. *Bakers Digest* 51: 58–65.

Jones, H. 1973. *Steam Engines.* London: Ernest Benn.

Jordan, C. F., and R. Herrera. 1981. Tropical rain forests: Are nutrients really critical? *The American Naturalist* 117: 167–180.

Joshi, Y. 2001. Heat out of small packages. *Mechanical Engineering* 123 (12): 56–58.

Joule, J. P. 1843. *The Mechanical Equivalent of Heat.* London: Taylor, 1850.

Kaiser, J. 2005. Mounting evidence indicts fine-particle pollution. *Science* 307: 1858–1861.

Kaler, J. B. 1992. *Stars.* New York: Scientific American Library.

Kalyoncu, R. S. 2000. Coal combustion products. In *Minerals Yearbook*, 20.1–20.5. U.S. Geological Survey. Washington, D.C.: U.S. Department of the Interior.

Kammen, D. M. 1995. Cookstoves for the developing world. *Scientific American* 273 (1): 72–75.

Kanamori, H., and E. Boschi, eds. 1983. *Earthquakes: Observation, Theory and Interpretation.* Amsterdam: North-Holland.

Kanamori, H., E. Hauksson, and T. Heaton. 1997. Real-time seismology and earthquake hazard mitigation. *Natrure* 390: 461–464.

Kanwisher, J. W., and S. H. Ridgway. 1983. The physiological ecology of whales and porpoises. *Scientific American* 246 (6): 110–120.

Kapitsa, P. 1976. Physics and the energy problem. *New Scientist* 72 (1021): 10–12.

Kasting, J. F., and D. H. Grinspoon. 1991. The faint young sun problem. In *The Sun in Time*, ed. C. P. Sonett, M. S. Giampapa, and M. S. Matthews, 447–462. Tucson: University of Arizona Press.

Kauffman, K. D. 1993. Why was the mule used in Southern agriculture? Empirical evidence of principal-agent solutions. *Explorations in Economic History* 30: 336–351.

Kaufmann, R. K. 1992. A biophysical analysis of the energy/real GDP ratio: Implications for substitution and technical change. *Ecological Economics* 6: 35–56.

Kê, B. 2001. *Photosynthesis: Photobiochemistry and Photobiophysics.* Boston: Kluwer.

Keeling, C. D. 1998. Reward and penalties of monitoring the Earth. *Annual Review of Energy and the Environment* 23: 25–82.

Keeney, D. R., and T. H. DeLuca. 1992. Biomass as an energy source for the Midwestern U.S. *American Journal of Alternative Agriculture* 7: 137–143.

REFERENCES

Keesing, F. 2000. Cryptic consumers and the ecology of an African savanna. *BioScience* 50: 205–215.

Keith, D. W., J. F. DeCarolis, D. C. Denkenberger, D. H. Lenschow, S. L. Malyshev, S. Pacala, and P. J. Rasch. 2004. The influence of large-scale wind power on global climate. *Proceedings of the National Academy of Sciences* 101: 16115–16120.

Keller, J. 2000. *Sprinkle and Trickle Irrigation*. Caldwell, N.J.: Blackburn Press.

Kelly, R. L. 1983. Hunter-gatherer mobility strategies. *Journal of Anthropological Research* 39: 277–306.

———. 1995. *The Foraging Spectrum: Diversity in Hunter-Gatherer Lifeways*. Washington, D.C.: Smithsonian Institution Press.

Kerr, R. A. 2006. Rising plumes in Earth's mantle: Phantom or real? *Science* 313: 1726.

Kessler, A. 1985. *Heat Balance Climatology*. Amsterdam: Elsevier.

Khazanov, A. M. 1984. *Nomads and the Outside World*. Cambridge: Cambridge University Press.

———. 2001. *Nomads in the Sedentary World*. London: Curzon.

Khazzoom, J. D. 1989. Energy savings from more efficient appliances: A rejoinder. *Energy Journal* 10: 157–166.

Kiehl, J. T., and K. E. Trenberth. 1997. Earth's annual global mean energy budget. *Bulletin of the American Meteorological Society* 78: 197–208.

Kim, S., and B. E. Dale. 2002. Allocation procedure in ethanol production system from corn grain. 1. System expansion. *International Journal of Life Cycle Assessment* 7 (4): 237–243.

King, Clarence D. 1948. *Seventy-five Years of Progress in Iron and Steel*. New York: American Institute of Mining and Metallurgical Engineers.

Kingsolver, J. G. 1985. Butterfly engineering. *Scientific American* 253 (2): 106–113.

Kirsch, S. 2005. *Proving Grounds: Project Plowshare and the Unrealized Dream of Nuclear Engineering*. New Brunswick, N.J.: Rutgers University Press.

Kittleman, L. R. 1979. Tephra. *Scientific American* 241 (6): 160–177.

Kleiber, M. 1932. Body size and metabolism. *Hilgardia* 6: 315–353.

———. 1961. *The Fire of Life*. New York: Wiley.

Klein, H. A. 1978. Pieter Bruegel the Elder as a guide to sixteenth-century technology. *Scientific American* 238 (3): 134–140.

Klemm, F. 1959. *A History of Western Technology*. Trans. D. W. Singer. London: Allen and Unwin. Reissued Cambridge, Mass.: MIT Press, 1964.

Klinge, H., W. A. Rodrigues, E. Brunig, and E. J. Fittkau. 1975. Biomass and structure in a Central Amazonian rain forest. In *Tropical Ecological Systems*, ed. F. Golley and E. Medina, 115–122. New York: Springer.

Koblents-Mishke, O. I., Y. G. Kabanova, and V. V. Volkovinskii. 1968. New data on the magnitude of primary production in the oceans. *Doklady Akademii Nauk SSSR Seria Biologicheskaya* 183: 1189–1192.

Kolb, A. 2001. Transport and communication in the Roman state. In *Travel and Geography in the Roman Empire*, ed. C. Adams and R. Laurence, 95–105. London: Routledge.

Kondratyev, K. I. 1988. *Climate Shocks: Natural and Anthropogenic*. New York: Wiley.

Kongshaug, G. 1998. *Energy Consumption and Greenhouse Gas Emissions in Fetilizer Production*. Paris: IFA.

Konopacki, S., H. Akbari, M. Pomerantz, S. Gabersek, and L. Gartland. 1997. *Cooling Energy Savings Potential of Light-Colored Roofs*. Report LNBL-39433. Berkeley, Calif.: Lawrence Berkeley National Laboratory.

Korgen, B. J. 1995. Seiches. *American Scientist* 83: 330–341.

REFERENCES

Kozłowski, J., and M. Konarzewski. 2005. West, Brown and Enquist's model of allometric scaling again: The same questions remain. *Functional Ecology* 19: 739–743.

Kreemer, C., W. E. Holt, and A. J. Haines. 2002. The global moment rate distribution within planetary boundary zones. In *Plate Boundary Zones*, ed. S. Stein and J. T. Fremueller, 173–190. Washington, D.C.: American Geophysical Union.

Krenz, J. H. 1976. *Energy Conversion and Utilization.* Boston: Allyn and Bacon.

Kudryavtsev, N. A. 1959. *Oil, Gas, and Solid Bitumens in Igneous and Metamorphic Rocks.* Leningrad: State Fuel Technical Press.

Kumar, S., and S. B. Hedges. 1998. A molecular timescale for vertebrate evolution. *Nature* 392: 917–920.

Kumar, S. N. 2004. Tanker transportation. In *Encyclopedia of Energy.* Vol. 6, 1–12.

Kunstler, J. H. 2005. *The Long Emergency: Surviving the End of the Oil Age, Climate Change, and Other Converging Catastrophes of the Twenty-first Century.* Boston: Atlantic Monthly Press.

Kurzweil, R. 1990. *The Age of Intelligent Machines.* Cambridge, Mass.: MIT Press.

Kvenvolden, K. A. 1993. Gas hydrates: Geological perspective and global change. *Reviews of Geophysics* 31: 173–187.

Kvist, A., A. Lindstrom, M. Green, T. Piersma, and G. H. Visser. 2001. Carrying large fuel loads during sustained bird flight is cheaper than expected. *Nature* 413: 730–732.

Lacey, J. M. 1935. *A Comprehensive Treatise on Practical Mechanics.* London: Technical Press.

Laherrère, J. H. 1996. Discovery and production trends. *OPEC Bulletin* 27 (2): 7–11.

———. 2000. Comments on "Global Natural Gas Perspectives." Review. ⟨http://www.hubbertpeak.com/laherrere/ngperspective/⟩.

———. 2001. *Estimates of Oil Reserves.* Laxenburg, Austria: International Institute of Applied Systems Analysis. ⟨http://www.oilcrisis.com/laherrere/⟩.

Lakatos, I., ed. 2001. *Recent Advances in Enhanced Oil and Gas Recovery.* Budapest: Akadémiai Kiadó.

Laloux, R., A. Falisse, and J. Poelaert. 1980. Nutrition and fertilization of wheat. In *Wheat*, 19–24. Basel: CIBA-Geigy.

Lamb, H. H. 1970. Volcanic dust in the atmosphere, with a chronology and assessment of its meteorological significance. *Philosophical Transactions of the Royal Society* A266: 425–533.

Landels, J. G. 1980. *Engineering in the Ancient World.* London: Chatto and Windus.

Landsberg, J. J. 1986. *Physiological Ecology of Forest Production.* New York: Academic Press.

Lane, F. C. 1934. *Venetian Ships and Shipbuilders of the Renaissance.* Baltimore, Md.: Johns Hopkins University Press.

Larson, W. E., F. J. Pierce, and R. H. Dowdy. 1983. The threat of soil erosion to long-term crop production. *Science* 219: 458–465.

Laub, B., and E. Venkatapathy. 2003. Thermal Protection System Technology and Facility Needs for Demanding Future Planetary Missions. Paper presented at the International Workshop on Planetary Probe Atmospheric Entry and Descent Trajectory Analysis and Science, Lisbon, October 6–9.

Lawlor, D. W. 2001. *Photosynthesis.* New York: Springer.

Lawrence, D. M. 2002. *Upheaval from the Abyss: Ocean Floor Mapping and the Earth Science Revolution.* New Brunswick, N.J.: Rutgers University Press.

Lawrence, W. 1964. *Home Fires Burning: The History of Domestic Heating and Cooking.* London: Routledge and Kegan Paul.

Lay, T., H. Kanamori, C. J. Ammon, M. Nettles, S. N. Ward, R. C. Aster, S. L. Beck et al. 2005. The Great Sumatra-Andaman earthquake of 26 December 2004. *Science* 308: 1127–1133.

REFERENCES

Layard, R. 2005. *Happiness: Lessons from a New Science.* New York: Penguin.

Layton, E. T. 1979. Scientific technology, 1845–1900: The hydraulic turbine and the origins of American industrial research. *Technology and Culture* 20: 64–89.

Layzer, D. 1988. Growth of order in the universe. In *Entropy, Information, and Evolution,* ed. B. H. Weber, D. J. Depew, and J. D. Smith, 23–39. Cambridge, Mass.: MIT Press.

Leach, E. R. 1959. Hydraulic society in Ceylon. *Past and Present* 15: 2–26.

Leach, G. 1975. *Energy and Food Production.* London: International Institute for Environment and Development.

Leckie, A. H., A. Millar, and J. E. Medley. 1982. Short- and long-term prospects for energy economy in steelmaking. *Ironmaking and Steelmaking* 9: 222–235.

Lee, C. A. 2004. Are Earth's core and mantle on speaking terms? *Science* 306: 64–65.

Lee, R., A. Krupnik, and D. Burtraw. 1995. *Estimating Externalities of Electric Fuel Cycles.* Washington, D.C.: McGraw-Hill/Utility Data Institute.

Lee, R. B. 1979. *The !Kung San: Men, Women and Work in a Foraging Society.* Cambridge: Cambridge University Press.

Lee, R. B., and R. Daly, eds. 1999. *Cambridge Encyclopaedia of Hunters and Gatherers.* Cambridge: Cambridge University Press.

Lee, R. B., and I. DeVore, eds. 1968. *Man the Hunter.* Chicago: Aldine.

Lefohn, A. S., J. D. Husar, and R. B. Husar. 1999. Estimating historical anthropogenic global sulfur emission patterns for the period 1850–1990. *Atmospheric Environment* 33: 3425–3444.

Legge, A. J., and P. A. Rowley-Conwy. 1987. Gazelle killing in Stone Age Syria. *Scientific American* 257 (2): 88–95.

Le Grand, H. E. 1988. *Drifting Continents and Shifting Theories.* Cambridge: Cambridge University Press.

Lehner, M. 1997. *The Complete Pyramids.* London: Thames and Hudson.

Lehtonen, A. et al. 2004. Biomass expansion factors (BEFs) for Scots pine, Norway spruce, and birch according to stand age for boreal forests. *Forest Ecology and Management* 188: 211–224.

Lelieveld, J. S., S. Lechtenböhmer, S. S. Assonov, C. A. M. Brenninkmeijer, C. Dienst, M. Fischedick, and T. Hanke. 2005. Low methane leakage from gas pipelines. *Nature* 434: 841–842.

Lenzen, M. 1999. Greenhouse gas analysis of solar-thermal electricity generation. *Solar Energy* 65: 353–368.

Lenzen, M., and C. Dey. 2000. Truncation error in embodied energy analyses of basic iron and steel products. *Energy* 25: 577–585.

Lenzen, M., and G. Treloar. 2002. Embodied energy in buildings: Wood versus concrete. Reply to Börjesson and Gustavsson. *Energy Policy* 30: 249–255.

Lepore, S. J. 1994. Crowding: Effects on human health and behavior. In *Encyclopedia of Human Behavior,* ed. V. S. Ramachandran. Vol. 2, 43–51. San Diego, Calif.: Academic Press.

Lepre, J. P. 1990. *The Egyptian Pyramids.* Jefferson, N.C.: McFarland.

Leser, P. 1931. *Entstehung und Verbreitung des Pfluges.* Münster: Aschendorff.

Lesser, I. O. 1991. *Oil, the Persian Gulf, and Grand Strategy.* Santa Monica, Calif.: Rand Corporation.

Lett, R. G., and T. C. Ruppel. 2004. Coal, chemical and physical properties. In *Encyclopedia of Energy.* Vol. 1, 411–423.

Lewis, G. N., and M. Randall. 1961. *Thermodynamics.* New York: McGraw-Hill.

Lewis, M. J. T. 1993. The Greeks and the early windmill. *History and Technology* 15: 141–189.

Leydon, K. 1987. So forecasting is easy! *Energy in Europe* 9: 17–25.

REFERENCES

Li, W. K. 2002. Macroecological patterns of phytoplankton in the northwestern North Atlantic Ocean. *Nature* 419: 154–157.

Liebig, J. von. 1843. *Die chemie in ihrer Anwendung auf Agricultur und Physiologie*. Braunschweig: F. Vieweg.

———. 1862. *Die Naturgesetze des Feldbaues*. Braunschweig: F. Vieweg.

Lieth, H., and R. H. Whittaker, eds. 1975. *Primary Productivity of the Biosphere*. New York: Springer.

Lifson, N., G. B. Gordon, and R. McClintock. 1955. Measurement of total carbon dioxide production by means of D_2O^{18}. *Journal of Applied Physiology* 7: 705–710.

Lindeman, R. 1942. The trophic-dynamic aspect of ecology. *Ecology* 23: 399–418.

Lindsay, J. 1974. *Blast-Power and Ballistics*. New York: Harper and Row.

Lindsay, R. B. 1975. *Energy: Historical Development of the Concept*. Stroudsburg, Pa.: Dowden, Hutchinson and Ross.

———, ed. 1976. *Applications of Energy: Nineteenth Century*. Stroudsburg, Pa.: Dowden, Hutchinson and Ross.

Lipfert, F. W., and S. C. Morris. 1991. Air pollution benefit-cost assessment. *Science* 253: 606.

Lipman, P. W., and D. R. Mullineaux, eds. 1981. *The 1980 Eruptions of Mount St. Helens*. Washington, D.C.: U.S. Geological Survey.

Liu, P. L., P. Lynett, H. Fernando, B. E. Jaffe, H. Fritz, B. Higman, R. Morton, J. Goff, and C. Synolakis. 2005. Observations by the international tsunami survey team in Sri Lanka. *Science* 308: 1595.

Lizot, J. 1977. Population, resources and warfare among the Yanomami. *Man* 12: 497–517.

Lobell, D. B., J. A. Hicke, G. P. Asner, C. B. Field, C. J. Tucker, and S. O. Los. 2002. Satellite estimates of productivity and light use efficiency in United States agriculture, 1982–1998. *Global Change Biology* 8: 722–735.

Lonsdale, W. M. 1990. The self-thinning rule: Dead or alive? *Ecology* 71: 1373–1388.

Loomis, S. H. 1995. Freezing tolerance of marine invertebrates. *Oceanography and Marine Biology: An Annual Review* 33: 373–350.

Lopreato, J. 1984. *Human Culture and Biocultural Evolution*. London: Unwin Hyman.

Lorenz, E. N. 1976. *The Nature and Theory of the General Circulation of the Atmosphere*. Geneva: World Meteorological Organization.

Lotka, A. J. 1922. Contribution to the energetics of evolution. *Proceedings of the National Academy of Sciences* 8: 147–151.

———. 1925. *Elements of Physical Biology*. Baltimore, Md.: Williams and Wilkins.

Lovelock, J. E. 1979. *Gaia*. Oxford: Oxford University Press.

Lovins, A. B. 1976. Energy strategy: The road not taken. *Foreign Affairs* 55 (1): 65–96.

———. 1988. Energy savings resulting from the adoption of more efficient appliances: Another view. *Energy Journal* 9: 155–162.

———. 2004. Energy efficiency: Taxonomic overview. In *Encyclopedia of Energy*. Vol. 2, 383–401.

Lowrance, R. R., B. R. Stinner, and G. J. House, eds. 1984. *Agricultural Ecosystems*. New York: Wiley.

Lowrie, A., and M. D. Max. 1999. The extraordinary promise and challenge of gas hydrates. *World Oil* 220 (9): 49–55.

Lui, L. 2004. Oil Sands Development and Research. Speech at TD Newcrest Oil Sands Forum, July 7. Calgary: Imperial Oil.

Luiten, E. E. M. 2001. *Beyond Energy Efficiency: Actors, Networks and Government Intervention in the Development of Industrial Process Technologies*. Utrecht, Netherlands: Universiteit Utrecht.

Lunt, R. R., and J. D. Cunic. 2000. *Profiles in Flue Gas Desulfurization*. New York: American Institute of Chemical Engineers.

Luo, Z. 1998. Biomass energy consumption in China. *Wood Energy News* 13 (3): 3–4.

Lyman, C. P. 1982. *Hibernation and Torpor in Mammals and Birds*. New York: Academic Press.

Mabro, R. 1992. OPEC and the price of oil. *Energy Journal* 13: 1–17.

MacArthur, R., and E. O. Wilson. 1967. *The Theory of Island Biogeography*. Princeton, N.J.: Princeton University Press.

Macdonald, A. M., and C. Wunsch. 1996. An estimate of global ocean circulation and heat fluxes. *Nature* 382: 436–439.

MacDonald, W. L. 1976. *The Pantheon: Design, Meaning, and Progeny*. Cambridge, Mass.: Harvard University Press.

Mace, G. M., P. H. Harvey, and T. H. Clutton-Brock. 1983. Vertebrate home-range size and energetic requirements. In *The Ecology of Animal Movement*, ed. I. R. Swingland and D. J. Greenwood, 33–53. Oxford: Clarendon Press.

Macedo, I. C., M. R. Leal, and E. A. Ramos da Silva. 2004. *Assessment of Greenhouse Gas Emissions in the Production and Use of Fuel Ethanol in Brazil*. 〈http://www.unica.com.br/i_pages/files/gee3.pdf〉.

Mach, E. 1896. *Die Prinzipien der Wärmelehre historisch-kritisch entwickelt*. Leipzig: J. A. Barth.

Machado-Moreira, C. A., F. Magalhães, A. C. Vimieiro-Gomes, N. R. Viana Lima, and L. O. Carneiro Rodrigues. 2005. Effects of heat acclimation on sweating during graded exercise until exhaustion. *Journal of Thermal Biology* 30: 437–442.

MacLean, H. L., and L. B. Lave. 1998. A life-cycle model of an automobile. *Environmental Science and Technology* 32: 322A–330A.

Maddison, A. 1995. *Monitoring World Economy 1820–1992*. Paris: OECD.

Mage, D., G. Ozolins, P. Peterson, A. Webster, R. Orthofer, V. Vandeweerd, and M. Gwynne. 1996. Urban air pollution in megacities of the world. *Atmospheric Environment* 30: 681–686.

Mahfoud, R. F., and J. N. Beck. 1995. Why the Middle East fields may produce oil forever. *Offshore* (April): 58–64+.

Major, J. K. 1980. Muscle power. *History Today* 30 (3): 26–30.

Makarieva, A. M., V. G. Gorshkov, and B. L. Li. 2005. Biochemical universality of living matter and its metabolic implications. *Functional Ecology* 19: 547–557.

Malhotra, V. M. 1999. Making concrete "greener" with fly ash. *Concrete International* 21 (5): 61–66.

Malmberg, T. 1980. *Human Territoriality*. The Hague: Mouton.

Mann, K. H. 1984. Fish production in open ocean ecosystems. In *Flows of Energy and Materials in Marine Ecosystems*, ed. M. J. R. Fasham, 435–438. New York: Plenum Press.

Mannion, A. M. 1999. Domestication and the origins of agriculture: An appraisal. *Progress in Physical Geography* 23: 37–56.

Marchaj, C. A. 2000. *Aero-Hydrodynamics of Sailing*. London: Adlard Coles Nautical.

Marchetti, C. 1977. Primary energy substitution models: On the interaction between energy and society. *Technological Forecasting and Social Change* 10: 345–356.

Marchetti, C., and N. Nakićenović. 1979. *The Dynamics of Energy Systems and the Logistic Substitution Model*. Laxenburg, Austria: International Institute of Applied Systems Analysis.

Marine drilling rigs 2000/2001. 2000. *World Oil* 221 (12) (Suppl.).

Mark, R. 1987. Reinterpreting ancient Roman structure. *American Scientist* 75: 142–150.

Markvart, T., ed. 2000. *Solar Electricity*. Wiley: New York.

Marland, G., T. Boden, and R. J. Andres. 2005. *Global CO₂ Emissions from Fossil-Fuel Burning, Cement Manufacture, and*

REFERENCES

Gas Flaring, 1751–2002. Oak Ridge, Tenn.: Oak Ridge National Laboratory.

Marquet, P. A., R. A. Quiñones, S. Abades, F. Labra, M. Tognelli, M. Arim, and M. Rivadeneira. 2005. Scaling and power-laws in ecological systems. *Journal of Experimental Biology* 208: 1749–1769.

Marshall, R. 1993. *Storm from the East: From Ghengis Khan to Khubilai Khan*. Berkeley: University of California Press.

Martin, D. H. 1998. *Federal Nuclear Subsidies: Time to Call a Halt*. ⟨http://www.cnp.ca/issues/nuclear-subsidies.html⟩.

Martin, P.-E. 1995. The external costs of electricity generation: Lessons from the U.S. experience. *Energy Studies Review* 7: 232–246.

Martin, T. C. 1922. *Forty Years of Edison Service, 1882–1922*. New York: N.Y. Edison Co.

Martinez-Alier, J. 1987. *Ecological Economics*. New York: Blackwell.

Masters, C. D., E. D. Attanasi, and D. H. Root. 1994. World petroleum assessment and analysis. In *Proceedings of the 14th World Petroleum Congress*, 529–541.

Matthews, E., R. Payne, M. Rohweder, and S. Murray. 2000. *Pilot Analysis of Global Ecosystems: Forest Ecosystems*. Washington, D.C.: World Resources Institute.

Maxwell, J. C. 1872. *Matter and Motion*. New York: D. van Nostrand.

Mayer, J. R. 1842. Die Mechanik der Wärme. *Annalen der Chemie und Pharmacie* 42: 233.

———. 1851. *Bemerkungen über das mechanische Aequivalent der Wärme*. Heilbronn, Germany: Johann Ulrich Landherr.

Maynard Smith, J. 1978. Optimization theory in evolution. *Annual Review of Ecology and Systematics* 9: 31–56.

Mazza, P., and R. Hammerschlag. 2004. *Carrying Energy Future: Comparing Hydrogen and Electricity for Transmission, Storage and Transportation*. Seattle: Institute for Lifecycle Environmental Assessment.

McBride, J. P., R. E. Moore, J. P. Witherspoon, and R. E. Blanco. 1977. *Radiological Impact of Airborne Effluents of Coal-Fired and Nuclear Power Plants*. Oak Ridge, Tenn.: Oak Ridge National Laboratory.

McCann, K., A. Hustings, and G. R. Huxel. 1998. Weak trophic interactions and the balance of nature. *Nature* 395: 794–798.

McClellan, J. E. III, and H. Dorn. 1999. *Science and Technology in World History: An Introduction*. Baltimore, Md.: Johns Hopkins University Press.

McCloy, S. T. 1952. *French Inventions of the Eighteenth Century*. Lexington: University of Kentucky.

McConnell, A. 2000. Extending the limits of human performance. *BMC News and Views* 2000: 1–3.

McCullough, D. R. 1973. Secondary production of birds and mammals. In *Analysis of Temperate Forest Ecosystems*, ed. D. E. Reichle, 107–130. New York: Springer.

McFate, K. L., ed. 1989. *Electrical Energy in Agriculture*. Amsterdam: Elsevier.

McGowan, J. G., and S. R. Connors. 2000. Windpower: A turn of the century review. *Annual Review of Energy and the Environment* 25: 147–197.

McGraw, M. G. 1982. Plant-site coal handling. *Electrical World* 196 (7): 63–93.

McInnes, W., ed. 1913. *The Coal Resources of the World*. Toronto: Morang.

McIntyre, R. S. 2001. *The Hidden Entitlements*. Washington, D.C.: Citizens for Tax Justice. ⟨http://www.ctj.org/pdf/hident.pdf⟩.

McKechnie, A. E., and B. O. Wolf. 2004. The allometry of avian basal metabolic rate: Good predictions need good data. *Physiological and Biochemical Zoology* 77: 502–521.

REFERENCES

McKelvey, V. E. 1973. Mineral resource estimates and public policy. In *United States Mineral Resources*, ed. D. A. Brobst and W. P. Pratt, 9–19. Washington, D.C.: U.S. Geological Survey.

McMahon, T. 1973. Size and shape in biology. *Science* 179: 1201–1204.

McMahon, T., and J. T. Bonner. 1983. *On Size and Life*. New York: Scientific American Library.

McManus, G. J. 1981. Inland's no. 7 start-up more than pushing the right buttons. *Iron Age* 224 (7): MP-7-MP-16.

McNab, B. K. 1963. Bioenergetics and the determination of home range size. *American Naturalist* 97: 133–140.

McNeill, W. H. 1989. *The Age of Gunpowder Empires, 1450–1800*. Washington, D.C.: American Historical Association.

Meadows, D. H., D. L. Meadows, J. Randers, and W. W. Behrens III. 1972. *Limits to Growth*. New York: Universe Books.

Meadows, D. H., J. Randers, and D. L. Meadows. 2004. *Limits to Growth: The 30-Year Update*. White River Junction, Vt.: Chelsea Green Publishing.

Meehan, B. 1977. Hunters by the seashore. *Journal of Human Evolution* 6: 363–370.

Meentemeyer, V., E. O. Box, and R. Thompson. 1982. World patterns and amounts of terrestrial plant litter production. *BioScience* 32: 125–128.

Megaw, J. V. S., ed. 1977. *Hunters, Gatherers and First Farmers Beyond Europe*. Leicester, UK: Leicester University Press.

Meier, A., and W. Huber. 1997. *Results from the Investigations of Leaking Electricity in the USA*. ⟨http://eetd.lbl.gov/EA/Reports/40909/⟩.

Meier, P. J. 2002. *Life-Cycle Assessment of Electricity Generation Systems and Applications for Climate Change Policy Analysis*. Madison: University of Wisconsin, Fusion Technology Institute. ⟨http://fti.neep.wisc.edu/pdf/fdm1181.pdf⟩.

Meier, P. J., and G. L. Kulcinski. 2002. *Life-Cycle Energy Requirements and Greenhouse Gas Emissions for Building-Integrated Photovoltaics*. Madison: University of Wisconsin, Fusion Technology Institute. ⟨http://fti.neep.wisc.edu/pdf/fdm1185.pdf⟩.

Meitner, L., and O. R. Frisch. 1939. Disintegration of uranium by neutrons: A new type of nuclear reaction. *Nature* 143: 239–240.

Mellars, P. A. 1985. The ecological basis of social complexity in the Upper Paleolithic of Southwestern France. In *Prehistoric Hunter-Gatherers: The Emergence of Cultural Complexity*, ed. T. D. Price and J. A. Brown, 271–297. New York: Academic Press.

Melville, H. 1851. *Moby-Dick, or the Whale*. New York: Harper.

Mendelssohn, K. 1974. *The Riddle of the Pyramids*. London: Thames and Hudson.

Menon, S., J. Hansen, L. Nazarenko, and Y. Luo. 2002. Climate effects of black carbon aerosols in China and India. *Science* 297: 2250–2252.

Menzies, R. J., R. Y. George, and G. T. Rowe. 1973. *Abyssal Environment and Ecology of the World Oceans*. New York: Wiley.

Merriam, M. F. 1978. Wind, waves, and tides. *Annual Review of Energy* 3: 29–56.

Merrill, A. L., and B. K. Watt. 1973. *Energy Value of Foods: Basis and Derivation*. U.S. Department of Agriculture.

Meyer, J. H. 1975. *Kraft aus Wasser: vom Wasserrad zur Pumpturbine*. Innertkirchen, Switzerland: Kraftwerke Oberhasli.

Miller, A. I. 1981. *Albert Einstein's Special Theory of Relativity*. Reading, Mass.: Addison-Wesley.

Miller, B. G. 2005. *Coal Energy Systems*. Amsterdam: Elsevier.

Miller, D. H. 1969. Development of the heat budget concept. *Yearbook of the Association of Pacific Coast Geographers* 30: 123–144.

REFERENCES

———. 1981. *Energy at the Surface of the Earth.* New York: Academic Press.

Miller, L., and B. C. Douglas. 2004. Mass and volume contributions to twentieth-century global sea level rise. *Nature* 428: 406–409.

Milliman, J. D., and J. P. Syvitksi. 1992. Geomoprhic/tectonic control of sediment discharge to the ocean: The importance of small mountainous rivers. *Journal of Geology* 100: 525–544.

Minchinton, W. 1980. Wind power. *History Today* 30 (3): 31–36.

Minchinton, W., and P. Meigs. 1980. Power from the sea. *History Today* 30 (3): 42–46.

Minetti, A. E. 2003. Efficiency of equine express postal systems. *Nature* 426: 785–786.

Minetti, A. E., and R. M. Alexander. 1997. A theory of metabolic costs of bipedal gaits. *Journal of Theoretical Biology* 186: 467–476.

Minetti, A. E., C. Moia, G. S. Roi, D. Susta, and G. Ferretti. 2002. Energy cost of walking and running at extreme uphill and downhill slopes. *Journal of Applied Physiology* 93: 1039–1046.

Minge-Klewana, W. 1980. Does labor time decrease with industrialization? A survey of time-allocation studies. *Current Anthropology* 21: 278–298.

Mintzer, I. M., A. S. Miller, and A. Serchuk. 1995. *The Environmental Imperative: A Driving Force in the Development and Deployment of Renewable Energy Technologies.* ⟨http://www.crest.org/repp_pubs/pdf/envImp.pdf⟩.

Mirowski, P. 1988. *More Heat Than Light.* New York: Cambridge University Press.

MIT (Massachusetts Institute of Technology). 2003. *The Future of Nuclear Power: An Interdisciplinary MIT Study.* Cambridge, Mass.: MIT.

Mitchell, B. 1992. *International Historical Statistics: Europe 1750–1988.* New York: Stockton Press.

Mithraratne, N., and B. Vale. 2004. Life cycle analysis for New Zealand houses. *Building and Environment* 39: 483–492.

Mobbs, P. 2005. Uranium supply and the nuclear option. *Oxford Energy Forum* 61: 1–5.

Molenaar, A. 1956. *Water Lifting Devices for Irrigation.* Rome: FAO.

Monod, T., ed. 1975. *Pastoralism in Tropical Africa.* Oxford: Oxford University Press.

Monteith, J. L. 1978. Reassessment of maximum growth rates for C_3 and C_4 crops. *Experimental Agriculture* 14: 1–5.

Montelli, R., G. Nolet, F. A. Dahlen, G. Masters, E. R. Engdahl, and S.-H. Hung. 2004. Finite-frequency tomography reveals a variety of plumes in the mantle. *Science* 303: 338–343.

Moore, D. W., and F. Newport. 1995. People throughout the world largely satisfied with personal lives. *Gallup Poll Monthly* 357: 2–7.

Moore, G. 1965. Cramming more components onto integrated circuits. *Electronics* 38 (8): 114–117.

Moore, J. G., W. R. Normark, and R. T. Holcomb. 1994. Giant Hawaiian underwater slides. *Science* 264: 46–47.

Moravec, H. P. 1999. *Robot: Mere Machine to Transcendent Mind.* New York: Oxford University Press.

Morgan, W. J. 1971. Convection plumes in the lower mantle. *Nature* 230: 42–43.

Morita, Z., and E. Toshihiko, eds. 2003. *An Introduction to Iron and Steel Processing.* Tokyo: Kawasaki Steel Twenty-first Century Foundation.

Moritz, L. A. 1958. *Grain-Mills and Flour in Classical Antiquity.* Oxford: Oxford University Press.

Morowitz, H. J. 1968. *Energy Flow in Biology.* New York: Academic Press.

Morrison, J. S., J. F. Coates, and N. B. Rankov. 2000. *The Athenian Trireme: The History and Reconstruction of an Ancient Greek Warship.* Cambridge: Cambridge University Press.

REFERENCES

Morrison, J. S., and T. R. Williams. 1968. *Greek Oared Ships, 900–322 B.C.* Cambridge: Cambridge University Press.

Mouton, F.-R. 2005. The Voluntary Standard and Data Gathering Exercise. Paper presented at the Joint OPEC/World Bank Workshop on Global Gas Flaring Reduction, Vienna, June 30–July 1.

MSHA (Mine Safety and Health Administration). 2000. *Injury Trends in Mining.* ⟨http://www.msha.gov/mshainfo/FactSheets/mshafct2.htm⟩.

Muller, G., and K. Kauppert. 2004. Performance characteristics of water wheels. *Journal of Hydraulic Research* 42: 451–460.

Müller, W. 1939. *Die Wasserräder.* Detmold, Germany: Moritz Schäfer.

Murdock, G. P. 1967. Ethnographic atlas. *Ethnology* 6: 109–236.

Murphy, J. B., and R. D. Nance. 2004. How do supercontinents assemble? *American Scientist* 92: 324–333.

Murra, J. V. 1980. *The Economic Organization of the Inka State.* Greenwich, Conn.: JAI Press.

Murthy, V. R., W. van Westrenen, and Y. Fei. 2003. Experimental evidence that potassium is a substantial radioactive heat source in planetary cores. *Nature* 423: 163–165.

Muybridge, E. 1887. *Animal Locomotion: An Electrophotographic Investigation of Consecutive Phases of Animal Movements.* Philadelphia: University of Pennsylvania Press.

MVMA (Motor Vehicle Manufacturers Association). 2000. *Facts and Figures.* Detroit: MVMA.

M. W. Kellogg Co. 1998. *First of a New Generation.* Houston: M. W. Kellogg Co..

Myneni, R. B., R. R. Nemani, and S. Running. 1997. Estimation of global leaf area index and absorbed PAR using radiative transfer models. *Transactions on Geoscience and Remote Sensing* 35: 1380–1393.

Nabuurs, G. J., R. Päivinen, A. Pussinen, and M. J. Schelhaas. 2003. *Development of European Forests until 2050: A Projection of Forest Resources and Forest Management in 30 Countries.* European Forest Institute Research Report 15. Leiden, Netherlands: Brill.

Nadel, S. 2006. *Energy Efficiency Resource Standards: Experience and Recommendations.* ⟨http://aceee.org/pubs/e063.htm⟩.

Nader, L., and S. Beckerman. 1978. Energy as it relates to the quality and style of life. *Annual Review of Energy* 3: 1–28.

Nagy, K. A. 2005. Field metabolic rate and body size. *Journal of Experimental Biology* 208: 1621–1625.

Nagy, K. A., I. A. Girard, and T. K. Brown. 1999. Energetics of free-ranging mammals, reptiles, and birds. *Annual Review of Nutrition* 19: 247–277.

NAS (National Academy of Sciences). 1978. *Nutrient Requirements of Horses.* Washington, D.C.: National Academies Press.

———. 1980. *Firewood Crops.* Washington, D.C.: National Academies Press.

NASA (National Aeronautics and Space Administration). 2005. *Langley 8-year Surface Radiation Budget.* NASA Langley Atmospheric Sciences Data Center. ⟨http://eosweb.larc.nasa.gov/⟩.

Nataf, H.-C. 2000. Seismic imaging of mantle plumes. *Annual Review of Earth and Planetary Sciences* 28: 391–417.

Natural Resources Canada. 2000. *1997 Survey of Household Energy Use.* Ottawa: Natural Resources Canada.

Naville, E. 1908. *The Temple of Deir el Bahari. Part VI.* London: Egyptian Exploration Fund.

NBS (National Bureau of Statistics). 2000. *China Statistical Yearbook 2000.* Beijing: China Statistics Press.

NEAA (Netherlands Environmental Assessment Agency). 2005. *HYDE: The History Database of the Global Environment.* ⟨http://www.mnp.nl/hyde/⟩.

Needham, J., ed. 1954–1986. *Science and Civilisation in China.* 7 vols. Cambridge: Cambridge University Press.

REFERENCES

———. 1964. *The Development of Iron and Steel Industry in China*. Cambridge: W. Heffer.

———. 1965. *Science and Civilisation in China*. Vol. 4, Pt. 2: *Physics and Physical Technology*. Cambridge: Cambridge University Press.

———. 1971. *Science and Civilisation in China*. Vol. 4, Pt. 3: *Civil Engineering and Nautics*. Cambridge: Cambridge University Press.

Nef, U. 1932. *The Rise of the British Coal Industry*. London: Routledge.

Nehring, R. 1978. *Giant Oil Fields and World Oil Resources*. Santa Monica, Calif.: Rand Corporation.

Nelson, D. L., and M. M. Cox. 2000. *Lehninger Principles of Biochemistry*. 3d rev. ed. New York: Worth.

Nemani, R. R., C. D. Keeling, H. Hashimoto, W. M. Jolly, S. C. Piper, C. J. Tucker, R. B. Myneni, and S. W. Running. 2003. Climate-driven increases in global terrestrial net primary production from 1982 to 1999. *Science* 300: 1560–1563.

NEOP (Near Earth Object Program). 2007. *Discovery Statistics*. ⟨http://neo.jpl.nasa.gov/stats⟩.

Neuburger, A. 1930. *The Technical Arts and Sciences of the Ancients*. London: Methuen.

Neumann, J. von. 1955. Can we survive technology? *Fortune* 51 (6): 106–107+.

New data lift world oil reserves by 27 percent. 1987. *Oil and Gas Journal* 85 (52): 33–37.

Newhall, C. G., and S. Self. 1982. The volcanic explosivity index (VEI): An estimate of explosive magnitude for historical volcanism. *Journal of Geophysical Research* 87: 1231–1238.

NHTSA (National Highway and Transport Safety Administration). 2005. *Domestic Passenger Car Fleet Average Characteristics*. ⟨http://www.nhtsa.dot.gov/cars/rules/CAFE/DomesticCarFleet.htm⟩.

Ni, S., H. Kanamori, and D. Helmberger. 2005. Energy radiation from the Sumatra earthquake. *Nature* 434: 582.

Nicholaides, J. J. et al. 1985. Agricultural alternatives for the Amazon Basin. *BioScience* 35: 279–285.

Nicholls, D. G., and S. J. Ferguson. 2002. *Bioenergetics*. San Diego, Calif.: Academic Press.

Niklas, K. J., and B. J. Enquist. 2001. Invariant scaling relationships for interspecific plant biomass production rates and body size. *Proceedings of the National Academy of Sciences* 98: 2922–2927.

NIRS (Nuclear Information and Resource Service). 1999. *Background on Nuclear Power and the Kyoto Protocol*. ⟨http://www.nirs.org/⟩.

Nitschke, W., U. Mühlenhoff, and U. Liebl. 1998. Evolution. In *Photosynthesis: A Comprehensive Treatise*, ed. A. Raghavendra, 285–304. Cambridge: Cambridge University Press.

NOAA (National Oceanic and Atmospheric Administration). 2005. International Tsunami Information Centre. ⟨http://www.tsunamiwave.info/⟩.

Noddack, W. 1937. Der Kohlenstoff im Haushalt der Natur. *Angewandte Chemie* 50: 505–510.

Norberg, U. M. 1990. *Vertebrate Flight*. Berlin: Springer.

Norgan, N. G., A. Ferro-Luzzi, and J. V. Durnin. 1974. The energy and nutrient intake and the energy expenditure of 204 New Guinean adults. *Philosophical Transactions of the Royal Society* B268: 309–348.

Norton, I. O. 2000. Global hotspot reference frames and plate motion. In *The History and Dynamics of Global Plate Motions*, ed. M. A. Richards, R. G. Gordon, and R. D. van der Hilst, 339–357. Washington, D.C.: American Geophysical Union.

NRC (National Research Council). 1971. *Atlas of Nutritional Data on United States and Canadian Feeds*. Washington, D.C.: National Academies Press.

———. 1986. *Electricity in Economic Growth*. Washington, D.C.: National Academies Press.

REFERENCES

———. 1994. *Solar Influences on Global Change*. Washington D.C.: National Academies Press.

———. 2004. *The Hydrogen Economy: Opportunities, Costs, Barriers, and R&D Needs*. Washington, D.C.: National Academies Press.

Nriagu, J. O., R. D. Coker, and L. A. Barrie. 1991. Origin of sulphur in Canadian Arctic haze from isotopic measurements. *Nature* 349: 142–145.

NTSG (Numerical Terradynamic Simulation Group). 2004. *Four Years of MOD17 Annual NPP*. ⟨http://images.ntsg.umt .edu/⟩.

NTTL (Nebraska Tractor Test Laboratory). 2001. ⟨http:// tractortestlab.unl.edu/publications.htm⟩.

———. 2005. *Tractor Test Reports*. ⟨http://tractortestlab.unl .edu/testreports.htm⟩.

O'Brien, W. J., H. I. Bowman, and B. I. Evans. 1990. Search strategies of foraging animals. *American Scientist* 78: 152–160.

Odell, P. R. 1984. The oil crisis: Its nature and implications for developing countries. In *The Oil Prospect*, ed. P. R. Odell and B. Mossavar-Rahmani, 33. Ottawa: Energy Research Group.

———. 1992. Global and regional energy supplies. *Energy Policy* 20 (4): 284–296.

———. 1999. *Fossil Fuel Resources in the 21st Century*. London: Financial Times.

———. 2004. *Why Carbon Fuels Will Dominate the 21st Century's Global Energy Economy*. Brentwood, UK: Multi-Science Publishing.

Odum, H. T. 1957. Trophic structure and productivity of Silver Springs, Florida. *Ecological Monographs* 27: 55–112.

———. 1971. *Environment, Power, and Society*. New York: Wiley.

———. 1973. Energy, ecology, economics. *Ambio* 2: 220–227.

———. 1975. Energy analysis and net energy. In *Report of the NSF-Stanford Workshop on Net Energy Analysis*, 90–115. Stanford, Calif.: Institute for Energy Studies.

———. 1983. *Systems Ecology*. New York: Wiley.

———. 1988. Self-organization, transformity, and information. *Science* 242: 1132–1139.

———. 1996. *Environmental Accounting: EMERGY and Environmental Decision Making*. New York: Wiley.

———. 2000. *Emergy of Global Processes*. Gainseville, Fla.: Center for Environmental Policy.

Odum, H. T., and E. C. Odum. 1981. *Energy Basis for Man and Nature*. New York: McGraw-Hill.

———. 2000. *Modeling for All Scales: An Introduction to System Simulation*. San Diego, Calif.: Academic Press.

Odum, H. T., and R. C. Pinkerton. 1955. Time's speed regulator: The optimum efficiency for maximum power output in physical and biological systems. *American Scientist* 43: 331–343.

OECD (Organisation for Economic Co-operation and Development). 1982. *The Energy Problem of the Agro-Food Sector*. Paris: OECD.

———. 2006. *Economic Policy Reforms: Going for Growth 2006*. Paris: OECD.

Oftedal, O. T. 1984. Milk consumption, milk yield and energy output at peak lactation: A comparative review. *Symposia of the Zoological Society of London* 51: 33–85.

Øhlenschlaeger, K. 1997. The trend toward larger wind turbines. *WindStats* 10 (4). BTM Consult, Denmark.

Oki, T. 1999. The global water cycle. In *Global Energy and Water Cycles*, ed. K. A. Browning and R. J. Gurney, 10–29. Cambridge: Cambridge University Press.

Okigbo, B. N. 1984. *Improved Production Systems as an Alternative to Shifting Cultivation*. FAO Soils Bulletin 53. Rome: FAO.

Olah, G. A., A. Goeppert, and G. K. S. Prakash. 2006. *Beyond Oil and Gas: The Methanol Economy.* Weinheim, Germany: Wiley-VCH.

Oleson, J. P. 1984. *Greek and Roman Mechanical Water-Lifting Devices: The History of a Technology.* Toronto: University of Toronto Press.

Ollier, C. D., and M. J. F. Brown. 1971. Erosion of a young volcano in New Guinea. *Zeitschrift für Geomorphologie* 15: 12–28.

Olson, J. S., J. A. Watts, and L. J. Allison. 1983. *Carbon in Live Vegetation of Major World Ecosystems.* Oak Ridge, Tenn.: Oak Ridge National Laboratory. ⟨http://cdiac.ornl.gov/ftp/ndp017/table.html⟩.

O'nions, R. K., P. J. Hamilton, and N. M. Evensen. 1980. The chemical evolution of the Earth's mantle. *Scientific American* 242 (5): 120–131.

Oreskes, N. 1999. *The Rejection of Continental Drift: Theory and Method in American Earth Science.* New York: Oxford University Press.

———, ed. 2001. *Plate Tectonics: An Insider's History of the Modern Theory of the Earth.* Boulder, Colo.: Westview Press.

Orme, B. 1977. The advantages of agriculture. In *Hunters, Gatherers and First Farmers Beyond Europe*, ed. J. V. S. Megaw, 41–49. Leicester, UK: Leicester University Press.

ORNL (Oak Ridge National Laboratory). 2006. *NPP Database.* ⟨http://www-eosdis.ornl.gov/NPP/npp_home.html⟩.

Orr, J. C., V. J. Fabry, O. Aumont, L. Bopp, S. C. Doney, R. A. Feely, A. Gnanadesikan et al. 2006. Anthropogenic ocean acidification over the twenty-first century and its impact on calcifying organisms. *Nature* 437: 681–686.

Orr, L., and Govindjee. 2006. *Photosynthesis and the Web.* Urbana-Champaign: University of Illinois. ⟨http://photoscience.la.asu.edu/photosyn/photoweb/⟩.

Ostrander, C. E. 1980. Energy use in agriculture poultry. In *Handbook of Energy Utilization in Agriculture*, ed. D. Pimentel, 379–392. Boca Raton, Fla.: CRC Press.

Ostwald, W. 1892. Studien zur Energetik. II. Grundlinien in der allgemeinen Energetik. *Berichte über die Verhandlungen der Königlich Sächsischen Gesellschaft der Wissenschaften zu Leipzig* 44: 211–237. Trans. in *Applications of Energy: Nineteenth Century*, ed. R. B. Lindsay. Stroudsburg, Pa.: Dowden, Hutchinson and Ross, 1976.

———. 1909. *Energetische Grundlagen der Kulturwissenschaften.* Leipzig: Alfred Kröner.

Pacala, S., and R. Socolow. 2004. Stabilization wedges: Solving the climate problem for the next 50 years with current technologies. *Science* 305: 968–972.

Pacey, A. 1990. *Technology in World Civilization: A Thousand-Year History.* Cambridge, Mass.: MIT Press.

Pallé, E., P. R. Goode, P. Montañes-Rodriguez, and S. E. Koonin. 2004. Changes in Earth's reflectance over the past two decades. *Science* 304: 1299–1301.

Palmer, M. R., and G. G. J. Ernst. 1998. Generating of hydrothermal megaplumes by cooling of pillow basalts and mid-ocean ridges. *Nature* 393: 643–647.

Panter-Brick, C., R. H. Layton, and P. Rowley-Conwy, eds. 2001. *Hunter-Gatherers: An Interdisciplinary Perspective.* Cambridge: Cambridge University Press.

Parker, B. F., ed. 1991. *Solar Energy in Agriculture.* Amsterdam: Elsevier.

Parker, D. 2004. Large-scale warming is not urban. *Nature* 432: 290.

Parkins, W. E. 2006. Fusion power: will it ever come? *Science* 311: 1380.

Parsons, T. R. et al. 1984. *Biological Oceanographic Processes.* Oxford: Pergamon Press.

Partl, R. 1977. *Power from Glaciers: The Hydropower Potential of Greenland's Glacial Waters.* Laxenburg, Austria: International Institute of Applied Systems Analysis.

Paton, J. 1890. Gas and gas-lighting. In Encyclopædia Britannica, 9th Ed., Vol. 10, 87–102. Chicago: R. S. Peale.

REFERENCES

Patton, J. F. 1997. Measurement of oxygen uptake with portable equipment. In *Emerging Technologies for Nutrition Research*, ed. S. J. Carlson-Newberry and R. B. Costello, 297–414. Washington, D.C.: National Academies Press.

Paul, E. A., and F. E. Clark. 1989. *Soil Microbiology and Biochemistry*. New York: Academic Press.

PDVSA (Petroleos de Venezuela SA). 2001. *PDVSA Orimulsion*. ⟨http://www.pdvsa.com⟩.

Peachey, B. 2005. *Strategic Needs for Energy Related Water Use Technologies*. ⟨http://www.aeri.ab.ca/sec/new_res/docs/EnergyINet_and_Water_Feb2005.pdf⟩.

Peacock, S. M. 2003. Thermal structure and metamorphic evolution of subducting slabs. In *Inside the Subduction Factory*, ed. J. Eiler, 7–22. Washington, D.C.: American Geophysical Union.

Peart, R. M., and R. C. Brook, eds. 1992. *Analysis of Agricultural Energy Systems*. Amsterdam: Elsevier.

Pedley, T. J., ed. 1977. *Scale Effects in Animal Locomotion*. London: Academic Press.

Peixoto, J. P., and A. H. Oort. 1992. *Physics of Climate*. New York: American Institute of Physics.

Pekáry, T. 1968. *Untersuchungen zu den römischen Reichsstrassen*. Bonn: R. Habelt.

Pellett, P. L. 1990. Protein requirements in humans. *American Journal of Clinical Nutrition* 51: 723–737.

Penner, P., J. Kurish, and B. Hannon. 1980. *Energy and Labor Cost of Coal Electric Fuel Cycles*. Urbana: University of Illinois Energy Research Group.

Perez, R., S. Valenzuela, V. Merino, I. Cabezas, M. Garcia, R. Bou, and P. Ortiz. 1996. Energetic requirements and physiological adaptation of draught horses to ploughing work. *Animal Science* 63: 343–351.

Perkins, D. H. 1969. *Agricultural Development in China, 1368–1968*. Chicago: Aldine.

Perlin, J. 1999. *From Space to Earth: The Story of Solar Electricity*. Ann Arbor, Mich.: Aatec Publications.

———. 2004. Wood energy, history of. In *Encyclopedia of Energy*. Vol. 6, 499–507.

Perrodon, A. 1985. *Histoire des grandes découvertes pétrolières*. Paris: Elf Aquitaine.

Perry, D. A. 1994. *Forest Ecosystems*. Baltimore, Md.: Johns Hopkins University Press.

Peschek, G. A., ed. 1999. *The Phototrophic Prokaryotes*. New York: Kluwer/Plenum.

Petchers, N. 2003. *Combined Heating, Cooling and Power Handbook*. Lilburn, Ga.: Fairmont Press.

Peterson, T. C. 2003. Assessment of urban versus rural in situ surface temperatures in the contiguous United States: No difference found. *Journal of Climate* 16: 2941–2959.

Peterson, T. C., and T. W. Owen. 2005. Urban heat island assessment: Metadata are important. *Journal of Climate* 18: 2637–2646.

Petit, J. R., D. Raynaud, J. Jouzel, and S. Duparcq. 1999. Climate and atmospheric history of the past 420,000 years from the Vostok ice core, Antarctica. *Nature* 399: 429–436.

Pettenkofer, M., and C. Voit. 1866. Untersuchungen über den Stoffverbrauch des normalen Menschen. *Zeitschrift für Biologie* 2: 459–573.

Phalen, R. F. 2002. *The Particulate Air Pollution Controversy*. Hingham, Mass.: Kluwer.

Phillips, K. J. H. 1992. *Guide to the Sun*. Cambridge: Cambridge University Press.

Phillips, R. E., and S. H. Phillips, eds. 1984. *No-Tillage Agriculture: Principles and Practices*. New York: Van Nostrand Reinhold.

Phoenix, G. K., W. K. Hicks, S. Cinderby, J. C. Kuylenstierna, W. D. Stock, F. J. Dentener, K. E. Giller et al. 2006. Atmospheric nitrogen deposition in world biodiversity hotspots: The need for a greater global perspective in assessing N deposition impacts. *Global Change Biology* 12: 470–476.

Piers, L. S., B. Diffey, M. J. Soares, S. L. Frandsen, L. M. McCormack, M. Latschini, and K. O'Dea. 1997. The validity of predicting the basal metabolic rate of young Australian men and women. *European Journal of Clinical Nutrition* 51: 333–337.

Piers, L. S., and P. S. Shetty. 1993. Basal metabolic rates of Indian women. *European Journal of Clinical Nutrition* 47: 586–591.

Piggott, S. 1983. *The Earliest Wheeled Transport.* Ithaca, N.Y.: Cornell University Press.

Pilkey, O. H., and J. A. G. Cooper. 2004. Society and sea level rise. *Science* 303: 1781–1782.

Pimentel, D. 1991. Ethanol fuels: Energy security, economics, and the environment. *Journal of Agriculture and Environmental Ethics* 4: 1–13.

———. 2003. Ethanol fuels: Energy balance, economics, and environmental impacts are negative. *Natural Resources Research* 12: 127–134.

———, ed. 1980. *Handbook of Energy Utilization in Agriculture.* Boca Raton, Fla.: CRC Press.

Pimentel, D., L. E. Hurd, A. C. Bellotti, M. J. Forster, L. N. Oka, O. D. Sholes, and R. J. Whitman. 1973. Food production and energy crisis. *Science* 182: 443–449.

Pinker, R. T., B. Zhang, and E. G. Dutton. 2005. Do satellites detect trends in surface solar radiation? *Science* 308: 850–854.

Piper, E. 1993. *The Exxon Valdez Oil Spill: Final Report, State of Alaska Response.* Anchorage: Alaska Department of Environmental Conservation.

Piperno, D. R., E. Weiss, I. Holst, and D. Nadel. 2004. Processing of wild cereal grains in the Upper Palaeolithic revealed by starch grain analysis. *Nature* 430: 670–673.

Pitzer, K. S. 1995. *Thermodynamics.* New York: McGraw-Hill.

Plank, C. J., and E. J. Rosinski. 1964. *Catalytic Cracking of Hydrocarbons with a Crystalline Zeolite Catalyst Composite.* U.S. Patent 3,140,249, July 7, 1964. Washington, D.C.: USPTO. ⟨http://www.uspto.gov⟩.

Platt's Oil Guide to Specifications. 2005. ⟨http://www.emis .platts.com/thezone/ guides/platts/oil/content.html⟩.

Plumptree, A. J., and S. Harris. 1995. Estimating the biomass of large mammalian herbivores in a tropical montane forest: A method of faecal counting that avoids a "steady state" system. *Journal of Applied Ecology* 32: 111–120.

Polis, G. A., and K. Winemiller. 1996. *Food Webs: Integration of Patterns and Dynamics.* New York: Chapman and Hall.

Pollack, H. N., and D. S. Chapman. 1977. The flow of heat from the Earth's interior. *Scientific American* 237 (2): 60–76.

Pollack, H. N., S. J. Hurter, and J. R. Johnson. 1993. Heat flow from the Earth's interior: Analysis of the global data set. *Reviews of Geophysics* 31: 267–280.

Pond, W. G. 1991. *Pork Production Systems: Efficient Use of Swine and Feed Resources.* New York: Van Nostrand Reinhold.

Poole, H. 1910. *The Calorific Power of Fuels.* New York: Wiley.

Post, D. M., M. L. Pace, and N. G. Hairston. 2000. Ecosystem size determines food-chain length in lakes. *Nature* 405: 1047–1049.

Post, W., A. W. King, and S. D. Wullschleger. 1997. Historical variations in terrestrial biospheric carbon storage. *Global Biogeochemical Cycles* 11: 99–109.

Poten and Partners. 1993. *World Trade in Natural Gas and LNG: 1985–2010.* New York: Poten and Partners.

PowerLight Corp. 2005. *Bavaria Solarpark.* ⟨http://www .powerlight.com/ bavaria/index.shtml⟩.

Poynting, J. H. 1885. On the transfer of energy in the electromagnetic field. *Philosophical Transactions of the Royal Society* 175: 343–349+.

REFERENCES

Prentice, A. M. 1984. Adaptations to long-term low energy intake. In *Energy Intake and Activity*, ed. E. Pollitt and P. Amante, 3–31. New York: Alan R. Liss.

Price, T. D., and J. A. Brown, eds. 1985. *Prehistoric Hunter-Gatherers*. Orlando, Fla.: Academic Press.

Prigogine, I. 1947. *Etude thermodynamique des phenomenes irreversibles*. Paris: Dunod.

Primack, R., and R. Corlett. 2005. *Tropical Rain Forest: An Ecological and Biogeographical Comparison*. Malden, Mass.: Blackwell.

Prins, H. H. T., and J. Reitsma. 1989. Mammalian biomass in an African equatorial forest. *Journal of Animal Ecology* 58: 851–861.

Priscu, J. C., C. H. Fritsen, E. E. Adams, S. J. Giovannoni, H. W. Paerl, C. P. McKay, P. T. Doran et al. 1998. Perennial Antarctic lake ice: An oasis for life in a polar desert. *Science* 280: 2095–2098.

Protzen, J.-P. 1993. *Inca Architecture and Construction at Ollantaytambo*. Oxford: Oxford University Press.

Pryor, F. L. 1983. Causal theories about the origin of agriculture. *Research in Economic History* 8: 93–124.

Psenner, R., and B. Sattler. 1998. Life at the freezing point. *Science* 280: 2073–2074.

Putnam, P. C. 1954. *Energy in the Future*. New York: Van Nostrand.

PVPRS (PV Power Resource Site). 2001. *PV History*. ⟨http://www.pvpower.com⟩.

Radick, R. R. 1991. The luminosity variability of solar-type stars. In *The Sun in Time*, ed. C. P. Sonett, M. S. Giampapa, and M. S. Matthews, 787–808. Tucson: University of Arizona Press.

Radler, M. 2002. Worldwide reserves increase as production holds steady. *Oil and Gas Journal* 100 (52): 113–115.

Rafiqul, I., C. Weber, B. Lehmann, and A. Voss. 2005. Energy efficiency improvements in ammonia production: Perspectives and uncertainties. *Energy* 30: 2487–2504.

Raghavendra, A. S. 1998. *Photosynthesis: A Comprehensive Treatise*. New York: Cambridge University Press.

Rakov, V. A., and M. A. Uman. 2003. *Lightning: Physics and Effects*. Cambridge: Cambridge University Press.

Ramage, J. 1983. *Energy: A Guidebook*. Oxford: Oxford University Press.

Ramanathan, V. 1998. Trace-gas greenhouse effect and global warming. *Ambio* 27: 187–197.

Ramanathan, V., B. R. Barkstrom, and E. F. Harrison. 1989. Climate and the Earth's energy budget. *Physics Today* 42 (5): 22–32.

Ramanathan, V., P. J. Crutzen, J. T. Kiehl, and D. Rosenfeld. 2001. Aerosols, climate, and the hydrological cycle. *Science* 294: 2119–2124.

Ramankutty, N., and J. A. Foley. 1999. Estimating historical changes in global land cover: Croplands from 1700 to 1992. *Global Biogeochemical Cycles* 13: 997–1027.

Ramelli, A. 1588. *Le diverse et artificiose machine*. Trans. M. Teach Gnudi. Baltimore, Md.: Johns Hopkins University Press, 1976.

Ramírez, A. R. 2005. *Monitoring Energy Efficiency in the Food Industry*. Utrecht, Netherlands: Universiteit Utrecht.

Ramirez-Zea, M. 2002. *Validation of Three Predictive Equations for Basal Metabolic Rate*. Report commissioned by FAO for the joint FAO/WHO/UNU Expert Consultation on Energy in Human Nutrition. Rome: FAO.

Rampino, M. R., and S. Self. 1992. Volcanic winter and accelerated glaciation following the Toba super-eruption. *Nature* 359: 50–52.

Rankine, W. J. M. 1859. *A Manual of the Steam Engine and Other Prime Movers*. London: R. Griffin.

REFERENCES

Rapp, A. 1960. Recent developments of mountain slopes in Karkevagge and surroundings, northern Scandinavia. *Geografiska Annaler* 42: 71–200.

Rappaport, R. A. 1968. *Pigs for the Ancestors.* New Haven, Conn.: Yale University Press.

Ratcliffe, M. 1985. *Liquid Gold Ships: A History of the Tanker, 1859–1984.* London: Lloyd's of London Press.

Rawitscher, M., and J. Mayer. 1977. National outputs and energy inputs in seafood. *Science* 198: 261–264.

Raynaud, D., J. Jouzel, J. M. Barnola, J. Chappellaz, and R. J. Delmas. 1993. The ice record of greenhouse gases. *Science* 259: 926–934.

Reagan, D. P., and R. B. Waide, eds. 1996. *The Food Web of a Tropical Rain Forest.* Chicago: University of Chicago Press.

Reaven, S. J. 1984. The concept of net energy. *Explorations in Knowledge* 1: 191–231.

Reed, C. A., ed. 1977. *Origins of Agriculture.* The Hague: Mouton.

Rees, M. 2003. *Our Final Hour.* New York: Basic Books.

Reich, P. B., M. G. Tjoelker, J.-L. Machado, and J. Oleksyn. 2006. Universal scaling of respiratory metabolism, size and nitrogen in plants. *Nature* 439: 457–461.

Reichle, D. E., ed. 1981. *Dynamic Properties of Forest Ecosystems.* Cambridge: Cambridge University Press.

Renne, P. R., and A. R. Basu. 1991. Rapid eruption of the Siberian Traps flood basalts at the Permo-Trisassic boundary. *Science* 253: 176–179.

Rennie, D. W., B. G. Covino, M. R. Blair, and K. Rodahl. 1962. Physical regulation of temperature in Eskimos. *Journal of Applied Physiology* 17: 326–332.

Reshetnikov, A. I., N. N. Paramonova, and A. A. Shashkov. 2000. An evaluation of historical methane emissions from the Soviet gas industry. *Journal of Geophysical Research* 105: 3517–3529.

Reynolds, J. 1970. *Windmills and Watermills.* London: Hugh Evelyn.

Reynolds, T. S. 1979. Scientific influences on technology: The case of the overshot waterwheel, 1752–1754. *Technology and Culture* 20: 270–295.

———. 1983. *Stronger Than a Hundred Men: A History of the Vertical Water Wheel.* Baltimore, Md.: Johns Hopkins University Press.

Reynolds, W. C. 1974. *Energy from Nature to Man.* New York: McGraw-Hill.

Ricci, P. F., and L. S. Molton. 1986. Health risk assessment: Science, economics, and law. *Annual Review of Energy* 11: 77–94.

Richards, M. A., R. G. Gordon, and R. D. van der Hilst, eds. 2000. *The History and Dynamics of Global Plate Motions.* Washington, D.C.: American Geophysical Union.

Richerson, P. J., R. Boyd, and R. L. Bettinger. 2001. Was agriculture impossible during the Pleistocene but mandatory during the Holocene? A climate change hypothesis. *American Antiquity* 66: 387–411.

Richter, C. F. 1935. An instrumental earthquake magnitude scale. *Bulletin Seismological Society of America* 25: 1–32.

Ricklefs, R. E. 1974. Energetics of reproduction in birds. In *Avian Energetics,* ed. R. A. Paynter, 152–292. Cambridge, Mass.: Nuttall Ornithological Club.

Ridley, B. K. 1979. *The Physical Environment.* Chichester, UK: Ellis Horwood.

Rifkin, J. 2002. *The Hydrogen Economy: The Creation of the Worldwide Energy Web and the Redistribution of Power on Earth.* New York: Tarcher.

Rigby, D. 1985. *Persistent Pastoralists.* London: Zed Books.

Riley, G. A. 1944. The carbon metabolism and photosynthetic efficiency of the Earth as a whole. *American Scientist* 32: 129–134.

REFERENCES

Rindos, D. 1984. *The Origins of Agriculture: An Evolutionary Perspective*. Orlando, Fla.: Academic Press.

Ritchie, D., and A. E. Gates. 2001. *Encyclopedia of Volcanoes*. New York: Facts on File.

Ritchie-Calder, P. 1979. Sunshine out of cucumbers: An energy perspective. *Environmental Science and Technology* 13: 1068–1073.

Roaf, S., and M. Hancock, eds. 1992. *Energy Efficient Building: A Design Guide*. New York: Halsted Press.

Robbins, C. T. 1983. *Wildlife Feeding and Nutrition*. New York: Academic Press.

Robertson, L. E. 2002. Reflections on the World Trade Center. *The Bridge* 32 (1): 5–10.

Robinson, D. A., and G. Kukla. 1984. Maximum surface albedo of seasonally snow-covered lands in the Northern Hemisphere. *Journal of Climate and Applied Meteorology* 24: 402–411.

Robock, A., and C. Oppenheimer, eds. 2003. *Volcanism and the Earth's Atmosphere*. Washington, D.C.: American Geophysical Union.

Rochereau, S. P. 1980. The energy requirements for inshore and offshore fishing crafts: The case of the Northeast fishery. In *Handbook of Energy Utilization in Agriculture*, ed. D. Pimentel, 441–446. Boca Raton, Fla.: CRC Press.

Rockwell, T. 1992. *The Rickover Effect: How One Man Made a Difference*. Annapolis, Md.: Naval Institute Press.

Rogin, L. 1931. *The Introduction of Farm Machinery*. Berkeley: University of California Press.

Rogner, H.-H. 1997. An assessment of world hydrocarbon resources. *Annual Review of Energy and the Environment* 22: 217–262.

———. 2000. Energy resources. In *World Energy Assessment: Energy and the Challenge of Sustainability*, ed. J. Goldemberg, 135–171. New York: United Nations Development Programme.

Rojey, A. 1996. *Natural Gas: Production Processing Transport*. Paris: Editions Technip.

Rose, D. J. 1974. Nuclear eclectic power. *Science* 184: 351–359.

———. 1986. *Learning about Energy*. New York: Plenum Press.

Rose, W. I., and C. A. Chesner. 1990. Worldwide dispersal of ash and gases from Earth's largest known eruption, Toba, Sumatra, 75 ka. *Global and Planetary Change* 89: 269–275.

Rosen, M. A. 2004. Exergy analysis of energy systems. In *Encyclopedia of Energy*. Vol. 2, 607–621.

Rosen, R. D. 1999. The global energy cycle. In *Global Energy and Water Cycles*, ed. K. A. Browning and R. J. Gurney, 1–9. Cambridge: Cambridge University Press.

Rosenberg, N. J., S. B. Verma, and B. L. Blad. 1983. *Microclimate*. New York: Wiley.

Rosenfeld, A. H., and D. Hafemeister. 1988. Energy-efficient buildings. *Scientific American* 258 (4): 78–85.

Ross, D. 1979. *Energy from Waves*. Oxford: Pergamon Press.

Rouse, J. E. 1970. *World Cattle*. Norman: University of Oklahoma Press.

Rowley-Conwy, P. 2001. Time, change and the archeology of hunter-gatherers: How original is the "Original Affluent Society"? In *Hunter-Gatherers: An Interdisciplinary Perspective*, ed. C. Panter-Brick, R. H. Layton, and P. Rowley-Conwy, 39–72. Cambridge: Cambridge University Press.

Roxburgh, S. H., S. L. Berry, T. N. Buckley, B. Barnes, and M. L. Roderick. 2005. What is NPP? Inconsistent accounting of respiratory fluxes in the definition of net primary production. *Functional Ecology* 19: 378–382.

Roy, J., B. Saugier, and H. A. Mooney, eds. 2001. *Terrestrial Global Productivity*. San Diego, Calif.: Academic Press.

Royal Dutch-Shell. 1983. *Petroleum Handbook*. Amsterdam: Elsevier.

REFERENCES

Rubin, E. S., M. R. Taylor, S. Yeh, and D. A. Hounshell. 2003. Experience Curves for Environmental Technology and Their Relationship to Government Actions. Paper presented at EXCETP-6 Workshop, Paris, January. ⟨http://www.iea.org/textbase/work/2003/extool-excetp6/III-rubin.pdf⟩.

Rubner, M. 1883. Über den einfluss der Körpergrösse auf Stoff- und Kraftwechsel. *Zeitschrift für Biologie* 19: 535–562.

———. 1902. *Die Gesetze des Energieverbrauchs bei der Ernährung.* Leipzig: F. Deuticke.

Ruddle, K. 1974. *The Yukpa Cultivation System.* Berkeley: University of California Press.

Ruddle, K., and G. Zhong. 1988. *Integrated Agriculture-Aquaculture in South China.* Cambridge: Cambridge University Press.

Rudin, A. 2004. How Greater Efficiency Increases Resource Use. Paper presented to the North Central Sociological Association, Cleveland, Ohio, April 2.

Ruff, L. J. 1996. Large earthquakes in subduction zones: Segment interaction and recurrence times. In *Subduction Top to Bottom*, ed. G. E. Bebout, D. W. Scholl, S. H. Kirby, and J. P. Platt, 91–104. Washington, D.C.: American Geophysical Union.

Rühlmann, G. 1962. *Kleine Geschichte der Pyramiden.* Dresden: Verlag der Kunst.

Running, S. W., R. R. Nemani, F. A. Heinsch, M. Zhao, M. Reeves, and H. Hashimoto. 2004. A continuous satellite-derived measure of global terrestrial production. *BioScience* 54: 547–560.

Russell, N. J., and T. Hamamoto. 1998. Psychrophiles. In *Extremophiles: Microbial Life in Extreme Environments*, ed. K. Horikoshi and W. D. Grant, 25–45. New York: Wiley-Liss.

Ruth, M. 1993. *Integrating Economics, Ecology, and Thermodynamics.* Dordrecht, Netherlands: Kluwer.

Ruth, M., and T. Harrington. 1998. Dynamics of material and energy use in U.S. pulp and paper manufacturing. *Journal of Industrial Ecology* 1: 147–186.

RWEDP (Regional Wood Energy Development Programme in Asia). 1997. *Regional Study of Wood Energy Today and Tomorrow.* ⟨http://www.rwdep.org/fd50.html⟩.

———. 2000. *Wood Energy Database.* ⟨http://www.rwedp.org/d_consumption.html⟩.

Ryder, H. W. et al. 1976. Future performance in footracing. *Scientific American* 234 (6): 109–119.

Sacher, E. 1881. *Grundzüge einer Mechanik der Gesselschaft.* Jena, Germany: Gustav Fischer.

Sagan, C., and C. Chyba. 1997. The early faint sun paradox: Organic shielding of ultraviolet-labile greenhouse gases. *Science* 276: 1217–1221.

Sagan, C., and G. Mullen. 1972. Earth and Mars: Evolution of atmospheres and surface temperatures. *Science* 177: 52–56.

Sage, R. F., and R. K. Monson. 1999. C_4 *Plant Biology.* San Diego, Calif.: Academic Press.

Sahlins, M. 1972. *Stone Age Economics.* Chicago: Aldine.

Sakharov, A. 1983. The danger of thermonuclear war. *Foreign Affairs* 61: 1001–1016.

Salter, S. H. 1974. Wave power. *Nature* 249: 720–724.

Salzman, P. C., ed. 1981. *Contemporary Nomadic and Pastoral Peoples: Asia and the North.* Williamsburg, Va.: College of William and Mary Press.

———. 2004. *Pastoralists: Equality, Hierarchy, and the State.* Boulder, Colo.: Westview.

Saris, W. H., M. A. van Erp Baart, F. Brouns, K. R. Westerterp, and F. Ten Hoor. 1989. Study on food intake and energy expenditure during extreme sustained exercise: The Tour de France. *International Journal of Sports Medicine.* 10: S26–S31.

Saunders, I., and A. Young. 1983. Rates of surface process on slopes, slope retreat and denudation. *Earth Surface Processes and Landforms* 8: 473–501.

REFERENCES

Schaefer, J. T., D. L. Kelly, and R. F. Abbey. 1980. Tornado track characteristics and hazard probabilities. In *Wind Engineering*, ed. J. E. Cermak, 95–109. Oxford: Pergamon Press.

Schaller, G. B. 1972. *The Serengeti Lion*. Chicago: University of Chicago Press.

Schidlowski, M. 1991. Quantitative evolution of global biomass through time: Biological and geochemical constraints. In *Scientists on Gaia*, ed. S. H. Schneider and P. J. Boston, 211–222. Cambridge, Mass.: MIT Press.

Schimmoller, B. K. 2000. Fluidized bed combustion. *Power* 104 (9): 36–42.

Schipper, L., and M. Grubb. 2000. On the rebound? Feedback between energy intensities and energy uses in IEA countries. *Energy Policy* 28: 367–388.

Schipper, L., and S. Meyers. 1992. *Energy Efficiency and Human Activity: Past Trends, Future Prospects*. Cambridge: Cambridge University Press.

Schmid, P. E., M. Tokeshi, and J. M. Schmid-Araya. 2000. Relation between population density and body size in stream communities. *Science* 289: 1557–1560.

Schmidt, M. J. 1996. Working elephants. *Scientific American* 274 (1): 82–87.

Schmidt-Nielsen, K. 1972. Locomotion: Energy cost of swimming, flying and running. *Science* 177: 222–228.

———. 1984. *Scaling: Why Is Animal Size So Important?* Cambridge: Cambridge University Press.

Schmitt, R. W. 1999. The ocean's response to the freshwater cycle. In *Global Energy and Water Cycles*, ed. K. A. Browning and R. J. Gurney, 144–154. Cambridge: Cambridge University Press.

Schmittner, A. 2005. Decline of the marine ecosystem caused by a reduction in the Atlantic overturning circulation. *Nature* 434: 628–632.

Schofield, W. N., C. Schofield, and W. P. T. James. 1985. Basal metabolic rate in man: Survey of the literature and computation of equations for prediction. *Human Nutrition Clinical Nutrition* 39 (Suppl. 1): 1–96.

Scholander, P. F., H. T. Hammel, J. S. Hart, D. H. LeMessurier, and J. Steen. 1958. Cold adaptation in Australian Aborigines. *Journal of Applied Physiology* 14: 605–615.

Scholz, C. H. 1998. Earthquakes and friction laws. *Nature* 391: 37–42.

Schott, D., ed. 1997. *Energie und Stadt in Europa.: von der vorindustriellen "Holznot" bis zur Ölkrise der 1970er Jahre*. Stuttgart: F. Steiner.

Schrödinger, E. 1944. *What Is Life?* Cambridge: Cambridge University Press.

Schulze, E.-D., J. Lloyd, F. M. Kelliher, C. Wirth, C. Rebmann, B. Lühker, M. Mund et al. 1999. Productivity of forests in the Eurosiberian boreal region and their potential to act as a carbon sink: A synthesis. *Global Change Biology* 5: 703–722.

Schurr, S. H. 1984. Energy use, technological change, and productive efficiency: An economic-historical interpretation. *Annual Review of Energy* 9: 409–425.

Schurr, S. H., and B. C. Netschert. 1960. *Energy in the American Economy, 1850–1975*. Baltimore, Md.: Johns Hopkins University Press.

Schweimer, G. W., and M. Levin. 2000. *Life Cycle Inventory for the Golf A4*. 〈http://www.volkswagen-environment.de/_download/sachbilanz_golf_a4_englisch.pdf〉.

Sciubba, E. 2004. Exergoeconomics. In *Encyclopedia of Energy*. Vol. 2, 577–591.

Sciubba, E., and S. Ulgiati. 2005. Energy and exergy analyses: Complementary methods or irreducible ideological options? *Energy* 30: 1953–1988.

Sclater, J. G., C. Jaupart, and D. Galson. 1980. The heat flow through oceanic and continental crust and the heat loss of the Earth. *Reviews of Geophysics and Space Physics* 18: 269–311.

Scudder, T. 1976. Social anthropology and the reconstruction of prehistoric land use in tropical Africa: A cautionary case study

REFERENCES

from Zambia. In *Origins of African Plant Domestication*, ed. J. R. Harlan, J. M. J. de Wet, and A. B. L. Stamler, 357–381. The Hague: Mouton.

Scurlock, J. M. O., G. P. Asner, and S. T. Gower. 2001. *Global Leaf Area Index Data from Field Measurements, 1932–2000.* ⟨http://www.daac.ornl.gov/vegetation/lai_des.html⟩.

Seaborg, G. T. 1972. Opening address. In *Peaceful Uses of Atomic Energy*, 29–35. Vienna: International Atomic Energy Agency.

Seavoy, R. E. 1986. *Famine in Peasant Societies.* New York: Greenwood Press.

Seebohm, M. E. 1927. *The Evolution of the English Farm.* London: Allen and Unwin.

Selby, M. J. 1985. *Earth's Changing Surface.* Oxford: Clarendon Press.

Sellin, H. J. 1983. The large Roman water-mill at Barbegal (France). *History of Technology* (Royal Society) 8: 91–109.

SEPCo (Shell Exploration and Production Company). 2004. *Shell in the Gulf of Mexico.* ⟨http://www.shellus.com/sepco/⟩.

———. 2006. *Ursa Tension Leg Platform.* ⟨http://www.shellus.com/sepco/⟩.

Service, E. R. 1979. *The Hunters.* London: Prentice-Hall.

Service, R. F. 2004. The hydrogen backlash. *Science* 305: 958–961.

Sexton, A. H. 1897. *Fuel and Refractory Materials.* London: Vlackie.

Shadwick, R. E. 2005. How tunas and lamnid sharks swim: An evolutionary convergence. *American Scientist* 93: 524–531.

Shaeffer, R. E. 1992. *Reinforced Concrete: Preliminary Design for Architects and Builders.* New York: McGraw-Hill.

Shannon, C. E., and W. Weaver. 1949. *The Mathematical Theory of Communication.* Urbana: University of Illinois Press.

Shapouri, H., J. A. Duffield, and M. Wang. 2004. *The 2001 Net Energy Balance of Corn-Ethanol.* U.S. Department of Agriculture.

Sharma, H. D. 1990. *Energy Alternatives: Benefits and Risks.* Waterloo, Canada: University of Waterloo Press.

Sharp, N. C. C. 1997. Timed running speed of a cheetah (*Acinonyx jubatus*). *Journal of Zoology* 241: 493–494.

Sheehan, G. W. 1985. Whaling as an organizing focus in Northwestern Eskimo society. In *Prehistoric Hunter-Gatherers*, ed. T. D. Price and J. A. Brown, 123–154. Orlando, Fla.: Academic Press.

Sheinbaum, C., and L. Ozawa. 1997. Energy use and CO_2 emissions for Mexico's cement industry. *Energy* 23: 725–732.

Shelton, J., and A. B. Shapiro. 1976. *The Woodburners Encyclopedia: An Information Source of Theory, Practice and Equipment Relating to Wood as Energy.* Waitsfield, Vt.: Crossroads Press.

Shen, T. H. 1951. *Agricultural Resources of China.* Ithaca, N.Y.: Cornell University Press.

Shiklomanov, I. A. 2003. *World Water Resources at the Beginning of the Twenty-First Century.* Cambridge: Cambridge University Press.

Shu, D.-G., H.-L. Luo, M. S. Conway, X.-L. Zhang, S.-X. Hu, L. Chen, J. Han et al. 1999. Lower Cambrian vertebrates from South China. *Nature* 402: 42–46.

Sidney, K., E. L. Hemingway, and L. H. Berkshire. 1976. *Basic Estimated Capital Investment and Operating Costs for Coal Strip Mines.* U.S. Bureau of Mines.

Sieferle, R. P. 2001. *The Subterranean Forest: Energy Systems and the Industrial Revolution.* Trans. M. P. Osmann. Cambridge: White Horse Press.

Sigurdsson, H., S. Carey, W. Cornell, and T. Pescatore. 1985. The eruption of Vesuvius in A.D. 79. *National Geographic Research* 1: 332–387.

Silberbauer, G. B. 1981. *Hunter and Habitat in the Central Kalahari Desert.* Cambridge: Cambridge University Press.

REFERENCES

Sillitoe, P. 2002. Always been farmer-foragers? Hunting and gathering in the Papua New Guinea Highlands. *Anthropological Forum* 12 (1): 45–76.

Silva, M., and J. A. Downing. 1995. The allometric scaling of density and body mass: A nonlinear relationship for terrestrial mammals. *American Naturalist* 145: 704–727.

Silver, C. 1976. *Guide to the Horses of the World.* Oxford: Elsevier Phaidon.

Simakov, S. N. 1986. *Forecasting and Estimation of the Petroleum-Bearing Subsurface at Great Depths.* Leningrad: Nedra.

Simkin, T. 1993. Terrestrial volcanism in space and time. *Annual Review of Earth and Planetary Sciences* 21: 427–452.

Simkin, T., L. Siebert, and R. Blong. 2001. Volcano fatalities: Lessons from the historical record. *Science* 291: 255.

Singer, C., E. J. Holmyard, A. R. Hall, and T. I. Williams, eds. *A History of Technology.* 1954–1978. 7 vols. Oxford: Clarendon Press.

Singer, J. D., and M. Small. 1972. *The Wages of War, 1816–1965: A Statistical Handbook.* New York: Wiley.

Singerman, P. 1991. *An American Hero: The Red Adair Story.* Thorndike, Me.: Thorndike Press.

Singh, R. P., ed. 1986. *Energy in Food Processing.* Amsterdam: Elsevier.

Sitwell, N. H. 1981. *Roman Roads of Europe.* New York: St. Martin's Press.

Skilton, C. P. 1947. *British Windmills and Watermills.* London: Collins.

Sleep, N. H. 2005. Evolution of continental lithosphere. *Annual Review of Earth and Planetary Sciences* 33: 369–393.

Slesser, M. 1978. *Energy in the Economy.* London: Macmillan.

Slicher van Bath, B. H. 1963. *The Agrarian History of Western Europe, A.D. 500–1850.* London: Arnold.

Slingo, J. 1982. A study of the earth's radiation budget using a general circulation model. *Quarterly Journal of the Royal Meteorological Society* 108: 379–405.

Smeaton, J. 1759. An experimental enquiry concerning the natural power of water and wind to turn mills, and other machines, depending on a circular motion. *Philosophical Transactions of the Royal Society of London* 51: 100–174.

Smil, V. 1976. *China's Energy.* New York: Praeger.

———. 1983. *Biomass Energies.* New York: Plenum Press.

———. 1985. *Carbon–Nitrogen–Sulfur.* New York: Plenum Press.

———. 1987. *Energy, Food, Environment: Realities, Myths, Options.* Oxford: Oxford University Press.

———. 1991. *General Energetics: Energy in the Biosphere and Civilization.* New York: Wiley.

———. 1992. Agricultural energy costs: National analyses. In *Energy in Farm Production*, ed. R. C. Fluck, 85–100. Amsterdam: Elsevier.

———. 1994. *Energy in World History.* Boulder, Colo.: Westview Press.

———. 1999a. Crop residues: Agriculture's largest harvest. *BioScience* 49: 299–308.

———. 1999b. Nitrogen in crop production: An account of global flows. *Global Biogeochemical Cycles* 13: 647–662.

———. 2000a. *Cycles of Life.* Rev. ed. New York: Scientific American Library.

———. 2000b. *Feeding the World: Challenge for the 21st Century.* Cambridge, Mass.: MIT Press.

———. 2001. *Enriching the Earth: Fritz Haber, Carl Bosch, and the Transformation of World Food Production.* Cambridge, Mass.: MIT Press.

———. 2002. *The Earth's Biosphere: Evolution, Dynamics, and Change.* Cambridge, Mass.: MIT Press.

REFERENCES

———. 2003. *Energy at the Crossroads: Global Perspectives and Uncertainties.* Cambridge, Mass.: MIT Press.

———. 2004a. *China's Past, China's Future: Energy, Food, Environment.* London: RoutledgeCurzon.

———. 2004b. Land requirements of energy systems. In *Encyclopedia of Energy.* Vol. 3, 613–622.

———. 2004c. War and energy. In *Encyclopedia of Energy.* Vol. 6, 363–371.

———. 2005a. *Creating the Twentieth Century: Technical Innovations of 1867–1914 and Their Lasting Impact.* New York: Oxford University Press.

———. 2005b. The next 50 years: Fatal discontinuities. *Population and Development Review* 31: 201–236.

———. 2006. *Transforming the Twentieth Century: Technical Innovations and Their Consequences.* New York: Oxford University Press.

Smil, V., P. Nachman, and T. V. Long II. 1983. *Energy Analysis in Agriculture: An Application to U.S. Corn Production.* Boulder, Colo.: Westview Press.

Smith, C. 1999. *The Science of Energy: A Cultural History of Energy Physics in Victorian Britain.* Chicago: University of Chicago Press.

Smith, D. R. 1987. The wind farms of the Altamont Pass area. *Annual Review of Energy* 12: 145–183.

Smith, E. 1857. Inquiries into the quantity of air inspired throughout the day and night under the influence of exercise, food, medicine, temperature. *Philosophical Magazine* 14: 546–572.

Smith, H. 1982. Light quality photoperception and plant strategy. *Annual Review of Plant Physiology* 33: 481–518.

Smith, H. O., R. Friedman, and J. C. Venter. 2003. Biological solutions to renewable energy. *The Bridge* 33 (2): 36–40.

Smith, K. 1993. Fuel combustion, air pollution exposure and health: The situation in developing countries. *Annual Review of Energy and the Environment* 18: 529–566.

Smith, N. 1980. The origins of the water turbine. *Scientific American* 242 (1): 138–148.

Smith, W. B., P. D. Miles, J. S. Vissage, and S. A. Pugh. 2004. *Forest Resources of the United States, 2002.* Washington, D.C.: USDA Forest Service.

Snow, C. P. 1959. *The Two Cultures.* Cambridge: Cambridge University Press.

So, J. K. 1980. Human biological adaptation to arctic and subarctic zones. *Annual Review of Anthropology* 9: 63–82.

Socolow, R. H. 1977. The coming age of conservation. *Annual Review of Energy* 2: 239–289.

Soddy, F. 1912. *Matter and Energy.* New York: Henry Holt.

———. 1926. *Wealth, Virtual Wealth and Debt: The Solution of the Economic Paradox.* London: Allen and Unwin.

———. 1933. *Money versus Man: A Statement of the World Problem from the Standpoint of the New Economics.* New York: E. P. Dutton.

Soden, B. J., R. T. Wetherald, G. L. Stenchikov, and A. Robock. 2002. Global cooling after the eruption of Mount Pinatubo: A test of climate feedback by water vapor. *Science* 296: 727–733.

Solanki, S. K., I. G. Usoskin, B. Kromer, M. Schüssler, and J. Beer. 2004. Unusual activity of the Sun during recent decades compared to the previous 11,000 years. *Nature* 431: 1084–1087.

SolarPACES. 2005. *Solar Power Tower.* ⟨http://www.solarpaces.org⟩.

Solomon, A. M., I. C. Prentice, R. Leemans, and W. P. Cramer. 1993. The interaction of climate and land use in future terrestrial carbon storage and release. *Water Air and Soil Pollution* 70: 595–614.

REFERENCES

Solow, R. M. 1957. Technical change and the aggregate production function. *Review of Economics and Statistics* 39: 312–320.

Sonntag, R. E., C. Borgnakke, and G. J. Van Wylen. 2003. *Fundamentals of Thermodynamics*. New York: Wiley.

Sørensen, B. 1980. *An American Energy Future*. Golden, Colo.: Solar Energy Research Institute.

———. 2004. *Renewable Energy: Its Physics, Engineering, Environmental Impacts, Economics and Planning*. San Diego, Calif.: Academic Press.

Sorrell, S. 2004. *The Economics of Energy Efficiency: Barriers to Cost-Effective Investment*. Cheltenham, UK: E. Elgar.

SPE (Society of Petroleum Engineers). 1991. *Horizontal Drilling*. Richardson, Tex.: Society of Petroleum Engineers.

Speakman, J. R. 2005. Body size, energy metabolism and lifespan. *Journal of Experimental Biology* 208: 1717–1730.

Spence, K. 2000. Ancient Egyptian chronology and the astronomical orientation of pyramids. *Nature* 408: 320–324.

Spencer, J. E. 1966. *Shifting Cultivation in Southeastern Asia*. Berkeley: University of California Press.

Spiegler, K. S. 1983. *Principles of Energetics*. Berlin: Springer.

Spotila, J. R., and D. M. Gates. 1975. Body size, insulation and optimum body temperatures of homeotherms. In *Perspectives of Biophysical Ecology*, ed. D. M. Gates and R. B. Schmerl, 291–301. New York: Springer.

Spreng, D. T. 1978. *On Time, Information, and Energy Conservation*. Oak Ridge, Tenn.: Institute for Energy Analysis.

Spruytte, J. 1983. *Early Harness Systems*. Trans. M. L. Littauer. London: J. A. Allen.

SRI (Stanford Research Institute). 1971. *Patterns of Energy Consumption in the United States*. Menlo Park, Calif.: SRI.

Stallo, J. B. 1900. *Concepts and Theories of Modern Physics*. New York: D. Appleton.

Stanford, C. B., and H. T. Bunn, eds. 2001. *Meat-Eating and Human Evolution*. Oxford: Oxford University Press.

Stanhill, G. 1976. Trends and deviations in the yield of the English wheat crop during the last 750 years. *Agro-Ecosystems* 3: 1–10.

———, ed. 1984. *Energy in Agriculture*. Berlin: Springer.

Starbuck, A. 1878. *History of the American Whale Fishery*. Waltham, Mass.: A. Starbuck.

Starr, C. 1969. Social benefit versus technological risk. *Science* 165: 1232–1238.

Starr, C., R. Rudman, and C. Whipple. 1976. Philosophical basis for risk analysis. *Annual Review of Energy* 1: 629–662.

Stein, S., and J. T. Freymueller, eds. 2002. *Plate Boundary Zones*. Washington, D.C.: American Geophysical Union.

Steinhart, J. S., and C. E. Steinhart. 1974. Energy use in the U.S. food system. *Science* 184: 307–316.

Stephens, D. W., and J. R. Krebs. 1987. *Foraging Theory*. Princeton, N.J.: Princeton University Press.

Stern, D. I. 2004. Energy and economic growth. In *Encyclopedia of Energy*. Vol. 2, 35–51. Amsterdam: Elsevier.

Stern, R. 2006. Oil market power and United States national security. *Proceedings of the National Academy of Sciences* 103: 1650–1655.

Stetter, K. O. 1996. Hyperthermophilic prokaryotes. *FEMS Microbiology Reviews* 18: 149–158.

———. 1998. Hyperthermophiles: isolation, classification and properties. In *Extremophiles: Microbial life in extreme environments*, ed. K. Horikoshi and W. D. Grant, 1–24. New York: Wiley.

Stevens, P., ed. 2000. *The Economics of Energy*. Cheltenham, UK: E. Elgar.

Stevenson, R. D. 1985. Body size and limits to the daily range of body temperature in terrestrial ectotherms. *American Naturalist* 125: 102–117.

REFERENCES

Stewart, B. A., and T. A. Howell, eds. 2003. *Encyclopedia of Water Science*. New York: Marcel Dekker.

Stockhuyzen, F. 1963. *The Dutch Windmill*. New York: Universe Books.

Stodolsky, F., A. Vyas, R. Cuenca, and L. Gaines. 1995. *Life-Cycle Energy Savings Potential from Aluminium-Intensive Vehicles*. Argonne, Ill.: Argonne National Laboratory.

Stopes, M. 1935. On the petrology of banded bituminous coals. *Fuel* 14: 4–13.

Storey, K. B., and J. M. Storey. 1988. Freeze tolerance in animals. *Physiological Reviews* 68: 27–84.

Stout, B. A. 1990. *Handbook of Energy in World Agriculture*. Amsterdam: Elsevier.

Stout, B. A., J. L. Butler, and E. E. Garett. 1984. Energy use and management in U.S. agriculture. In *Energy and Agriculture*, ed. G. Stanhill, 175–176. New York: Springer.

Straker, E. 1931. *Wealden Iron*. Reissued Newton Abbot, UK: David and Charles, 1969.

Stuart, J. S. 2001. A near-Earth asteroid population estimate from the LINEAR survey. *Science* 294: 1691–1693.

Syncrude Canada. 2006. ⟨http://www.syncrude.com/⟩.

Syvitski, J. P. M., C. J. Vörösmarty, A. J. Kettner, and P. Green. 2005. Impact of humans on the flux of terrestrial sediment to the global coastal ocean. *Science* 308: 376–380.

Szekely, J. 1987. Can advanced technology save the U.S. steel industry? *Scientific American* 257 (1): 34–41.

Taha, H. 2004. Heat islands and energy. In *Encyclopedia of Energy*. Vol. 3, 133–143.

Tanaka, J. 1980. *The San Hunter-Gatherers of the Kalahari*. Tokyo: University of Tokyo Press.

Tanner, A. H. 1998. *Continuous Casting: A Revolution in Steel*. Fort Lauderdale, Fla.: Write Stuff Enterprises.

Tanno, K., and G. Willcox. 2006. How fast was wild wheat domesticated? *Science* 311: 1886.

Tansley, A. G. 1935. The use and abuse of vegetational concepts and terms. *Ecology* 16: 284–307.

Tappen, M. 2001. Deconstructing the Serengeti. In *Meat-Eating and Human Evolution*, ed. C. B. Stanford and H. T. Bunn, 13–29. Oxford: Oxford University Press.

Tapponnier, P., X. Zhiqin, F. Roger, B. Meyer, N. Arnaud, G. Wittlinger, and Y. Jingsui. 2001. Oblique stepwise rise and growth of the Tibet Plateau. *Science* 294: 1671–1677.

Taverner, M. R., and A. C. Dunkin, eds. 1996. *Pig Production*. Amsterdam: Elsevier.

Taylor, G. K., R. L. Nudds, and A. L. R. Thomas. 2003. Flying and swimming animals cruise at a Strouhal number tuned for high power efficiency. *Nature* 425: 707–711.

Taylor, N. A. S. 2006. Ethnic differences in thermoregulation: Genotypic versus phenotypic heat adaptation. *Journal of Thermal Biology* 31: 90–104.

Teal, J. M. 1962. Energy flow in the salt marsh ecosystem of Georgia. *Ecology* 43: 614–624.

Technocracy. 1937. *Energy Certificate*. ⟨http://www.technocracy.org/⟩.

Teller, E. 1979. *Energy from Heaven and Earth*. San Francisco: W. H. Freeman.

Testart, A. 1982. The significance of food storage among hunter-gatherers: Residence patterns, population densities, and social inequalities. *Current Anthropology* 23: 523–537.

TGVweb. 2000. Under the hood of a TGV. ⟨http://www.trainweb.org/tgvpages/motrice.html⟩.

Thackeray, M. M. 2004. Batteries, transportation applications. In *Encyclopedia of Energy*. Vol. 1, 127–139.

Thekaekara, M. P. 1977. Solar irradiance, total and spectral. In *Solar Energy Engineering*, ed. A. A. M. Sayigh, 37–59. New York: Academic Press.

REFERENCES

Theobald, D. W. 1966. *The Concept of Energy*. London: Spon.

Thieme, H. 1997. Lower Paleolithic hinting spears from Germany. *Nature* 385: 807–810.

Thirring, H. 1958. *Energy for Man*. Bloomington: Indiana University Press.

Thomas, J., ed. 1979. *Energy Analysis*. Boulder, Colo.: Westview Press.

Thomas, L. P. 2002. *Coal Geology*. New York: Wiley.

Thomson, K. S. 1987. How to sit on a horse. *American Scientist* 75: 69–71.

Thomson, W. 1853. On the dynamical theory of heat. *Royal Society of Edinburgh Transactions* 20: 261–298+.

Thorne, J., and M. Suozzo. 1997. *Leaking Electricity Estimates*. ⟨http://www.sciencenews.org/sn_arc97/10_25_97/bob1a.htm⟩.

Tillman, D. A. 1978. *Wood as an Energy Resource*. New York: Academic Press.

Tilton, J. E., and B. J. Skinner. 1987. The meaning of resources. In *Resources and World Development*, ed. D. J. McLaren and B. J. Skinner, 13–27. Chichester, UK: Wiley.

Tiratsoo, E. N. 1986. *Oilfields of the World*. Houston: Gulf Publishing.

———, ed. 1980. *Natural Gas*. Houston: Gulf Publishing.

Tolbert, N. E. 1997. The C_2 oxidative photosynthetic carbon cycle. *Annual Review of Plant Physiology and Plant Molecular Biology* 48: 1–25.

Tomlinson, G. H. 1990. *Effects of Acid Deposition on the Forests of Europe and North America*. Boca Raton, Fla.: CRC Press.

Tompkins, P. 1971. *Secrets of the Great Pyramid*. New York: Harper and Row.

Tonon, S., M. T. Brown, F. Luchi, A. Mirandola, A. Stoppato, and S. Ulgiati. 2006. An integrated assessment of energy conversion processes by means of thermodynamics, economic and environmental parameters. *Energy* 31: 149–163.

Torr, G. 1964. *Ancient Ships*. Chicago: Argonaut Publishers.

Torrey, V. 1976. *Wind-Catchers: American Windmills of Yesterday and Tomorrow*. Brattleboro, Vt.: Stephen Greene Press.

Total. 2005. *Factbook, 1998–2005*. ⟨http://www.total.com/2005-factbook/corporate/corporate_1.htm⟩.

Transocean. 2003. *Firsts and World Records*. ⟨http://www.deepwater.com/FactsandFirsts.cfm⟩.

Travis, C., and E. L. Etnier, eds. 1982. *Health Risks of Energy Technologies*. Boulder, Colo.: Westview Press.

Treloar, G. J. 1997. Extracting embodied energy paths from input-output tables: Toward an input-output-based hybrid energy analysis method. *Economic Systems Research* 9: 375–392.

Tresouthwick, S. W., and A. Mishulovich. 1991. Energy and environmental considerations for the cement industry. In *Energy and the Environment in the 21st Century*, ed. J. W. Tester, 397–404. Cambridge, Mass.: MIT Press.

Trinkhaus, E. 2005. Early modern humans. *Annual Review of Anthropology* 34: 207–230.

Truesdell, C., and S. Bharatha. 1977. *The Concepts and Logic of Classical Thermodynamics as a Theory of Heat Engines, Rigorously Constructed upon the Foundation Laid by S. Carnot and F. Reech*. New York: Springer.

Tucker, V. A. 1973. Bird metabolism during flight: Evaluation of a theory. *Journal of Experimental Biology* 58: 689–709.

———. 1975. The energetic cost of moving about. *American Scientist* 63: 413–419.

Turkenburg, W. C. 2000. Current status and potential future costs of renewable energy technologies (table 7.25). In *World Energy Assessment: Energy and the Challenge of Sustainability*, ed. J. Goldemberg. New York: United Nations Development Programme.

Turner, W. C. 2005. *Energy Management Handbook*. New York: Marcel Dekker.

REFERENCES

Tyedmers, P. 2004. Fisheries and energy use. In *Encyclopedia of Energy*. Vol. 2, 683–693.

Ubbelohde, A. R. 1954. *Man and Energy*. London: Hutchinson.

UNDP (United Nations Development Programme). 2001. *Human Development Report*. ⟨http://hdr.undp.org/reports/global/2001/en/⟩.

———. 2005. *Human Development Report*. ⟨http://hdr.undp.org/reports/global/2005/⟩.

Unger, R. 1984. Energy sources for the Dutch Golden Age. *Research in Economic History* 9: 221–253.

Unger, T. A. 1996. *Pesticide Synthesis Handbook*. Park Ridge, N.J.: Noyes Publications.

UNO (United Nations Organization). 1956. World energy requirements in 1975 and 2000. In *Proceedings of the International Conference on the Peaceful Uses of Atomic Energy*. Vol. 1, 3–33.

———. 1990. *Energy Statistics Database*. ⟨http://unstats.un.org/unsd/energy/edbase.htm⟩.

———. 2000. *Energy Statistics Database*. ⟨http://unstats.un.org/unsd/energy/edbase.htm⟩.

———. 2006. *Cities, Population Density*. ⟨http://unstats.un.org/unsd/cdb/cdb_series_xrxx.asp?series_code=14722⟩.

Urbanski, T. 1967. *Chemistry and Technology of Explosives*. Oxford: Pergamon Press.

Urry, D. W. 2004. *What Sustains Life? Consilient Mechanisms for Protein-Based Machines and Materials*. New York: Springer.

U.S. Congress. 1974. *Non-Nuclear Energy Research and Development Act of 1974*. Public Law 93–577. Washington, D.C.: Government Printing Office.

USBC (U.S. Bureau of the Census). 1975. *Historical Statistics of the United States: Colonial Times to 1970*. ⟨http://www.census.gov/prod/www/abs/statab.html⟩.

———. 2006. *Statistical Abstract of the United States*. ⟨http://www.census.gov/compendia/statab/⟩.

USDA (U.S. Department of Agriculture). 1959. *Changes in Farm Production and Efficiency*. Washington, D.C.: USDA.

———. 1980. *Energy and U.S. Agriculture: 1974 and 1978*. Washington, D.C.: USDA.

———. 2006. *USDA Food and Nutrient Database for Dietary Studies*. ⟨http://www.ars.usda.gov/Services/docs.htm?docid=7673⟩.

USDOE (U.S. Department of Energy). 1997. *Energy, Environmental, and Economic (E3) Handbook*. ⟨http://www.eia.doe.gov/⟩.

———. 2000. *Energy Use: Steel Industry Analysis Brief*. ⟨http://www.eia.doe.gov/emeu/mecs/iab/steel/⟩.

———. 2001. *Price-Anderson Act*. ⟨http://www.ans.org/pi/ps/docs/ps54-bi.pdf⟩.

US EIA. 2006. *Electric Power Annual 2005*. Washington, D.C.: EIA. ⟨http://www.eia.doe.gov/cneaf/electricity/epa/epa+sum.html⟩.

USGS (U.S. Geological Survey). 2000. *World Petroleum Assessment*. ⟨http://pubs.usgs.gov/dds/dds-060/⟩.

———. 2001. *The World Coal Quality Inventory (WoCQI)*. ⟨http://energy.er.usgs.gov/coal_quality/wocqi/index.html⟩.

———. 2003. *Moving Slabs*. ⟨http://pubs.usgs.gov/gip/dynamic/slabs.html⟩.

———. 2004. *Estimated Use of Water in the United States in 2000*. ⟨http://pubs.usgs.gov/circ/2004/circ1268/⟩.

———. 2005. *National Earthquake Information Center*. ⟨http://earthquake.usgs.gov/regional/neic/⟩.

———. 2006. *Cement Statistics and Information*. ⟨http://minerals.usgs.gov/minerals/pubs/commodity/cement/⟩.

Usher, A. P. 1954. *A History of Mechanical Inventions*. Cambridge, Mass.: Harvard University Press.

USPIRG (U.S. Public Interest Research Group). 2004. *Reel Danger: Power Plant Mercury Pollution and the Fish We Eat*. ⟨http://www.uspirg.org/home/reports/report-archives/⟩.

REFERENCES

Utley, F. 1925. *Trade Guilds of the Later Roman Empire.* London: London School of Economics.

Valk, M., ed. 1995. *Atmospheric Fluidized Bed Coal Combustion.* Amsterdam: Elsevier.

Van Deman, E. B. 1934. *The Building of the Roman Aqueducts.* Washington, D.C.: Carnegie Institution.

Van der Linden, S. 2006. Bulk energy storage potential in the USA: Current developments and future prospects. *Energy* 31: 3446–3457.

Van der Post, L., and J. Taylor. 1984. *Testament to the Bushmen.* New York: Viking.

Van Dyke, K. 2000. *Drilling Fluids.* Austin, Tex.: PETEX. ⟨http://www.utexas.edu/cee/petex/⟩.

Van Loon, H. W. 1935. *Ships and How They Sailed the Seven Seas, 5000 B.C.–A.D. 1935.* New York: Simon and Schuster.

van Winkle, T. L., J. Edeleanu, E. A. Prosser, and C. A. Walker. 1978. Cotton versus polyester. *American Scientist* 66: 280–290.

Venkataraman, C., G. Habib, A. Eiguren-Fernandez, A. H. Miguel, and S. K. Friedlander. 2005. Residential biofuels in South Asia: Carbonaceous aerosol emissions and climate impacts. *Science* 307: 1454–1456.

Verbraeck, A., ed. 1976. *The Energy Accounting of Materials, Products, Processes and Services.* Rotterdam: TNO (Netherlands Institute for Applied Scientific Research).

Verhoogen, J. 1980. *Energetics of the Earth.* Washington, D.C.: National Academies Press.

Vernadsky, V. I. 1926. *Biosfera.* Leningrad: Nauchnoe khimiko-tekhnicheskoye izdatel'stvo.

Vesey-Fitzgerald, B. 1946. *The Book of the Horse.* London: Nicholson and Watson.

Vestas. 2007. ⟨http://www.vestas.com/vestas/global/en/⟩.

Videler, J. J. 1993. *Fish Swimming.* New York: Chapman and Hall.

Videler, J. J., E. J. Stamhuis, and G. D. E. Povel. 2004. Leading-edge vortex lifts swifts. *Science* 306: 1960–1962.

Vimal, O. P., and M. S. Bhatt. 1989. *Wood Energy Systems.* Delhi: K. L. Publications.

Viswanath, R., V. Wakharkar, A. Watwe, and V. Lebonheur. 2000. Thermal performance challenges from silicon to systems. *Intel Technology Journal* Q3: 1–16.

Voeikov, A. I. 1884. *Klimaty zemnogo shara.* St. Petersburg: Izdatel'stvo kartograficheskogo zavedeniya.

Vogel, H. U. 1993. The great well of China. *Scientific American* 268 (6): 116–121.

Volcano Live. ⟨http://www.volcanolive.com/⟩.

von Braun, W., and F. I. Ordway. 1975. *History of Rocketry and Space Travel.* New York: Crowell.

Von Moltke, A., C. McKee, and T. Morgan, eds. 2004. *Energy Subsidies: Lessons Learned in Assessing their Impact and Designing Policy Reforms.* Nairobi: UNEP.

von Tunzelmann, G. N. 1978. *Steam Power and British Industrialization to 1860.* Oxford: Clarendon Press.

Voznesenskaya, E. V., V. R. Franceschi, O. Kiirats, H. Freitag, and G. E. Edwards. 2001. Kranz anatomy is not essential for terrestrial C_4 plant photosynthesis. *Nature* 414: 543–546.

Vringer, K., and K. Blok. 2000. The energy requirement of cut flowers and consumer options to reduce it. *Resources, Conservation and Recycling* 28: 3–28.

Wailes, R. 1975. *Windmills in England: A Study of Their Origin, Development and Future.* London: Architectural Press.

Walsberg, G. E. 2000. Small mammals in hot deserts: Some generalizations revisited. *BioScience* 50: 109–120.

Wang, M., and Y. Ding. 1998. Fuel-saving stoves in China. *Wood Energy News* 13 (3): 9–10.

Warburton, M. 2001. Barefoot running. *Sportscience* 5 (3): 1–4.

REFERENCES

Ward, C. R., ed. 1984. *Coal Geology and Coal Technology*. Melbourne: Blackwell.

Ward, S., and S. Day. 2001. Cumbre Vieja volcano: Potential collapse and tsunami at La Palma, Canary Islands. *Geophysical Research Letters* 28: 3397–3400.

Ward, S., U. Möller, J. M. V. Rayner, D. M. Jackson, D. Bilo, W. Nachtigall, and J. R. Speakman. 2001. Metabolic power, mechanical power, and efficiency during wind tunnel flight by the European starling *Sturnus vulgaris*. *Journal of Experimental Biology* 204: 3311–3322.

Wark, K. W., and D. E. Richards. 1999. *Thermodynamics*. 6th ed. New York: McGraw-Hill.

Warneck, P. 2000. *Chemistry of the Natural Atmosphere*. San Diego, Calif.: Academic Press.

Warren-Rhodes, K., and A. Koenig. 2001. Escalating trends in the urban metabolism of Hong Kong, 1971–1997. *Ambio* 30: 429–438.

Waring, R. H., and S. W. Running. 1998. *Forest Ecosystems: Analysis at Multiple Scales*. 2d ed. San Diego, Calif.: Academic Press.

Waterbury, J. 1979. *Hydropolitics of the Nile Valley*. Syracuse, N.Y.: Syracuse University Press.

Watt, B. K., and A. L. Merrill. 1963. *Composition of Foods*. Handbook 8. U.S. Department of Agriculture.

Watters, R. F. 1971. *Shifting Cultivation in Latin America*. Rome: FAO.

WBCSD (World Business Council for Sustainable Development). 2004. *Mobility 2030: Meeting the Challenges of Sustainability*. Geneva: WBCSD.

WCD (World Commission on Dams). 2000. *Dams and Development*. ⟨http://www.dams.org/report/⟩.

WCI (World Coal Institute). 2005. *The Coal Resource: A Comprehensive Overview of Coal*. London: WCI.

———. 2006. *Coal Facts 2006*. London: WCI.

Weber, B. H., D. J. Depew, and J. D. Smith, eds. 1988. *Entropy, Information, and Evolution*. Cambridge, Mass.: MIT Press.

WEC (World Energy Council). 1998. *Survey of Energy Resources*. ⟨http://www.worldenergy.org/wec-geis/publications/reports/ser/overview.asp⟩.

———. 2001. *Survey of Energy Resources*. ⟨http://www.worldenergy.org/wec-geis/publications/reports/ser/overview.asp⟩.

———. 2004. *Survey of Energy Resources*. ⟨http://www.worldenergy.org/wec-geis/publications/reports/ser/overview.asp⟩.

———. 2006. *Energy Efficiencies: Pipe-Dream or Reality?* ⟨http://www.worldenergy.org/wec-geis/global/downloads/statements/stat2006.pdf⟩.

WEC/IIASA. 1998. *Global Energy Perspectives*. Cambridge: Cambridge University Press.

Wegener, A. 1924. *The Origin of Continents and Oceans*. London: Methuen.

Weimerskirch, H., O. Chastel, C. Barbraud, and O. Tostain. 2003. Frigatebirds ride high on thermals. *Nature* 421: 333–334.

Weimerskirch, H., J. Martin, Y. Clerquin, P. Alexandre, and S. Jiraskova. 2001. Energy saving in flight formation. *Nature* 413: 697–698.

Weinberg, A. 1994. *The First Nuclear Era*. Washington, D.C.: American Institute of Physics.

Weindruch, R., and R. S. Sohal. 1997. Caloric intake and aging. *New England Journal of Medicine* 337: 986–994.

Wesley, J. P. 1974. *Ecophysics*. Springfield, Ill.: C.C. Thomas.

West, G. B., and J. H. Brown. 2005. The origin of allometric scaling laws in biology from genomes to ecosystems: Towards a quantitative unifying theory of biological structure and organization. *Journal of Experimental Biology* 208: 1575–1592.

REFERENCES

West, G. B., J. H. Brown, and B. J. Enquist. 1997. A general model for the origin of allometric scaling laws in biology. *Science* 276: 122–126.

———. 1999. A general model for the structure and allometry of plant vascular systems. *Nature* 400: 664–667.

———. 2000. The origin of universal scaling laws in biology. In *Scaling in Biology*, ed. J. H. Brown and G. B. West, 87–112. New York: Oxford University Press.

Westerterp, K. R., B. Kayser, F. Brouns, J.-P. Herry, and W. H. M. Saris. 1992. Energy expenditure climbing Mt. Everest. *Journal of Applied Physiology* 73: 1815–1819.

Westerterp, K. R., B. Kayser, L. Wouters, J.-L. Le Trong, and J.-P. Richalet. 1994. Energy balance at high altitude of 6,542 m. *Journal of Applied Physiology* 77: 862–866.

Westheimer, F. H. 1987. Why nature chose phosphates. *Science* 235: 1173–1177.

Westoby, M. 1984. The self-thinning rule. *Advances in Ecological Research* 14: 167–225.

Weyand, P. G., and J. A. Davis. 2005. Running performance has a structural basis. *Journal of Experimental Biology* 208: 2625–2631.

Weyand, P. G., D. B. Sternlight, M. J. Bellizzi, and S. Wright. 2000. Faster top running speeds are achieved with greater ground forces, not more rapid leg movements. *Journal of Applied Physiology* 89: 1991–1999.

Whipp, B. J., and S. A. Ward. 1992. Will women soon outrun men? *Nature* 355: 25.

White, C. R., and R. S. Seymour. 2003. Mammalian basal metabolic rate is proportional to body mass. *Proceedings of the National Academy of Sciences* 100: 4046–4049.

———. 2005. Allometric scaling of mammalian metabolism. *Journal of Experimental Biology* 208: 1611–1619.

White, J. 1985. The thinning rule and its application to mixtures of plant populations. In *Studies on Plant Demography*, ed. J. White, 291–309. London: Academic Press.

White, K. D. 1970. *Roman Farming*. London: Thames and Hudson.

———. 1984. *Greek and Roman Technology*. Ithaca, N.Y.: Cornell University Press.

White, L. 1978. *Medieval Religion and Technology*. Berkeley: University of California Press.

Whitman, W. B., D. C. Coleman, and W. J. Wiebe. 1998. Prokaryotes: The unseen majority. *Proceedings of the National Academy of Sciences* 95: 6578–6583.

Whittaker, R. H., and G. E. Likens. 1975. The biosphere and man. In *Primary Productivity of the Biosphere*, ed. H. Lieth and R. H. Whittaker, 305–328. New York: Springer.

Whitton, B. A., and M. Potts. 2000. *The Ecology of Cyanobacteria: Their Diversity in Time and Space*. Boston: Kluwer.

Widdowson, E. M. 1983. How much food does man require? In *Nutritional Adequacy, Nutrient Availability and Needs*, ed. J. Mauron, 11–25. Basel: Birkhäuser.

Wielicki, B. A., T. Wong, R. P. Allan, A. Slingo, J. T. Kiehl, B. J. Soden, C. T. Gordon et al. 2002. Evidence for large decadal variability in the tropical mean radiative energy budget. *Science* 295: 841–844.

Wielicki, B. A., T. Wong, N. Loeb, P. Minnis, K. Priestley, and R. Kandel. 2005. Changes in Earth's albedo measured by satellite. *Science* 308: 825.

Wier, S. K. 1996. Insight from geometry and physics into the construction of Egyptian Old Kingdom pyramids. *Cambridge Archaeological Journal* 6: 150–163.

Wild, M., H. Gilgen, A. Roesch, A. Ohmura, C. N. Long, E. G. Dutton, B. Forgan et al. 2005. From dimming to brightening: Decadal changes in solar radiation at Earth's surface. *Science* 308: 847–850.

Williams, E. D. 2004. Energy intensity of computer manufacturing: Hybrid assessment combining process and economic input-output method. *Environmental Science and Technology* 38: 6166–6174.

REFERENCES

Williams, E. D., R. U. Ayres, and M. Heller. 2002. The 1.7 kilogram microchip: Energy and material use in the production of semiconductor devices. *Environmental Science and Technology* 36: 5504–5510.

Williams, T., G. Deskins, S. L. Ward, and M. Hightower. 2001. *Sound Coil-Tubing Drilling Practices.* ⟨http:// www.maurertechnology.com/NewsEvents/publications/ CTDmanual.pdf⟩.

Williams, T. I. 1987. *The History of Invention: From Stone Axes to Silicon Chips.* New York: Facts On File.

Williams, T. M. 1999. The evolution of cost efficient swimming in marine mammals: Limits to energetic optimization. *Philosophical Transactions of the Royal Society* B354: 193–201.

Willrich, M., and T. B. Taylor. 1974. *Nuclear Theft: Risks and Safeguards.* Cambridge, Mass.: Ballinger.

Wilson, A. M. 1999. Windmills, cattle and railroad: The settlement of the Llano Estacado. *Journal of the West* 38 (1): 62–67.

Wilson, A. M., M. P. McGuigan, A. Su, and A. J. van den Bogert. 2001. Horses damp the spring in their step. *Nature* 414: 895–899.

Wilson, C. 1990. *The Gothic Cathedral: The Architecture of the Great Church, 1130–1530.* London: Thames and Hudson.

Wilson, C. L. 1980. *Coal-Bridge to the Future: Report of the World Coal Study (WOCOL).* Cambridge, Mass.: Ballinger.

Wilson, D. G. 2004. *Bicycling Science.* Cambridge, Mass.: MIT Press.

Wilson, J. T. 1963. A possible origin of the Hawaiian islands. *Canadian Journal of Physics* 41: 863–870.

Wilson, P. N. 1956. *Watermills: An Introduction.* London: Society for Protection of Ancient Buildings.

Winterhalder, B., and E. A. Smith, eds. 1981. *Hunter-Gatherer Foraging Strategies.* Chicago: University of Chicago Press.

Wirsenius, S. 2000. *Human Use of Land and Organic Materials: Modeling the Turnover of Biomass in the Global Food System.* Göteborg, Sweden: Chalmers University of Technology.

Wischmeier, W. H., and D. D. Smith. 1978. *Predicting Rainfall Erosion Losses.* Washington, D.C.: U.S. Department of Agriculture.

Wittmuss, H., L. Olson, and D. Lane. 1975. Energy requirements for conventional versus minimum tillage. *Journal of Soil and Water Conservation* 30: 72–75.

WNA (World Nuclear Association). 2006a. *Energy Analysis of Power Systems.* London: WNA.

———. 2006b. *Nuclear Power in the World Today.* ⟨http:// www.world-nuclear.org/info/inf01.htm⟩.

Wolf, B. O., and G. E. Walsberg. 2000. The role of the plumage in heat transfer processes in birds. *American Zoologist* 40: 575–584.

Wölfel, W. 1987. *Das Wasserrad: Technik und Kulturgeschichte.* Wiesbaden, Germany: U. Pfriemer.

Wolff, A. R. 1900. *The Windmill as Prime Mover.* New York: Wiley.

Wood, B., and M. Collard. 1999. The human genus. *Science* 284: 65–71.

Wood, W. 1922. *All Afloat.* Glasgow: Brook & Company.

Woodall, F. P. 1982. Water wheels for winding. *Industrial Archaeology* 16: 333–338.

World Bank. 2001. *World Development Report 2000/01.* New York: Oxford University Press.

Worrell, E., and C. Galitsky. 2003. *Profile of the Petroleum Refining Industry in California.* Berkeley, Calif.: Lawrence Berkeley National Laboratory.

Wrangham, R. W., J. H. Jones, G. Laden, D. Pilbeam, and N. Conklin-Brittain. 1999. The raw and the stolen. *Current Anthropology* 40: 567–594.

REFERENCES

WRI (World Resources Institute). 2005. *Earth Trends.* ⟨http://www.wri.org/⟩.

Wulff, H. E. 1966. A postscript to Reti's notes on Juanelo Turriano's water mills. *Technology and Culture* 7: 398–401.

Wunsch, C. 2002. What is the thermohaline circulation? *Science* 298: 1179–1180.

Wyndham, C. H. 1969. Adaptation to heat and cold. *Environmental Research* 2: 442–469.

Xiong, J., W. M. Fischer, K. Inoue, M. Nakahara, Carl E. Bauer. 2000. Molecular evidence for theearly evolution of photosynthesis. *Science* 289: 1724–1729.

Yachandra, V. K., V. J. DeRose, M. J. Latimer, I. Mukerji, K. Sauer, and M. P. Klein. 1993. Where plants make oxygen: A structural model for the photosynthetic oxygen-evolving manganese cluster. *Science* 260: 675–679.

Yang, Q., ed. 1997. *Geology of Fossil Fuels, Coal: Proceedings of the 30th International Geological Congress.*

Yesner, D. R. 1980. Maritime hunter-gatherers: Ecology and prehistory. *Current Anthropology* 21: 727–750.

Young, H. E. 1979. Forest Biomass a Renewable Source of Energy. Paper prepared for the UNITAR Conference on Long-Term Energy Resources, Montreal.

Zahedi, A. 2003. *The Engineering and Economics of Solar Photovoltaic Energy Systems.* Melbourne: New World Publishing.

Zebroski, E., and M. Levenson. 1976. The nuclear fuel cycle. *Annual Review of Energy* 1: 101–130.

Zemansky, M. W., and R. H. Dittman. 1981. *Heat and Thermodynamics.* New York: McGraw-Hill.

Zens, M. S., and C. O. Webb. 2002. Sizing up the shape of life. *Science* 295: 1475–1476.

Zhang, Q. 2004. Residential energy consumption in China and its comparison with Japan, Canada, and USA. *Energy and Buildings* 36: 1217–1225.

Zhang, X., M. A. Friedl, C. B. Schaaf, A. H. Strahler, and A. Schneider. 2004. The footprint of urban climates on vegetation phenology. *Geophysical Research Letters* 31: doi:10.1029/2004GL020137. ⟨http://www.agu.org/pubs/crossref/2004/2004GL020137.shtml⟩.

Zierenberg, R. A., M. W. W. Adams, and A. J. Arp. 2000. Life in extreme environments: Hydrothermal vents. *Proceedings of the National Academy of Sciences* 97: 12961–12962.

Zvelebil, M. 1986. Postglacial foraging in the forests of Europe. *Scientific American* 254 (5): 104–115.

REFERENCES

NAME INDEX

NAME INDEX

Subject Index

SUBJECT INDEX

SUBJECT INDEX

SUBJECT INDEX